STEEL FIBER CONCRETE

US–Sweden joint seminar (NSF–STU) Stockholm, June 3–5, 1985

Editors

SURENDRA P. SHAH

Professor of Civil Engineering, Northwestern University, Evanston, Illinois, USA

and

ÅKE SKARENDAHL

Åke Skarendahl AB, Djursholm, Sweden

ELSEVIER APPLIED SCIENCE PUBLISHERS
LONDON and NEW YORK

ELSEVIER APPLIED SCIENCE PUBLISHERS LTD
Crown House, Linton Road, Barking, Essex IG11 8JU, England

Sole Distributor in the USA and Canada
ELSEVIER SCIENCE PUBLISHING CO., INC
52 Vanderbilt Avenue, New York, NY 10017, USA

WITH 61 TABLES AND 280 ILLUSTRATIONS

British Library Cataloguing in Publication Data

Steel fiber concrete: US–Sweden joint
seminar (NSF–STU), Stockholm, June 3–5,
1985.

1. Reinforced concrete, Fiber
I. Shah, Surendra P. II. Skarendahl, Åke
624.1′8341 TA444

Library of Congress Cataloging in Publication Data

Steel fiber concrete.

Bibliography: p.
Includes index.
1. Reinforced concrete, Fiber—Congresses.
2. Gunite–Congresses. I. Shah, S. P. (Surendra P.)
II. Skarendahl, Åke
TA444.S73 1986 620.1′37 86-16223

ISBN 1-85166-043-7

Printed in Great Britain by Galliard (Printers) Ltd, Great Yarmouth

CONTENTS

PREFACE

Although the concept of reinforcing brittle materials with fibers is quite old, the recent interest in reinforcing portland cement based materials with randomly distributed fibers was spured by pioneering research on steel fiber reinforced concrete conducted in the United States in the 1960's. Since then there has been substantial research and development activities throughout the world. It has been established that the addition of randomly distributed steel fibers to brittle cement based matrices can increase, by an order of magnitude, their cracking resistance (fracture thoughness, ductility, impact resistance, fragmentation and spalling resistance). Since fibers can be premixed in a conventional manner or fabricated in a fashion similar to fiber-glass production or sprayed on site, the concept of steel fiber concrete has added an extra dimension to concrete construction.

The currently successful applications of steel fiber concrete include: (1) shotcrete applications, e.g. for tunnel linings and repair, (2) pavements and overlays, (3) precast products, (4) protective concrete structures and (5) furnace linings with stainless steel fibers. Considerable work has been done in Sweden on thin walled precast products made with steel fibers and the interest for this product is growing in the United States. Different methods of spraying steel fiber concrete have been developed in Sweden, while considerably more experience in cast-in-place steel fiber concrete for pavement has been gained in the United States.

To evaluate the experience gained in the United States and Sweden in the field of steel fiber concrete, a joint US-Sweden seminar was held on June 3-5 1985 in Stockholm, Sweden. Each participant presented a contribution based on their expertise. Their papers are contained in these proceedings.

The scope of the seminar was threefold: (1) to review and discuss gained experience from practical applications to identify possibilities and problems, (2) to present and evaluate current research activities and (3) to identify needed research areas and possible future joint research efforts.

The seminar was sponsored by the US National Science Foundation (Dr Christine Glenday) and the National Swedish Board for Technical Development - STU (Professor Carl E Brinck). CBI (Swedish Cement and Concrete Research Institute) participated in the organization of the seminar and is publishing these proceedings. Additional financial support was obtained from Ribbon Technology Corporation, ABV, Besab, Ekebro AB, Svenskt Stål AB and Stabilator AB.

The positive and enthusiastic support from the participants and the sponsors in carrying out the seminar is gratefully acknowledged.

Stockholm, June 1985

Surendra P Shah　　　*Åke Skarendahl*

Editors

STEEL FIBER CONCRETE
US-SWEDEN joint seminar
(NSF-STU)
Stockholm 3-5 June, 1985
S P Shah and Å Skarendahl,
Editors

STEEL FIBER REINFORCED SHOTCRETE
A State-of-the-Art Report

by V Ramakrishnan, Prof, South Dakota School of Mines and Technology, Rapid City, South Dakota, USA

ABSTRACT

Steel fiber reinforced shotcrete (SFRS) was first experimented in U.S.A. in 1971, and the first major field application was in 1973. Since that time there has been extensive use of this new promising material in all advanced countries of the world.

This report traces the history of development of SFRS and describes the laboratory research and successful field applications. SFRS has found its largest application in mining operations, for forming linings in various railway, road, and water tunnels. Other applications include rock slope stabilization work, brick bridge arch strengthening, dome structures, canal linings, and repair and rehabilitation of water front structures and other deteriorated concrete structures. The most successful applications are when it has been used in lieu of mesh reinforced shotcrete. Materials, steel fibers, additives, admixtures, mixing and placing techniques are described. Factors affecting rebound are discussed. Physical and elastic properties, particularly the improvements achieved in shotcrete performance due to the addition of steel fibers are described. Other typical properties including, shrinkage, impact resistance, toughness index, fatigue, pullout strength, bond, durability, permeability and porosity are discussed.

V. Ramakrishnan is Professor of Civil Engineering at South Dakota School of Mines and Technology, Rapid City, South Dakota, U.S.A. He is chairman of ACI Committee 214, a member of Concrete Material Research Council, and Committee on Awards for Papers, Shotcrete (506), Fiber Reinforced Concrete (544) and Research (123). He is also Chairman of Transportation Research Board Committee on Mechanical Properties of Concrete. He has authored 2 books and over 130 publications in the field of concrete structures and concrete technology. Dr. Ramakrishnan has received numerous awards including the Presidential Award as Outstanding Professor 1980 from SDSM&T for excellence in teaching, research and community service.

INTRODUCTION

Steel Fiber Reinforced Shotcrete (SFRS) is defined as a mortar or concrete containing discontinuous discrete steel fibers which are pneumatically projected at high velocity onto a surface (2).

The addition of steel fibers to Shotcrete has considerably improved the flexural strength, impact strength, direct tensile strength, shock resistance, ductility, and failure toughness. The increase in strength depends on adjustments made to the basic conventional mix so that fibers can be added, and the quantity, the type, size, shape, and aspect ratio of fibers used.

It is suggested that fiber shotcrete can be used to line irrigation channels where cavitation and erosion damage is a problem, to line tunnels, and stabilize rock slopes. Because of its high energy absorption ability it could also be used in earthquake areas. In addition, SFRS has attractive potential in shotcrete tanks, swimming pools, steel pipe coatings, refractory coatings, shell roofs, elevated tanks, repair and rehabilitation of waterfront substructures, and other structural applications. (1 to 8)

BACKGROUND

Research: The early experiments with SFRS was carried out by Lankard at the Battelle Laboratories in 1971 (9). Additional work was done under the direction of Poad by the U.S. Bureau of Mines in an investigation of new and improved methods of using shotcrete for underground support (10).

In 1978 Ramakrishnan et.al. (4) carried out a comparative evaluation of various types of fibers and found that hooked end fibers provided better phycial properties than straight fibers for a given volume concentration of fiber. This is because of the better end anchorage provided by the hooked end fibers. These fibers provide a higher effective aspect ratio than equivalent length straight fibers. The fibers used were two types of collated fibers with deformed ends, straight fibers, and cut steel tire cord. Three different fiber concentrations were used at 0.6, 1.0, and 1.3 percent. The dry-process sanded mix shotcrete with an assumed water-cement ratio of 0.46 and 5 percent entrapped air was used in this investigation. One major finding is that collated fibers with hooked ends could be effectively used in shotcreting. Excellent distribution was noted from the x-ray photos.

Morgan (5) carried out a laboratory study to assess the performance of hooked steel fibers in "dry" shotcrete mixes. He investigated five different shotcrete mixes; all mixes were supplied as dry, premixed, bagged materials. There were no significant problems of balling or blockages in the equipment or segregation in the test panels, even at 1.5 percent steel fiber concentration. It was concluded that for practical SFRS applications steel fiber additions to the premixed shotcrete of the order of 0.7 to 1.0 percent by volume of the total shotcrete mix; equivalent to 55 kg/m^3 (92 lb/yd^3) to 78 kg/m^3 (132 lbs/yd^3) of hooked 30 mm long x 0.40 mm diameter steel fibers, should prove adequate. This study also demonstrated that SFRS can be applied with conventional plain shotcrete equipment and the supply of dry, premixed, prebagged steel fiber shotcrete for dry mix applications.

Recently Krantz (1) had conducted an experimental investigation of SFRS in the U.S. Bureau of Mines Lake Lynn Laboratory Experimental Mine using a variety of pneumatic guns and gunning crews. The performance of both wet and dry mixes were studied. Flexural strength, compressive strength, rebound, permeability, porosity, dust loading, shooting rate, crew requirements, and cost factors were evaluated. Results indicated that all of the commercially available SFRS materials tested can provide beneficial sealant, spall prevention or roof stability control attributes for underground mining environments when applied by an experienced crew. Wet-mix with silica fume was tested. Silica fume increased the cohesiveness of the mix and it was possible to place a thickness of more than 76 mm (3 in.) without sagging or slumping. The silica fume gave a smooth overall finish to the shotcrete and even the protruding steel fibers appeared to have a paste coating on them.

Wet-mixes containing steel fibers, water reducers (mainly of the lignosulfonate or hydroxylated carboxylic acid type), accelerators, superplasticizers or high range water reducers, and polymer latex additives, were investigated. The wet-mix shotcrete with water reducers tested in this investigation had a slump ranging from 89 to 127 mm (3.5 to 5 in.) and provided an increased pumpability. The wet-mix containing silica fume essentially needed superplasticizer and these mixes took on a silky, almost lustrous feel and could be pumped for great distances without clogging the placement and/or delivery hoses. In adding silica fume and superplasticizer, a sensitive balance must be maintained. On one occasion the addition of too much superplasticizer in a wet-mix with silica fume resulted in a total loss of cohesiveness and the initiation of mix segregation, a condition known as total collapse. This mix had to be discarded and cleaned out of the shotcrete pump system. SFRS with polymer latex addition was investigated. The polymer latex used was a styrene-butadiene polymeric emulsion in whch the polymer comprised 47 $\pm$ 1 percent by weight and water made up the balance. The polymer contained 64 $\pm$ 2 percent styrene and 36 $\pm$ 2 percent butadiene. The prebagged shotcrete materials were shot using a rotary gun. Considerable difficulty was experienced in obtaining a consistent latex-shotcrete flow from the nozzle. The polymer latex caused partial blockage of the nozzle water ring assembly by partially hydrating the cement fraction and forming a sticky residue within the orifices. The gunned material had a very high slump and would not be suitable for application on a vertical surface. The polymer specimens were seriously laminated. A decrease in the 28-day compressive and flexural strength were observed for SFRS compared to non-polymer SFRS, owing to the mix hydration problems. However, the permeability reduction achieved by the polymer latex specimens when compared to that with non-polymer specimens was impressive even in specimens with lamination zones. This reduced permeability would lead to improved freezing and thawing durability.

The investigation showed that SFRS had the highest strength parameters when compared with shotcrete containing other fibers and plain shotcrete. The wet-mix shotcrete with steel fibers, silica fume and superplasticizer showed the highest compression and flexural strengths.

Applications: SFRS had its first practical application in 1972(19) for lining of a tunnel audit at the Ririe Dam, Idaho. Since that time there has been considerable use of SFRS in most of the industrialized nations like Sweden, Germany, England, Norway, Canada, Australia, Finland, Switzerland, Poland and Japan (2). So far the largest volume of use for SFRS was in support of underground openings. It has been used for linings in various road, railway and water tunnels. It has been used to rehabilitate deteriorating old tunnels on the Canadian Pacific Railway main trans-Canada line and to control water-flow and ice formation and stabilize rock scaling (11) and to line exploratory adits in a slaking cretaceous shale and line drainage tunnels in a rock slide area in two British Columbia hydro-electric projects (12).

SFRS has been used in Sweden to line and stabilize a gravity ore transfer shaft (20) at the Bolidens Gruv AB mine. It has also been used in the lining of an oil storage cavern in Sweden.

In U.S., SFRS has been used for lining a 61/m (200 ft.) length of the subway tunnel by the Metropolitan Atlanta Rapid Transit Authority (22), and for lining the enlarged underground rooms at the U.S. Bureau of Mines experimental mine at Bruceton, Pa. (24). It has been used to provide good fireproofing protection for urethane foam (23) and to coat bulk-heads, seals and stoppings formed by Bernold Steel (25).

Morgan and McAskill (3) reported the extensive use of SFRS to line several kilometers of new tunnels constructed through Rocky Mountain by the British Columbia Railway in 1981-83. This project has shown that SFRS can be successfully used to line underground openings by using conventional shotcrete equipment and procedures. Uniformity in the quality of shotcrete was achieved by the use of weight batched bulk bin bag materials. This pre-dry-bagging facilitated efficient winter shotcrete construction when outside temperature dropped as low as -30° C (-22° F). The application of shotcrete to follow the contours of the jointed and blocky rock closely, thus minimizing the volume of shotcrete required, was possible because of the use of SFRS instead of mesh reinforced shotcrete. It is claimed that this project has demonstrated that SFRS can provide a viable technical, economic, and practical alternative to conventional mesh reinforced shotcrete for lining of underground openings.

SFRS has also been used extensively in numerous rock slope stabilization projects. Kaden reported (19,26) a large application of SFRS near Little Goose Dam along the Snake River in the state of Washington. Using SFRS, a portion of rock surface at an oil refinery at the west coast of Sweden was stabilized with a layered construction of plain shotcrete 5 to 10 mm (0.2 to 0.4 in.) thick followed by 30 mm (1.2 in.) of SFRS covered with a top layer of 5 to 10 mm (0.2 to 0.4 in.) of plain shotcrete (27).

Special SFRS mixes have been used in construction and repairs of refractories (18). In recent years, SFRS has been used in various new types of structures like domes and barrel-vault structures using a process in which an inflated membrane is sprayed with a polyurethane foam which creates the form for the application of SFRS (17). This technique has been used for building farm storage sheds, commercial offices, industrial warehouses, residential complexes and military hardened shelters. These structures are very efficient thermally and can support heavy roof loads (13 to 16).

In England SFRS has been used to strengthen brick arches under three rail bridges (2). A flash coat of 13 mm (0.5 in.) was used to cover exposed fibers. Traffic interruption is minimized because the scaffolding required for mesh installation is eliminated. In Japan, vehicular tunnel linings damaged by rock pressure have been repaired with SFRS (2).

In Sweden a lighthouse damaged by freezing and thawing and the interior of a 50 m (164 ft.) tall concrete chimney were repaired with SFRS. It has also been used in Australia to repair an eroded roof in a concrete bunker used for absorbing energy from impacting projectiles; relining a steel bin used for aggregate storage and lining curved sections of a stormwater drain (2). SFRS has been used for sandwich wall construction and coal mine strengthening and sealing of stoppings in England, resurfacing of a rocket flame deflector at Cape Canaveral, Florida, and constructing boat hulls (2).

Ironman (21) has reported the use of SFRS and robots in Norway. The production of high quality wet-mix shotcrete and its consequent overwhelming acceptance in Norway resulted from the combined use of microsilica, superplasticizer and steel fibers. Since the introduction of steel fibers the number of special applications for shotcrete has increased. Because of greater impermeability SFRS is favored for jobs involving tanks, pools, and marine structures.

STEEL FIBERS

Steel fibers for use in shotcrete are available in a number of shapes, sizes and metal types. Currently they are produced by three different processes. 1. A sheet of metal is cut or slitted, producing a square or rectangular fiber. 2. Cold-drawn wire is chopped to specific length. Some are collated with water-soluble glue into bundles of 10 to 30 fibers to facilitate handling and to increase their bond and anchorage parameters. 3. Melt-extracted fibers are produced by a process whereby a rotating, cooled disc with indentations of the size of fiber is dipped in the surface of a molten pool of high quality metal.

Many different types of fibers, with round, rectangular and crescent shaped cross sections, are commercially available. They range in ultimate strength from 345 to 2070 MPa (50,000 to 300,000 psi). Fiber sizes range from 13 x 0.25 mm to 64 x 0.76 mm (0.5 x .01 in. to 2.5 x 0.03 in.). However, most successful shotcrete applications have utilized fibers with length of 13 to 30 mm (0.5 to 1.2 in.). In one investigation, brass plated cut steel wire tire cord with a tensile strength of 2758 MPa (400,000 psi) has been successfully used. Fibers with hooked or deformed ends could be used in smaller quantities (4) because they develop higher pullout resistance. Fibers with large surface area, square or rectangular as compared to round, have more concrete bonding area. And fibers with pitted surface have a greater surface area.

For high temperature application up to 1650° C (3000° F), like in refractory shotcretes, stainless steel fibers are used. Corrosion of steel fibers in shotcrete with a high water-to-cement ratio may cause deterioration. The free moisture in wet concrete provides an aqueous medium which facilitates transport of chlorides toward the metal. It also increases the electrical conductivity of the material, thus aiding any tendency for electrochemical corrosion. However, in actual field cases, corrosion of carbon steel fibers have been found to be minimal. When fibers were exposed on a surface, they showed evidence of corrosion; however, internal fibers showed no corrosion

A conventional numerical parameter describing a fiber is its aspect ratio, defined as the fiber length divided by diameter (or equivalent diameter in the case of non-round fibers). Typical aspect ratios used ranged from 30 to 150.

Most of the early research on SFRS was conducted using relatively high fiber loadings and using steel fibers of relatively small diameter and relatively high aspect ratios. Later research (4) showed that by using hooked fibers with a high equivalent aspect ratio, considerably smaller fiber quantities can be used to achieve the same strength characteristics. Fiber loadings in construction projects have typically ranged from 0.5 to 2.0 percent by volume: 47 to 157 kg/m^3 (66 to 265 lb/yd.^{3}). Higher percentages of fibers have been generally used with straight fibers. All these quantities refer to fibers added to the shotcrete mix; the actual volume of fibers in the in-place shotcrete may be smaller depending on the degree of rebound of steel fibers relative to the other shotcrete materials.

ADDITIVES AND ADMIXTURES

A variety of additives and admixtures are added to shotcrete materials, particularly in the wet mix to improve strength, adhesiveness, cohesiveness, and freezing and thawing and abrasion-resistance characteristics, to inhibit corrosion and to reduce rebound. Major additives and admixtures used are briefly discussed.

Silica Fume: Condensed silica fume and ferrosilicon dust (essentially the same material but with slightly different chemical compositions) are by-products of silicon metal and ferrosilicon alloy manufacturing. The light grey and extremely fine powder was once discarded as waste material; research and testing have turned it into a valuable product, sometimes costing more than that of cement. Silica fume is highly pozzolonic and can be used in shotcrete as a cement replacement. When silica fume is added, a superplasticizer has to be added to achieve the required workability.

Water Reducers: Water reducers are used to improve the wet-mix shotcrete workability, cohesiveness in the plastic state, and pumpability.

Accelerators: Shotcrete accelerators are used to shorten product set time, thereby reducing any sagging or sloughing tendencies when thick applications are to be made in a single pass.

Superplasticizers: Known as high-range water reducers, superplasticizers are chemically distinct from normal plasticizers or water reducers. They are mainly sulfonated melamine formaldehyde condensates or sulfonated naphthalene formaldehyde condensates or modified lignosulfonates. They are used either to increase the strength or to increase the workability considerably without reduction in strength.

Polymer Latex Additives: These additives could be added to the dry-mix shotcrete to impart special desired properties, like adhesion improvement, permeability reduction, resistance to chloride attack, increased durability in freezing and thawing conditions, impact resistance, steel protection, and strength improvement.

Admixtures containing calcium chlorides, fluorides, sulfites, sulfates, and nitrates should never be used in SFRS, because these chemicals could promote corrosion of the steel fibers.

MIX DESIGN AND MATERIALS

Cements: Normal Type I portland cement has been used in most SFRS applications. In tunnels and marine structures where a high early strength development was needed to resist the effects of blasting or wave action at an early age, type III cements have been used. Type V sulfate resisting cements have been used in sulfate bearing soil or groundwater conditions. Cement contents used have ranged from 390 kg/m^3 (658 lb/yd.^{3}) to as much as 558 kg/m^3 (940 lb/yd.^{3}) in mortar mixes. When a 10 mm (3/8 in.) maximum size aggregate was used in SFRS projects, typically the mixes contained 445 kg/m^3 (750 lb/yd.^{3}) cement. However, the cement content of the in-place shotcrete will always be higher than the as-batched cement content, because of the higher degree of rebound of coarse aggregates and sand particles than cement. This is particularly true in the case of dry-mix process, shotcrete sprayed overhead.

Aggregate Gradation: The ACI Standard Specification for Materials, Proportioning and Application of Shotcrete (28) lists three desirable combined gradation limits for shotcrete aggregates. Gradation No. 1 refers to a mortar mix containing no coarse aggregates.

The concrete materials used in much of the early research was a cement-sand mortar containing no coarse aggregates. They required higher cement contents, particularly in wet mix shotcrete applications with consequent higher shrinkages and creep. These mixes had higher porosity than mixes containing coarse aggregate particles.

ACI Gradation No. 2 refers to a 10 mm (3/8 in.) maximum size aggregate gradation as shown in Figure 1. This is the gradation most used recently in SFRS projects. A typical mix contained between 20 to 35 percent by weight of the total combined aggregates of 10 to 5 mm (3/8 to

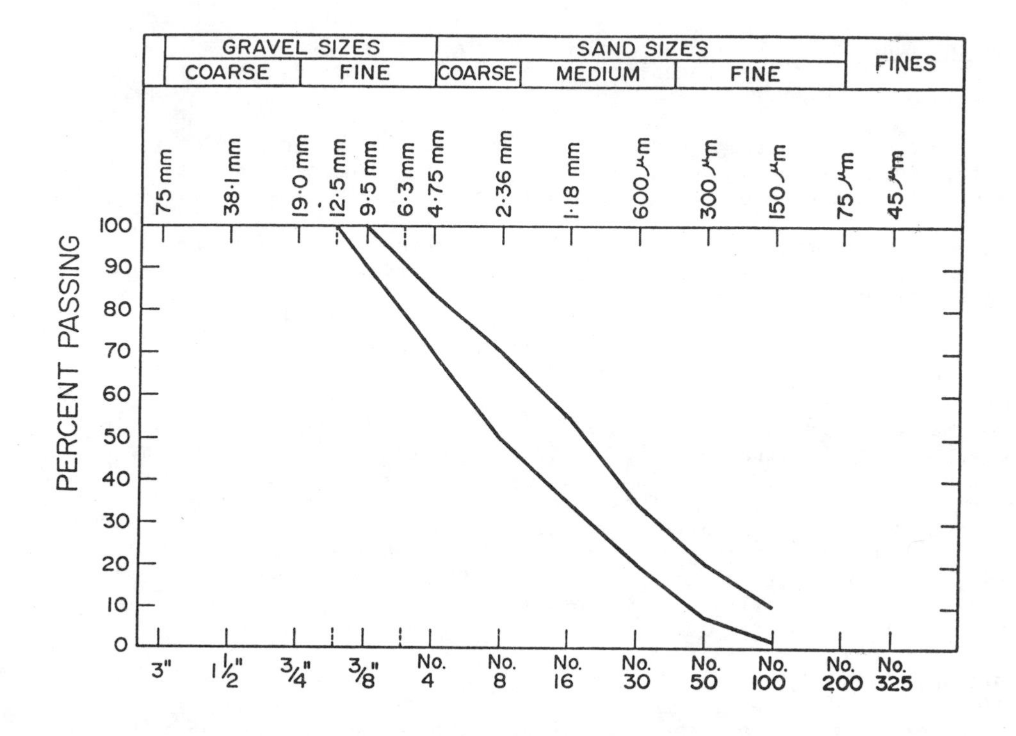

FIG. 1. Gradation No. 2, ACI506.2-77

1/16 in.) coarse aggregate. For overhead application, the mixes are proportioned to the finer side of the gradation envelope; to the middle of the gradation envelope for vertical applications, and to the coarser side of the gradation envelope for downward application.

Gradation No. 3 refers to a 20 mm (3/4 in.) maximum size aggregate. This is seldom used in SFRS mixes.

BATCHING, MIXING AND PLACING

SFRS has two primary application techniques - wet mix and dry mix. In the dry process, the ingredients are volume or weight batched and mixed dry or prebagged. The cement, damp sand and coarse aggregate can be mixed in a ready mix concrete truck with steel fibers then being added to the truck. Alternatively, the shotcrete materials can be mixed in a central mix plant or mobile site batch plant (2). Different techniques have been used for adding the fibers. These include dispensing the fibers through a shaker or vibrating screen so that fiber-balls (clusters of fibers) are not dispensed into shotcrete materials. Fibers have been discharged onto conveyor belts and fed in with aggregates or added through special dispensors (3). Caution should be exercised when batching and mixing loose, bulk fibers to avoid the formation of fiber balls. When collated fibers, bundled together with a water soluble glue, are used in the dry process, they can be added directly to the mixes after the aggregate has been added. Complete separation of these fibers has been observed in the sprayed concrete (4). In the dry-mix process, it has been found that a good electrical ground to the gun and nozzle dramatically reduces the fiber clumping and plugging that might otherwise occur (1).

The prebagged SFRS has been widely used in British Columbia (3). The cement, completely dry aggregates and steel fibers are weighed, batched and mixed in a dry bag plant and then discharged into paper bags or bulk synthetic cloth bags. Paper bags have generally been supplied in 23 kg (50 lb.) or 40 kg (88 lbs.) sizes. Bulk bin bags have been supplied with weights varying from 1134 kg (2500 lb.) to 1814 kg (4000 lb.). The dry materials are then discharged into premoisturizer in which water is added to bring the moisture content of SFRS into a 3 to 6 percent by weight range before discharging into shotcrete pot. Bulk bin bags have the advantage of requiring little manpower. Once supported above the premoisturizer hopper they are essentially self discharging.

In repair or rehabilitation work, small volume work or where intermittent placement is involved or in mines where a mixer and bulk materials would aggravate space problems, it is often more economical and convenient to use prebagged SFRS supply. For proper control of setting time, prebagged material is preferable because time of contact between cement and moisture prior to discharge into shotcrete pot is kept to a minimum. The time of set of shotcrete mixes is greatly influenced by the time of prehydration of the cement prior to shooting.

Conventional dry process shotcrete equipment with no modifications has been successfully used to place SFRS with different types and sizes of fibers (4). Some modifications can be made to reduce plugging by eliminating restrictions such as 90 degree elbows or abrupt changes in hose size. It is recommended that the hose diameter should be two times the fiber length (2). Large hose size 50 mm (2 in.) diameter and up works better. Some contractors have reported a greater rate of wear with the rubber hoses used to convey the SFRS mixes, compared to plain shotcrete.

Wet-Mix Process: In wet process shotcrete, procedures for batching, mixing, and placing SFRS are essentially the same as those used for steel fiber reinforced concrete. Good guidance is given in the ACI State-of-the-Art Report on Fiber Reinforced Concrete (17). Details about some special equipment are given in references 1 and 2.

It is reported that a high percent of wet-mix SFRS is used in Norway (21).

It is recommended that due to the high cement factors used in SFRS, the delivery and discharge time-lapse for wet-mix should not exceed 20 minutes, when the concrete is hauled in nonagitating trucks, nor 40 minutes when the concrete is hauled in truck mixers or truck agitators. Retempering after 30 minutes of initial mixing operation is not advised.

Placing: The placement of SFRS is basically the same as applying plain shotcrete. Special precautions should be used when dry mix shotcrete with the steel fibers are sprayed in multiple layered applications to avoid laminations (lenses of poorly cemented material). In an experimental project (4) inspection of the cut specimens showed some laminations when the panels were built up to less than full depth initially and additional shotcrete was applied a few minutes later to obtain the required depth. Premature uniaxial compressive, flexural, and impact failures occurred along these weak layers.

In SFRS, some steel fibers may be exposed at the surface or protruding. These exposed fibers may corrode and stain the surface even though they do not cause structural weakness. To avoid this, a finish coat of plain shotcrete of the same mix but excluding fibers should be applied to cover exposed and protruding fibers. The nominal thickness of this finish coat should be 13 mm (0.5 in.) and it should be applied as a final pass before the main body of the shotcrete has stiffened.

REBOUND

The pneumatic application of SFRS causes a relatively high percentage of the high velocity aggregate and fibers to bounce or rebound off the fresh rock surface until a layer of cohesive material has accumulated there and can cushion the impact. This rebound material is waste and it cannot be reused or reprocessed. This material must be discarded.

For plain shotcrete, the rebound is 15 to 30 percent for sloping and vertical walls and 25 to 50 percent for overhead work. No reliable values are available for SFRS. Some researchers (4,2,30,31) have reported that SFRS showed less total rebound than plain shotcrete and others have reported no difference (29). However it has been observed that a greater percentage of fibers than aggregate rebound from the wall (2).

Parker (29) has reported extensive work on rebound. He reported fiber retention of 44 to 88 percent for coarse aggregate mixtures gunned on to vertical panels and an average rebound of 18.3 percent and 17.7 percent for plain and fiber shotcretes respectively. He used high-speed photography to observe the shotcrete airstreams. The photography showed that many of the fibers were in the outer portion of the airstream and that many of them were blown away radially from near the point of intended impact shortly before or after they hit. He also showed that the rebound process differed during establishment of an initial critical thickness and subsequent spraying into the fresh shotcrete. The following mixer conditions could reduce the rebound: higher cement content, more fines in the mixture, smaller maximum size aggregate, proper moisture content of the aggregate, a finer gradation, and inclusion of fly ash or silica fume.

For dry mix process the most effective means to reduce rebound include reduction of air pressure, use of more fines, use of shorter fibers, predamping and shooting at the wettest stable consistency.

PHYSICAL PROPERTIES

Test Procedures: The reported properties for SFRS are generally measured by tests advocated in ACI Committee 544 Report (32) and ASTM Standard tests. Currently ACI Committee 506 Shotcreting is preparing a report on in-place evaluation of shotcrete.

Compressive Strength: The compressive strength of SFRS is governed by the compressive strength of the shotcrete matrix. A perusal of the literature (4) shows that inclusion of fibers in shotcrete does not show appreciable increase in compressive strength. However, the mode of failure is changed, particularly for high-strength shotcrete and concretes, from a brittle, sudden and explosive failure to a more ductile failure. The descending portion of the post-peak stress-strain curve is much flatter.

The 28-day compressive strengths reported have varied from 29 MPa to 51.7 MPa (4200 psi to 7500 psi).

Flexural Strength: A significant difference in the performance of plain shotcrete and SFRS is shown in the static flexure test and the load-deflection curves. These curves indicate a ductile behavior. They also show the advantages of fibers in shotcrete in obtaining higher load-carrying cpacity, higher ultimate strength, higher toughness, and excellent energy absorbing characteristics. The apparent energy absorbed is represented by the total area under the load-deflection curve.

Placement of shotcrete tends to orient the fibers in a plane parallel to the surface being shot. This orientation is beneficial to the structural properties of the shotcrete layer, particularly when thin sections are applied.

Two values of the flexural strengths are significant: 1. The first crack flexural strength, recognized as the point at which the load-deformation curve deviates from the linearity; and 2. The ultimate flexural strength at the point of maximum load (Modulus of rupture). For plain shotcrete and for low aspect ratio or low volume concentrations of fiber (particularly for straight fibers), the beams fail at the first crack load without any appreciable increse in the strength after the first crack. Typical 28-day flexural strengths as determined from beam specimens vary from about 5.5 MPa (800 psi) to about 10.3 MPa (1500 psi) with the average near 6.9 to 7.6 MPa (1000 to 1100 psi) (2).

Typical load-deflection curves for 4 different types of steel fibers at 1 percent concentration are compared with the curve for the plain shotcrete in Figure 2 and the effect of different fiber concentrations is shown in Figure 3. The fibers compared are 1. straight fibers 0.25 x 0.56 x 25.4 mm (0.01 x 0.2 x 1.00 in.), 2. hooked and collated low carbon steel fibers 30 mm (1.18 in.) long and 0.50 mm (0.02 in.) diameter; 3. low carbon steel fibers 30 mm (1.18 in.) long and .40 mm (0.016 in.) diameter; and 4. brass plated cut steel tire cord with a ultimate tensile strength of 2758 MPa (400,000 psi), straight fibers 18 mm (0.71 in.) long and 0.25 mm (0.01 in.) diameter (4). The flexural strength increase with increasing fiber volume, higher aspect ratio, and increased pull-out resistance of the fibers (hooked fibers). The hooked fibers and cut steel tire cord fibers have shown better ability as crack arrestors. The cracks were prevented from propagating until the composite ultimate stress was reached. The mode of failure was simultaneous yielding of the fibers and the matrix. During the test one could actually hear the popping sound of the fibers failing in tension and see the broken fibers indicating fiber failure in tension. The deformed ends contributed significantly to the increase in the bond between the fiber and the matrix. The mode of failure for the straight fiber specimens was strictly a bond failure. It was observed after testing that the fibers had pulled out from the matrix. All of the fiber shotcrete specimens except the straight fiber shotcrete specimens displayed substantial post-crack load carrying capacity. The ultimate failure load and the first crack load are one and the same for the plain and straight fiber specimens indicating sudden brittle failure.

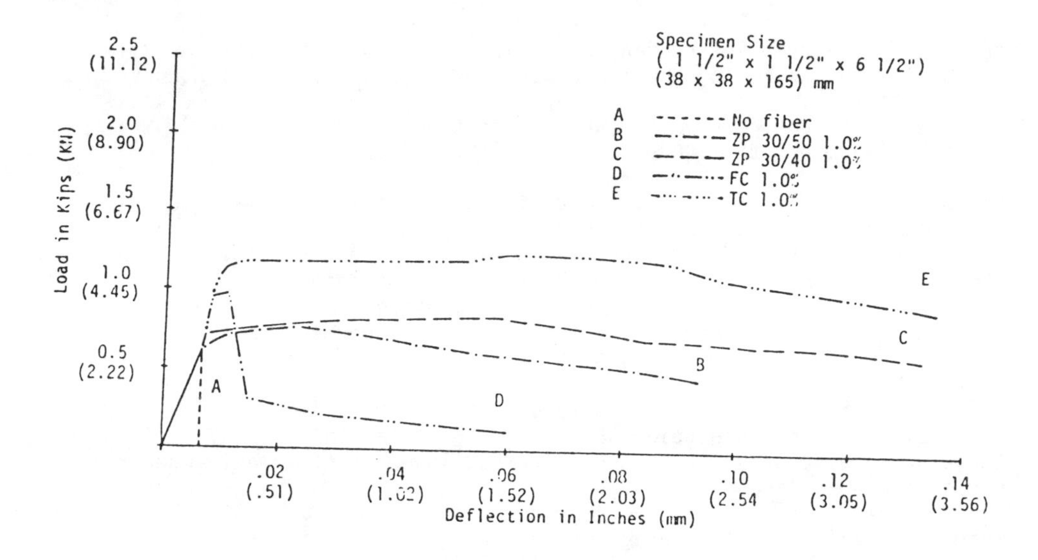

FIG. 2. Load-deflection curves for different fiber types.

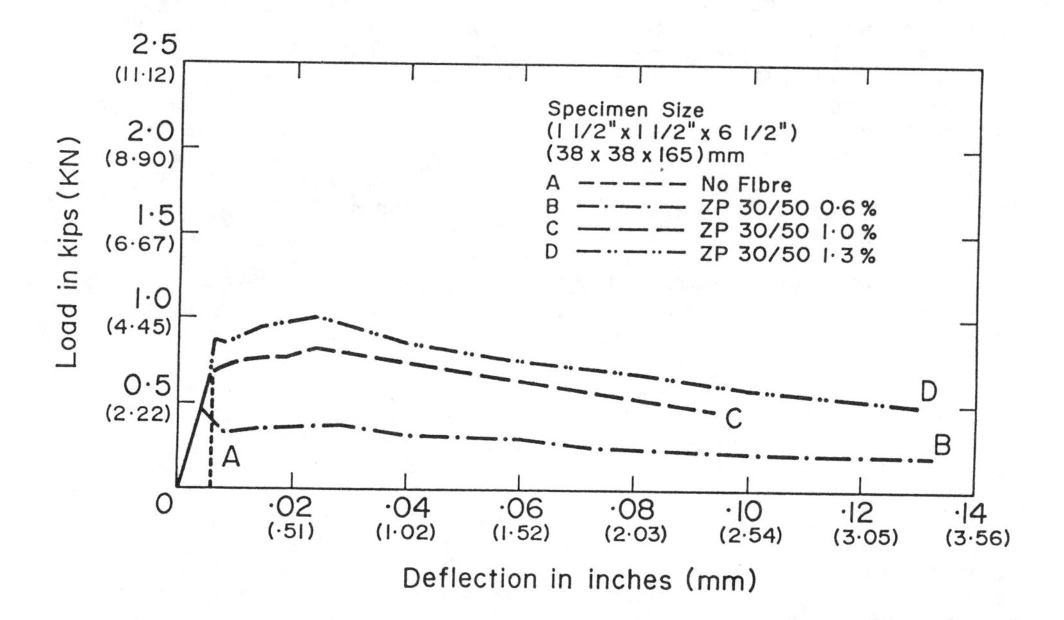

FIG. 3. Load-deflection curves for different fiber contents.

Toughness Index: Energy absorption or toughness of shotcrete is increased considerably by the addition of steel fibers. The toughness index is taken as the measure of the amount of energy required to deflect the beam in the modulus of rupture test, a given amount compared to the energy required to bring the beam to the point of first crack (32). It is calculated as the area under the load-deflection curve up to 1.9 mm (0.075 in.), divided by the area under the load-deflection curve of the fibrous beam up to the first crack strength (proportional limit, defined as first deviation from linear). A new ASTM test procedure for the determination of toughness (33) index has been published and this procedure is slightly different from that of ACI procedure. However, all the values reported so far are obtained by the ACI test procedure. For plain shotcrete, which sustains no post first-crack load, this value is 1.0. The toughness index of SFRS increases with increased fiber loading and aspect ratio of fiber. The toughness index is also affected by the aggregate size.

Typical toughness index values for SFRS range from 4 to 10. This parameter indicates the real benefits of adding steel fibers in shotcrete.

Impact Strength: The ACI Committee 544 has published a test procedure (32) for measuring impact resistance. It involves dropping a 4.54 kg. (10 lb.) hammer 457 mm (18 in.) onto a 64 mm (2.5 in.) diameter ball which rests on a 152 mm (6 in.) diameter by 64 mm (2.5 in.) high shotcrete core. The number of blows to first crack, as well as the number of blows to cause disruption of the specimens are measured. Though the impact test is empirical and the test results have a high coefficient of variation, it gives a useful indication of the impact resistance of shotcrete.

The number of blows required to cause failure at 28 days ranges from 100 to 500 or more depending upon the fiber amount, length and type as shown in Figure 4. Plain shotcrete specimens fail at from 10 to 40 blows (2,4). It was found that hooked fiber shotcrete had a tremendous ability to absorb impact load, and this ability increased dramatically as the fiber concentration increased. A significant contribution to shotcrete due to the addition of steel fibers is the ability to control cracking and deformation under impact loads.

It was noted during testing that those specimens with large laminations tended to rupture at the laminations which reduced the effective thickness of the test specimens and failure occurred at a lower number of impacts (4).

Shrinkage: A knowlege of shrinkage strains is important when SFRS is used in repair and rehabilitation work. Some typical curves are shown in Figure 5. The addition of fibers reduces the drying shrinkage values and the reduction is proportional to the quantity of fibers added (4).

Elastic Properties: The dynamic and static modulus of elasticity values follow the same pattern as the compressive strength values for SFRS. The addition of steel fibers does not appreciably influence the elastic properties (4).

The measured pulse velocities varied from 4.22 km/sec (13,000 ft/sec) to 4.59 km/sec (15,000 ft/sec) which means that all these shotcretes could be classified as of excellent quality. The measured pulse veloci-

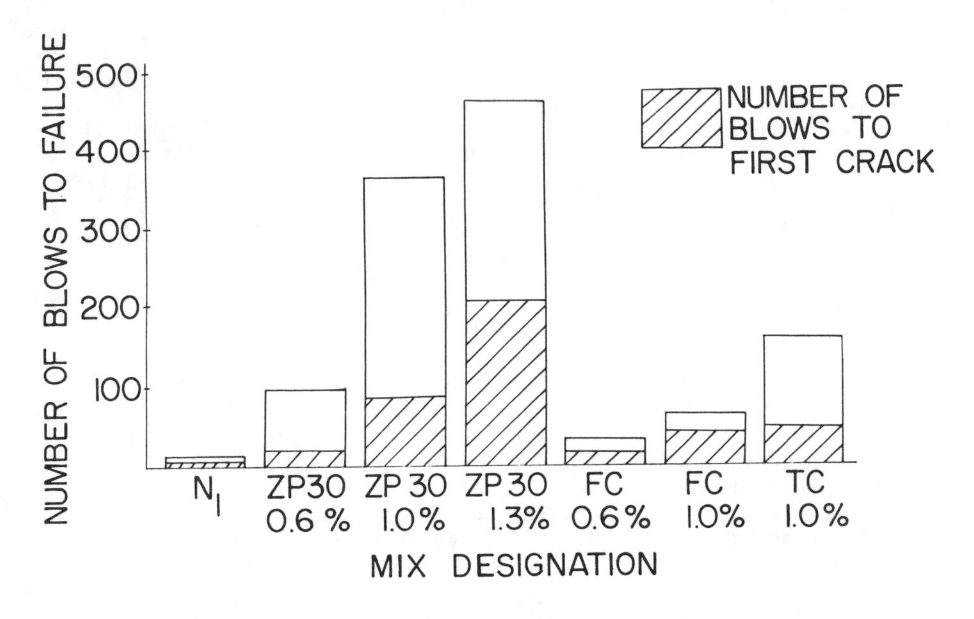

FIG. 4. Impact resistance for different types of fibers.

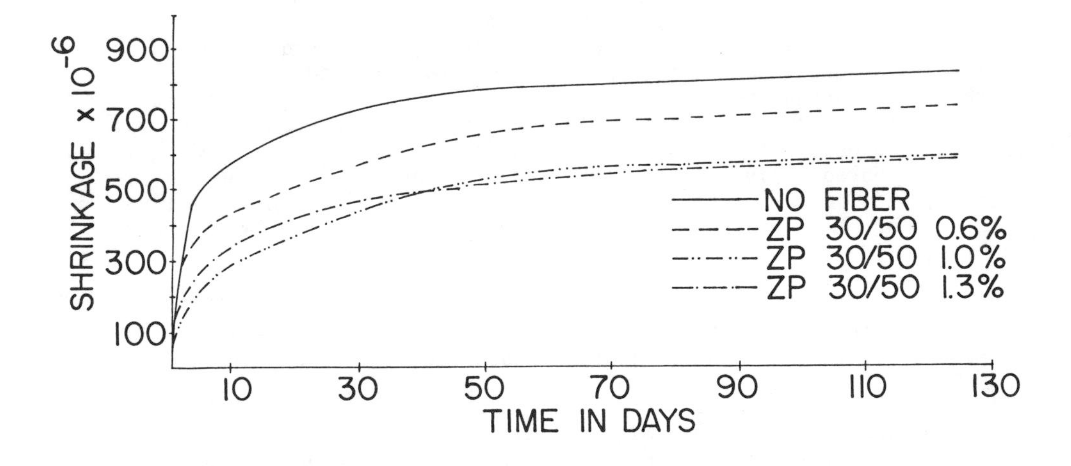

FIG. 5. Shrinkage curves for different fiber contents. Time reckoned since end of wet curing at the age of 28-days.

ties and the elastic properties were proportional to the compressive strengths of shotcrete (4).

Pullout Strength: The quality of in-place shotcrete is evaluated by the pullout strength test which uses pullout anchors which are embedded in the shotcrete as it is gunned. The pullout anchors are discs about 25 mm (1.0 in.) in diameter, embedded about 30 mm (1.25 in.) deep. For SFRS, a pullout strength of 12.4 MPa (1800 psi) has been reported (2).

Bond Strength: The reported bond strengths for SFRS vary from 1 MPa (145 psi) to 3.7 MPa (540 psi) on a rough-surfaced granite grain (2). These tests were done by pulling off a 610 x 610 mm (2 x 2 ft) steel plate embedded in a flat shotcrete layer and calculating the bond strength.

Strain-to-Failure: Significantly higher strain-to-failure in SFRS, measured in a rapid load flexural test, has been reported (19). The measured strains at 28 days for SFRS were nearly twice that of plain shotcrete.

Fatigue Strength: Fatique properties have not been reported for SFRS. However, fatigue performance characteristics are available for steel fiber reinforced concrete. The addition of steel fibers considerably increases fatigue life and decreases the crack width under fatigue loading.

Durability: There is paucity of information regarding durability of SFRS. However considerable amount of information is available for steel fiber reinforced concretes with similar mix proportions and fiber types.

Permeability and Porosity: Krantz reported (1) permeability and porosity values for SFRS. These properties are the primary factors in determining the durability of SFRS, which involves its alkali-acid reactivity, leaching characteristics, resistance to chloride or sulfate attack, reinforcement steel corrosion, and freezing and thawing characteristics. The shotcrete specimens exhibited a general trend of increasing permeability with decreasing compressive strength. Permeability also increases with an increase in water-cement ratio. Details of the test methods and test results are available in Reference 1.

REFERENCES

(1) Krantz, Gary W., "Selected Pneumatic Gunite for Use in Underground Mining: A Comparative Engineering Analysis," Bureau of Mines Information Circular, 1984; 1C8984, U.S. Department of the Interior.

(2) ACI Committee 506, "State-of-the-Art Report on Fiber Reinforced Shotcrete", Concrete International: Design & Construction, V. 6, No. 12, December 1984, pp. 15-27.

(3) Morgan, Dudley R. and McAskill, Neil, "Rocky Mountain Tunnels Lined with Steel Fiber Reinforced Shotcrete", Concrete International: Design & Construction, V. 6, No. 12, December 1984, pp. 38-38.

(4) Ramakrishnan, V.; Coyle, W.V.; Dhal, Linda Fowler; and Schrader, Ernest K., "A Comparative Evaluation of Fiber Shotcrete", Concrete International: Design & Construction, V. 3, No. 1, January 1981, pp.56-69.

(5) Morgan, Dudley R., "Steel Fiber Shotcrete - A Laboratory Study", Concrete International: Design & Construction, V. 3, No. 1, January 1981, pp. 70-74.

(6) Shotcrete for Ground Support, Sp-54, American Society of Civil Engineers/ American Concrete Institute, Detroit, 1977, 766 pp.

(7) Ramakrishnan, V.; Coyle, W.V.; Kulandaisamy, V.; and Schrader, Ernest K., " Performance Characteristics of Fiber Reinforced Concretes with Low Fiber Contents", ACI Journal, Proceedings V. 78, No. 5, Sept.-Oct 1981, pp. 384-394.

(8) Ramakrishnan, V.; Brandshaug, Terje; Coyle, W.V.; and Schrader, Ernest K., " A Comparative Evaluation of Concrete Reinforced with Straight Steel Fibers and Fibers with Deformed Ends Glued Together into Bundles", ACI Journal, Proceedings, V. 77, No. 3, May-June 1980, pp. 135-143.

(9) Lankard, D.R., "Field Experiences with Steel Fibrous Concrete", presented at American Ceramic Society Meeting, Chicago, Apr. 26, 1971.

(10) Poad, M.E.; Serbousek, M.O.; and Goris, J., "Engineering Properties of Fiber-Reinforced and Polymer-Impregnated Shotcrete", Report of Investigations No. 8001, U.S. Bureau of Mines, Washington, DC 1975, 25pp.

(11) Van Rysaryk, R., "Tunnel Repairs", Construction West, September 1979.

(12) Little, T., "An Evaluation of Steel Fibre Reinforced Shotcrete", 36th Canadian Geotechnical Conference, Vancouver, British Columbia, June 1983.

(13) Williamson, G.R., et. al., "Inflation/Foam/Shotcrete System for Rapid Shelter Construction", CERL Technical Report No. M-215, U.S. Army Construction Engineering Research Laboratory, Champaign, May 1977.

(14) Wilkinson, Bruce M., "Foam Domes, High Performance Environmental Enclosures", Concrete Construction, V. 23, No. 7, July 1978, PP. 405-406.

(15) "Shotcrete and Foam Insulation Shaped over Inflated Balloon Form", Concrete Construction, V. 27, No. 6, June 1982, pp. 511-513.

(16) Nelson, K.O., and Henager, C.H., "Analysis of Shotcrete Domes Loaded by Deadweight", Preprint No. 81-512, ASCE Convention (St. Louis, Oct. 1981), American Society of Civil Engineers, New York, 1981.

(17) ACI Committee 544, "State-of-the-Art Report on Fiber Reinforced Concrete", Report ACI 544-1R-82, Concrete International: Design & Construction, May 1982.

(18) Glassgold, I. Leon, "Refractory Shotcrete-Current State-of-the-Art", Concrete International: Design & Construction, V. 3, No. 1, Jan. 1981, pp. 41-49.

(19) Kaden, Richard A., "Fiber-Reinforced Shotcrete. Ririe Dam and Little Goose (CPRR) Relocation", Shotcrete for Ground Support, SP-54, American Concrete Institute/American Society of Civil Engineers, Detroit, 1977, pp. 66-68.

(20) Sandell, Bertill, "Steel Fiber Reinforced Shotcrete (Stalifiberar-merad Sprutbeton)", Proceedings, Information-Dagen 1977, Cement-Och Betonginstitutet, Stockholm, 1977, pp. 50-75.

(21) Ironman, Ralph, "Shotcreting by Robot-Wet Mix Process Gains Favor in Norway", Concrete Products, 1984, pp. 28 and 34.

(22) Rose, D.C., et. al., " The Atlanta Research Chamber, Applied Research for Tunnels", Report No. UMTA-GA-060-0007-81-1, U.S. Department of Transportation, Washington, DC, March 1981.

(23) Warner, B.L., "Evaluation of Materials for Protecting Existing Urethane Foam in Mines", Report No. ORF 75-76 (NTIS PB 254 682), U.S. Bureau of Mines, Washington, DC, Sept. 1974.

(24) Chronis, N.P., "Three Innovations in Mine Expansion Tested at Bruceton Experimental Mine", Coal Age, V. 80, No. 4, Apr. 1975.

(25) Murphy, E.M., "Steel Fiber Shotcrete in Mines", Concrete Construction, V. 20, No. 10, Oct. 1975, pp. 443-445.

(26) Kaden, R.A., "Slope Stabilized with Steel Fibrous Shotcrete", Western Construction, Apr. 1974, pp. 30-33.

(27) Malmberg, B., and Ostfyord, S., " Field Test of Steel Fibre Reinforced Shotcrete at Scan-Raff, Brofyorden", Fiberbetong, Norforskc Projckt Committcc for FRC Matcrial Dclvapportcr, Cement-Och Betonginstitutet, Stockholm, 1977, pp. y1-y16.

(28) ACI Committee 506, "Standard Specification for Materials, Proportioning and Application of Shotcrete", ACI 506.2-77, ACI Manual of Concrete Practice, Part 5, 1982.

(29) Parker, H.W.; Fernandez, G.; and Lorig, L.J., "Field-Oriented Investigation of Conventional and Experimental Shotcrete for Tunnels", Report No. FRA.OR & D 76-06, Federal Railroad Administrastion, Washington, DC, Aug. 1975, 628pp.

(30) Ryan, T.F., "Steel Fibers in Gunite, An Appraisal" , Tunnels and Tunneling (London), July 1975, pp. 74-75.

(31) Henager, C.H., "The Technology and Uses of Steel Fibrous Shotcrete: A State-of-the-Art Report", Battelle-Northwest, Richland, Sept. 1977, 60pp.

(32) ACI Committee 544, "Measurement of Properties of Fiber Reinforced Concrete", ACI544.2R.78, ACI Manual of Concrete Practice, Part 5, 1982.

(33) American Society for Testing and Materials (ASTM) C-1018-84-Test Method for Flexural Toughness of Fiber-Reinforced Concrete (Using Beam with Third-Point Loading), 1916 Race St., Philadelphia, PA 19103.

STEEL FIBER CONCRETE
US-SWEDEN joint seminar
(NSF-STU)
Stockholm 3-5 June, 1985
S P Shah and Å Skarendahl,
Editors

SYSTEM BESAB FOR HIGH STRENGTH STEEL FIBER REINFORCED SHOTCRETE

by Nils-Olov Sandell, Man Dir and Börje Westerdahl, MSc, AB BESAB, Göteborg, Sweden

---

ABSTRACT

BESAB, a contracting and machine manufacturing company, present their technique for production of high-strength steel fiber reinforced shotcrete, i.e. a composite material where the relationship between load and deformation enables the load to be increase well in excess of the load at which the matrix cracks. The necessity of a high value of the parameter w · l/d is indicated; where w = weight % fiber, l = length of fiber, and d = diameter of fiber.

The method of feeding the steel fiber to the spray nozzle in a separate hose is presented, a method which may be utilized for both dry and wet spraying. Nozzles developed for the two methods are shown.

Results from large-scale load trials done with system BESAB in co-operation with Royal Swedish Fortifications Administration are presented.

The results of several contract operations are presented.

Nils-Olov Sandell, Managing Director, AB BESAB, Göteborg

Börje Westerdahl, MSc, AB BESAB, Göteborg

INTRODUCTION

BESAB is a machine manufacturing contracting company primarily active within the fields of rock reinforcement, foundation reinforcement, concrete demolition and facade renovation. It was activity in the fields of rock reinforcement and facade renovation that led to a growing interest in fiber reinforcement of cement based matrices. Initially, reinforcement involved thin layers of wire netting, but this proved difficult and expensive to apply, and in many cases the technical results were inadequate.

During the early 70´s we first began experimenting with fiber reinforcement. The uncertainties of alkali resistance in glass fiber favoured the use of steel fiber for which well documented evidence concerning the combination of steel and cement was already available.

Steel fiber reinforced shotcrete is not a uniform material, and the main parameters which govern the characteristics of this composite material are the characteristics of the fiber and concrete matrix, and the adhesion between the two.

If we concentrate on the fiber we can see that there are three basic methods of producing steel fiber on the market today:

- by cutting cold drawn wire
- by slitting steel sheet
- by extracting them from a pool of molton steel (melt-extraction).

Fiber manufactured from drawn wire has a tensile strength of about 1.200 MN/m^2. Fiber made by slitting steel sheet about 600 MN/m^2.

A fracture in a fiber reinforced composite occurs when the initial crack, or some other form of crack, expands to the extent that the fibers bridging the crack are pulled out or breaked. A variety of stress and deformation relationship are possible, depending on the characteristics of the matrix, the fiber, etc. Fig 1 describes load and deformation relationships for flexural tested beams. A comparison may be made between a matrix reinforced wich 4.3 weight % fiber of drawn wire (d = 0.35 mm, l = 35 mm, l/d = 100) and the same matrix reinforced with 7.2 weight % fiber made by slitting sheet (b = 0.6 mm, t = 0.3 mm and l = 18 mm; l/d = 32.6).

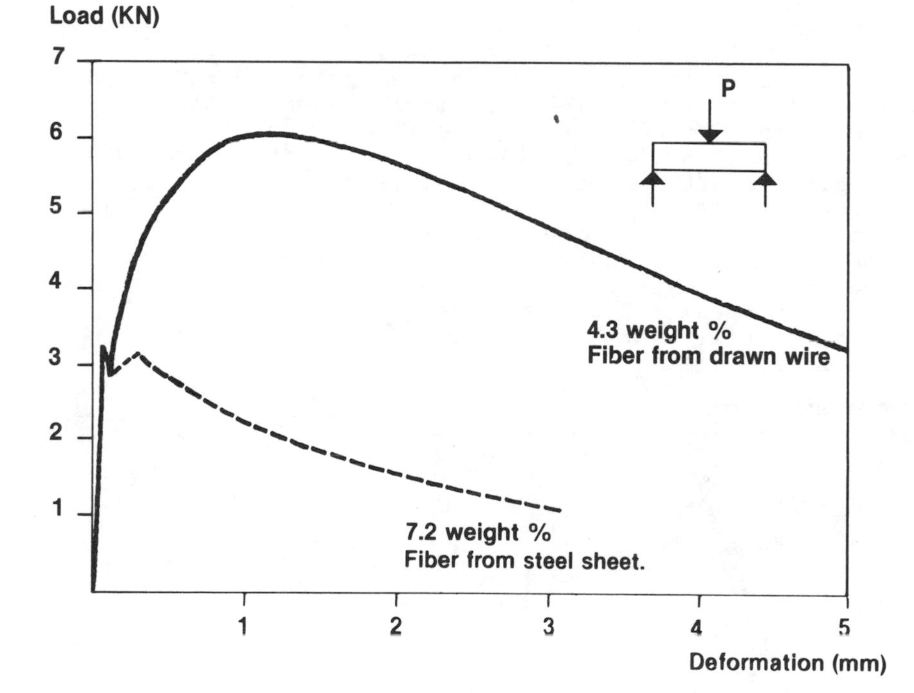

FIG 1. Comparative flexural tests on shotcrete reinforced with steel fibers made from drawn wire and steel sheet. The tests carried out by the National Swedish Institute for Materials Testing.

Steel fiber shotcrete can be made with about half the content of steel sheet fiber shown in Fig 1 without the aid of a special technique. The wet mix method enables the fiber to be mixed in the mixer and pumped with conventional pumps. BESAB have carried out a number of contracts in this manner. If the material characteristics achieved with this method are found to be satisfactory then there is no cause for further worry. However, we realize that in practice improved material characteristics are often required. The remainder of this article will therefore be concerned with the technique that has been developed for the production of high-strength steel fiber shotcrete, implying a shotcrete where the relationship between load and deformation enables load to be considerably increased beyond the load at which the matrix cracks. The curve for fiber made from drawn wire in Fig 1 is an example of high-strength steel fiber shotcrete.

THE DEVELOPMENT OF THE BESAB SYSTEM

We first began experimenting with steel fiber about ten years ago. We acquired 100 kilos of fiber made of wire originally intended for use in tyres. It was brass coated; d = 0.25 mm, l = 25 mm, and l/d = 100. The experiment was conducted with dry spraying equipment, and we ran into difficulties immediately with the notorious "balling problem" in our conventional free-fall mixers. The problem was not diminished when we attempted to spray the mixture with the aid of a feed-wheel gun. These first trials to get the feel of the thing resulted in a search for documented evidence in relevant literature on the subject. We came across an ACI publication, SP44, Steel Fiber Reinforced Concrete. In professor C.D. Johnston´s article we found a relationship which later became the basis of our present innovation. This is shown in Fig 2 below.

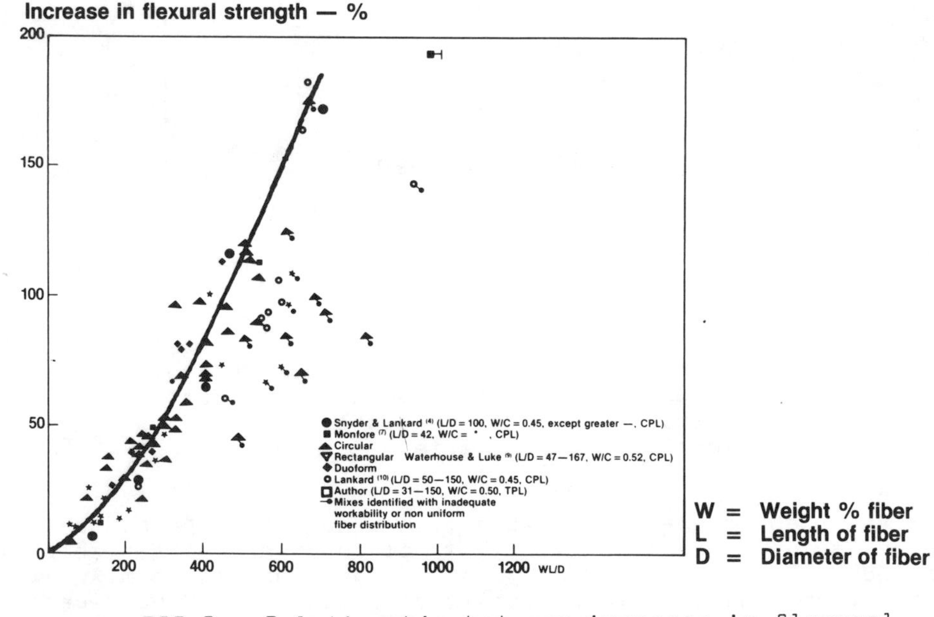

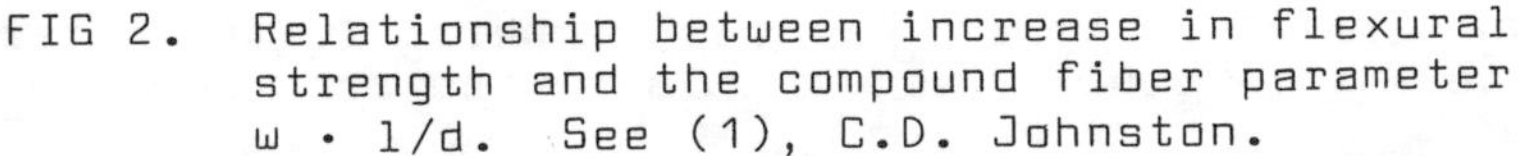
FIG 2. Relationship between increase in flexural strength and the compound fiber parameter w · l/d. See (1), C.D. Johnston.

Our first target was to increase the flexural strength of the composite material. The relationship in Fig 2 and the result of our experiments convinced us that we would not succeed without the innovation of a new technique.

Fracture characteristics were also crucial. Here the flexural toughness based on area under the load-deformation curve could be of great assistance.

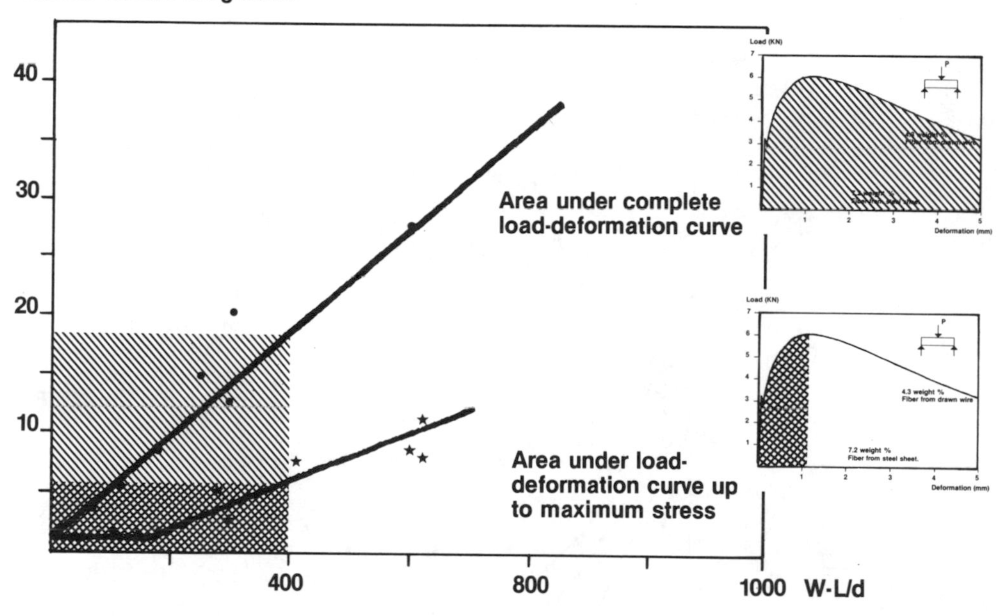

FIG 3. Influence of w · l/d on flexure toughness based on area under load-deformation curve. See (2).

When we processed the curve in Fig 2 with an economic parameter we established that this new technique was economically feasible; see Fig 4.

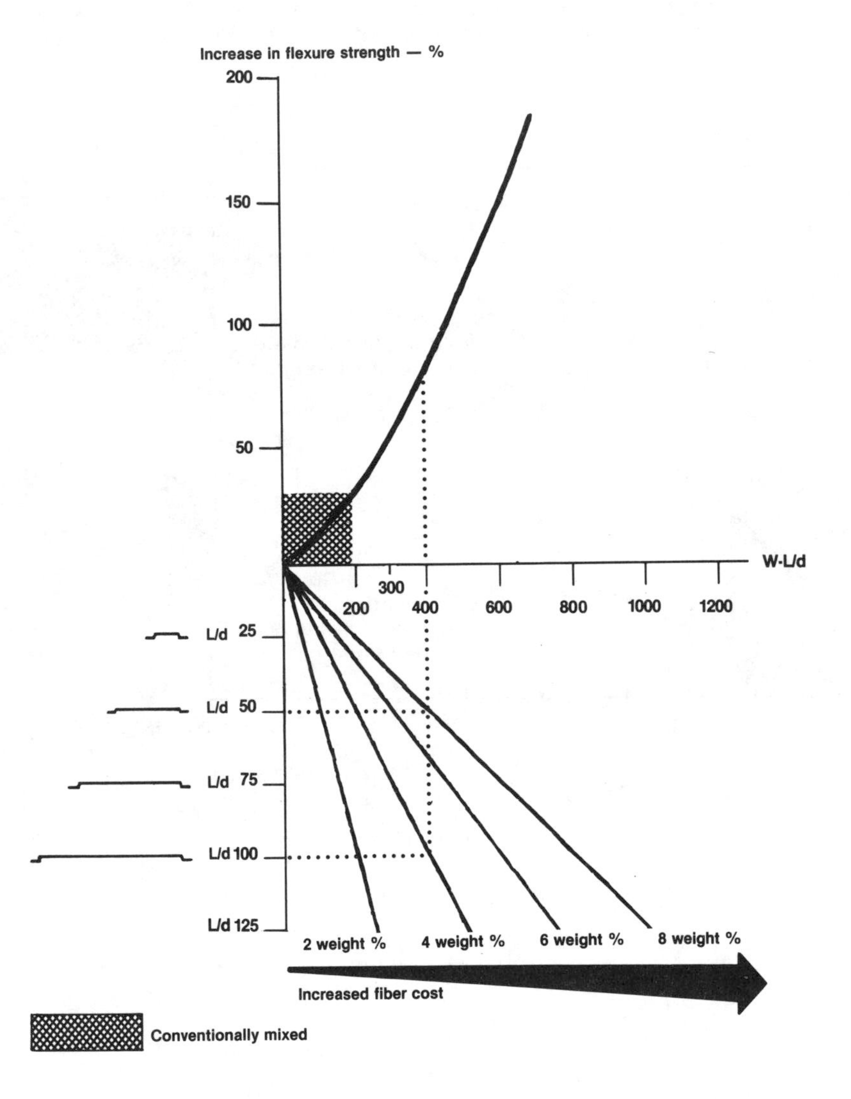

FIG 4.

THE BESAB SYSTEM

The BESAB system was developed for the production of high-strength steel fiber shotcrete. This required the inclusion of a high weight % of long thin fibers, i.e. a high value of the parameter w . l/d, which trials had shown to be incompatible with conventional mixer and spraying methods.

A new technique was envisaged where the fiber was supplied in a separate hose to the spray nozzle prior to being mixed with the concrete.

The feeder equipment comprised a sloping, rotating drum which broke up the incoming fiber mass with the aid of special spikes. Gravitation then fed the loose fibers through an adjustable grille aperture into the front section of the drum where controlled feeding took place; see Fig 5.

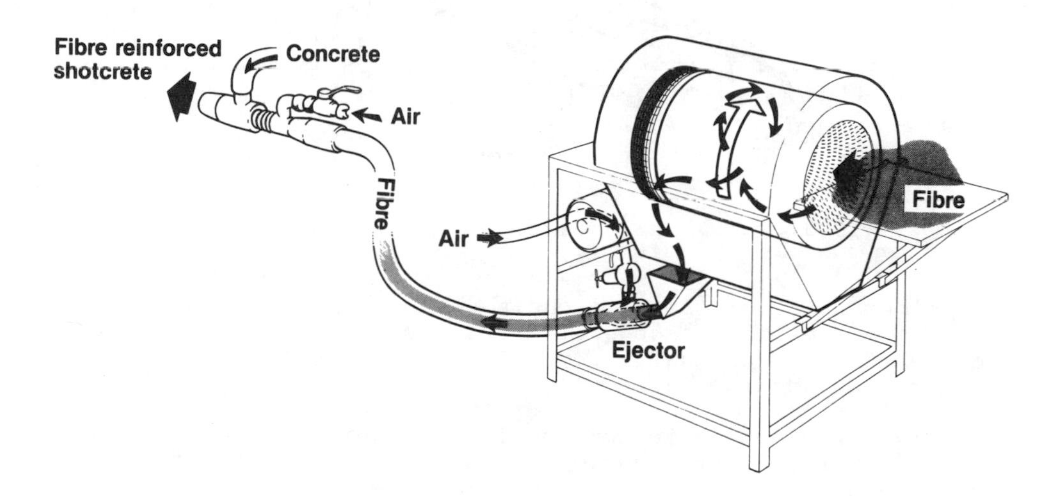

FIG 5. System BESAB for the production of high-strength steel fiber shotcrete.

By reducing or increasing the size of the aperture the quantity of fiber could be regulated. The fiber falls in a regular flow to the receiving funnel and is then blown via a ring-ejector into the spray nozzle. The ejector was specially designed for conveying fiber.

The primary air-hole accomodates supersonic velocities, providing long and reliable conveyance with a minimum of air consumption. The quantity of fiber can be adjusted to correspond to the various capacities of spray units and the feeder equipment may be used for all types of fiber on the market, including punched, wave shaped, hooked, and smoth fibers, etc. Feeder precision is excellent for the full range of fibers.

The equipment may be used for both wet and dry spraying methods. Spray nozzles have also been designed for both methods.

The design of the wet spraying method enables the fiber to be centrally supplied and coated with concrete when it emerges from the aperture, see Fig 6 below.

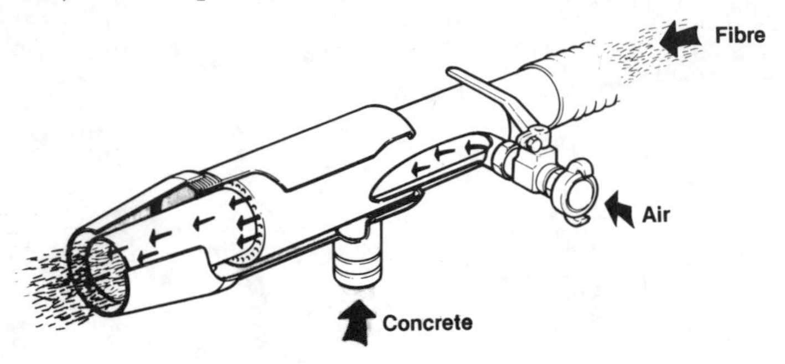

FIG 6. Wet spray nozzel for system BESAB.

A minimum of fiber waste is achieved and the fiber has a consolidating effect on the concrete mass when it is applied to the form surface, reducing or completely avoiding the use of an accelerator addmixture.

The dry spray nozzle is constructed so that the dry concrete is mixed with the fiber first, water added and mixing carried out prior to ejection through the nozzle; see Fig 7 below.

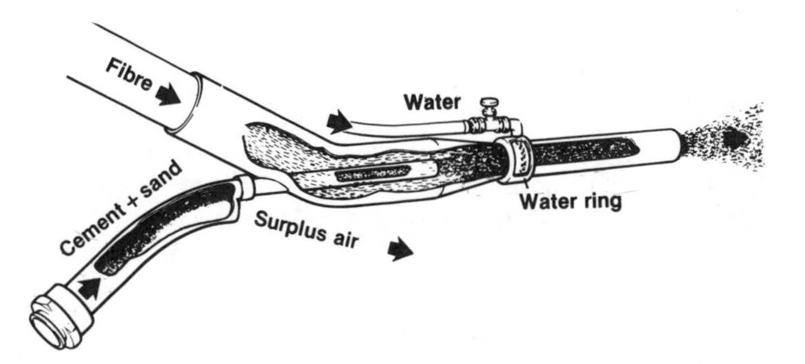

FIG 7. Dry spray nozzle for system BESAB.

The dry spray method requires a large volume of air to facilitate conveyance of the dry concrete, increasing fiber spillage at the point of application. Excess air is therefore extracted at the dry spray nozzle prior to the mixing of fiber and concrete.

The spray nozzles are currently the object of further experimentation in co-operation with Chalmers University of Technology. High-speed film is used to study the mixing process. Fig 8 shows one such experiment.

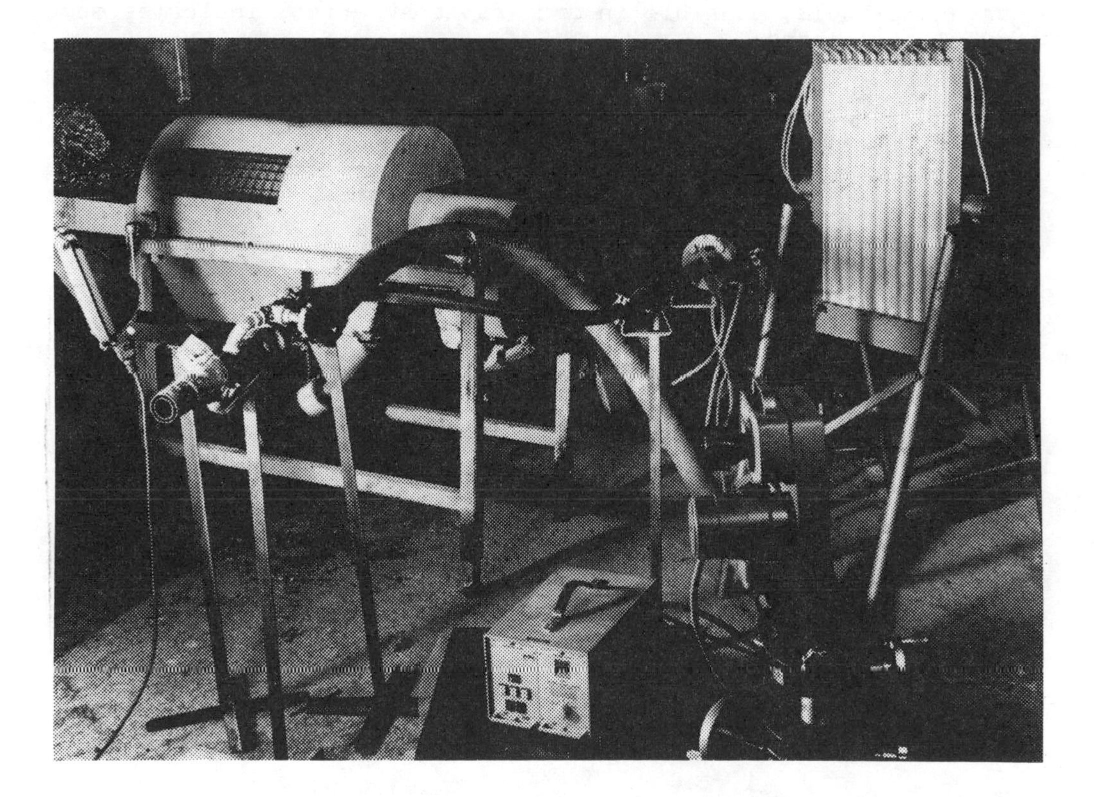

FIG 8. High speed film on nozzle function.

LARGE-SCALE LOAD TRIALS

In co-operation with the Royal Swedish Fortifications Administration large-scale load trials have been carried out under the supervision of Dr. Jonas Holmgren. A detailed report is included in Dr. Holmgren´s report "Bolt Anchored Steel Fiber Reinforced Shotcrete Linings". The experiments showed that even a small quantity of steel fiber (approx. 1.5 weight %)ensures superior rock reinforcement than can be achieved with conventional net reinforcement. A 5 weight % steel fiber shotcrete reinforcement, where l/d = 129, gives the same reinforcement as a moment adjusted system, a system which is neither technically or economically possible in rock caverns. Fig 9 indicates some results attained.

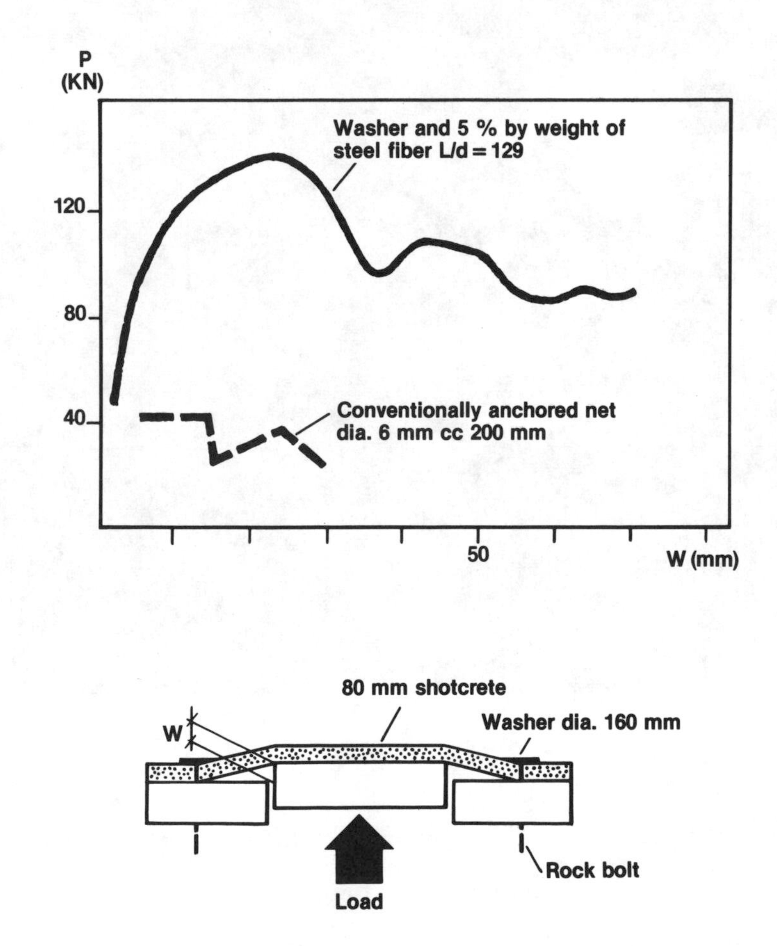

FIG 9. Steel fiber reinforcement compared with conventional net reinforcement in shotcrete.

The large-scale trials indicate several advantages with the use of high-strength steel fiber shotcrete. The fiber spins with a definite resistance in the matrix, making a possible fracture supple with respect to both bending and shearing. Several shear tests were carried out with small crached test samples. The cracks were widened to 3 mm, a shear force applied and the results registered on a stress-displacement diagram as shown in Fig 10 below.

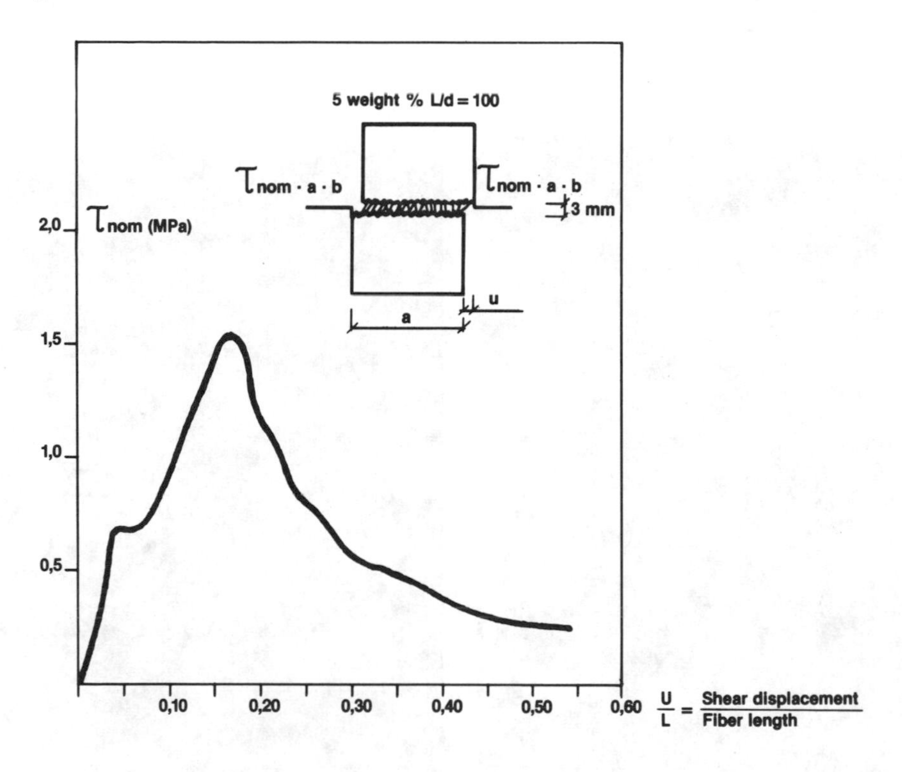

FIG 10. Shear stress versus displacement for small scale shear tests.

The diagram is characterized by a pronounced peak and a long "tail", desplaying substantial plasticity after maximum loading. A steel fiber shotcrete enables a shearing strength five times greater than conventionally reinforced concrete.

Advanced reinforcement system where the joint effect of rock bolts and shotcrete is used is easier to design when high-strength steel fiber shotcrete is used.

The main results from the large-scale load trials were

1. It is possible to produce steel fiber reinforced shotcrete linings which are at least equally strong and ductile in bending as conventionally reinforced ones.

2. Cheaper washers can be used for the anchorage of the rock bolts than when mesh is used.
3. The shear performance of a steel fiber reinforced lining seems to be superior to that of a conventionally reinforced one also after cracking.

EXAMPLES OF FINISHED WORK

Rock reinforcement

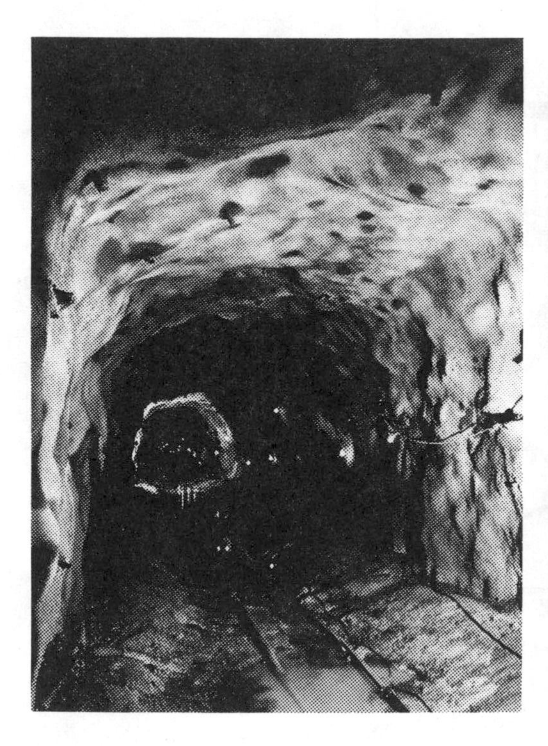

Emergency cooler tunnel at Ringhals nuclear power station commissioned in 1977 by the National Swedish Power Administration. Steel fiber reinforced shotcrete used in conjunction with rock bolts.

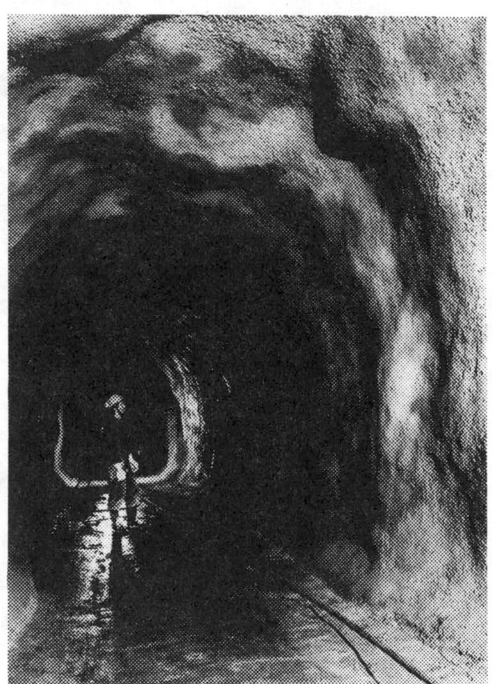

During 1981 a tunnel to a power station plant in Tollered, Sweden, was reinforced. The commissioner was Viskans Power Station.

In addition to spot reinforcement of weak sections throughout the tunnel a pressure tunnel was constructed with steel fiber shotcrete. The fiber reinforcement replaced a powerful conventional reinforcement.

Robot spraying of steel fiber reinforced shotcrete in a rock cavern for oil storage, 1978.

Fiber reinforced shotcrete has a very large energy absorbtion capacity. Steel fiber reinforced shotcrete has been used in a number of mines in Sweden. At the Boliden Mine Ltd Company mine near Kristineberg the material was used to line and stabilize a gravity ore transfer shaft which was deteriorating from the impact of the ore. The shaft was filled with ore so that the top surface of the ore became the working platform for the shotcreating operation.

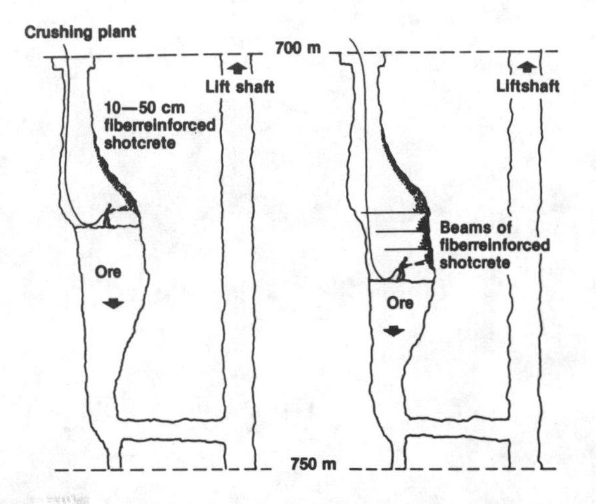

FIG 13. Technique for lining of ore shaft at Boliden´s mine, Sweden.

Render-bearing plate of 20 mm thick steel fiber reinforced shotcrete. The plate is sprayed on the curtain wall and anchored with fastening at cc 400 mm in the concrete wall and the framing of joists.

Steel fiber reinforced render on additional insulation.

REFERENCES

(1) ACI Publication SP-44."Fiber Reinforced Concrete", 1974.

(2) Hannant, D.J., "Fibre Cements and Fibre Concretes", John Wiley and Sons Ltd., 1978.

(3) Holmgren, J., "Bolt Anchored Steel Fibre Reinforced Shotcrete Linings".

(4) Sandell, B., "Våtsprutad stålfiberarmerad betong", Statens Råd för Byggnadsforskning. Rapport R 87:1978.

(5) Sandell, B., "Fiberarmerad puts", Statens Råd för Byggnadsforskning. Rapport R 1:1982.

(6) ACI Committee 506. State of the Art Report on Fiber Reinforced Shotcrete. Concrete International, Dec. 1984.

STEEL FIBER CONCRETE
US-SWEDEN joint seminar
(NSF-STU)
Stockholm 3-5 June, 1985
S P Shah and Å Skarendahl,
Editors

LACK OF PRACTICABLE STANDARDS AND TEST-METHODS RESTRICT THE DEVELOPMENT OF STEEL-FIBER SHOTCRETE

by Contr. Mgr Stig Östfjord, Stabilator AB, Gothenburg, Sveden.

ABSTRACT

Fiber cement composites represent one of major development in the construction industry in recent times.Fibers unique properties in controlling cracking and deformation are unquestionable. Another advantage is the (SFRS)-materials possibility to mould into intricate shapes. In spite of the fact that this has been known for quite a long time, still no standards or codes of practice for safe design stresses, deformation or durability reqvirements exist for the use of steel-fiber-reinforced-shot-shotcrete in building industry. This paper want to emphasize the urgent need of existence of practicable standards and test-methods by which the (SRFS) material should be studied without incorrectly favouring one type of material or design at the expence of another.

Stig Östfjord is working as a Contract Manager at Stabilator AB, a contracting company specialized in foundation engineering and underground works. He was a member of the Nordforsk project comitté for FRC-material 1974-1977. He has also been in charge of several contracts where steel-fiber-reinforced shotcrete has been used in full production, where one of these is the sloop-protection at Brofjorden Refinery which is still the largest structure made in Sweden with SFRC-material.

INTRODUCTION

During the last ten years the use of fiber-reinforced-shotcrete has increased, and in many cases with great success.
On the other hand - the number of structures where expected properties of the final product not have been achieved, has also increased.

There is many reasons for these failures to occur but it is obvious that lack of basic principals as "codes of good practice" and test-methods is essential.

Important factors with great influence on the final product is also what process (wet or dry), cement content and aggregate grading is used in combination with the wide variety of admixtures and steel-fibers available on the market.

Based on results from our last ten years of shotcreting with fibre, where wet shotcreting process with fiber is concentrated to the last 3 years, has it been obvious that the reqvired mixdesign, with high content of cement and fines, which is essential for the fiber-reinforced-shotcrete to obtain bond between fibers and matris, asks for our very special attention.

Also a study of the failures casued by wrong technic used by the crew in production is of most importance in order to seperate these from failures in material or mechanical properties.

Fig 1 A. Stabilator shotcrete system with robot is used for the application of steel-fiber-reinforced shotcrete at Kalhäll, Sweden

Fig 1 B. The delivery of shotcrete is done by transmixers from a ready-mix-plant

MATERIAL

In most of our shotcrete-production the standard Portland cement has been used together with Silica fume.In addition to this additive as water-reducer and superplasticiser in dry powder form has been added to the mix in different dosages.

The aggregate gradation confirm on the fine side to the gradation envelope specified in ACI 506, gradation no 2.

Steel-fibers which has been used is both EE-fibers 19 mm Ø 0,6 and straight fibers in various length up to 60 mm with a diameter of 0,5 - 0,6 mm.

WET PROCESS COMPARED WITH DRY

Most steel-fiber-reinforced-shotcrete placed to date has been with the dry porcess.
With the dry process, the reqvired high content of cement and fines in the mix for steel-fiber-shotcrete, has made it very difficult to achieve an acceptable even wetting of the dry mixed material in the nozzle.

There is a great risk that the result ends up in inhomogeneaus shotcrete and high rebound.

With the wet process, where all materials including water has been premixed and only the air and if necessary the accelerator are added in the nozzle, the rebound could be reduced to less then half and a throghly mixed shotcrete is produced.

The difficult part with the wet process is to get a pumpable mix with a consistence, which without accelerator can be shotcreated on a overhead or vertical surface to an acceptable thickness (30-50 mm) and with high capacity. This is most important for permanent works where accelerator in general is not accepted.

ACCERELATOR

The fact that use of accerelator is not accepted in permanent works, is specially a problem for the wet-process. Most of the pumps used for wet-shotcreting today require such a consistence for pumping that when it is shotcreted on a vertical or overhead surface, it does not stick to the surface unless accerelator is added.

If the problem with pulsation can be solved the piston-type of concrete pumps are able to handle the low slump mix, which is reqvired to get the shotcrete to stick to the surface without accelerator.

Using the right gradation of aggregate and a not too slow cement, the use of accerelator with the dry process can almost be avoided or heavily reduced.

On the other hand in addition to the wettingproblem even with normal rebound, which is at least twice the quantaty for wet-shotcrete, the costfactor for wasted material has to be encountered.

ADMIXTURES

Together with the wet shotcrete-process large number of admixtures have been introduced to the shotcrete process, where water-reducer (WR) and concrete-plasticiser (SP) together with Silika fume hase become the most freqvently used.

The advantage to use admixtures is out of question, but there is a great risk of missusing these. The diagram in Figure no 2, shows som of the testresults from a testprogramme made at our workingsite Bolmen where the influence of admixtures in different dosage has been studied.

The results shows that what the shotcrete gain in toughness by using long steel-fibers can be reduced to a minimum by too high or too low dosage of admixture. Also a wrong combination of admixtures gives considerable reduction in strength and toughness.

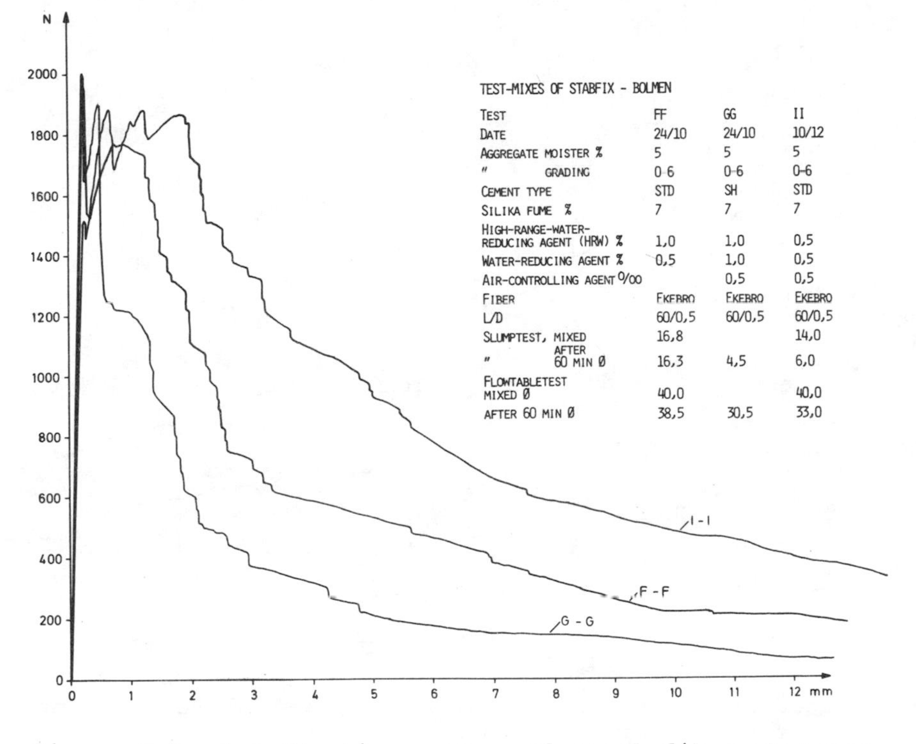

TEST-MIXES OF STABFIX - BOLMEN

TEST	FF	GG	II
DATE	24/10	24/10	10/12
AGGREGATE MOISTER %	5	5	5
" GRADING	0-6	0-6	0-6
CEMENT TYPE	STD	SH	STD
SILIKA FUME %	7	7	7
HIGH-RANGE-WATER-REDUCING AGENT (HRW) %	1,0	1,0	0,5
WATER-REDUCING AGENT %	0,5	1,0	0,5
AIR-CONTROLLING AGENT 0/00		0,5	0,5
FIBER	EKEBRO	EKEBRO	EKEBRO
L/D	60/0,5	60/0,5	60/0,5
SLUMPTEST, MIXED	16,8		14,0
" AFTER 60 MIN Ø	16,3	4,5	6,0
FLOWTABLETEST MIXED Ø	40,0		40,0
AFTER 60 MIN Ø	38,5	30,5	33,0

Fig no 2 Load-deflection curves of steel-fiber shotcrete with different dosage of admixtures.

FIBER-LENGTH AND SHAPE

That the fiber-lenght is essential, and with no doubts the most important factor for fiber-reinforced shotcrete, is unquestionable why the trend in Sweden and Norway, to use short fibers is strange.

At our worksite in Bolmen we have tested different lenght of fibers which can be studied in Figure no 3 which shows the effect of 1% fiber (by volume) with different lenght and the result is confirming the importance of the fiberlength.

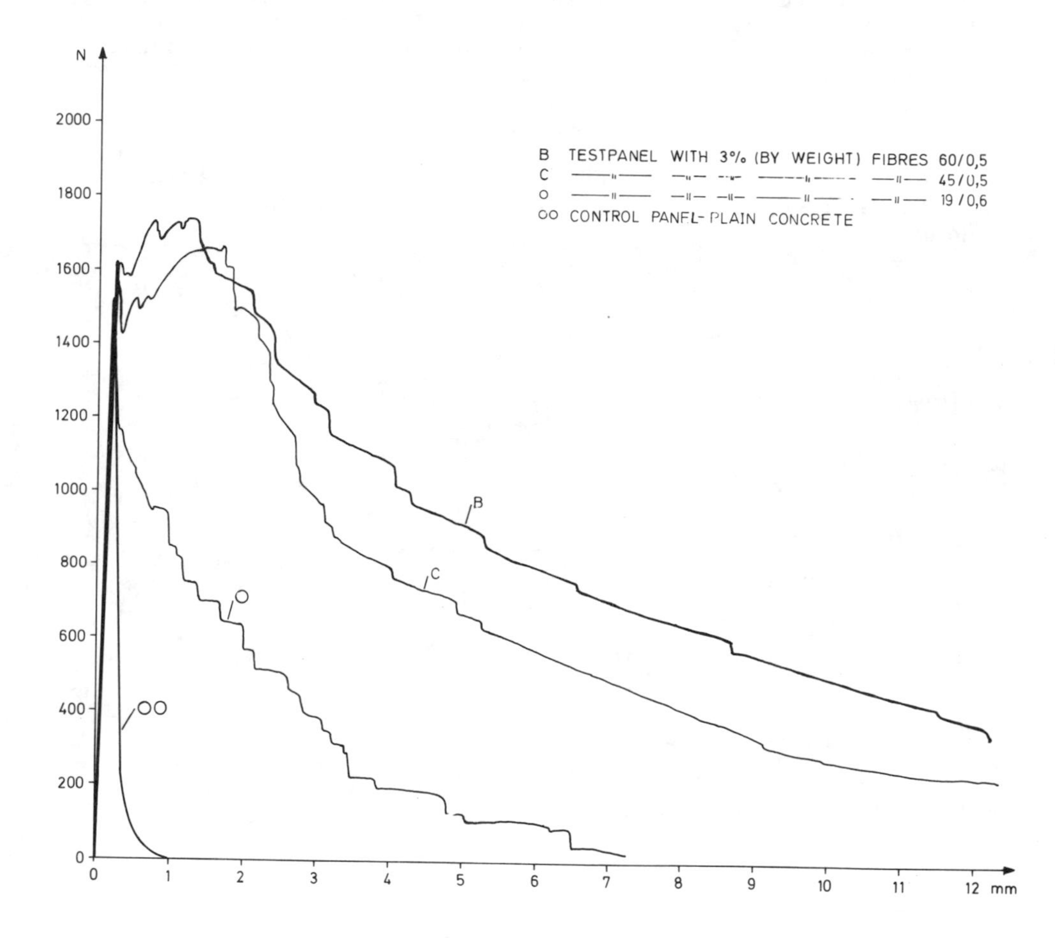

Fig no 3. Load-deflection curves of steel-fiber shotcrete with different length of fibers.

To show that the lenght is more important than the quantity of fibers, 1% of EE-fibers (L= 19 mm) were added into a mix where1% 60 mm long fibers already were mixed in.

The results indicate very clear that double guantity in weight of fibers does not give any notable increas of strength, compare diagram A and B in Figure no 4.

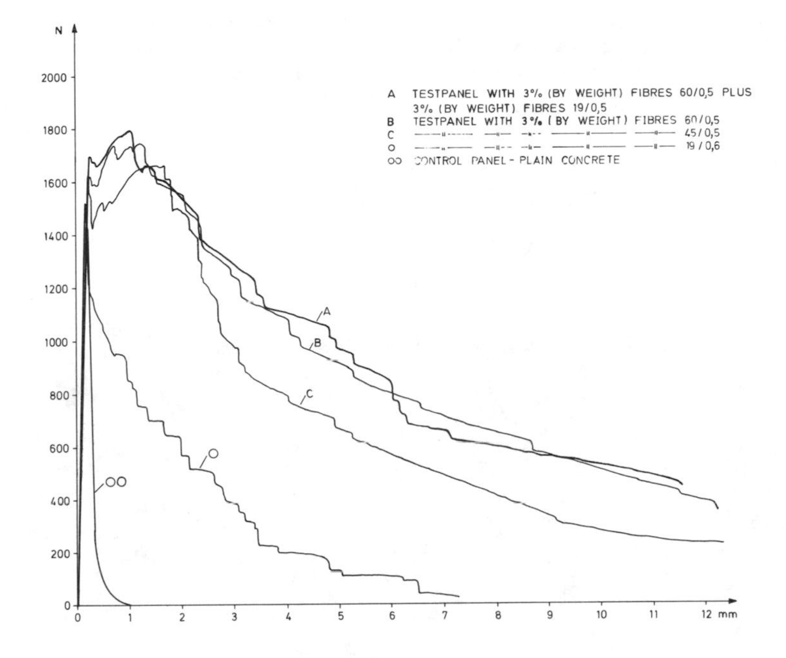

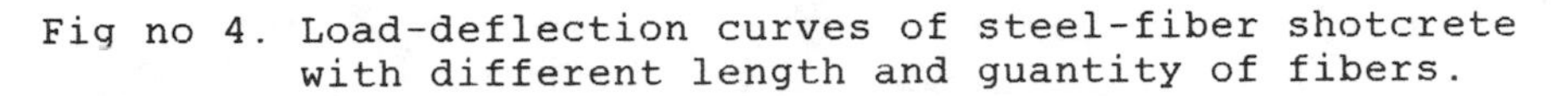
Fig no 4. Load-deflection curves of steel-fiber shotcrete with different length and quantity of fibers.

FIBER TYPES

Fibers made of steel is available in various shapes and sizes. We have in our production been using straight fibers with a length range from 19 mm to 60 mm.
We have for minor works also used anchored fibers, and for testpurpose the waved one.

Steel-fibers with bent or deformed ends have a high pullout resistant and can be used in smaller quantaties than straight fibers to achieve the same properties.

Test - methods

The testing criterion in Bolmen has been based on load-deflection - diagrames. The specimen chosen for all mixes has been 500 x 100 x 25 mm which has been tested in third point loading testing machine with a major span of 300 mm.

The constant rate of crosshead motion has been 0,3 mm/min, and the load-deflection diagram has been monitored by a plotter. The principle of testing can be seen in Figure no 5.

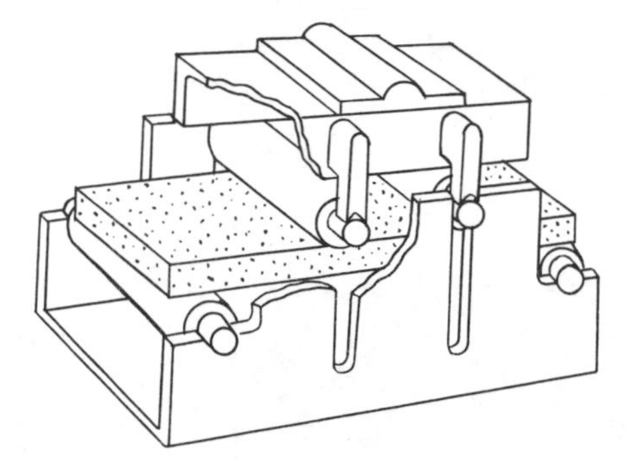

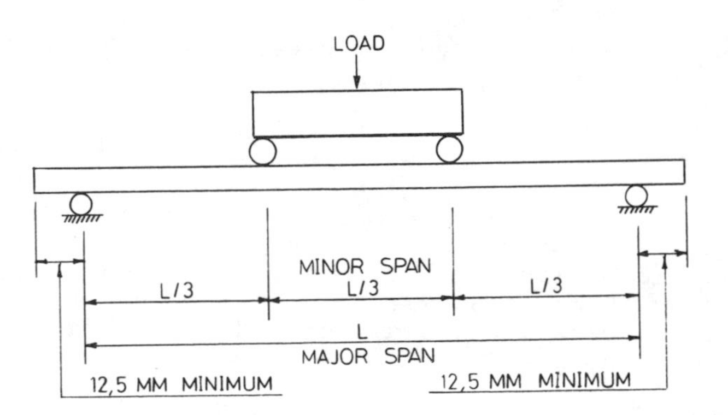

Fig 5. Loading arrangements and testing configuration for 3 RD point loading test.

During the test the distance between the midpoint and the first crack has been measured. This measurment has shown that the optimum load-deflection is achieved when the first crack is close to the mid-point, of the specimen, which indicate that the method with a notch in the mid-point of the specimen, can be one way of getting the most relavant testing-results out of the fiber-reinforced material.

SUMMARY

From fullscale production and research, it has been shown that the use of steel-fibers in reinforced shotcrete has the following benefits in flexural behavior.

- increases the moment capacity

- increases the ductility

- increases the tensile strength of material

- improves crack control

- increases the stiffness

It has also been shown that the material properties of fiber-reinforced shotcrete are influenced to a large extent by the type, volume percentage (se Figures 6, 7 and 8), aspect ratio, nature of deformation and orientation of the fibres.

For the beneficial use of fiber-reinforced shotcrete it is important that designers realize the importance of proper testing procedures when specifying a toughness index, tensile and compressive strength.

If the specimen size and testing procedure selected not is related to the structure size and fibre length, it can mislead the designers and cause failures and slow down the developing process of fiber-reinforced shotcrete applications due to that expected benefits not can be proved.

The most freqvent used tests in Sweden today is compressive and flexural strength-test, with no request on a load-deflextion diagram, which in most of the cases where fiber-reinforced shotcrete is used, is the most important criterion.

As a start, it could be of great importance to get a testing-method, simular to the one, which is specified in a Rilem draft of recommendations published by Rilem Technical Comitte 49 TFR accepted in Sweden.

From a contractors point of wiew it is essential that the testing-procedure is construed in a way which gives the contractor more than one option to produce the final product.

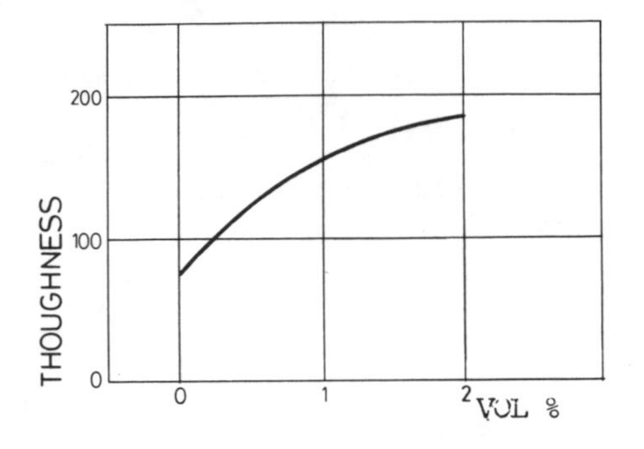

Fig no 6. The relation between thougness and fiber ratio.

STRENGTH TEST DATA ARE SHOWN IN THE FOLLOWING TABLE.

S.F. RATIO VOL %	BENDING STRENGTH	TENSILE STRENGTH	IMPACT RESISTANCE
0	1	1	1
1 - 2	2 - 4	1.4 - 2.7	3 - 12

Fig no 7. Strength test data are shown in the following table.

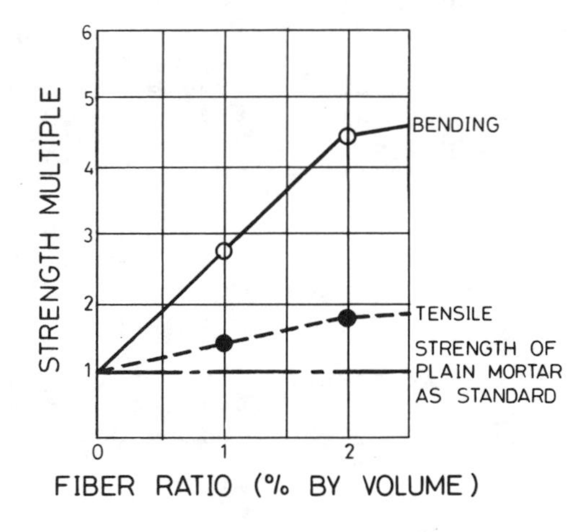

Fig no 8. Relationship between strength of fiber-mortar and fiber ratio.

STEEL FIBER CONCRETE
US-SWEDEN joint seminar
(NSF-STU)
Stockholm 3-5 June, 1985
S P Shah and Å Skarendahl,
Editors

WHY WET PROCESS STEEL FIBER REINFORCED SHOTCRETE?

by Ole Arnfinn Opsahl, MSc (Eng), Karl-Eric Buhre, MSc (Eng) and Rolf Hörnfeldt, MSc (Eng) , presented by Rolf Hörnfeldt, the General Contractor ABV, Solna, Sveden.

ABSTRACT

The development of shotcrete technology from the dry to the wet process, with or without steel fiber, and from manual spraying to robot controlled equipment has opened up new areas of use. The wet process with steel fibers gives concrete that not only has improved fracture toughness but is also more homogeneous and resistant. High quality shotcrete (100 MPg 14000 psi) can be made to meet any demands necessary. The current range of applications includes:

- reinforcement of tunnels and underground storage facilities
- stabilization of rock and earth slopes
- repair of concrete structures

Wet-process shotcrete has numerous advantages, including:

- cleaner working conditions
- safer working conditions
- higher capacity
- wide range of applications
- economy
- high quality
- durability
- water tightness
- toughness

Ole Arnfinn Opsahl, M.Sc. - Civil Engineering - Norwegian Institute of Technology, University of Trondheim.
Robocon.

Karl-Erik Buhre, M.Sc. - Civil Engineering - Royal Institute of Technology in Stockholm.
Robocon System AB.

Rolf Hörnfeldt recieved his M.Sc. degree in civil engineering at the Royal Institute of Technology in Stockholm 1961. He is now Technical Manager and responsibel for business and technical development in civil engineering at the general contractor ABV. Earlier he worked as project manager for bridge and harbour projects in Sweden and in GDR.

---

1. HISTORY AND THE PRESENT

According to the American Concrete Institute, shotcrete is concrete or mortar pneumatically projected at high velocity on to a surface (ACI 506.2-77).

Shotcrete consists of cement, aggregates, water and appropriate additives, i.e. the same materials as ordinary concrete.

Experiments with sprayed concrete started in the U.S.A. around 1910. The product was introduced under the name of Gunite by the Cement Gun Company of Allentown, Pennsylvania. Today the term "Gunite" is mainly used for dry-process shotcrete containing fine aggregates. Shotcrete was introduced in Scandinavia around the 1920s.

Dry-process shotcrete used to be the only method in use for quality reasons. Today, however, modern concrete technology, using water reducing additives, silica, and steel fiber, has made it possible to produce wet-process shotcrete with compressive strength from 25 to 100 MPa.

Improvements in rock tunneling methods more than before require efficient support systems during tunneling. As a result, wet-process shotcrete has been the only method used in Norway in the last few years.

Replacing mesh reinforcement with fibers was an important step towards reducing the construction time for shotcreting. Together with high nominal capacity (10-14 m^3/hour) and reduced rebound (5-10%) the delay due to support work can be reduced considerabely.

Continued technical development has opened up new applications for wet-process steel reinforced shotcrete. This technology is increasingly used for the stabilization of earth and rock slopes and also for strengthening and reparing concrete structures.

In Norway practically all shotcreting is done by the wet process today. In Sweden and Finland the wet process is starting to take over the dryprocess market. On the European continent, except in Italy, the dry method is used. The same has been true of the U.S.A. until today, but there are signs of a change to the wet method. This is especially significant on the west coast, where shotcrete to some extent is used in construction.

2. WET SHOTCRETE TECHNOLOGY

2.1 Applications

Wet shotcrete with fiber reinforcement may be used in all types of shotcreting work. Its high capacity, 10-14 m^3 per hour when robot is used, makes the method especially useful for shotcrete lining for ground support. This includes both temporary and permanent lining of tunnels, rock storage, rock slopes, earth slopes etc.

Fig. 1. Wet-process steel fiber reinforced shotcreting by robot rig

Wet-process steel fiber reinforced shotcrete should be used where high strength, durability, wear resistance, and crack distribution are required.

Where only small quantities are needed the advantages of the wet process are less conspicuous. Still it may well be competitive with the dry process thanks to less rebound, higher quality, cleaner working conditions, and high production stability.

2.2 Materials and production

The materials used for production of wet-process shotcrete are aggregate, cement, silica, steel fiber, plasticizers and accelerators.

Cement

All types of cement used for regular concrete are normally suitable for shotcrete. For example, if early strength is required, a rapid hardening PC should be prefered. In aggressive environments silica or combination of and special cement and silicas should be used (sulphate resisting or low alkali types).

Aggregate

Wet-process shotcrete requires well graded aggregates in order to achive good pumpability. This also means there is less waste, because well graded aggregate gives less rebound.

Steel fiber

The fiber length most commonly used with the wet process today is 18 mm. This represents a compromise between fiber handling characteristics and fiber efficiency. However, it is possible to use longer fibers in the wet process, up to 35 mm, if appropriate measures are taken.

Admixtures

Plasticizers and superplasticizers are used to give a suitable consistence at low W/C-ratios. Accelerator is used to achieve quick set and enable the application of thicker layers.

Concrete production

Most concrete is produced at ready-mix plants, carried to the site, and then loaded into the hopper of the pump (see Figs 1 and 2).

Fig. 2. Principles of wet-process shotcrete equipment.

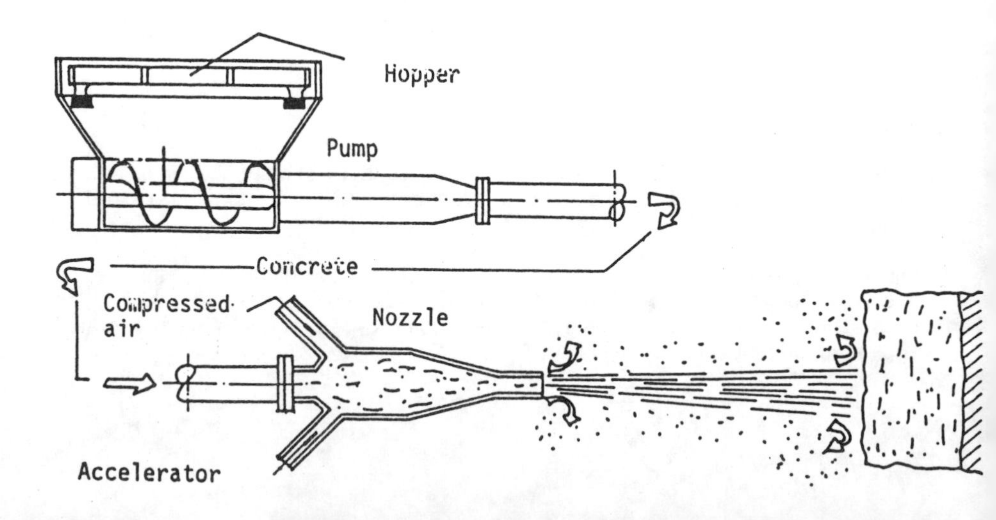

Rebound

One of the big advantages of wet-process shotcreting is the low rebound. The amount of rebound depends on the consistence of the concrete, the use of accelerators, spraying technique, and last but not least aggregate grading. In wet-process shotcreting a rebound of between 5 and 10 percent by weight is experienced while in the dry process rebound can be as high as 50 percent.

Working conditions

The wet process gives cleaner working conditions than the dry process. Measurements by the Construction Industry's Organization for Working Environment Safety and Health, performed for the Swedish State Power Board at its Messaure worksite, yielded the following values in connection with wet process shotcreting:

Table 1. Working conditions

Measuring point	Total dust content (mg/m^3)	Respirable dust (mg/m^3)	Respirable quartz dust (mg/m^3)
Pump operator	-	1.9	0.09
Spraying operator	-	2.1	0.09
Rig	7.2	-	-
Threshold limiting value	10.0	5.0	0.1

Results of measuring dust contamination during shotcreting in tunnel. The table shows the means for two shifts.

3. PROPERTIES

3.1 Mechanical properties

Given appropriate personnel, materials and equipment, wet-process concrete at the current state of research and development gives concrete of high quality.

Shotcrete is usually sprayed in thin layers, often in heavily ventilated tunnels. To improve the mechanical properties of the concrete and avoid cracking it is often advisable to use steel fiber reinforcement. Membrane and water curing should also be considered especially for repair work.

If appropriate care is taken, it is no problem to produce high strength concrete with steel fiber reinforcement giving it better tensile and toughness characteristics (see Figs. 3 and 4). As Fig. 4 shows, the tensile properties in particular are affected by the curing conditions.

Compressive strength

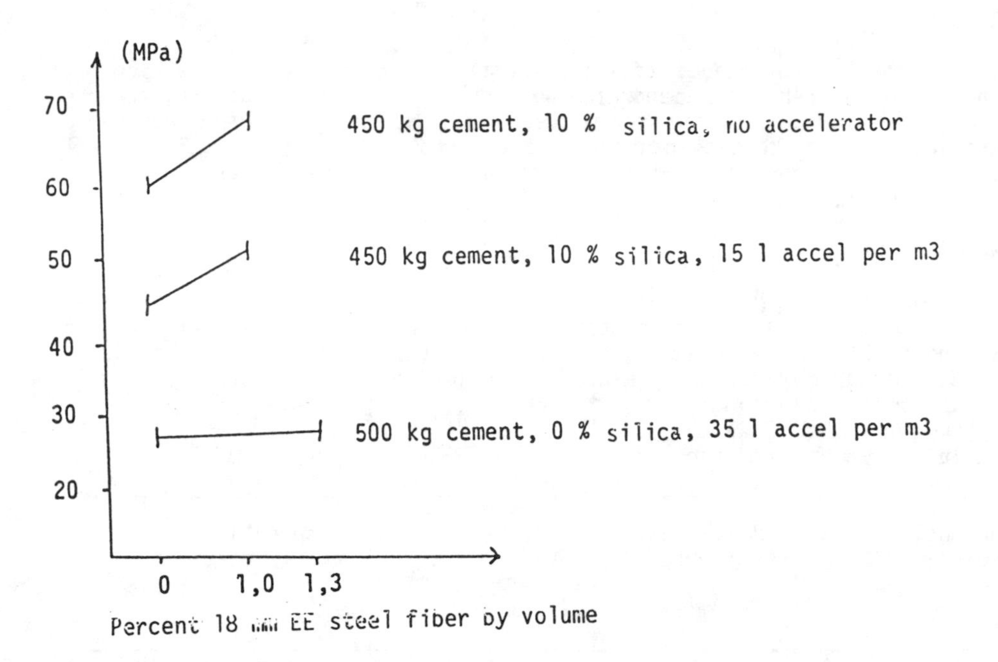

Fig. 3. Typical compressive strength of wet-process shotcrete with various fiber contents. Values from drill cores 60 mm in diameter, ø/h = 1,0.

Tensile strength

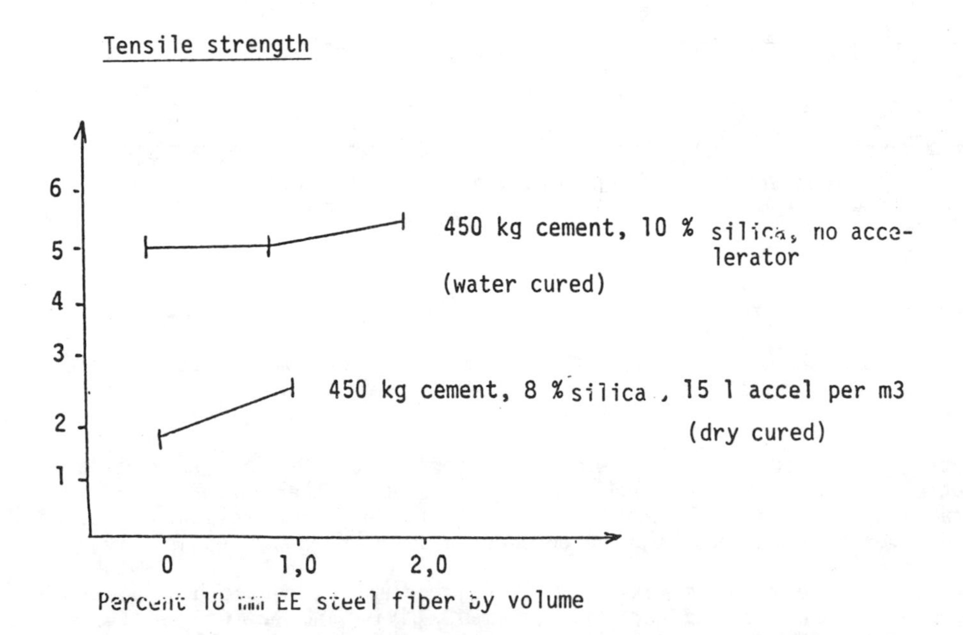

Fig. 4. Uniaxial tensile strength of wet-process shotcrete with various fiber and silica contents. Values from sawn prisms 100 x 100 x 500 mm, dry and water cured specimens.

Wet-process steel fiber reinforced shotcrete has high flexural strength and increased fracture energy, as shown in Figs 5a and b.

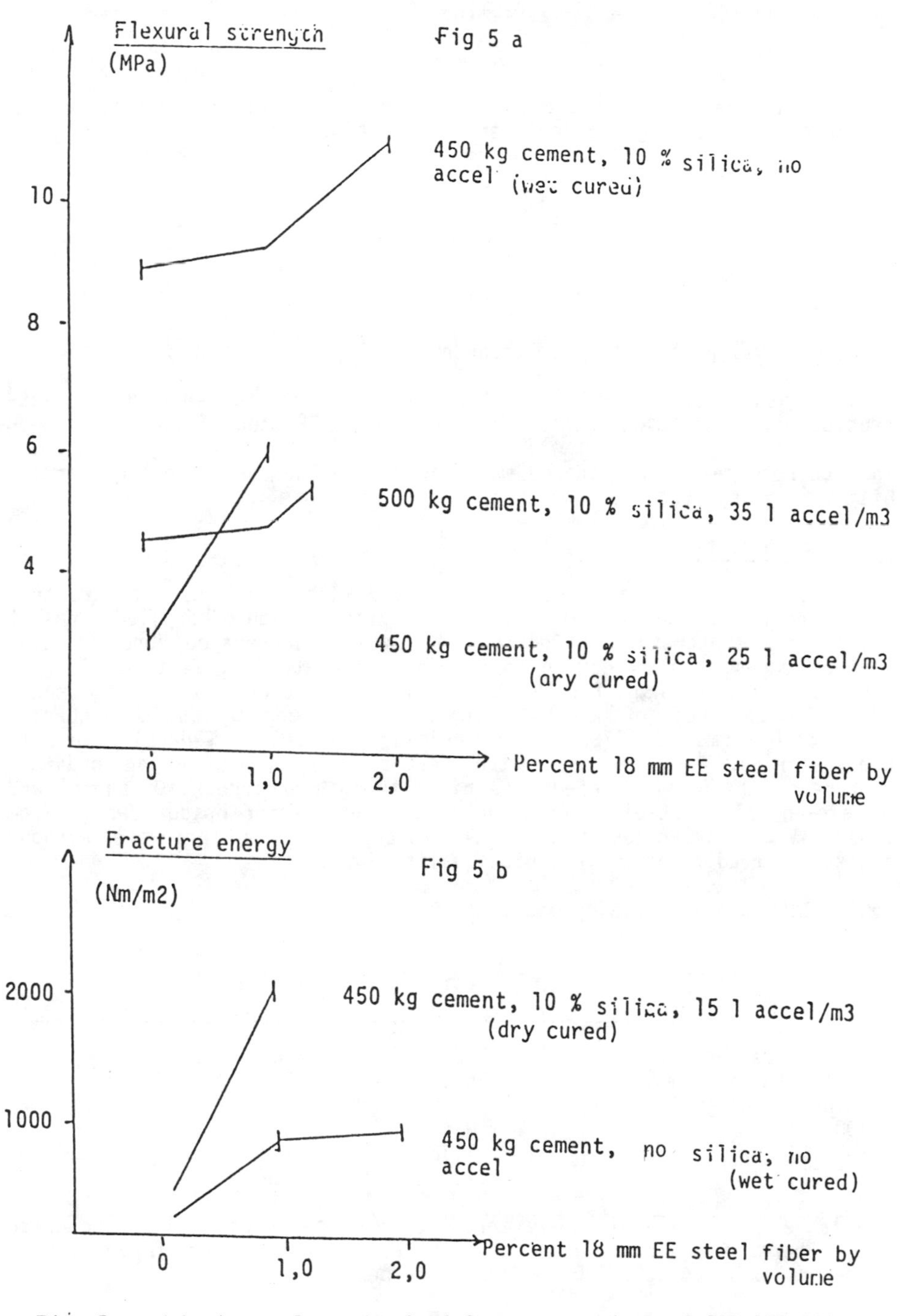

Figs 5a and b show values obtained fram sawn prisms of 100x100x500 mm. Ductility or toughness is the property most affected by the fiber reinforcement.

3.2 PERMEABILITY, AIR-VOID CHARACTERISTICS

Permeability

Tests have shown that concretes containing silica additives have a very low water permeability. Table II shows typical test results.

Table II: Permeability

	k (m/s)
Shotcrete, 45 MPa compressive strength (35 l acc/m³)	2.10^{-14} - 1.10^{-13}
Shotcrete, 75 MPa compressive strength	$< 10^{-15}$

Table II shows the coefficient of permeability of drilled cores of wet-process silica shotcrete containing 1.0 % 18 mm EE steel fiber by volume.

The concrete may be considered watertight when the coefficient of permeability k is less than 10^{-12} m/s.

Air-void characteristics

Good air-void characteristics were achieved without the use of airentraining agents, as was proved in the quality inspection of drilled cores of wet-process shotcrete with silica. Air voids in sprayed concrete have sharper edges than is common in ordinary concrete (5). Testing of frost resistance due to ASTM 666-77 gave durability factors (DF) in the range 60-80 for shotcrete of 60-70 MPa compressive strength. Testing according to Swedish norms SS 137225 using de-icing chemicals (3% NaCL) have just been started. Today we have the results from tests on an ordinary shotcrete (the Bolmen project), a high strength shotcrete (Offshore) and an air-entrained steel fiber reinforced concrete for bridge decks. From table IV it is clear that a freeze-thaw durable shotcrete without admixtures need to be of very high quality.

Table III. Examples of air-void characteristics

	Test 1	Test 2
Air voids measured direct (%)	3.6	4.0
Specific surface (mm²/mm³)	36.1	31.0
Spacing factor (mm)	0.16	0.13
A 100 (%)	0.47	0.48
A 200 (%)	1.18	1.23
A 300 (%)	1.57	1.78

The table shows air-void characteristics of wet-process silica shotcrete with fibers. A_{100}, A_{200} and A_{300} refer to air-void volume in voids with diameters of up to 100, 200 and 300 um respectively.

Table 1V

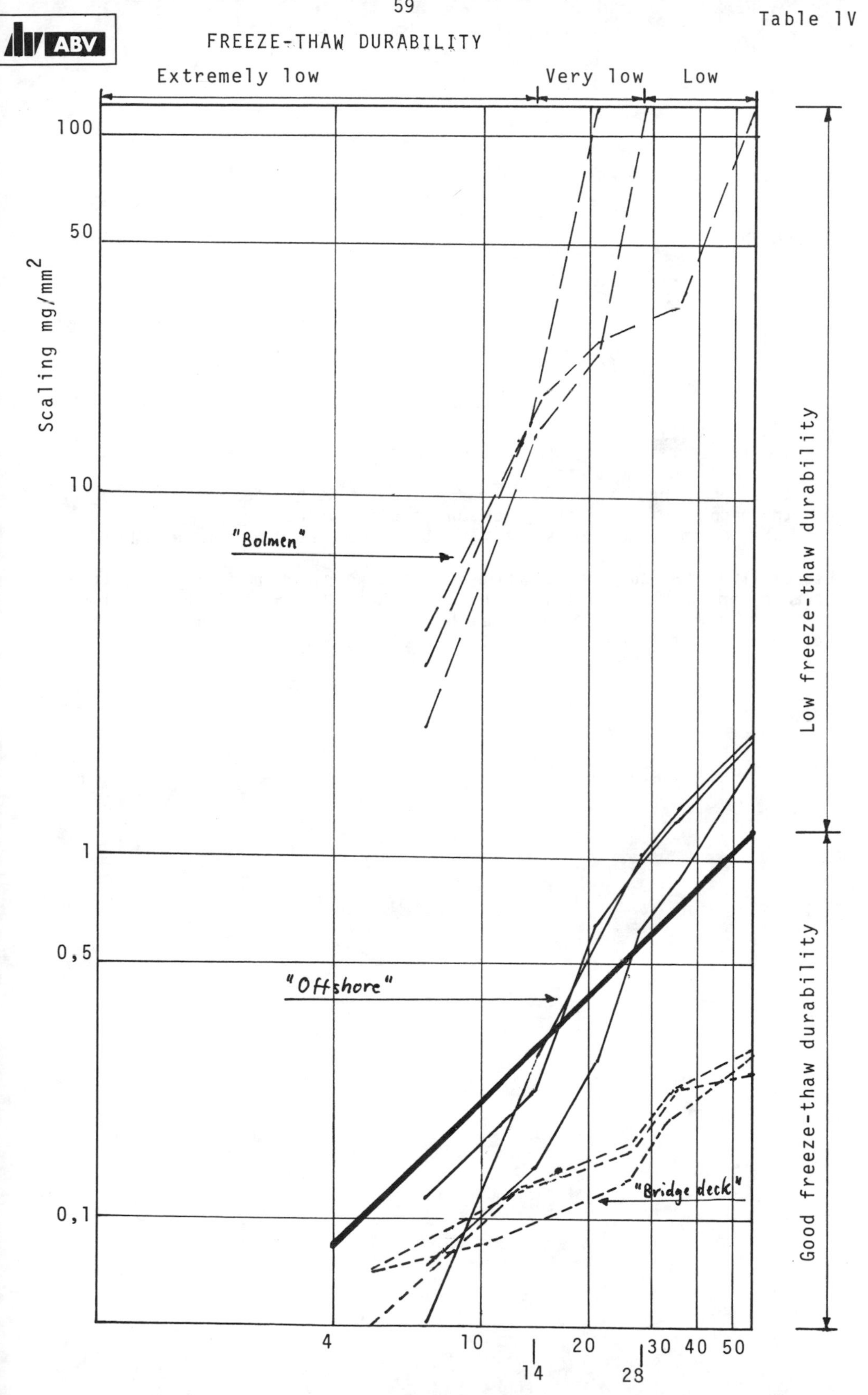

ABV
FREEZE-THAW DURABILITY
Extremely low
Very low
Low
Scaling mg/mm2
100
50
10
1
0,5
0,1
"Bolmen"
"Offshore"
"Bridge deck"
4
10
14
20
28
30
40
50
Low freeze-thaw durability
Good freeze-thaw durability

4. GROUND SUPPORT

4.1 Temporary support

When tunneling in poor quality rock and similar cases where shotcreting is part of the work cycle, support work is a critical activity in terms of time. The advantages of steel fiber reinforced shotcrete are as follows:

- quick setting up of shotcrete rig
- high capacity
- the reinforcement is applied at the same time as the concrete, i.e. quick reinforcement effect
- time-consuming fixing of mesh reinforcement is avoided
- the problem of shotcreting mesh on wire is avoided
- safe working condition since the robot is operated by remote control

Especially quick support is achieved by using rapid hardening PC, accelerator, and a concrete temperature of between 20 and 30°C. Fig. 6 shows the development of splitting tensile strength bond strength and for wet-process steel fiber reinforced shotcrete.

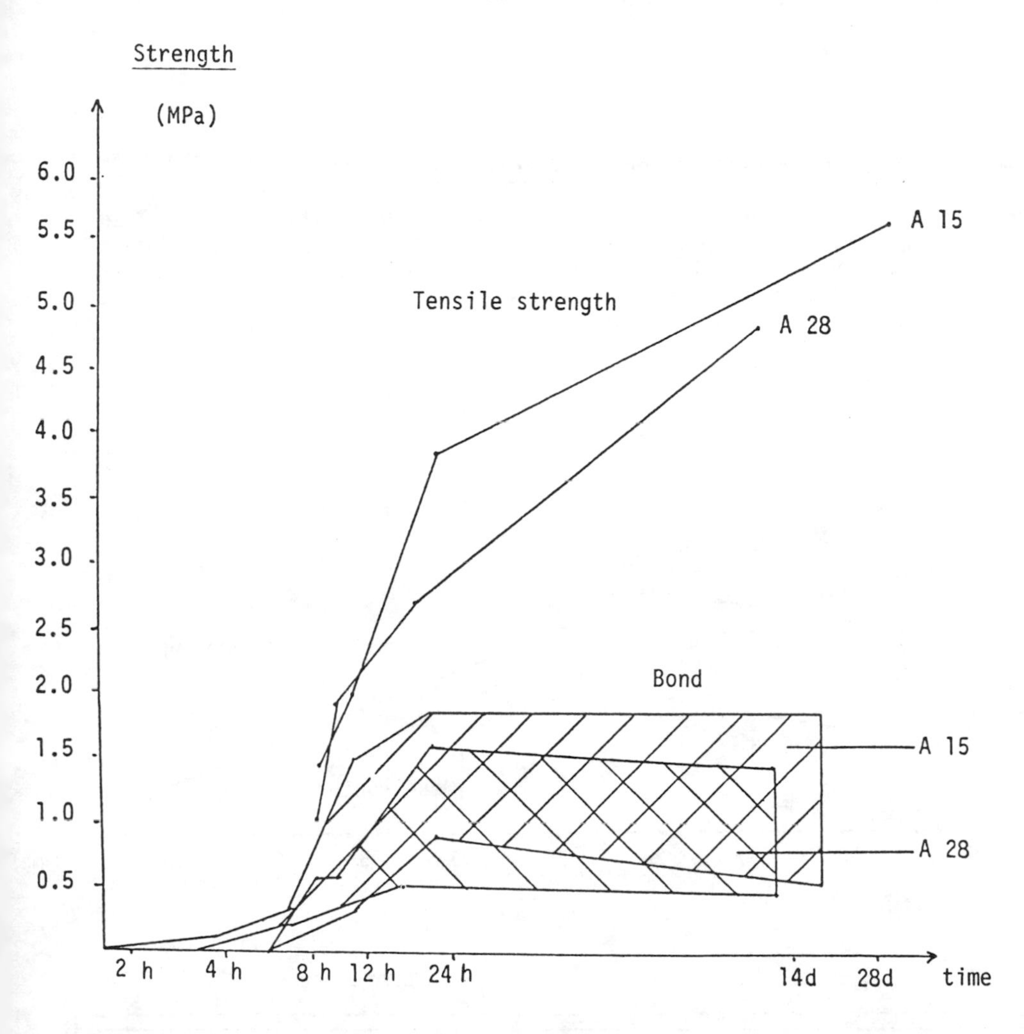

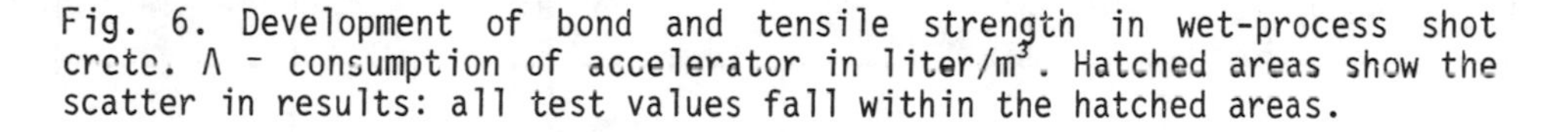

Fig. 6. Development of bond and tensile strength in wet-process shot crete. A - consumption of accelerator in liter/m^3. Hatched areas show the scatter in results: all test values fall within the hatched areas.

4.2 Permanent lining

Permanent lining is required to restrain or avoid

- o rock falls and slides
- o later scaling
- o water leakage

The durability of the concrete is a vital factor for permanent linings. Steel fiber reinforced wet-process shotcrete today can meet high durability requirements.

4.3 Combined effects of fiber reinforced shotcrete and bolting

Steel fiber reinforced shotcreting combined with bolting gives load bearing capacity and toughness equal to or greater than those of mesh reinforced concrete (see Fig. 7). In poor quality rock, steel fiber reinforced shotcreting should be combined with systematic bolting and big blocks should be bolted separately.

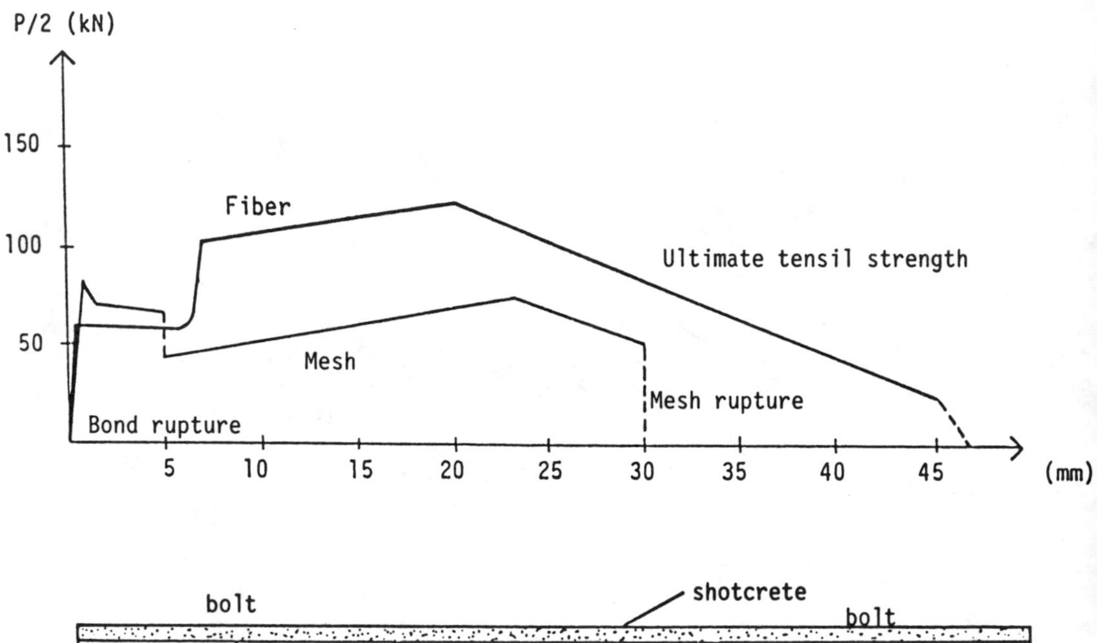

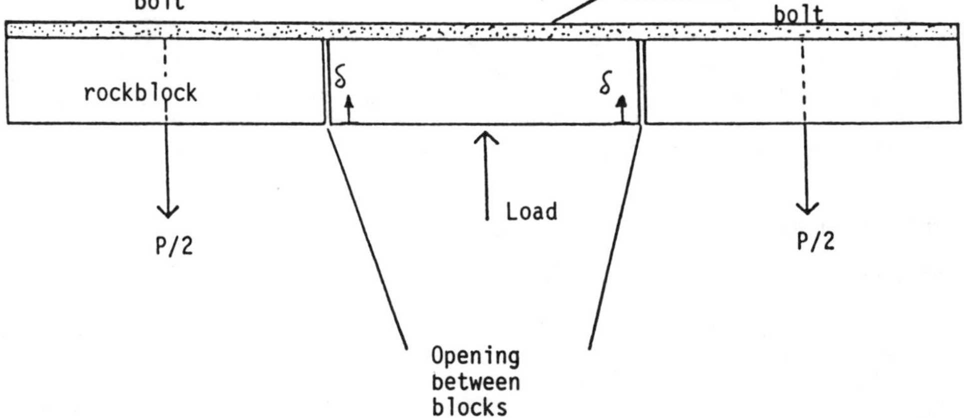

Fig. 7. Load and deflection curves from tests due to falling-block model (3) and (6).

5. MARKET

Shotcreting used to be confined to ground support. The rapid technical development of shotcreting, in particular wet-process steel fiber reinforced shotcrete, has opened up a wider range of applications and new markets:

- o temporary and permanent lining of tunnels
- o temporary and permanent lining of rock storage facilities
- o temporary and permanent stabilization of rock slopes
- o temporary and permanent stabilization of earth slopes
- o reinforcement of excavations
- o repair of concrete structures, e.g. bridges, docks, silos etc.
- o construction

6. PROJECTS

Today has a total of 70 - 80 000 m^3 of steel fiber reinforced shotcretes been sprayed with the Robocon system all over the world.

Among many projects and test:

Swedish State Power Board

For some years the Swedish State Power Board has been testing steel fiber reinforced shotcrete on full scale at one of its power plant construction sites. At the Vietas and Forsmark sites, rigs are used for spraying concrete using the Robocon technique. Temporary and final underground support, volume 5 - 6 000 m^3, strength 5 - 8 000 psi.

Sydvatten AB, Malmö, Sweden

ABV has introduced the Robocon technique using wet-process shotcrete at the Bolmen Tunnel projekt, Stage 1. After full-scale testing in 1983, the wet-process spraying technique is now in use. ABV has successively used:

1. ready-mixed dry mortar from Norway
2. site-mixed mortar
3. mortar from external ready-mix producer

ABV started by using manual equipment but changed over to a Robocon robot rig in March this year. Temporary and final underground support, strength 5 000 psi.

Statoil, Norway

Geological conditions in the Kårstø Tunnel required extensive safety lining during construction. The Robocon technique was used here too. It was found to be the best and most economical solution to the problem of achieving a safe tunnel. Temporary and final underground support, volume 8 000 m^3,strength 4 - 5 000 psi.

Inicel, Ecuador

At the Agoyan hydro power project steel fiber reinforced shotcrete is in use for ground support, both underground and on slopes. Volume 5 000 m^3 (until today), strength 5 - 6 000 psi.

NET, Ireland

A prill tower (fertilizer) was repaired in Arklow and 100 m^3 high strength shotcrete, 12 - 14 000 psi, was sprayed in one week. The result was very satisfactory and total crackfree.

Norwegian Contractors, Norway

Steel fiber reinforced shotcrete was used as a cover for tendon anchorage on Statfjord C 100 m^3 and on Gullfaks A 80 m^3, strength 8 000 psi (bending 1 500 - 3 000 psi).

US Bureau of Mines, USA

Several test in coal mines has been performed with high strength shotcrete, compressive strength up to 15-000 psi and flexural strength up to 2 000 psi.

Peoples republic of China

For a hydro power project at Lubrige a Robocon remote control shotcreting Rig (Robot) has been delivered and the first tests are running.

7. THE FUTURE

Demands for better and safer working conditions and demands for high quality and for more durable concrete structures will require the continued development of concrete technology.

New and supplemented standards for shotcrete workmanship and quality control are essential in order to further the rational use of shotcrete.

REFERENCES

1. Opsahl, O A: "Bruk av silika i sprøytebetong" NIF-kurs: Silika i Betong 21.23 april 1982, Fagernes

2. Mohn S og Pedersen J: "Tidligfasthet i fiberarmert sprøytebetong" Hovedoppgave ved Inst for byggn.matr. läre NHT. BML-rapport nr 83.412, 12-12-1983

3. Opsahl, O A: "Steel-fibre reinforced shotcrete for rock support" NTNF-report 1053.09511, Royal Norwegian Council for Scientific and Industrial Research Oslo - sept 1983

4. Opsahl, O A: "Reparasjoner med fiberarmert sprøtebetong: Utforelse og materialkvalitet" Teknologidagarna i Nord-Norge 1982: Tillstandskontroll og vedlikehold av marine konstruksjoner. Narvik 10-12 mars

5. ASTM C 457: Standard Recommended Practise for Micro scopical Determination of Air-Void Content and Parameters of Air-Void System in Hardened Concrete

6. Holmgren, J: "Punch-Loaded shotcrete lining on hard rock" 1979. Fortifikationsförvaltningens Forskningsbyrå Rapport nr 121:6

STEEL FIBER CONCRETE
US-SWEDEN joint seminar
(NSF-STU)
Stockholm 3-5 June, 1985
S P Shah and Å Skarendahl,
Editors

USE OF STEEL FIBER REINFORCED CONCRETE IN BRIDGE DECKS AND PAVEMENTS

by George C Hoff, Dr, Mobil Research and Development Corp, Dallas, Texas, USA

ABSTRACT

Steel fiber reinforced concrete has been used throughout the world in bridge decks, floor slabs, pavements and other horizontal constructions. In general, the performance has been satisfactory while allowing for a substantial reduction in concrete thickness (40 to 50 percent) for given loads when compared to conventionally designed flat work. Overlays can often be kept to 50 mm (2 in.). Because of very small crack openings, durability is also improved by preventing the ingress of deteriorating substances. Joint spacings can be significantly increased thus reducing the number of joints needing future maintenance. Impact resistance to heavy wheel loads is greatly improved, Numerous applications of steel fiber reinforced concrete in horizontal construction are cited along with some performance information. An extensive bibliography is provided.

Dr. George C. Hoff is a Senior Associate Engineer for Mobil Research and Development Corp., Dallas, Texas. He is a member of the ACI Board of Directors, Secretary of ACI 223, and a member of several other ACI technical committees. Dr. Hoff is a Fellow of the ACI and a Registered Professional Engineer. He has authored or co-authored over 75 technical papers and 2 books.

INTRODUCTION

The use of steel fiber reinforced concretes for horizontal construction or "flat work" has been actively pursued in North America and Japan since the early 1960's. More recently this application has become more common in the United Kingdom, Australia and Europe. The most reported use of steel fiber reinforced concrete has been in the areas of pavements,* bridge decks,** and floors[3,20,110,133]. The following sections of this paper briefly describe a number of the actual uses of steel fiber reinforced concrete in horizontal construction. As the use of this material became more prevalent in the late 1970's and early 1980's, documented information on actual jobs occurred less frequently in the literature. An extensive reference list is included for readers who wish to obtain more detail on these applications.

---

* References: 4, 5, 9, 10, 14, 17, 32, 33, 34, 35, 41, 42, 43, 44, 45, 48, 52, 54, 55, 62, 64, 67, 71, 74, 77, 79, 80, 81, 90, 91, 93, 95, 99, 100, 101, 115, 116, 119, 120, 122, 124, 126, 129, 130, 131, 136, 137, and 138.

** References: 23, 106, 107, 141

BRIDGE DECKS

It has been suggested that fibrous concrete could be applied to a number of different situations within bridge construction.[7] These include permanent soffit formwork as precast lightweight panels, total depth construction, wearing surfaces, composite construction combined with conventional reinforcement, expansion joint nosings and curb edgings where high impact and abrasion resistance requirements occur, and shotcrete repairs to existing structures. Wearing surfaces have been the most common application.

A number of fiber-reinforced bridge deck surfacings have been constructed in the United States since 1972. The following sections briefly describe some of these placements.

BRIDGE DECK OVERLAYS

Winona, Minnesota. The first job was a fibrous concrete overlay of a precast bridge in Winona, Minnesota in August, 1972.[106,107] Severe scaling of the surface of the 9.1 by 29 m (30 by 95 ft) precast deck necessitated the repair. The surface scale was removed to a depth of 38 mm (1-1/2 in.) and a 76 mm (3 in.) steel fiber concrete overlay placed over the deck with no joints. Thicknesses varied from 64 to 102 mm (2.5 to 4.0 in.). The fiber content was 118 kg/m^3 (200 lb/yd^3) of 0.25 mm (0.010 in.) diameter by 12.7 mm (0.5 in.) long fibers. The cement content was 445 kg/m^3 (752 lb/yd^3). Reports after one year[7] indicated that the overlay was successfully combating the effects of salt scaling and studded tires.

New Cumberland, Pennsylvania. Also in August, 1972, the deck of a steel truss bridge in New Cumberland, Pennsylvania was overlayed with steel fiber concrete. This deck, which was 47.2 m (155 ft) long and 12.2 m (40 ft) wide, had been frequently overlayed with asphaltic concrete. After removal of this overlay and any unsound concrete, a steel fiber overlay was placed to a depth of 51 to 127 mm (2 to 5 in.) depending on how much old concrete had been removed. A cement paste bonding agent was used to fully bond the overlay. The overlay was done in one placement with no joints. The fiber content was 118 kg/m^3 (200 lb/yd^3) of 0.25 by 0.56 by 25 mm (0.010 by 0.022 by 1.0 in.) steel fibers. The cement content was was 445 kg/m^3 (752 lb/yd^3). Traffic was allowed on this overlay after 7 days. After one year of service the overlay had no major defects however a few cracks had developed.[39] Most of these developed at areas where the concrete overlay varied abruptly in thickness. The cracks measured from 0.25 to 0.75 mm (0.01 to 0.03 in.) in width.[39] The overlay was still in service in 1984 and was experiencing an average daily traffic of 13,700 vehicles.

Cedar Rapids, Iowa. Early in October, 1972, one lane of a wood truss railroad viaduct on First Avenue in Cedar Rapids, Iowa, was replaced with an unbonded fibrous concrete overlay. The deck had its old asphaltic concrete removed down to the wood plank decking. Double polyethylene sheeting served as the bond breaker. The fiber concrete slab placed was 49 m (152 ft) long by 3.4 m (11 ft) wide by 76 mm (3 in.) thick and contained no joints. The fiber content was 89 kg/m^3 (150 lb/yd^3) of 0.64 mm (0.025 in.) diameter by 64 mm (2.5 in.) long fibers. Traffic was allowed on the overlay after four days. After three years of 30,000 vehicles per day the overlay was essentially crack free.[1,36]

New York City, New York. The Dyckman Street Bridge in New York City was overlayed in March, 1973, using 254 to 305 mm (10 to 12 in.) of steel fiber concrete which was bonded to the old deck.[64] The overlay was four lanes wide and 76 m (250 ft) long and contained 102 by 305 mm (4 by 12 in.) steel mesh at the midpoint of the overlay section. Steel fiber contents varied and were either 89 or 118 kg/m^3 (150 or 200 lb/yd^3) of 0.64 mm (0.025 in.) diameter by 64 mm (2.5 in.) long fibers.

Pittsburgh, Pennsylvania. In July, 1973, a five-lane prestressed box beam bridge over Turtle Creek in Pittsburgh, Pennsylvania, was overlayed with 76 mm (3-in.) of steel fiber concrete.[64] The 12.2 by 18.3 m (40 by 60-ft) overlay was bonded to the old deck using an epoxy bonding agent. The mixture had a fiber content of 118 kg/m^3 (200 lb/yd^3) of 0.25 by 0.56 by 25 mm (0.010 by 0.022 by 1.0 in.) fibers. Some shrinkage cracks appeared on the surface after curing but these did not open in the ensuing years. As of 1984, some polishing and wear are evident due to an average daily traffic of 26,500 vehicles. Surface crazing and spalling exist in a 0.9 by 2.7 m (3 by 9 ft) area along one curb line. The overlay is considered quite satisfactory, however.[36]

Jefferson, Iowa (Greene County). Another steel fiber concrete overlay was placed on a bridge on Highway E 53 near Jefferson, Iowa (Greene County) in September, 1973.[14,64] The 49 m (160 ft) long, 76 mm (3 in.) thick overlay was fully bonded to the old bridge deck using a cement paste bonding agent and contained no joints. The fiber content was 95 kg/m^3 (160 lb/yd^3) of 0.25 by 0.56 by 25 mm (0.010 by 0.022 by 1.0 in.) fibers. As of 1983, the overlay had become completely unbonded but showed no signs of distress. It is still performing satisfactorily.[36]

Hudson, New York. During October, 1973, a 8.5 by 109 m (28 by 356 ft) steel fiber reinforced overlay averaging 89 mm (3.5 in.) thick was placed on Bridge No. 2 over the Roeliff Jansen Kill on New York State Road 9-G near Hudson, New York. An existing asphalt overlay on the badly cracked concrete deck of this four-lane bridge was removed prior to construction of the overlay. Two different fiber contents of 59 and 89 kg/m^3 (100 and 150 lb/yd^3) were used over the period of the job. Fiber size was 0.64 mm (0.025 in.) diameter by 64 mm (2.5 in.) long. The cement content varied from 348 to 445 kg/m^3 (588 to 752 lb/yd^3). At that time, the bridge sustained traffic of approximately 2300 vehicles a day of which 10 percent were trucks. Performance after one year was excellent with no cracks occurring.[67]

Winchester, Virginia. During June, 1974, the State of Virginia constructed six new deck sections on two bridges on U. S. Route 7 Bypass at Berryville, Virginia. A two-course, bonded construction technique was used in the deck construction. This involved the casting of an initial layer of concrete to a level near the top reinforcement and then following this with a high quality 51 mm (2 in.) thick wearing course. Steel fiber reinforced concrete was one of three materials being evaluated for this wearing course. Two test sections 11.6 by 15.8 m (38 by 52 ft) were prepared using a concrete containing 445 kg/m^3 (752 lb/yd^3) of cement and 95 kg/m^3 (160 lb/yd^3) of 0.41mm (0.016 in.) diameter by 25 mm (1 in.) long steel fibers. The performance of these decks was planned to be compared to that of single lift decks made with high quality portland cement concrete and latex modified concrete.

Rome, Georgia. In June, 1974, a 7.3 by 183 m (24 by 600 ft) long bridge deck in Rome, Georgia, was overlayed with 51 mm (2 in.) of steel fiber reinforced concrete in an attempt to stiffen the bridge against the vibrations caused by the passage of heavily loaded gravel and lumber trucks. The mixture contained 104 kg/m^3 (175 lb/yd^3) of 0.25 by 0.64 by 25 mm (0.01 by 0.22 by 1.0 in.) steel fibers and 473 kg/m^3 (800 lb/yd^3) of cement. While the stiffness of the structure was not significantly improved, the performance of the overlay in accomodating the large oscillatory deflections of the deck without cracking has been extraordinary. After one year service very few cracks could be found in an overlay that visibly oscillated and damped out over the length of the deck with each passing truck.

Houston Texas: In April, 1975 the Texas Highway Department conducted a field test overlay of a bridge deck at the Interstate and Highway 610 Loop. Three test sections, each 21 by 23 m (70 by 75 ft), were cast, one each, with steel fiber reinforced concrete, glass fiber reinforced concrete and conventionally reinforced concrete. The overlays varied in thickness from 32 to 70 mm (1 1/4 to 2 3/4 in.). The steel fiber content was 71 kg/m^3 (120 lb/yd^3) of 0.25 by 0.56 by 25 mm (0.010 by 0.022 by 1.0 in.) fibers. The cement content was 389 kg/m^3 (658 lb/yd^3). No information was available for the glass fiber additions other than the fibers were 25 mm (1.0 in.) long. No performance information is available

FULL DEPTH DECKS AND ORTHOTROPIC DECKS

No full depth construction of fiber reinforced bridge decks has been attempted although an analysis of such a construction was made for the Pennsylvania Department of Transportation.[7] For a span of 2.3 m (7-1/2 ft), and using an allowable stress of 6.9 MPa (1000 psi) in flexure, it was found that a bridge deck 152 mm (6 in.) thick made entirely of steel fiber reinforced concrete had the same strength as a standard 216 mm (8-1/2 in.) thick reinforced concrete deck.

Zollo[141] investigated the feasibility of steel fiber reinforced concrete overlays for orthotropic bridge deck type loadings by evaluating flexural specimens of composite beams consisting of (1) concrete cast over steel plates, and (2) concrete cast over floor grating. The concrete contained 2 percent by volume of 0.15mm (0.006 in.) diameter by 13mm (1/2 in.) long steel fibers. He concluded that the use of steel fiber reinfroced concrete in composite with steel plates and gratings is a structurally flexible system that is highly crack resistant. Tests under static and fatigue loading suggested that the composite system exhibits substantial material toughness which is highly desireable for bridge deck wearing surfaces. Zollo[141] concluded that no appreciable reduction in design stress is necessary to account for fatigue loading conditions when fiber concrete is used. The nominal overlay on the plate and grating specimens was 25 mm (1 in.).

PAVEMENTS AND OVERLAYS

Fiber reinforced concrete pavements and overlays have been used in locations which are urban,[4,5,32,33,35,45,52,55] rural,[62,85] industrial,[10,93,95,106,111] airports,* residential[32,138] and other specialized applications. The following sections review some of the more publicized of these many applications.

Residential:

Cedar Rapids, Iowa. Two residential streets in Cedar Rapids, Iowa, were overlayed with steel fiber reinforced concrete during October 1972.[32,138] A 54 m (178 ft) long by 8.5 m (28 ft) wide (curb to curb) partially bonded, no joint, overlay was placed on Danbury Street, NE. The old 178 mm (7-in.) thick reinforced concrete street was badly cracked and spalled. The preparation of the surface for the overlay involved simply broom cleaning and wetting. The thickness varied from 64 to 102 mm (2-1/2 to 4 in.) depending on the amount of deterioration of the underlying pavement. The concrete cement content was 500 kg/m^3 (846 lb/yd^3) with a fiber content of 104 kg/m^3 (175 lb) of 0.41mm (0.016 in.) diameter by 25 mm (1-in.) long steel fibers. The aggregate was 9.5mm (3/8-in.) maximum size. A few months after placing, the overlay had a few cracks, but they were not opening up.[119] The street was eventually resurfaced because some of the residents complained about abrasions suffered by playing children when they fell on the pavement. These complaints were never substantiated, however.

The second placement was a 61 m (200 ft) long by 7.3 m (24 ft) wide partially bonded overlay placed on 5th Avenue between 16th and 17th Streets in Cedar Rapids.[32,138] The old surface was brick covered with a partially removed asphaltic concrete overlay. The center of the street contained exposed rail tracks. The street received the same preplacement treatment as Danbury Street (above). the overlay thickness was 64 mm (2.5 in.). The concrete mixture cement content was 500 kg/m^3 (846 lb/yd^3) with a fiber content of 118 or 148 kg/m^3 (200 or 250 lb/yd^3) of 0.41mm (0.016 in.) diameter by 25 mm (1-in.) long steel fibers. Traffic was allowed on the finished overlay seven days after construction. Two transverse

---

* References: 6, 32, 34, 41, 42, 43, 44, 48, 67, 74, 90, 91, 93, 115, 136

cracks developed early in the curing history over an old sewer construction area. Other than that, the performance has been judged excellent[67] and it is still in service today as an arterial street.[88]

Fayetteville, Arkansas. In 1972, a short section of city street was overlayed with 64 mm (2.5 in). of steel fiber reinforced concrete. The street was 8.8 m (29 ft) wide and had five 12.2 m (40 ft) sections placed on it. The concrete contained 59 kg/m^3 of 0.30 mm (0.012 in.) diameter by 19 mm (0.75 in.) long fibers. Longitudinal centerline joints of the original pavement did reflect through the overlay. After 10 years of service at least two of the five sections had developed some failures.[88]

Rural:

Ashland, Ohio. The first experimental project to evaluate the use of steel fiber reinforced concrete in new pavement construction was completed during August 1971 and opened to traffic in 1973 as the entrance to a truck-weighing station of Interstate 71, north of Ashland, Ohio.[85] The 152 m (500 ft) long by 4.9 m (16 ft) wide by 102 mm (4 in.) thick pavement was placed, with no joints, on a 127 mm (5 in.) asphaltic concrete base. Ends were tapered to 229 mm (9 in.) over a 2.1 m (7 ft) length. Doweled expansion joints were installed at either end. The concrete had a cement content of 500 kg/m^3 (846 lb/yd^3) and a fiber content of 157 kg/m^3 (265 lb/yd^3) of 0.25 by 0.56 by 25 mm (0.010 by 0.022 by 1 in.) steel fibers. The maximum aggregate size was 9.5 mm (3/8 in.).

Although the pavement was not officially opened to traffic for two years, for more than one year it was used as a storage area and sustained an indeterminate amount of light and heavy load bearing vehicles. The day following the placement, a transverse crack occurred 81 m (265 ft) from one end or approximately at the midpoint of the slab.[85] The crack varied in width from 3.2 to 12.7 mm (0.125 to 0.500 in.). Four months later a second transverse crack occurred 30 m (100 ft) from the end and 50 m (165 ft) from the first crack. This crack only opened slightly. A third transverse crack occurred some time in 1973 and was 30 m (98 ft) from the other end of the pavement. An irregular longitudinal crack, hairline in nature, was also observed over a 12 m (39 ft) length at the center of the pavement beginning approximately 30 m (100 ft) from the end of the pavement.

At the time the ramp was put into service, the daily traffic passing over the pavement was approximately 2400 to 3600 trucks per day.[85] This traffic produced spalling in the areas of the three transverse cracks and prompted the replacement of 4.6 m (15 ft) sections in these areas with full-depth portland cement concrete.[67] In 1976, the entire project was resurfaced. The performance of this pavement suggested that future designs for this type of pavement include jointing details and reduced slab lengths.[85]

Greene County, Iowa. The most ambitious early project was an experimental fibrous concrete overlay project placed during September and October 1973, on a 4.8 km (3 mile) section of Highway E53 near Jefferson, Iowa, in Greene County, Iowa.[62] The roadway selected to receive the fibrous concrete overlay a badly cracked pavement constructed originally in 1920 and 1921. The existing pavement was 5.5 m (18 ft) wide and was being widened an additional 0.6 m (2 ft) on each side. The total new width of 6.7 m (22 ft) was overlayed. The project included thirty-three 122 by 6.7 m (400 by 20 ft) sections of steel fiber reinforced concrete 51 and 76 mm (2 and 3 in.) thick, four 122 by 6.7 m (400 by 20 ft) sections of continuously reinforced concrete overlays 76 and 102 mm (3 and 4 in.) thick, and five 122 by 6.7 m (400 by 20 ft) sections of plain concrete and mesh reinforced concrete overlays 102 lnd 127 mm (4 and 5 in.) thick. Variables in the steel fiber reinforced concrete included:

a. Overlay thickness: 51 and 76 mm (2 and 3 in.).

b. Cement content: 355 and 444 kg/m^3 (600 and 750 lb/yd^3.

c. Fiber contents: 36, 59 and 95 kg/m^3 (60, 100, and 160 lb/yd^3).

d. Fiber sizes: 0.25 by 0.56 by 25 mm (0.010 by 0.022 by 1 in.), and 0.64 mm (0.025 in.) diameter by 64 mm (2.5 in.) long.

e. Type and condition of bond with the old slab: bonded, partially bonded, and unbonded.

Joint spacing on the fibrous sections was 12 m (40 ft.). The concrete used a 50/50 weight ratio of sand/coarse aggregate with the coarse aggregate being 9.5 mm (3/8 in.) maximum size. The amount of coarse aggregate used per cubic meter (yard) of concrete was greater than that used on any previous overlay project.

The Greene County project was an experiment in which all of the major mixture and overlay design variables were studied under the same loading and environmental conditions. The main shortcoming of the project was the light traffic count which was approximately 1000 vehicles per day. Early observations indicated that the use of debonding techniques had greatly minimized the formation of transverse cracks during the early life of the overlay. The 122 m (400 ft) long unbonded sections had exhibited fewer than two cracks (in some cases no cracks) after 9 months of services. Fully and partially bonded overlays had from 8 to 15 cracks per section.[67] The 76 mm (3 in.) thick pavements were performing significantly better than the 51 mm (2 in.) pavements[93] which were tending to deteriorate along the edges and joints for all bond conditions. These thin 51 mm (2-in.) overlays, when unbonded, were attempting to curl and warp. Bond was apparently being lost in the bonded sections. Cement content variations did not seem to affect performance but the 95 kg/m^3 (160 lb/yd^3) fiber content sections performed significantly better than the sections with lesser amount of fibers.[427] A survey in 1978 indicated that the higher fiber content concrete of a 76 mm (3 in.) overlay and a 76mm (3 in.) slab placed directly on the subgrade were still performing satisfactorily.[88]

Urban:

Detroit, Michigan (Eight Mile Road). In October 1972, a steel fiber reinforced overlay was placed on a section of M102 (Eight Mile Road) in Detroit, Mich.[4,5,32,33,35] This was the first use of a fibrous overlay on a heavily traveled urban roadway. The original reinforced concrete pavement, constructed in 1930, was 6.1 m (20 ft) wide and had a 3.6 m (12 ft) concrete widening lane placed in 1955 and new 1.8 and 3.0 m (6 and 10 ft) base course widening. The overlay was four lanes, 14.6 m (48 ft) wide, with normal slab lengths of 15 and 30 m (50 and 100 ft). Two slabs were made approximately 20 and 24 m (67 and 79 ft) long because of manhole covers. Pavement joints were sawed. The overlay was designed for a 76 mm (3 in.) minimum thickness and used concrete which contained 500 kg/m^3 (846 lb/yd^3) of cement and either 71 or 118 kg/m^3 (120 or 200 lb/yd^3) of 0.25 by 0.56 by 25 mm (0.010 by 0.022 by 1 in.) steel fibers . The aggregate used was 25A slag. The concrete was placed using a conventional batch plant and a slipform paver. The old concrete surface was simply swept and wet down prior to the placement of the overlay thus resulting in a partially bonded condition.

Heavy traffic in the area required opening the lanes two days after placement.[4] The construction was done under adverse weather conditions at curing temperatures considerably lower than desired. Through the first three months, performance of the 118 kg/m^3 (200 lb) fiber content concrete sections was adequate, but the outside lanes of the 71 kg/m^3 (120 lb) fiber content concrete sections developed some serious problems. Several reflective cracks developed over joints and cracks in the old pavement and closely spaced fractures developed in localized areas. In many locations, the planned overlay thickness was not achieved, with thicknesses occurring as thin as 32 mm (1-1/4 in.).[4] After nine months, the deterioration significantly progressed.[5] The overlay had become completely unbonded and some edge warping was occurring. Approximately 3/4 of the overlay was then removed. The problems were attributed to insufficient thickness and incomplete curing of the overlay prior to opening to traffic.[5] The remainder of the pavement was 76 mm (3 in.) thick and contained the higher fiber contents. In May 1976, that portion of the pavement was considered as failed but was not replaced until 1981. Some of the failures in this pavement occurred over joints in the base pavement that were not matched in the fiber concrete resurfacing.[88]

<u>Calgary, Alberta, Canada</u>. Three projects involving the study of performance of pavements containing steel fibers under city bus traffic have been undertaken in Calgary. The first section was constructed in May 1973[52,55] and occupies a 55 m (180 ft) length of one lane of an existing two-lane 7.3 m (24 ft) asphalt concrete road which forms a city bus route through the University of Calgary campus in Canada. The fiber reinforced test slabs were placed directly on a granular subbase and were subjected to approximately 1000 loadings by busses each week. The test section consisted of fifteen 3.6 m (12 ft) wide by 3.4 m (11 ft) long slabs ranging from 76 to 178 mm (3 to 7 in.) in thickness with fiber contents of 0, 40 and 79 kg/m^3 (0, 67, or 133 lb/yd^3) of 0.25 mm (0.01 in.) diameter by 19 mm (0.75 in.) long steel fibers. The fibrous concrete mixtures contained 296 kg/m^3 (500 lb/yd^3) of cement, 118 kg/m^3 (200 lb/yd^3) of fly ash and 12.7 mm (1/2 in.) maximum size aggregate. Companion plain concrete sections used the same aggregate but only 325 Kg/m^3 (550 lb/yd^3) of cement. Fiber contents were varied at 1/2 or 1 percent by volume. The performance of this pavement was reported after 1 and 4 years[52,55] at which time the test section had experienced more than 200,000 bus passes. From these observations, Johnston[55] concluded that for pavement of constant thickness, the 1 percent fiber

content concrete has a life about four times that of the 1/2 percent fiber content concrete which in turn has a life about four times that of plain concrete. Also, the 1 percent fiber concrete has a life more than 12 times that of plain concrete. For equivalent performance in terms of cracking and, by implication, equivalent pavement life, the 1 percent fiber concrete can be used in pavements of less than 60 percent and possibly as little as 50 percent of the thickness needed for plain concrete. With 1/2 percent fiber concrete, the thickness could be reduced to 75 to 80 percent of that of a plain concrete pavement. He also noted that predictions of the performance of fiber concrete from relationships established for plain concrete using the standard flexural strength value as the measure of concrete quality are overly conservative.[55]

The second bus lane project[53] was constructed on Seventh Avenue in September 1976 and consisted of 12 steel fiber reinforced, 3 conventionally reinforced, and three unreinforced concrete slabs 7.6 m (25 ft) long in a 7.0 m (23 ft) wide two-lane section representing half the street width. Three types of steel fiber were used: 0.5 by 0.25 by 25 mm (0.02 by 0.01 by 1 in.) slit sheet; 32 mm (1.25 in.) long melt extracted fiber; and, 0.25 by 19 mm (0.01 by 0.75 in.) wavy cut fiber. All three types were used in amounts of 45 or 74 kg/m^3 (75 or 125 lb/yd^3). The mixture contained 326 kg/m^3 (550 lb/yd^3) of cement and 89 kg/m^3 (150 lb/yd^3) with 13mm (1/2 in.) aggregate. The performance of this pavement, after 32 months and approximately 100,000 bus passes plus an unknown mix of urban traffic, paralled that of the previous test section with the fiber reinforced concrete out performing the other types of sections. This study indicated that the high flexural strength of the fiber reinforced concrete is essential to good performance. It also indicated that with fiber reinforced concrete, greater joint spacings are possible.

The third bus lane project[53] was constructed on Sixth Avenue in October 1977 and consisted of a composite of 279 mm (11 in.) of lean concrete (econocrete) overlayed with 51 mm (2 in.) of steel fiber reinforced topping. The test section was 148 m (487 ft) long and 3 m (10 ft) wide and was constructed in six sections 22 to 30 m (72 to 100 ft) long. Three types of steel fibers were again used: 0.4 by 25 mm (0.016 by 1 in.) straight brass coated wire; 0.55 by 0.25 by 25 mm (0.022 by 0.01 by 1 in.) slit sheet; and, 0.55 by 0.25 by 13 mm (0.022 by 0.01 by 0.5 in.) slit sheet. The fiber content was 74 kg/m^3 (125 lb/yd^3). The mixture contained 326 kg/m^3 (550 lb/yd^3) of cement and 148 kg/m^3 (250 lb/yd^3) with 13mm (1/2 in.) aggregate. Some sections were monolithically cast while others had the overlay bonded to the lean concrete by epoxy. Johnston[53] reported that after 7 years of carrying 500 busses a day, the pavement is still serviceable although 9 of the 33 slabs exhibited significant damage amounting in most cases only to a single transverse crack. He attributed much of the damage to things other than traffic, however.

Great Britain (M10 Motorway). Gregory,[45] et al evaluated the potential of steel fiber reinforced concrete for thin overlays on heavily traffiked concrete roads in Great Britain. Overlays were constructed on a portion of a reinforced concrete carriageway on the M10 motorway. This motorway had been in service 15 years and had suffered some cracking. The overlay was constructed in two thicknesses, 60 and 80 mm (2-3/8 and 3-3/16 in.) using fiber contents of 1.3, 2.2, and 2.7 percent by weight of the 0.50 mm (0.02 in.) diameter by 38 mm (1-1/2 in.) long steel fiber. Sixteen test sections were placed in two lanes, some of which were bonded to the original pavement and some which were only partially bonded. The total length of the test overlay was 200 m (655 ft.) Expansion joints were provided in the overlay at intervals of 37 m (120 ft) to coincide with those in the existing pavement. Transverse contraction joints were formed at 12 m (40 ft) intervals to provide test slabs of a reasonable length enabling all the variables to be covered. No dowelbars or tie bars were used in any of the overlay joints.

One week after placing and prior to opening the overlay to traffic, some hairline cracking developed in one lane and this was attributed to shrinkage and thermal movements.[45] Numerous reflection cracks also occurred and appeared to be independent of whether the overlay was bonded or partially bonded. Traffic was allowed on the overlay after it was from 7 to 11 days old. After 11 days of traffic, numerous reflection cracks had occurred along with a few new cracks. Most remained as hairline cracks. The total length of longitudinal cracking in the overlay exceeded that of the original pavement and longitudinal cracks tended to be more pronounced in the partially bonded sections. Between 11 weeks and 6 weeks after opening the road to traffic, there was little further change in the crack pattern. Gregory, et al[45] found no clear relationship in their work between fiber content and resistance to cracking but concluded that the thinner bonded sections performed better than the others.

Industrial:

Niles, Michigan. Luke, et al[75] have described a number of full thickness slabs and overlays of steel fiber reinforced concrete placed at Niles, Michigan. The first full thickness slab was placed in September 1968 and was 102 mm (4 in.) thick. It was placed as one lane of a road into an industrial plant. The other lane was 190 mm (7-1/2 in.) of conventional concrete. After four years and eight months, the conventionally reinforced concrete lane had three transverse cracks across the complete width. The steel fiber concrete section had no cracks even though trucks routinely cut the corners of this lane. The steel fibers used were both flat and round being 25 mm (1 in.) long with a 0.30 mm (0.012 in.) diameter or equivalent diameter. In June 1970, a section of road 3.6 m (12 ft) wide and 49 mm (160 ft) long leading out of a ready-mix concrete plant was then constructed using steel fiber reinforced concrete. The road would carry the ready-mix trucks at axial loads of 7260 and 8160 kg (16,000 and 18,000 lbs) with a total weight of 15,240 kg (50,000 lb) passing over the road each day.

The road was divided into 13 test sections of different lengths, thickness, mixture proportions, fiber reinforcements, and different laminations and overlays.[75] Test section lengths varied from 0.6 to 6 m (2 to 20 ft) with slab thicknesses varying from 51 to 102 mm (2 to 4 in.) Overlays and toppings were 51 mm (2 in.) thick. Fiber contents were varied at 109 or 118 kg/m^3 (185 or 200 lb/yd^3). Fiber dimensions included diameters of 0.25, 0.38 and 0.41 mm (0.010, 0.015 and 0.016 in.) and lengths of 19 and 25 mm (3/4 and 1 in.) Loads were allowed on the new concrete 64 hours after placement. After 2 years, the only section containing no cracks was 102 mm (4 in.) thick and contained 118 kg/m^3 (200 lb/yd^3) of 0.41 mm (0.016 in.) diameter by 25 mm (1 in.) long fibers. All sections made at thicknesses of 76 mm (3 in.) or less experienced varying degrees of cracking with most of it beginning after one year of service. A 51 mm (2 in.) overlay on 152 mm (6 in.) of conventional concrete had only 3 hairline cracks after 2 years.

Japan(Kashima Works). Nishioka, et al[84] reported on the use of a steel fiber reinforced pavement in the Kashima Works yard of Sumitoma Metals in Japan. The road was routinely used by fork lift trucks with a gross weight of 47 tonnes (52 tons) and the heavy wheel loading had badly damaged conventional pavements. A steel fiber reinforced pavement for similar types of loadings was also constructed at the Homestead Works of US Steel in Pittsburgh, Pennsylvania, in 1969.[111] The Japanese pavement was done in two sections each 15 m (50 ft) long, 4 m (13.6 ft) wide, and 150 mm (6 in.) thick. The joint interval was varied from 5 m (16.4 ft) in one section to 15 m (49.2 ft) in the other section. A 200 mm (8 in.) thick conventional pavement with a 5 m (16.4 ft) joint interval was also placed as a control section. The concrete had a cement content of 222 kg/m^3 (375 lb/yd^3), a water-cement ratio of 0.464 and a steel fiber content of 131 kg/m^3 (222 lb/yd^3) of 0.25 by 0.56 by 25 mm (0.010 by 0.022 by 1 in.) long fibers.

Loading tests were performed one month after the concrete was placed with slab stresses being determined from strains obtained from molded wire strain gages.[84] The maximum bending stresses measured on the surface at the corner edge of the slab at the point of severest loading were 3.75 and 3.31 MPa (545 and 480 psi), respectively, for the steel fiber concrete and the conventional concrete. The measured flexural strength of both these concretes was 6.86 and 4.48 MPa (995 and 650 psi), respectively. An examination of the cracking pattern in the slabs after one year showed that in the conventional concrete slab, there was a large crack which went through the thickness near one corner and 10 rather small cracks of 1.8 to 4.9 m (6 to 16 ft) in length. There was only one hairline crack in the center of a section of fiber reinforced pavement with the 15 m (49.2 ft) joint spacing. This was attributed to shrinkage and the long joint spacing. Nishioka, et al[84] concluded that the joint interval of steel fiber reinforced concrete pavement could be longer than that of conventional concrete, and that the life of the pavement became remarkedly increased in proportion to the volumetric percentage of reinforcing fibers.

<u>Midlothian, Texas (Gifford-Hill Quarry)</u>. Buckley[10] evaluated four test slabs containing glass fibers and one control slab containing steel reinforcing bars on a road used at a quarry of the Gifford-Hill Cement Company, Midlothian division. Each slab was 4.9 m (16 ft) wide and 6.1 m (20 ft) long placed on a subgrade whose modulus of subgrade reaction "K" was conservatively estimated to be 2760 MPa (400 kips/in.^{2}) The slabs contained the following variables:

<u>Slab 1</u>. 203 mm (8 in.) thick, reinforced with No. 3 bars on 610 mm (24 in.) centers.

<u>Slab 2</u>. 152 mm (6 in.) thick, glass fiber reinforced.

<u>Slab 3</u>. 152 mm (6 in.) thick, glass fiber reinforced.

Slab 4. 102 mm (4 in.) thick, glass fiber reinforced.

Slab 5. 102 mm (4 in.) thick, glass fiber reinforced on 1 in. of recycled rubber.

The concrete for the glass fiber reinforced slabs used 25 mm (1 in.) long glass fibers, a cement content of 334 kg/m^3 (564 lb/yd^3) concrete, 13 mm (1/2 in.) maximum size aggregate, sand, and water. Glass fiber contents varied from 1.34 volume percent for slab 5 to 0.97 volume percent for slabs 2, 3, and 4. The water-cement ratios also varied being 0.800, 0.688, 0.688, and 0.700, respectively, for slabs 2, 3, 4, and 5. The control slab used a cement content of 278 kg/m^3 (470 lb/yd^3), a water-cement ratio of 0.65, and 38 mm (1 1/2 in.) maximum size aggregate.

Loads were applied by a Euclid rock hauler which, when loaded, carried a front axle load of 19,500 kg (43,000 lb) or 9773 kg (21,550 lb) per front wheel and a rear axle load of 38,640 kg (85,200 lb) or 9660 kg (21,300 lb) per rear wheel. Only the 203 mm (8 in.) glass fiber reinforced slab performed better than the conventionally reinforced slab. The other slabs had become so badly cracked they had to be removed. After one year's observation, minor growth of the crack patterns was continuing with the glass reinforced concrete slab performing somewhat better than the conventional reinforced slab.[10]

Vicksburg, Mississippi. (Waterways Experiment Station). Two experimental pavements were placed on a street at the Waterways Experiment Station (WES), Vicksburg, Mississippi, in 1972 and 1973. The first of these was placed in June 1972 and was 30 m (100 ft) long, 6.1 m (20 ft) wide, and 127 mm (5 in.) thick placed on rough graded soil. The slab contained no joints and was placed using transit-mix trucks, hand labor, and a vibrating screed. The concrete used 486 kg/m^3 (822 lb/yd^3) of Type I-P cement (containing approximately 15 percent fly ash) and 68 kg/m^3 (115 lb/yd^3) of 0.41 mm (0.016 in.) diameter by 19 mm (3/4 in.) long steel fibers. The maximum aggregate size was 9.5 mm (3/8 in.). After twelve years of service with the first year subjecting the pavement to heavy truck and construction equipment loads and the later years involving only light truck and car traffic, the slab remains crack free.

The second WES pavement was constructed in July 1973 and was 305 m (1000 ft) long, 7.3 m (24 ft) wide, and 102 mm (4 in.) thick.[93] It was constructed as one 7.3 m (24 ft) wide lane without joints using slipform paving techniques. The pavement was permitted to crack where necessary to reduce tensile stresses induced during curing. The ends of this pavement were adjacent to the fiber slab described above and a new 152 mm (6-in.) concrete pavement. A 25 mm (1 in.) expansion joint was provided at both ends of the pavement. The pavement was placed on a clay-gravel subbase. In places, the grade of the placement approached 12 percent and this presented a few problems with the slipform paving operation proceeding downhill. The concrete mixture had a cement content of 500 kg/m^3 (846 lb/yd^3) and a fiber content of 118 kg/m^3 (200 lb/yd^3) of 0.25 by 0.56 by 25 mm (0.010 by 0.022 by 1 in.) steel fibers. The maximum aggregate size was 9.5 mm (3/8 in.).

Immediately after placing some shrinkage cracking or crazing occurred in this second pavement because application of curing compound did not keep up with the paver. The problem was accentuated by high ambient temperatures 35°C (95°F), the high cement content (500 kg/m^3 (846 lb/yd^3)), the rough finishing technique, and the downhill paving. A distinct crack pattern developed in the pavement the night after placing. Seven cracks ranging in spacing from 19.2 to 73.1 m (63 to 240 ft) occurred. The average length between cracks was 38.5 m (126.4 ft). The cracks widened considerably more over a period of a few days and then stabilized at between 4.8 to 10.2 mm (0.19 to 0.40 in.) except for normal thermal opening and closing. These cracks from the inception were not hairline cracks held tightly closed by fibers such as is the case for load-induced cracks. No additional cracking has been observed in the pavement. The traffic load is light, however, but does include an occasional truck loading. The pavement was still in service in 1984.

Airports:

Numerous airport uses of steel fiber reinforced concrete* and several experimental test slabs for aircraft type loadings[41,43,44,93] have been reported. Most of the airport uses have been overlays. The experimental slabs involved both full depth slabs and overlays.

Vicksburg, Mississippi. (Waterways Experiment Station). Two experimental fibrous concrete test pavements consisting of a slab-on-grade and an overlay were constructed at WES during June 1971-March 1972 as part of a test section designed to study the effects of multiple-wheel heavy gear load aircraft such as the Boeing 747 and the military C-5A transport on keyed longitudinal construction joints.[41,43,44,93] The steel fiber rein-

* References 15, 24, 25, 29, 32, 67, 74, 90, 91, 93.

forced slab on grade was located at the east end of a 98 m (320 ft) long test track. The slab was 7.6 m (25 ft)long and 15.2 m (50 ft) wide resting a 102 mm (4 in.) thick sand filler course having a modulus of subgrade reaction of 5.65 kg/cm^3 (42 lb/in^3). The 152 mm (6 in.) slab was thickened to 229 mm (9 in.) by a uniform taper over 1181 mm (30 in.) at the transverse edges. No provision for load transfer to other adjacent slabs was made. A plain concrete nonreinforced test section 15.2 m (50 ft) square consisting of four equal individual slabs 254 mm (10 in.) thick was also constructed on a 102 mm (4 in.) thick sand filter course having a modulus of subgrade reaction of 16.8 kg/cm^3 (125 lb/in^3) and was used for making comparisons of performance between the fibrous concrete and conventional concrete pavements. The concrete mixture had a cement content of 500 kg/m^3 (846 lb/yd^3) and a fiber content of 148 kg/m^3 (250 lb/yd^3) of 0.41 mm (0.016 in.) diameter by 25 mm (1 in.) long steel fibers. The maximum aggregate size was 9.5 mm (3/8 in.).

The construction of both slabs used manual techniques because of their small size. The steel fiber reinforced concrete slab contained no joints.

Loads were applied 73 days after placing the concrete by using a WES load cart which simulated one 12-wheel gear loading with 13,605 kg (30,000 lb) per wheel. The cart was driven back and forth along five evenly spaced parallel lines across an area 6.1 m (200 in.) wide on the pavements.

The 152 mm (6 in.) thick fibrous concrete slab on a weak subgrade was about half the design thickness of the 254 mm (10 in.) thick plain concrete slab on a medium strength subgrade. The 152 mm (6 in.) thick fibrous concrete slab developed its first visible crack at 350 traffic loadings and the second visible crack at 700 traffic loadings. The 254 mm (10 in.) thick plain concrete slab developed the first crack at less than 40 traffic loadings and was in a shattered condition after 700 loadings. After 950 loadings this pavement was completely failed due to structural cracking and spalling. The progression of cracking in the fibrous slab was gradual. At 8735 traffic loadings, many hairline-width cracks had developed but the pavement was in excellent condition.

The completely shattered plain concrete pavement was then cleaned and moistened and covered with a no-joint fibrous overlay 102 mm (4 in.) thick. The overlay concrete was similar to that of the fibrous slab except that the steel fibers were 0.25 by 0.56 by 25 mm (0.010 by 0.022 by 1 in.) in size. The overlay was loaded in the same manner as the slab and was 28 days old when loading began. The first crack formed at 900 traffic loadings and was a reflection of the longitudinal joint in the base pavement. The second and third cracks appeared at 1400 and 2600 loadings, respectively. Other cracks tended to form gradually under traffic. After 6900 traffic loadings, only one crack was classified as a working crack.[44]

Between March 1972 and May 1973, two additional fibrous concrete test sections were constructed and tested at WES as a part of a series of full-scale pavement tests designed to study the effects of chemical stabilization, insulating materials, and fibrous concrete on pavement performance.[93] The first pavement consisted of 178 mm (7 in.) of steel fiber reinforced concrete on a 508 mm (20 in.) layer of lean clay encased in a waterproof membrane over a heavy clay subgrade. The second pavement consisted of 102 mm (4 in.) of the same concrete on a 432 mm (17 in.) layer of clay gravel stabilized with a percent portland cement over a heavy clay subgrade. Each pavement was composed of two 7.6 by 15.2 m (25 by 50 ft) slabs divided by a construction joint. The concrete used three different fibers. Round, 0.41 mm (0.016 in.) diameter by 25 mm (1 in.) long fibers were used for the full depth pavement. One-half of the second pavement used 19 mm (3/4 in.) long deformed fibers having a 0.41 mm (0.016 in.) diameter while the other half used flat fibers of 0.25 by 0.36 by 19 mm (0.010 by 0.014 by 3/4 in.) long cross-section. Test traffic was applied in a similar manner as that described above for the first two fibrous concrete test sections except that both a 90.7 and 108.9 tonne (100 and 120 ton) twin-tandem assembly equipped with 49 x 17 tires with a ply rating of 26 were used. These represented on twin-tandem component of a Boeing 747 assembly.

One lane of the full thickness, 178 mm (7 in.) steel fiber reinforced concrete slab initially failed after 1000 coverages of the 90.7 tonne (100 ton), dual tandem assembly. Although the slab was not completely bisected by a crack, it was felt that the transverse crack and permanent deformation along the longitudinal construction joint was as detrimental as a continuous crack across the slab. The pavement was considered to have reached the shattered slab condition at 1800 coverages. Complete failure occurred at 3000 coverages. The other lane of this pavement was loaded using the 108.9 tonne (120 ton) dual-tandem assembly. Initial failure in this case occurred at 200 coverages. The slab was considered completely shattered at 650 coverages and completely failed at 1010 coverages. Failure of the pavement under both 90.7 and 108.9 tonne (100 and 120 ton) traffic was characterized by multiple cracking. The cracks did not spall but did ravel around the edges and widen as traffic was applied. The pavement also experienced a maximum permanent deformation of approximately 17.8 mm (0.7 in.) in both lanes.

The 102 mm (4 in.) thick pavement over the clay-gravel and cement stabilized clay subgrade was loaded in the same manner as the 178 mm (7 in.) thick pavement. Under the 90.7 tonne (100 ton) loading, initial failure of the first lane occurred at 500 coverages. At 1200 coverages, it had reached the shattered slab condition. Some of the longitudinal cracks had begun to widen and transverse cracks connecting the longitudinal cracks had begun to widen and transverse cracks connecting the longitudinal cracks were beginning to occur. Complete failure occurred at 1770 coverages. When the second lane was loaded with the 108.9 tonne (120 ton), dual-tandem assembly, initial failure was noted at 150 coverages. Permanent deformations of 12.7 mm (0.5 in.) had also occurred at this point. The shattered slab condition occurred at 400 coverages with complete failure occurring at 740 coverages. As in the case of the 178 mm (7 in.) slab, the 102 mm (4 in.) slab failure was characterized by multiple cracking and large permanent deformations. It was noted that the pavements behaved more like flexible that rigid pavements.[93]

Tampa, Florida. During February 1972, two fibrous concrete overlay test pavements were constructed at Tampa International airport, Florida.[91] The overlays were constructed on a taxiway parallel to one of the primary runways. The existing taxiway pavement was constructed in 1965 and opened to traffic in January 1966. Two thicknesses of overlay were used. A 102 mm (4 in.) overlay was situated in the center of the taxiway and was formed by the construction of two 7.6 m (25 ft) wide paving lanes. The paving lanes were constructed so that the center of the paving lanes coincided with the longitudinal construction joints in the base pavement. The longitudinal construction joint was a butt-type joint without load-transfer capabilities. No transverse joints were formed in the overlay. The ends of the section coincided with transverse contraction joints so that one transverse contraction joint and one longitudinal construction joint in the base pavement were spanned by each lane of the overlay. Bituminous transition overlays were constructed around the overlay.

A 152 mm (6 in.) overlay spanned the entire width of the taxiway. The section was formed by constructing three 7.6 m (25 ft) wide paving lanes. The longitudinal construction joints in the overlay matched the longitudinal construction joints in the base pavement. Vertical faces were formed along the edges of the paving lanes with no provisions for load transfer. No transverse joints were formed in the overlay, therefore, all transverse contraction joints in the base pavement were spanned by the overlay. One expansion joint in the base pavement located 23 m (75 ft) from the south end of the section was spanned by the overlay. The ends of the overlay coincided with transverse contraction joints in the base pavement. Bituminous transition sections are constructed on each end of the overlay.

A minimum of surface preparation was performed, and no attempt was made to either bond the fibrous concrete to the base pavement or to completely eliminate the bond between the overlay and base pavement. The concrete mixture had a cement content of 306 kg/m^3 (517 lb/yd^3), a pozzolan (fly ash) content of 133 kg/m^3 (225 lb/yd^3), and a fiber content of 118 kg/m^3 (200 lb/yd^3) of 0.25 by 0.56 by 25 mm (0.010 by 0.022 by 1 in.) steel fibers. The maximum aggregate size was 19 mm (3/4 in.).

These overlays were partially bonded but no reflection cracking was observed prior to opening for traffic. After traffic was applied, reflection cracking occurred. The cracks in the overlays coincide with cracks or joints in the base pavement. Those cracks corresponding to base pavement joints did widen sufficiently to cause failure (bond or fracture) of the fibers across the cracks. Those cracks corresponding to base pavement cracks did not widened significantly thus indicating that, at least at early ages of the pavement, the fibers across the crack were still effective.[93] The overlays were still in service in 1984.[88]

Lockbourne AFB, Ohio. Two slabs were placed at Lockbourne AFB, Ohio, in July 1970.[93] The first slab was a paving apron on grade that was approximately 10.6 by 14 m (35 ft by 46 ft) by 152 mm (6 in.) thick with a 102 by 127 mm (4 by 5 in.) leave-out in the center. The second slab was approximately 1.5 by 6.7 m (5 by 22 ft) by 152 mm (6 in.) thick placed on grade on a taxiway. The slabs were constructed over 229 mm (9 in.) of a lean concrete with a cement content of 222 kg/m^3 (376 lb/yd^3) and granular base course. The fiber reinforced concrete mixture had a cement content of 429 kg/m^3 (725 lb/yd^3) and a fiber content of 106 kg/m^3 (180 lb/yd^3) of 0.25 by 0.56 by 25 mm (0.010 by 0.022 by 1 in.) steel fibers. The maximum aggregate size was 9.5 mm (3/8 in.). The 152 mm (6 in.) fibrous concrete slabs and the 229 mm (9 in.) lean mixture concrete was separated by 0.10 mm (4 mil) polyethylene sheeting thus producing an unbonded overlay.

Cracks developed at each corner of the leave-out in the parking apron but were arrested in growth by the fibers at approximately 0.9 to 1.2 m (3 to 4 ft) from the corners. Without the fibers, the cracks probably would have propagated to a free edge.[93] No distress was observed in the taxiway slab, but an adjacent 7.6 by 3.6 m (25 by 12 ft) by 381 mm (15 in.) thick plain concrete slab had developed a longitudinal crack and a number of short transverse cracks.[88]

Cedar Rapids, Iowa. In September 1972, a steel fiber reinforced concrete overlay was placed on the main taxiway at the Cedar Rapids, Iowa, airport.[32] Deteriorated concrete and asphalt patches in the old taxiway were removed prior to the overlay operation. The overlay was 26 m (86 ft) long, 23 m (75 ft) wide, and varied in thickness from 25 to 102 mm (1 to 4 in.) across the width. The base concrete had longitudinal joints at 3.8 m (12.5 ft) intervals and transverse joints spaced at 6.1 m (20 ft.) The overlay pavement had no joints except a longitudinal construction joint in the center. It was placed using a Fomaco bridge machine. The fiber reinforced concrete mixture for the east half of the overlay had a cement content of 445 kg/m^3 (752 lb/yd^3) and a fiber content of 118 kg/m^3 (200 lb/yd^3) of 0.41 mm (0.016 in.) diameter by 25 mm (1.0 in.) long steel fibers. The same mixture but with a fiber content of 89 kg/m^3 (150 lb/yd^3) of 0.64 mm (0.025 in.) diameter by 64 mm (2.5 in.) long steel fibers was used for the west half.

The completed overlay was opened to aircraft traffic seven days after construction. In 1975, the overlay had reflective cracks from the longitudinal construction and contraction joints and transverse construction joints in the base pavement.[93]

Detroit, Michigan. A steel fiber reinforced concrete aircraft parking apron slab (on grade) was placed in July 1971 at the Detroit airport.[74] One of the problems with pavements at this airport was the drain boxes around which conventional portland cement concrete continues to crack providing a repetitive repair problem. The slab placed was in a gate area used by 747 aircraft. It was 6.1 by 9.1 m (20 by 30 ft) by 203 mm (8 in.) thick and used concrete containing 0.41 mm (0.016 in.) diameter by 25 mm (0.010 in.) long steel fibers. The adjacent slabs were 305 mm (12 in.) thick so the base course for the steel fiber reinforced concrete slab had to be built up 102 mm (4 in.) The fibrous slab was tied to adjacent slabs with deformed rebars installed by drilling and grouting in adjacent slabs. Observations at nine months indicated that the slab was performing satisfactorily.[74]

New York, New York (JFK Airport). During May 1974, an unbonded steel fiber reinforced concrete overlay was placed on the end of a main runway at John F. Kennedy (JFK) International Airport, New York, New York.[67] Construction consisted of the removal of the old asphalt concrete wearing course, construction of a new 51 mm (2 in.) thick asphalt leveling course, placement of a double thickness of 0.15 mm (6 mil) thick polyethylene, and then construction of the steel fiber reinforced concrete overlay. The overlay was 53 by 37 m (175 by 120 ft) by 140 mm (5.5 in.) thick and used both keyed and doweled construction joints. The fiber reinforced concrete mixture had a cement content of 445 kg/m^3 (752 lb/yd^3) and a fiber content of 104 kg/m^3 (175 lb/yd^3) of 0.64 mm (0.025 in.) diameter by 64 mm (2.5 in.) long steel fibers. A 1975 survey indicated satisfactory performance and the repaired area was enlarged at that time. It is still in service although there is some cracking and shattered corners at intersections.[88]

Las Vegas, Nevada. The largest fibrous concrete paving project at that time was done at the Las Vegas McCarran International Airport in early 1976 with the placement of a 288 by 183 m (945 by 600 ft) by 152 mm (6 in.) thick overlay on a transit parking apron and connecting taxiway.[29] A problem for overlaying the existing pavement existed because existing cargo buildings were already 0.6 to 0.9 m (2 to 3 ft) below the ramp. A conventional concrete overlay would add 381 to 406 mm (15 to 16 in.) to this difference. The selection of fiber reinforced concrete allowed this thickness to be reduced to 152 mm (6 in.).

The overlay was placed in alternating 7.6 m (25 ft) strips using slipform paving techniques. Construction joints were spaced 15.2 m (50 ft) apart. The concrete mixture contained 95 kg/m^3 (160 lb/yd^3) of 0.25 by 0.56 by 25 mm (0.010 by 0.022 by 1 in.) steel fibers. The pavement was placed in 20 working days. The overlay was designed for a 20-year service life.

The overlay presently has a number of corner breaks (0.9 to 1.5 m (3 to 5 ft)) and tight longitudinal cracks near the panel centerline. Some of these corner cracks have faulted and are developing spalling. The repair is still in service.[88]

A new construction of an aircraft parking ramp was completed at the same airfield in December 1979. A 178 mm (7 in.) steel fiber concrete pavement was placed over 51 mm (2 in.) of new asphaltic concrete which was on top of 305 mm (12 in.) of conventional concrete. The ramp configuration was irregular but encompassed 72,850 m^2 (87,120 yd^2). It was constructed in 7.6 m (25 ft) strips with 15.2 m (50 ft) between sawed joints. The concrete mixture had a cement content of 385 kg/m^3 (650 lb/yd^3), a pozzolan (fly ash) content of 149 kg/m^3 (252 lb/yd^3), and a fiber content of 50 kg/m^3 (85 lb/yd^3) of 0.51 mm (0.020 in.) diameter by 51 mm (2.0 in.) long deformed steel fibers. A total of 656 tonnes (723 tons) of fibers was used.

As in the 1976 pavement, corner breaks are present in at least 10 percent of the panels. About one-half of the transverse contraction joints have not functioned properly even though the joints were sawed 76 mm deep in the 178 mm pavement. This lack of joint opening has resulted in wide open joints at those that have functioned properly. This wide opening has produced joint seal failures.[88]

Reno, Nevada. A steel fiber reinforced concrete overlay was completed in May 1975 at the Cannon International Airport, Reno.[15] The 102 mm (4 in.) overlay was placed on 29,300 m^2 (35,000 yd^2) of the hardstand at the passenger debarking area at the air terminal building. The concrete mixture had a cement content of 389 kg/m^3 (658 lb/yd^3), a pozzolan (fly ash) content of 128 kg/m^3 (216 lb/yd^3), and a fiber content of 118 kg/m^3 (200 lb/yd^3) of 0.25 by 0.56 by 25 mm (0.010 by 0.022 by 1 in.) steel fibers. The maximum aggregate size was 9.5 mm (3/8 in.). The concrete was initially spread to grade with a Blaw-Knox spreader, vibrated with a mobile three-bar roller-vibrator, dressed with bull floats, scored with a rake, and then sprayed with curing compound. Sawed joints and grooves were placed in the overlay directly above joints and grooves in the original paving. The original paving had joints forming 6.1 by 7.6 m (20 by 25 ft) slabs with a groove cut down the center of the 7.6 m (25 ft) side thus resulting in 3.8 by 6.1 m (12.5 by 20 ft) subsections.

The bonding apparently failed at many of the corners before it had developed properly due to extreme daily temperature fluctuations, thus resulting in some curling and warping at the corners. Later cracks developed parallel.to joints at a distance of 460 to 610 mm (18 to 24 in.) from the joint thus suggesting the limit of the debonding that occurred. These cracks remained tight and free from spalls but other cracks in the overlay are requiring maintenance.[88]

The second project built at Cannon International was in 1980. It was a new 203 mm (8 in.) fiber reinforced taxiway pavement on a aggregate base and subbase. The taxiway construction employed 7.6 by 12.2 m (25 by 40 ft) sections. A 1982 survey indicated that in several of the sections, there is a longitudinal crack at the approximate centerline but that this crack was tight.[88]

Norfolk Naval Air Station, Virginia. Several large placements of steel fiber reinforced concrete have been placed at this location since December 1977 principally for use in aircraft parking aprons. The principal reason for the use is to reduce apron overlays from a conventional requirement of 254 mm (10 in.) of unreinforced concrete to 127 mm (5 in.) and limit new construction to pavement thicknesses of 178 mm (7 in.). Improved durability is also a consideration.

The first large placement was in 1979 and used a joint spacing of 7.6 by 7.6 m (25 by 25 ft). All overlay concrete was placed on a 38 mm (1.5 in.) bituminous bond breaker and leveling course. The area overlayed was 29,275 m^2 (35,015 yd^2). The mixture used a cement content of 355 kg/m^3 (600 lb/yd^3), a pozzolan (fly ash) content of 148 kg/m^3 (250 lb/yd^3), and a fiber content of 95 kg/m^3 (160 lb/yd^3). The fiber size was 0.25 by 0.56 by 25 mm (0.010 by 0.022 by 1 in.). The largest overlay was completed in 1982 on 74,580 m^2 (89,200 yd^2) of apron using the same overlay thickness, fiber and mixture as in 1979. A total of 943 tonnes (1040 tons) of steel fiber was used in that application. At the time of this paper, another paving job was begun at that site involving 574 tonnes (633 tons) of steel fibre.

Although these overlays have only been in short while, they have exhibited a minimum of cracking and no spalling.[88]

Taoyuan Air Base, Taiwan. This project was completed in June, 1984 and involved a 152 mm (6 in.) bonded overlay over 55,000 m^2 (65,780 yd^2) of an old concrete pavement. All joints in the old pavement were matched by the overlay. Approximately 800 tonnes (882 tons) of 0.89 mm (0.035 in.) equivalent diameter by 51 mm (2 in.) fully crimped steel fibers were used. No performance data is yet available.

OTHER PAVEMENT APPLICATIONS

Fort Hood, Texas (Tank Parking Area). The first full-scale nonexperimental steel fiber reinforced concrete overlay was completed in March 1974 at Fort Hood, Texas.[130,131] The area overlayed was a maintenance hardstand for tanks, tank retrievers, and other tracked vehicles. The existing pavement consisted of 127 to 178 mm (5 to 7 in.) of asphaltic concrete on a stabilized limestone base. Because of severe wear and rutting caused by the action of the tracks, the flexible pavement surface has required replacement every three to four years. The steel fiber concrete overlay was selected in hopes to extend that maintenance period.

The concrete mixture had a cement content of 307 kg/m^3 (519 lb/yd^3), a pozzolan (fly ash) content of 137 kg/m^3 (231 lb/yd^3), and a fiber content from 95 to 118 kg/m^3 (160 to 200 lb/yd^3) of 0.25 mm (0.010 in.) diameter by 12.7 mm (0.50 in.) long fibers. The maximum aggregate size was 9.5 mm (3/8 in.). Approximately 26,755 m^2 (32,000 yd^2) of 51 mm (4 in.) thick overlay were machine placed in 6.1 m (20 ft) wide lanes with several thousand additional square meters (yards) being hand placed. Joints were sawed every 15.2 m (50 ft.). During an early evaluation, only a minimum amount of cracking had appeared in the overlay.[131] This occurred a few days after placement and before traffic was allowed on the overlay. Only one shrinkage crack 2.4 to 3.6 m (8 to 12 ft) long had appeared on the several thousand square meters (yards) of overlay placed by hand. Several transverse shrinkage cracks appeared in the machine-placed areas, but the most common crack was a longitudinal one in the center of the paving lanes. Approximately 60 m (200 ft) of this type of cracking occurred for undetermined reasons. All the cracks were hairline and none appeared to be working cracks after the pavement was open to traffic.

A survey of the overlay after being opened to traffic for nine months revealed only one working crack.[130] It crossed a 3.6 m (12 ft) wide hand finished lane and opened up approximately 6.3 mm (1/4 in.). The hairline cracks that developed shortly after construction remained tightly closed. The overall performance of the overlay was satisfactory,[131] but softening of the underlying asphalt due to penetration of oil and gasoline through the joints and complaints from Army mechanics about fibers in their shoes and clothes, forced a resurfacing on the area in 1980.[88]

PATCHING

Luke[74] reported on 11 small patches that were installed along the key joint on a runway used by 747's at Chicago's O'Hare International Airport in October 1970. The runway was 305 mm (12 in.) thick and had cracks across it about 1.5 m (5 ft). The runway was reinforced with reinforcing steel every 152 mm (6 in.) in both directions of a depth of 152 mm (6 in.). The patches were cut out with a saw and an air hammer was used to remove the material. They were approximately 0.3 m (1 ft) wide by 0.9 to 3.4 m (3 to 11 ft) long and 76 to 152 mm (3 to 6 in.) deep and were blown out with air, wetted, and filled with steel fiber reinforced concrete containing 0.41 mm (0.016 in.) diameter by 25 mm (1 in.) long fibers. Similar repair work two years prior to then had been done with epoxy and not proved satisfactory. Luke[74] reported the steel fiber reinforced concrete patches were performing satisfactorily after two winters.

Walker and Lankard[124] reported on the use of precast steel fiber reinforced mortar slabs for patching the Prospect Expressway, South[124] and the Queens Midtown Tunnel in New York, New York.[129] The Queens Midtown tunnel patching was done in March 1971. The pavement in the truck lane of the westbound tube had several areas where the brick surface had settled as much as 76 mm (3 in.). Repair of these areas required removal of a 1.1 m (3.5 ft) square of the brick surface and the deteriorated concrete subbase. The repairs had to be completed between the hours of midnight and five a.m. of the same day. The precast slabs were 0.9 m (3 ft) square and 51 mm (2 in.) thick. The excavation was approximately 178 mm (7 in.) deep and was filled with freshly mixed steel fiber reinforced mortar to a depth of 127 mm (5 in.). The precast slabs were then set into the fresh mortar and adjusted to be flush with the adjacent wearing course. The void between the precast slab and the adjoining bricks was then filled with the freshly mixed fibrous mortar. The mortar for the slabs contained 572 kg/m^3 (972 lb/yd^3) of cement and 145 kg/m^3 (245 lb/yd^3) of 0.25 mm (0.010 in.) diameter by 25 mm (1 in.) long steel fibers. The subbase concrete mixture contained 478 kg/m^3 (808 lb/yd^3) of cement and 114 kg/m^3 (193 lb/yd^3) of 0.41 mm (0.016 in.) diameter by 25 mm (1 in.) long steel fibers, and a set accelerator. Traffic was allowed on the patches approximately four hours after the concrete was placed.

In October 1971, a similar technique was used to repair an area in the extreme right lane between 8th Avenue Exit and the Fort Hamilton Parkway on the Prospect Expressway South. A 1016 mm (40 in.) square of deteriorated concrete pavement was removed by saw cutting to a depth of 127 mm (5 in.) and removing the material with air hammers. The section was approximately 305 mm (12 in.) from the curb and abutted an expansion joint which was not replaced. The excavated patch area was filled to a depth of approximately 76 mm (3 in.) with a proprietary fast setting concrete mixture, the precast steel fiber reinforced mortar patch set in place and positioned, and the edge voids filled with the fast setting mixture. Traffic was resumed on the slab approximately 445 minutes after the work was done.

SUMMARY OF PERFORMANCE

The majority of the bridge decks, floors, pavements and overlays placed to date have had some visual examination to qualitatively assess their performance. Meaningful evaluations of performance have been made difficult, in most cases, by the relatively brief observation periods and the fact that very few projects studied major design variables systematically under identical load and environmental conditions. Even with only limited information available, some design procedures for both pavements and overlays have developed.[77,93,99,100,101] The analysis of the data has been based on several rather arbitrary, though conservative, decisions and assumptions. The existing design criteria can be considered as satisfactory but subject to change as additional performance data are accumulated.

Fibrous concrete used in horizontal construction has generally performed better than comparable plain concrete constructions having identical thickness, foundation conditions, and concrete flexural strength. The use of fibrous concrete will result in thinner pavements and offer an alternative design that has several advantages over conventional construction.[93]

Many of the experimental overlays placed to date have developed full-width transverse cracks within 24 to 36 hours after placing. Lankard and Walker[67] felt that this problem was related to the high cement content and relatively low aggregate content of the fibrous concrete used in the experiments. The drying shrinkage and heat release of these concretes is greater than for conventional concretes. Restrained shrinkage occurs in the overlay at a time when the bond between the fiber and matrix is inadequate and a crack is formed. Successful techniques for eliminating or minimizing this problem have included: (1) reducing the cement content and increasing the aggregate content of the concrete; (2) replacement of a portion of the cement with fly ash; and (3) using a shrinkage compensating concrete.

Slab movements due to temperature change must also be accomodated. The use of double polyethylene sheeting and bituminous leveling courses as bond breakers has been very successful in many of the field projects. Experience indicates that unless underlying joints are matched, some induced form of debonding will be necessary in steel fiber rein- forced concrete overlays to eliminate nonload related cracking, especially when overlaying existing concrete pavements.

The use of sliding dowels to join adjacent fibrous concrete overlay slabs or to joint fibrous concrete overlay slabs to existing pavement has been successful in sections as thin as 76 mm (3 in.).[67]

REFERENCES

1. "A Status Report on Fiber Reinforced Concretes," Concrete Construction, Vol 21, No. 1, Jan 1976, pp 13-16.

2. ACI Committee 544, "Listing of Fibrous Concrete Projects," Feb 15, 1978; Addendum 1, Feb 29, 1980; Addendum 2, Mar 25, 1981; Addendum 3, Sep 3, 1982, American Concrete Institute, Detroit, Michigan.

3. "Add Fiber Wires to the Mix for Better Warehouse Floors," Materials Handling News (UK), Oct 1973, pp 1-4.

4. Arnold, C. J., "Steel Fiber-Reinforced Concrete Overlay," Research Report No. R-878, Aug 1973, Michigan State Highway Commission, Lansing, Michigan.

5. Arnold, C. J. and Brown, M. G., "Experimental Steel Fiber-Reinforced Concrete Overlay," Research Report No. R-852, Apr 1973, Michigan State Highway Commission, Lansing, Michigan.

6. Batson, G. B., "Introduction to Fibrous Concrete," Proceedings, Conference M-28, Fibrous Concrete--Construction Material for the Seventies, Construction Engineering Research Laboratory, Champaign, Illinois, Dec 1972, pp 1-25.

7. Beckett, R. E., "Fibrous Concrete Bridge Decks," Civil Engineering (London), Vol 69, No. 814, May 1974, pp 32-35.

8. Beckett, R. E., "Highway Construction in Wirand Concrete," 2 E'me Symposium sur les Routes en Beton, Berne, 1973.

9. Brockenbrough, T. W., and Davis, C. F., "Progress Report No. 1 on Fiber Reinforced Concrete," Jan 1970, Deleware State Highway Department and University of Deleware, Newark, Del.

10. Buckley, E. L., "Accelerated Trials of Glass Fiber Reinforced Rigid Pavements," Construction Research Center, The University of Texas at Arlington, Texas, Apr 1974,.

11. Buckley, E. L., and Clines, W. A., "Rigid Pavements on Interstate Highway Using Glass Fiber Reinforcement," Research Report TR-5-74, Construction Research Center, The University of Texas at Arlington, Texas, Jul 1974, 25 pp.

12. Buckley, E. L., and Sriboonlue, W., "Investigation of the Use of E-Glass Fibers for Crack Control in Post-Tensioned Slabs-on-Ground Foundations," Research Report TR-1-76, Construction Research Center, The University of Texas at Arlington, Texas, 1976.

13. Cleary, M., "GRC in Highway Maintenace and Construction," The Institution of Highway Engineers, United Kingdom, Dec 1975.

14. "Concrete Means Better Pavement," Newsletter, American Concrete Paving Association, Oakbrook, Ill., Vol 9, No. 10, Oct 1973, pp 1-6.

15. Denson, R. H., "Report of Trip to Reno, Nevada to Observe Fiber-Reinforced Concrete Paving Project," Memorandum for Record, U. S. Army Engineer Waterways Experiment Station, Structures Laboratory, Vicksburg, Mississippi, 8 May 1975, 2 pp.

16. Dickerson, R. F., A New Pavement Material, Battelle Columbus Laboratories, Columbus, Ohio, Apr 1972, 20 pp.

17. Dickerson, R. F., "Steel Fiber Reinforced Concrete--A New Pavement Material," Roads and Streets, Vol 115, No. 7, Jul 1972, pp 68-71.

18. Dickerson, R. F., Reduced Maintenance Costs, Battelle Columbus Laboratories, Columbus, Ohio, Apr 1972, 20 pp.

19. Dickerson, R. F., "Wirand Concrete Reduced Maintenance Costs," American Association of State Highway Officials (AASHO), Phoenix, Arizona, Nov 1972.

20. "Experimental House Uses Fibrous Concrete Floor Slab," Engineering News-Record, Vol 193, 16 Jan 1975, p 12.

21. "Fiber Reinforcement Demonstrates Potential in Concrete Pavements," Roads (formerly Rural & Urban Roads), Vol 22, No. 7, July 1984, p 68.

22. "Fibre Reinforcing for Pavements," Civil Engineering (London), Dec 1982, pp 27-28.

23. "Fibrous Concrete--A New Dimension in Bridge Deck Construction," Concrete Construction, Vol 18, No. 7, Jul 1973, pp 321-324.

24. "Fibrous Concrete Cuts Airport Overlay to 6 In.," Engineering News-Record, Vol 196, No. 24, 10 Jun 1976, p 21.

25. "Fibrous Concrete Airport Pad Draws Worldwide Inquiries," Modern Concrete, Industry News, Vol 40, No. 12, Apr 1977, pp 9-10.

26. "Fibrous Concrete Experiments Show Promising Future as Paving Material,"NZ Concrete Construction, Vol 18, No. 1, Feb 1974, pp 32-33.

27. "Fibrous Concrete Makes Thinner Overlays Possible," World Construction, Vol 30, No. 11, Nov 1977, p 61.

28. "Fibrous Concrete Overlay," NZ Concrete Construction, Development News, Vol 20, No. 8, Sep 1976, p 34.

29. "Fibrous Concrete Overlay at Las Vegas Airport," Newsletter, American Concrete Paving Association, Oakbrook, Ill., Vol 12, No. 6, Jun 1976, p 2.

30. "Fibrous Concrete Overlay Gets The Big Test," Concrete Construction, Vol 19, No. 3, Mar 1974, pp 103-105.

31. "Fibrous Concrete Overlay-Texas Test Section," Highway Focus, Vol 8, No. 2, Apr 1976, pp 66-68.

32. "Fibrous Concrete--Pavement of Tomorrow," Newsletter, American Concrete Paving Association, Oakbrook, Ill., Vol 8, No. 10, Oct 1972, pp 1-6.

33. "Fibrous Concrete Placed 3 Inches Thick in Urban Highway Test," Roads and Streets, Vol 116, No. 2, Feb 1973, pp 59-62.

34. "Fibrous-Reinforced Concrete Performs Well in Airfield Pavement Tests," Concrete Construction, Vol 17, No. 3, Mar 1972, pp 119-120.

35. "Fibrous Concrete Tested on Highway," Engineering News-Record, Vol 190, No. 36, 19 Oct 1972, p 16.

36. Galinat, M. A., "An Alternative Bridge Deck Renewal System," Paper No. IBC-84-21, Presented at the International Bridge Conference, Pittsburgh, Pennsylvania, June, 1984, Mitchell Fibercon, Inc., Pittsburgh, Pennsylvania.

37. Galloway, J. W.,and Gregory, J. M., "Trial of a Wire-Fiber-Reinforced Concrete Overlay on a Motorway," TTRL Laboratory Report No. 764, Transport and Road Research Laboratory, Crowthorne, Berkshire, 1977, 22 pp.

38. "Glass-Fibre Cement Bridge," Concrete, Vol 10, No. 4, Apr 1976, p 13.

39. Gramling, W. L. and Nichols, T. H., "Steel Fiber Reinforced Concrete," Report 71-3, Pennsylvania Department of Transportation, Dec 1972; also Special Report 148, Transportation Research Board, Washington, D. C., 1974, pp 160-165.

40. Granchich, A., "Bearing Capacity and Durability of Pavements with a Thin Wire Concrete Surfacing," Inzhenyrske Stavby (Czechoslovakia), Vol 25, No. 1, 1977, pp 8-16.

41. Gray, B. H., "Fiber Reinforced Concrete--A General Discussion of Field Problems and Applications," Technical Manuscript M-12, Apr 1972, Construction Engineering Research Laboratory, Champaign, Illinois.

42. Gray, B. H., "Fiber Reinforced Concrete Pavement Performance Investigation," Highway Focus, Vol 4, No. 5, Oct 1972, pp 53-64.

43. Gray, B. H. and Rice, J. L., "Fibrous Concrete for Pavement Applications," Preliminary Report M-13, Apr 1972, Con- struction Engineering Research Laboratory, Champaign, Illinois.

44. Gray, B. H. and Rice, J. L., "Pavement Performance Applications," Proceedings, Conference M-28, Fibrous Concrete--Construction Material for the Seventies, Construction Engineering Research Laboratory, Champaign, Illinois, Dec 1972, pp 147-157.

45. Gregory, J. M., Galloway, J. W., and Raithby, K. D., "Full Scale Trials of a Wire-Fibre-Reinforced Concrete Overlay on a Motorway," Proceedings, RILEM Symposium, Fibre Reinforced Cement and Concrete, Construction Press, Lancaster, England, Sep 1975, pp 383-394.

46. Hanna, A. N., "Steel Fiber Reinforced Concrete Properties and Resurfacing Applications," Research and Development Bulletin RD049.01P, Portland Cement Association, Skokie, Illinois, 1977, 18 pp.

47. Hoff, G. C., Godwin, L. N., Saucier, K. L., Buck, A. D., Husbands, T. B. and Mather, K., "Identification of Candidate Zero Maintenance Paving Materials, Chapter 9, Fiber Reinforced Concrete," Report No. FHWA-RD-77-110, Vol 2, Final Report, Federal Highway Administration, Washington, D. C., May 1977, pp 264-439.

48. Hutchinson, R. L., "Performance of Concrete Pavements Sub- jected to Wide-Body Jet Aircraft Loadings," SP-51, Roadways and Airport Pavements, American Concrete Institute, Detroit, Michigan, 1975, pp 135-159.

49. Ibukiyama, S., "Pavement Applications (of Fiber Reinforced Concrete)" (in Japanese), Concrete Journal (Japan), Vol 15, No. 3, Mar 1977, pp 44-48.

50. Ibukiyama, S., Seto, K., and Kokubu, S., "Steel Fiber Reinforced Concrete Overlays on Asphalt Pavement," SP-81, Fiber Reinforced Concrete, International Symposium, American Concrete Institute, Detroit, Michigan, 1984, pp 351-373.

51. "Iowa's Concrete Overlay Tests May Solve Rehabilitation Riddles," Rural and Urban Roads, Vol 14, No. 8, Aug 1976, pp 40-46.

52. Johnston, C. D., "Steel Fiber Reinforced Concrete Pavement--Construction and Interim Performance Report," SP-51, Roadways and Airport Pavements, American Concrete Institute, Detroit, Michigan, 1975, pp 161-173.

53. Johnston, C. D., "Steel Fiber Reinforced Concrete Pavement Trials," Concrete International: Design & Construction, Vol 6, No. 12, Dec 1984, pp 39-43.

54. Johnston, C. D., "Steel Fiber Reinforced Mortar and Concrete--A Review of Mechanical Properties," Fiber Reinforced Concrete, SP-44, American Concrete Institute, Detroit, Michigan, 1974, pp 127-142.

55. Johnston, C. D., "Steel Fiber Reinforced Concrete Pavement--Second Interim Performance Report," Proceedings, RILEM Symposium, Fibre Reinforced Cement and Concrete, Construction Press, Lancaster, England, Sep 1975, pp 409-418.

56. Josifek, C. W., "The Largest Fibrous Concrete Aircraft Parking Ramp in the World-At McCarran International Airport," Bekaert Steel Wire Corp., Niles, Illinois, Apr 1980, 21 pp.

57. Josifek, C. W., "Technical Report--Fiber Concrete Aircraft Parking Ramp at McCarran International Airport," The International Journal of Cement Composites, Vol 2, No. 4, Nov 1980, pp 235-238.

58. Keeton, J. R., "Shrinkage-Compensating Cement for Airport Pavement, Phase 3, Fibrous Concretes, Addendum," Report No. CEL-TN-1561-ADD, Final Report, Jan 1979-May 1980, Civil Engineering Lab (Navy), Port Hueneme, California, Sep 1980, 28 pp.

59. Kobayashi, K., and Inoue, T., "Concrete Pavement Using Steel Fibers," (in Japanese), Concrete Journal (Japan), Vol 14, No. 6, Jun 1976, pp 50-54.

60. Kohno, K., Nakano, K., Nakahara, Y., and Yurugi, M., "Pavement Applications of Steel Fiber Reinforced Concrete," (in Japanese), Proceedings of the Symposium on Steel Fiber Reinforced Concrete, Tokyo, Nov 7, 1977, Japan Concrete Institute, Tokyo, Nov 1977, pp 156-159.

61. Kubota, K., "Steel Fiber Reinforced Concrete Pavement," (in Japanese), The Doboku-Seko (Japan), Vol 17, No. 14, Dec 1976, pp 65-70.

62. Knutson, M. J., "Green County, Iowa, Concrete Overlay Research Project," SP-51, Roadways and Airport Pavements, American Concrete Institute, Detroit, Michigan, 1975, pp 175-195.

63. Lankard, D. R. and Henager, C. H., "Condition of the Wirand Concrete Overlay Project in Greene County, Iowa, as of September 1977," Summary Report, BattelleColumbus Laboratories, Columbus, Ohio, 1977.

64. Lankard, D. R. and Walker, A. J., "Bridge Deck and Pavement Overlays with Steel Fibrous Concrete," Fiber Reinforced Concrete, SP-44, American Concrete Institute, Detroit, Michigan, 1974, pp 375-392.

65. Lankard, D. R. and Walker, A. J., "Field Work in New York City for the Triborough Bridge and Tunnel Authority--East Side Airlines Terminal," Battelle Columbus Laboratories, Columbus, Ohio, Jul 1971, 11 pp.

66. Lankard, D. R. and Walker, A. J., "Installation of Wirand Concrete Slab--Ohio Highway Experiment, Ashland, Ohio," Battelle Columbus Laboratories, Columbus, Ohio, Aug 1971, 10 pp.

67. Lankard, D. R. and Walker, A. J., "Pavement Applications for Steel Fibrous Concrete," Transportation Engineering Journal, American Society of Civil Engineers, Vol 101, No. TE1, Paper 11108, Feb 1975, pp 137-153.

68. Lankard, D. R. and Walker, A. J., "Wirand Field Trials -- Detroit Metropolitan Airport," Battelle Columbus Laboratories, Columbus, Ohio, Nov 1971, 9 pp.

69. Lankard, D. R. and Walker, A. J., "Wirand Field Trials: Queens Midteown Tunnel Patch, New York, N. Y.," Battelle Columbus Laboratories, Columbus, Ohio, Jul 1971, 10 pp.

70. Lankard, D. R. and Walker, A. J., "Wirand Field Trials -- Wirand Precast Slab Patching of New York City Expressway," Battelle Columbus Laboratories, Columbus, Ohio, Jul 1971, 11 pp.

71. Largest Fibrous Concrete Paving Project Solves Clumping Problem," Engineering News-Record, Vol 192, No. 15, 11 Apr 1974, pp 68-69.

72. Lott, J. L., and Abdel-Malek, R. A., "Evaluation of the Response of Fiber Concrete Pavement Systems," Presented at the InterSociety Conference on Transportation, 18-24 Jul 1976, Development Engineers, Naperville, Illinois.

73. Lowe, R. A., "Fiberous Concrete Construction at Reno and Las Vegas Airports (Abridgement)," Special Report 1975, Researching Airport Pavements, Transportation Research Board, Washington, D. C., 1978, pp 67-68.

74. Luke, C. E., "Driveway, Road and Airport Slabs," Proceedings, Conference M-28, Fibrous Concrete--Construction Material for the Seventies, Construction Engineering Research Laboratory, Champaign, Illinois, Dec 1972, pp 199-208; also Highway Focus, Vol 4, No. 5, Oct 1972, pp 65-70.

75. Luke, C. E., Waterhouse, B. L., and Wooldridge, J. F., "Steel Fiber-Reinforced Concrete Optimization and Applications," SP-44, Fiber Reinforced Concrete, American Concrete Institute, Detroit, Michigan, 1974, pp 393-413.

76. Marsden, W. A., "Bulk Steel Fibre Reinforced Concrete Has Arrived in Australia," Proceedings, Symposium on Fibrous Concrete, 16 April 1980, The Concrete Society (U. K.), The Construction Press, Lancaster, England, 1980, pp 128-136.

77. Marvin, E., "Fibrous Concrete Pavement Overlay Thickness Design," Technical Note, Construction Engineering Research Laboratory, Champaign, Illinois, Dec 1974.

78. Mayo, R. E., "Fibrous Concrete Overlay," The Military Engineer, No.432, Jul-Aug 1974, pp 242-243.

79. McDonald, A. R., "Wirand Concrete Pavement Trials," Proceedings, Conference M-28, Fibrous Concrete--Construction Material for the Seventies, Construction Engineering Research Laboratory, Champaign, Illinois, Dec 1972, pp 209-234.

80. Meyer, A. H., Ledbetter, W. B., Layman, A. H., and Saylak, D., "Reconditioning Heavy Duty Freeways in Urban Areas," Texas Transportation Insitute, Texas A&M University, College Station, Texas, Dec 1975.

81. Meyer, A. H., Ledbetter, W. B., and White, F. S., "Annotated Bibliography for Reconditioning Heavy Duty Freeways in Urban Areas," Texas Transportation Insitute, Texas A&M University, College Station, Texas, Mar 1976, pp 33-55.

82. Nichols, T. H., "Steel Fiber Reinforced Concrete - Research Project No. 71-3," Bureau of Materials and Research, Pennsylvania Department of Transportation, Dec 1972.

83. Nichols, T. H., "Steel Fiber Reinforced Concrete - Research Project No. 71-3(2)," Bureau of Materials and Research, Pennsylvania Department of Transportation, Sep 1974.

84. Nishioka, K., Kakima, N., Yamakawa, S., and Shirakawa, K., "Effective Applications of Steel Fibre-Reinforced Concrete," Proceedings, RILEM Symposium, Fibre Reinforced Cement and Concrete, Construction Press, Lancaster, England, Sep 1975, pp 425-433.

85. Ohio Department of Transportation, "Experimental Wirand Concrete Pavement Report 2, Ashland County-IR-71," Feb 1974.

86. Ohno, T., and Ibukiyama, S., "Steel Fiber Reinforced Concrete Pavement Testing at Kuroiso Bypass," (in Japanese), Concrete Journal (Japan), Vol 14, NO. 6, Jun 1976, pp 62-66.

87. Ohno, T., Kubota, K., and Kokubu, S., "Performance of Steel Fiber Reinforced Concrete Pavement (at Kuroiso Bypass, Route 4) Subjected to Traffic," (in Japanese), Proceedings of the Symposium on Steel Fiber Reinforced Concrete, Tokyo, Nov 7, 1977, Japan Concrete Institute, Tokyo, Nov, 1977, pp 144-147.

88. Packard, R. G., and Ray, G. K., "Performance of Fiber- Reinforced Concrete Pavement," SP-81, Fiber Reinforced Concrete, International Symposium, American Concrete Institute, Detroit, Michigan, 1984, pp 325-349.

89. Paillere, A. M. and Bonnet, G., "Incorporation of Metal Fibres in Concrete in New Roads--Laboratory Tests and Site Investigations at Egletons," Proceedings, Volume 2, RILEM Symposium, Fibre Reinforced Cement and Concrete, Construction Press, Lancaster, England, Sep 1975, pp 539-550.

90. Parker, F., Jr., "Construction of Fibrous Concrete Overlay: Tampa International Airport," Proceedings, Conference M-28, Fibrous Concrete--Construction Material for the Seventies, Construction Engineering Research Laboratory, Champaign, Illinois, Dec 1972, pp 177-197.

91. Parker, F., Jr., "Construction of Fibrous Concrete Overlay Test Slabs, Tampa International Airport, Florida," Report No. FAA-RD-72-119, Federal Aviation Administration, U. S. Department of Transportation, Washington, D. C., Oct 1972.

92. Parker, F., Jr., "Pavement Design: Prestressed, Steel Fibrous and Continuously Reinforced Concrete," Special Report 175, Transportation Research Board, Washington, D. C., 1978, pp 46-53.

93. Parker, F., Jr., "Steel Fibrous Concrete for Airport Pavement Applications," Technical Report S-74-12, U.S. Army Engineer Waterways Experiment Station, Vicksburg, Mississippi, Nov 1974.

94. Parker, F., Jr. and Rice, J. L., "Steel Fibrous Concrete for Airport Pavements," Proceedings, International Conference on Concrete Pavement Design, Purdue University, West Lafayette, Indiana, Feb 15-17, 1977, pp 541-555.

95. Pasko, T. J., Jr., "Pavement Applications of Steel Fiber Reinforced Concrete," Highway Focus, Vol 4, No. 5, Oct 1972, pp 71-91.

96. "Pioneer Overlay Job Uses Fiber-Reinforced Concrete," Civil Engineering, American Society of Civil Engineers, Vol 44, No. 1, Jan 1974, PP 38-39.

97. Popovics, S., "Polymer Pavement Concrete for Arizona; Study I," Final Report, Northern Arizona University, Flagstaff, Arizona, 1974.

98. "Research in Airport Pavements," Report No. SR-17S, National Academy of Sciences, Transportation Research Board, Washington, D. C., May 1977, 13 pp.

99. Rice, J. L., "Fibrous Concrete Pavement Design Summary," Final Report, No. CERL-TR-M-134, Construction Engineering Research Laboratory, Champaign, Illinois, Jun 1975.

100. Rice, J. L., "Pavement Design Considerations," Proceedings, Conference M-28, Fibrous Concrete--Construction Material for the Seventies, Construction Engineering Research Laboratory, Champaign, Illinois, Dec 1972, pp 159-176.

101. Rice, J. L., "Proposed Design Criteria for Fibrous Concrete Pavement," Preliminary Report, No. S-5, Construction Engineering Research Laboratory, Champaign, Illinois, Apr 1972.

102. Richter, H., Mandel, J., and Kirsten, E., "Steel-Fiber Reinforced Concrete and It's Use For Foundary Floors (In German)," Bauplanung-Bautechnik (Berlin), Vol 34, No. 10, Oct 1980, pp 460-462.

103. Roth, L., "Little Steel Fibers May Solve a Big Paving Problem," Michigan Contractor and Builder, Vol 68, No. 14, Jul 1974, pp 20-24.

104. Sakurai, H., "Overlay on Asphalt with Steel Fiber Reinforced Concrete," Composite Materials II, Proceedings of 2nd Japan- USSR Symposium on Composite Materials, Japan Society for Composite Materials, 1980, p 151.

105. Sakurai, H., Itoh, S., Ibukiyama, S., Okuno, M., and Mizusawa, S., "Steel Fiber Reinforced Concrete Overlay," (in Japanese), Proceedings of the Symposium on Steel Fiber Reinforced Concrete, Tokyo, Nov 7, 1977, Japan Concrete Institute, Tokyo, Nov, 1977, pp 148-151.

106. Sather, W. R., "First Fiber Concrete Placed in Minnesota," Construction Bulletin (USA), 3 Aug 1972, pp 3-5.

107. Sather, W. R., and Wilson, J. R., "A New Dimension in Bridge Deck Construction," Concrete Construction, Jul 1973, pp 321- 324.

108. Schrader, E. K., Design Methods for Pavements with Special Concretes," SP-81, Fiber Reinforced Concrete, International Symposium, American Concrete Institute, Detroit, Michigan, 1984, pp 197-212.

109. Schrader, E. K. and Munch, A. V., " Deck Slab Repaired by Fibrous Concrete Overlay," Proceedings, American Society of Civil Engineers, Vol 102, CO1, Mar 1976, pp 179-196.

110. Schrader, E. K. and Munch, A. V., "Fibrous Concrete Repair of Deck Slab at Libby Dam Visitor Facility, Libby, Montana," Walla Walla District Report, U. S. Army Corps of Engineers, Walla Walla, Washington.

111. Schwarz, A. W., "Steel Fiber Reinforced Concrete," U.S. Steel Corporation, New Product Development, Pittsburgh, Pennsylvania, 1975.

112. Spires, J. W., Romualdi, J. P., and Pichumani, R., "Analysis of Steel-Fiber Reinforced Concrete Warehouse Floor Slabs," Proceedings, American Concrete Institute Journal, Vol 74, No. 12, Dec 1977, pp 616-622.

113. "Steel Fibrous Concrete for Las Vegas Airport Job," Concrete Products, Vol 79, No. 8, Aug 1976, p 9.

114. "Steel Fibrous Concrete Overlay Get Good Reviews in Las Vegas," Transportation Research News, No. 66, Transportation Research Board, Washington, D. C., Sep-Oct 1976, pp 20-21

115. "Steel Fibres in Airport Runways," Concrete, Vol 6, No. 8, Aug 1972, pp 34-35.

116. "Steel Fibre Reinforced Concrete," Highways and Road Construction, Vol 41, Oct 1973, p 55.

117. "Steel Fibre Reinforced Concrete for Airfield Pavements," Report RB-23-1982, Cement Research Institute of India, New Delhi, India, Jul 1982, 17 pp.

118. Steinmetz, E., "Experience with the Construction of a Steel Fibre Reinforced Concrete Road," Proceedings, Conference on Fibre Reinforced Concrete and Other Fibre Reinforced Building Materials, Stevin Laboratory, Delft, Netherlands, Sep 1973, pp 181-185.

119. "Thin Concrete Overlay for Residential Streets," Better Roads, Vol 43, No. 4, Apr 1973, pp 24-26.

120. "Thin Concrete Overlays Get County Test," Better Roads, Vol 43, No. 11, Nov 1973, pp 20-22.

121. Tyson, S. S., "Two-Course Bonded Concrete Bridge Deck Construction in Virginia," Transportation Research Record, No. 652, 1977, pp 15-21.

122. "Unusual Construction Problems Overcome on Fibrous Concrete Pavement Research Project," Roads and Streets, Vol 117, No. 2, Feb 1974, pp 61-65.

123. Walker, A. J., "Chicago O'Hare International Airport Field Trial Wirand Concrete Overlay Patch Repair," Battelle Columbus Laboratories, Columbus, Ohio, Feb 1971, 17 pp.

124. Walker, A. J., "Wirand Field Trial--Wirand Precast Slab Patching of New York City Expressway," Battelle Columbus Laboratories, Columbus, Ohio, Oct 1971.

125. Walker, A. J. and Lankard, D. R., "Bridge Deck Rehabilitation with Steel Fibrous Concrete," Presented at the Third Annual World of Concrete International Exposition on Concrete Construction, New Orleans, Louisiana, 5-8 Jan 1977, Battelle Columbus Laboratories, Columbus, Ohio

126. Walker, A. J. and Lankard, D. R., "Field Work in New York City for the Tri-Borrough Bridge and Tunnel Authority--East Side Airlines Studies," Battelle Columbus Laboratories, Columbus, Ohio, Mar 1971.

127. Walker, A. J. and Lankard, D. R., "Wirand Concrete Field Placements, Cedar Rapids, Iowa, Summary Report," Battelle Columbus Laboratories, Columbus, Ohio, Feb 1973, 28 pp.

128. Walker, A. J. and Lankard, D. R., "Wirand Concrete Field Projects Contracted by the New York-New Jersey Port Authority, Jun 1975 to Sep 1976," ," Battelle Columbus Laboratories, Columbus, Ohio, Mar 30, 1977, 14 pp.

129. Walker, A. J. and Lankard, D. R., "Wirand Field Trials, Queens Midtown Tunnel Patch, New York, N. Y.," Battelle- Columbus Laboratories, Columbus, Ohio, Mar 1971.

130. Williamson, G. R., "Fort Hood Fibre Concrete Overlay," Proceedings, RILEM Symposium, Fibre Reinforced Cement and Concrete, Construction Press, Lancaster, England, Sep 1975, pp 453-459.

131. Williamson, G. R., "Technical Information Pamphlet on Fibrous Concrete Overlays--Fort Hood Project," Final Report No. CERL-TR-M-147, Construction Engineering Research Laboratory, Champaign, Illinois, Aug 1975.

132. "Wirand Concrete Data Pack for Highway Applications," Battelle Development Corporation, Columbus, Ohio, Dec 1973, 6 pp.

133. "Wire-Reinforced Precast Concrete Decking Panels," Precast Concrete, Vol 2, No. 12, Dec 1971, pp 703-708.

134. Yamakawa, S., Shirakawa, K., Nose, H., and Koyama, S., "Steel Fibre Reinforced Concrete Pavement Trials," (in Japanese), Proceedings of the Symposium on Steel Fiber Reinforced Concrete, Tokyo, Nov 7, 1977, Japan Concrete Institute, Tokyo, Nov, 1977, pp 152-155.

135. Yamakawa, S., et al., "Tested Pavements of Steel Fibre Reinforced Concrete," Journal of Pavement (Japan), Vol 13, No.9, Sep 1978, pp 1-5.

136. Yrjanson, W. A., "A Review of Field Applications of Fibrous Concrete," Special Report 148, Transportation Research Board, Washington, D. C., 1974, pp 69-79.

137. Yrjanson, W. A., "Pavement Applications of Fibrous Concrete," Proceedings, Conference M-28, Fibrous Concrete--Construction Material for the Seventies, Construction Engineering Research Laboratory, Champaign, Illinois, Dec 1972, pp 139-145; also Highway Focus, Vol 4, No. 5, Oct 1972, pp 92-98.

138. Yrjanson, W. A. and Halm, H. J., "Field Applications of Fibrous Concrete Pavements," American Concrete Paving Association, Oak Brook, Illinois, Nov 1973.

139. Yuhki, M., Hinuma, T., and Nakazaki, S., "Steel Fiber Reinforced Concrete Pavement," (in Japanese), The Doboku-Seko (Japan), Vol 18, No. 11, Aug 1977, pp 27-32.

140. Yuhki, M., and Nakazaki, S., "Trials of Steel Fibre Reinforced Concrete Paving," (in Japanese), Proceedings of the Symposium on Steel Fiber Reinforced Concrete, Tokyo, Nov 7, 1977, Japan Concrete Institute, Tokyo, Nov, 1977, pp 140-143.

141. Zollo, R. F., "Wire Fiber Reinforced Concrete Overlays for Orthotropic Bridge Deck Type Loadings," Proceedings, American Concrete Institute Journal, Vol 72, No. 10, Oct 1975, pp 576-582.

STEEL FIBER CONCRETE
US-SWEDEN joint seminar
(NSF-STU)
Stockholm 3-5 June, 1985
S P Shah and Å Skarendahl,
Editors

FIBER REINFORCED CONCRETE PAVEMENTS AND SLABS A State-of-the-Art Report

by Ernest K Schrader, Consulting Eng, Walla Walla, Washington, USA

ABSTRACT

The hardened properties of fiber reinforced concrete determined by tests in the lab resulted in its use with great expectations for pavements and slabs in the 1970's. Theoretically and in tests, it provided a tougher concrete with high flexural strength and fatigue endurance. This allowed reduced pavement thicknesses and lead to expected better performence. Practical construction problems were experienced in early field work concerning production rates, mix designs, workability, addition of fibers to the mix, and finishing. The industry suffered from this, but now knows how to deal with these issues. Similarily, there were disappointments when some slabs and pavements developed curl, problems at joints, cracking, and potentially hazardous exposed surface fibers. Even with these conditions, most fiber reinforced pavements are functioning and performing their intended purpose. Experiences where problems developed have allowed the industry to evaluate what occurred and address it in future work so that reoccurrences are eliminated or greatly minimized. The use of lower cement and water contents, larger aggregate, and high-range water reducers can be expected for fiber reinforced pavements and slabs in the future.

Ernest K. Schrader is a Consulting Engineer from Walla Walla, Washington, USA, specializing in concrete materials and construction. He has been involved with fiber reinforced concrete research, design, inspection, and construction for 17 years. He is a member of various ACI technical committees including Committee #544, Fibrous Concrete. Ernie has published numerous papers, including many on FRC and has received national awards from various professional organizations.

INTRODUCTION

During the late 1960's and early 1970's, fiber reinforced concrete was studied and tested extensively, and subsequently was used in a variety of demonstration projects with great enthusiasm and expectations. Laboratory data indicated that it could provide a major improvement and reduced thickness requirements for slabs, airfield pavements, overlays, and highways primarily because of improved fatigue endurance, flexural strength, toughness, impact resistance, and tensile strength. Because of its theoretical crack arrest capabilities, crack propagation was logically expected to be much less than for a noncomposite conventional slab without fibers.

Practical difficulties were experienced with early full-scale fiber reinforced concrete pavement projects from the standpoints of maintaining high productivity while adding the fibers into the mix; avoiding fiber balls or tangles; finishing surfaces without excessive fibers being exposed at the surface; and establishing appropriate mix designs. These problems can now be eliminated or minimized using techniques, equipment, and/or improved fibers developed as a result of this experience.

Although essentially all of the FRC overlays, slabs, and pavements that were placed during the first 5 to 10 years of the industry have performed their intended function, the crack-free extraordinary performance hoped for did not always result. In fact, curling of corners and edges (normally with some subsequent cracking) is common (1). Some transverse and longitudinal cracking also resulted, and reflective cracking in FRC overlays placed on old slabs has been experienced when the original cracks were active and no special precautions were taken.

Design charts and methods for pavement and overlay design that were developed and used for these early pavements appropriately addressed stresses from applied wheel loads. However, they did not account for internal stresses resulting from drying shrinkage, external and internally developed temperature changes, and restraint from various sources including jointing techniques that had been thought to be suitable. Observation and evaluation of the performance of these early pavements have provided the basis for current and future designs that can address both internally AND externally developed loads.

HISTORY

General

The paper presented by Hoff at this conference contains an excellent history of FRC pavements and overlays. Only the pertinent details of selected specific projects of major significance to the current state-of-the-art of FRC pavements, overlays, and slabs are discussed here.

Slabs, Overlays, and Floors

In 1971, an FRC sidewalk 6 feet (1.8 m) wide and 2.75 inches (62 mm) thick was placed at the University of Illinois campus in Champaign, Illinois. A slipform paver was used. One construction joint separated the two days of production which were 652 feet (199 m) and 488 feet (149 m) long. A 1.5 percent (volume) fiber content with 0.010 x 0.022 x 1.00 inch (0.25 x 0.56 x 25 mm) steel fiber typical of earlier work in the U. S. was used.

Flexural strengths in excess of 1000 psi (70 Kg/sq cm) resulted. Also typical of early fiber projects, the mix contained about 50 percent sand and 50 percent pea gravel (10 mm), and used fly ash (31 percent by weight) for a total cementitious volume of about 800 lbs/cy (474 Kg/cu m).

Detailed follow-up performance data was not published, but the project significantly influenced continued use of SFRC. It reportedly had little or no open cracking except for the construction joint which opened an estimated 1/2 inch (13 mm). The project was considered successful. In "hindsight," it demonstrated two problems that were overlooked at the time and became major factors in later pavements. (a) Without jointing to provide adequate relief, the ends of the placement had to contract inward by a significant amount to relieve shrinkage or internal stress which would develop if the movement were restrained. (b) The surface had exposed fibers which individually stained the concrete and represented a potential hazard to barefoot personnel.

In 1972, a thin SFRC overlay was placed on existing badly cracked conventional concrete in the severe exposed climate at Libby, Montana, USA (2). The project was probably the first FRC placement to be done by a construction contractor for a fixed fee without considering it to be a research or demonstration project. Because of the unevenness in the original surface, the thickness varied from about 0.75 to 2.75 inches (19 to 69 mm). The approximately 25 x 75 foot (7.6 x 22.9 m) surface contained three transverse joints. Placing was done with a vibrating screed. Hand finishing by wood and magnesium floats completed the work to a surface tolerance less than 1/8 inch per 10 feet (3 mm per 3 m).

The project demonstrated that nonworking cracks could be overlayed, but working joints and cracks would reflect through a thin overlay of FRC. It also showed that with attention to technique, the surface could be finished with essentially all fibers buried. The overlay was successful and has been in service in an extremely severe climate of heat, cold, snow, and rains. Substantial corner and edge curling occurred. This did not detract from the slab's performance, but it was an indication of problems or concerns that developed in subsequent slabs and pavements. Flexural strengths well in excess of 1000 psi (70 Kg/cm sq) were achieved. The mix was typical of that time with fly ash, cement, fiber, and aggregate proportions very similar to the University of Illinois sidewalk.

During the late 1970's and early 1980's, exterior slabs for heavy equipment, storage yards, and overlays were placed using SFRC with similar results to the Libby work. Thicknesses generally were on the order of 3 to 4 inches (76 to 102 mm) for new slabs and even less for overlays. In the United States, this work was primarily done on an individual and isolated case basis without detailed documentation or published reports. Despite a published guide for design (3), thicknesses and desired strength levels were typically judgemental values, and the slabs were often considered to be demonstration projects. A few examples are a fueling area for trucks and equipment in Lewiston, Idaho, warehouse floors for facilities belonging to fiber suppliers, and the first of several subsequent SFRC slabs for Dow at Midland, Michigan.

About 1980 the use of FRC for warehouse floors and slabs progressed to a somewhat more systematic and scientific approach in the U. S., but it was still influenced by judgement. Subsequent marketing efforts for warehouse floors in the U. S. were sporatic, and relatively little work was done. Meanwhile, the use of SFRC in warehouse and shop floors became reasonably common in Australia and parts of Europe where they had developed appropriate design thickness charts, jointing recommendations, and other necessary technical support. Their achievements are beginning to influence the U. S. and will probably result in increased usage there in the near future.

Examples in Europe are the heavily loaded NATO warehouses such as the Burnssum project in Holland (4). It economically used an SFRC floor in lieu of a double reinforced conventional concrete slab. This eliminated the conventional steel and also reduced the slab thickness from 7.9 inch (200 mm) to 7 inch (180 mm). The total area of SFRC surfaces for this use in Holland alone is 2.7 million sq. ft. (250,000 sq. m). The mix used a relatively low fiber concentration of 51 lb/cu. yd. (30 Kg/cu. m) achievable with the highly efficient bent end fiber made from high-strength wire. In addition to having lower fiber contents than were common in early slab and pavement work, these mixes also typically used less cement 573 lb/cu. yd. (340 Kg/cu. m), larger aggregate 1-1/4 inch (32 mm), no fly ash, and a lower water content achieved by the use of plasticizers and because of the lower cement content. These changes in approach to mix proportioning result in lower cost with reduced internal shrinkage and stresses.

An example in Australia is the recently completed Kidston Gold Mine shop and maintenance building floor for heavy equipment. The original design called for a 5.9-inch (150 mm) thick floor with heavy-duty F102 wire mesh. A contractor/engineer proposed FRC at 4.7-inch (120 mm) thick without mesh and with a better jointing system. The owner realized a savings of $25,000 (Aus). The contractor also completed the work faster. The hard troweled surface is excellent and will provide better resistance to damage by tracked equipment than the original conventional design.

In recent years, polypropylene fibers have gained wide usage for slabs in the U. S. Their rapid acceptance seems to be due to relatively low cost, ease of addition into the mix, lack of potential for rusting and injury that is perceived to be a problem with steel fibers, an ability to minimize or eliminate visible cracks in many instances, improved impact or spall resistance, and good marketing. However, flexural strengths are not changed significantly. The approach has been to add the fibers without reductions of thickness or redesign of the mix proportions.

Highways and Roads

In the early and mid-1970's, there was great anticipation for large amounts of highway paving with SFRC, but it did not develop. A series of test sections and small projects were completed, the most comprehensive being the Greene County project in Iowa. Various fiber contents, fiber types, and thicknesses were used there and most of the sections were very thin.

In general, early pavement projects did not demonstrate a noteworthy savings, a marked performance improvement, easier construction, or any other major overall advantage to conventional paving materials. At that time, the combined fiber supply capabilities of all manufacturers were unable to meet the demand for any major project with miles of road, so large projects which could have been a fairer evaluator never developed. Consequently, emphasis on highways shifted to airport pavements which were something the industry in the U. S. found easier to work with and supply.

It should also be noted that the "state-of-the-art" for mix design, optimum fiber content and shape, evaluating overall economy, designing for internal shrinkage stresses and curl, etc., was in its infancy when the road trials and early airport projects were done. They suffered from inefficiencies and inadequate design which can now be much better dealt with. With this better understanding and improved supply capabilities, a return to enthusiasm of SFRC for highway application is probable. The Ministry of Quebec Canada has just recently completed a major highway demonstration project, and SFRC is being considered for a major section of interstate in Montana, U.S.

In 1984, Schrader (5) demonstrated a procedure for incorporating SFRC with a no-slump, inexpensive, low shrinkage, roller compacted concrete to create a sandwich panel with SFRC on the top and bottom. Roller compacted concrete is a stable filler material separating the SFRC. It is placed while the SFRC mixes are sufficiently fresh to achieve a monolithic bond or laminate. The resulting section can be very deep and will carry approximately the same highway load and fatigue as if the entire section were made of just SFRC.

The expense of a full depth SFRC section is eliminated. The benefit from a curl standpoint is discussed later.

Airfields

The majority of pavement work with SFRC has been for airfield taxiways, runways, and parking aprons. This primarily resulted from work sponsored by the U. S. Army Corps of Engineers in 1971-1972 which demonstrated that a much thinner section of SFRC performed better after more passes of a C5A landing gear than a thicker section of conventional concrete (6, 7, 8).

The vast majority of SFRC airfield pavements have been constructed within the United States, with some work also completed in Europe, Australia, India, Taiwan, and Cuba. After discounting relatively minor demonstration projects and repairs, the U. S. work essentially consists of 22 pavements or overlays for parking aprons and taxiways at 9 differnt airfields. These projects are listed in Table 1. They are geographically located from throughout the United States.

Each of the 22 projects was thoroughly and systematically inspected in late 1982 and early 1983, and a detailed report of the findings was prepared (1). At that time, all of the placements were performing their intended function and very litle or no maintenance was required. However, concerns of exposed surface fibers, curling, and cracking were expressed by airport personnel. Since then, some routine corner maintenance has become necessary at one of the high-use pavements, and a few of them have developed cracking and curl which may require remedial work.

Two examples of SFRC airfield projects are the Las Vegas McCarran International Commerical Airport and the Norfolk Naval Military Air Station.

Two pavement projects have been completed at Las Vegas. These are approximately 8 and 5 years old. The first is a parking area for air freight. It is 6 inches (152 mm) thick and used 160 lbs/cy (95 Kg/cu. m) of straight 1-inch (25 mm) long mild steel fibers. The mix used cement and fly ash proportions of 600 and 250 lbs/cy (356 and 148 Kg/cu. m). The later project was for an apron at the passenger terminal. It had a depth of 8 inches (203 mm) and used similar proportions of cement and fly ash, but only 80 lbs/cy (47 Kg/cu. m) of bent end high-strength wire fibers that were 2 inches (50 mm) long. Flexural strengths were in excess of 1,000 psi (70 Kg/cm. sq.). Slipform paving equipment was used for the 25-foot-wide lanes with transverse joints spaced at 50 ft (15 m).

The Norfolk work consists of a series of separate pavement contracts which were 5 inches (127 mm) thick placed over an asphalt leveling course and 7 inches (187 mm) of old conventional concrete pavement. These placements are now 2 to 7 years old.

DESIGN METHODS

Overlays

Two approaches are available for overlay thickness design. The first uses formulas adapted from conventional concrete work but reduces the thickness requirements by mathematically taking advantage of the improved fatigue life of SFRC (6). However, the formulae do not directly address the suitability of the overlay to bridge cracks of different severities in the original concrete. Based on experience to date, the general industry opinion is that the "partial bond" overlay system should be used with caution and only after a thorough analysis of the existing pavement. Totally bonded (to good condition pavements) or debonded pavements appear to be better approaches to most future work.

The second overlay approach is to treat the existing pavement as a subgrade and determine its equivalent subgrade modulus. An SFRC pavement can then be designed as a new pavement over subgrade. Again, special attention must be given to the condition and activity of existing joints and cracks.

New Pavements

Suitably supported design charts are available (3, 4, 10) for determining the thickness of SFRC warehouse pavements. These charts address heavy storage loads and wheel loads from typical fork lifts. For lightly loaded warehouse slabs, the thickness is normally determined based on engineering judgement, experience, possible future loadings, adjacent floor levels and elevations, ceiling heights, and practical minimum placing depths.

Airfield slab thicknesses for new designs using SFRC can be obtained directly from charts developed for SFRC by Parker (7). They account for subgrade modulus, aircraft type and gross or gear loading, static flexural strength and traffic repetitions. These charts are similar in type to those developed and used for years by various agencies or authorities for conventional concrete. Schrader (9) has demonstrated a method for adapting charts for conventional concrete thickness design by adjusting the working stress levels to the appropriate values based on the fatigue life of SFRC and its static strength gain with maturity.

SPECIFICATIONS

SFRC pavments are typically designed for a static flexural strength specified in construction contracts to be achieved at a given age. Design charts assume a substantial improvement in fatigue life for the SFRC but the specifications and design procedures seldom address this or maturity as separate issues. Other important influencing factors such as shrinkage and impact resistance also are not directly addressed. SFRC pavement design charts were developed many years ago for the typical 1-inch (25 mm) long straight smooth fiber used at relatively high batch weights of fiber per cubic yard of concrete. A variety of significantly different fibers are now available which were not the basis for the design curves. Some of them perform better and others worse in different tests.

Specifications are needed that assure that the fibers used provide an SFRC with ALL of the material properties on which the design was based. Attempts to deal with variations in fiber types have been made in some of the more recent projects. With improved awareness of the difference in performance of various fibers, and with more use of fatigue, toughness index, and impact strength tests, better and more equitable specifications can be anticipated in the future.

A performance specification taking into account static flexural strength, fatigue life, impact resistance, toughness, compressive strength, and possibly shear strength is desirable. Unfortunately, many pavement projects do not allow enough time for complete testing with the particular aggregates, fiber proportion and type, cementitious materials, and admixtures at each project. Some of this testing is also quite expensive.

However, a knowledgeable and experienced engineer with expertise and experience in the field of SFRC pavements, construction, design, and testing can prepare fair specifications that would use a "modified performance" approach. This would require specific mix designs to be prepared and submitted prior to construction, with at least some of the more pertinent tests being performed with the particular aggregates proposed for the project to demonstrate compliance with design criteria. Experience and data from previous testing could be substituted for the more complicated tests such as fatigue if they were for similar mixes and fibers. Alternatively, testing during design could be undertaken to preselect specific acceptable fibers and mix designs. This approach is developing now and can be expected for most future projects.

CONSTRUCTION

Mix Design

In order to perform satisfactorily, fibrous concrete must have correct mix proportions. This is true of conventional concrete also, but it is especially critical with fibrous concrete. There are two aspects to be considered when speaking of correct mixes: (a) Obtaining the strength and basic engineering properties needed to meet structural design requirements of strength, fatigue endurance, etc., and (b) constructability or ability to be mixed, transported, placed, and finished. Success depends on satisfactorily meeting both of these criteria.

Selecting mix proportions was first done on a trial and error basis and is still often done that way with experience being used to get a reasonable starting point. A systematic mix proportioning procedure has been developed and used very successfully (2) which optimizes the aggregate blend and minimizes cement contents while assuring an adequate paste conent which is typically higher for SFRC.

Initially, very high cementitious factors on the order of 900 lbs/cy (534 Kg/cu. m) were used in SFRC pavements, along with 3/8-inch (10 mm) aggregate and fiber contents in the range of 160 to 200 lbs/cy (95 to 119 Kg/cu. m).

Gradually, the aggregate sizes increased to a more typical 3/4-inch (19 mm) with fiber contents decreasing to about 85 lbs/cy (50 Kg/cu. m) of the efficient high strength bent end fiber which is roughly equivalent to 140 lbs/cy (83 Kg/cu. m) of the smaller smooth fiber of mild strength steel.

Pavement mixes generally have low slumps and low water/cement ratios on the order of 0.39 to 0.43 (without the use of superplasticizers). The industry has typically achieved this by use of high cement factors and small aggregate. Until recently, the detrimental internal curl stresses and high shrinkage associated with high cement factors have not been appreciated. The use of superplasticizers, better attention to gradation, and larger aggregate have become recognized methods of decreasing cement factors and improving SFRC field performance in future pavements. Larger aggregates to about 1-1/2 inch (381 mm) are suitable for paving mixes with large fibers having lengths greater than the largest aggregate dimension, and mixes of this type can be expected in the future. This also should permit a reduction in the fiber dosage on a volume basis per unit of concrete because it will result in less mortar or paste.

Equipment

Batching and mixing equipment used in pavements and overlays has varied from sophisticated fully automated on-site batch and mixing plants to manual off-site plants for batching while mixing is done in transit trucks. Either system will work if it is properly monitored, calibrated, and utilized. Either system can result in problems if it is not properly done.

Early projects suffered from problems with fiber addition. It slowed the production of concrete to undesirable and expensive levels and it often resulted in fiber balling (fibers tangling together). The early fibers were individual elements usually shipped in containers of about 40 lbs (18 Kg). They were manually dumped onto shaker screens that would sift the fibers so that they could fall onto a feed belt as individual elements instead of tangled clumps. These problems have been overcome in current projects by improvements in fibers which include heavier fibers with greater stiffness (length/diameter ratio), shorter fibers with resulting greater stiffness, deformed ends to offset the loss of bond strength resulting from the shortened length, and collated fibers which enter the mix as fiber groups glued together so that they don't tangle during addition but will later separate within the mix into individual fibers.

For small projects, these fibers can be dumped into the back of a transit truck and mixed. Large projects such as airfield pavements requiring high production and short delivery times use conventional on-site mix plants with conveyors to charge the fibers onto the aggregate as it enters the mix.

Central mix plants handle the SFRC as they would any conventional paving mix. Dump trucks haul it to the placement and slipform pavers place it. High frequency-low amplitude vibration has been found to work best. Adequate vibrators are essential to any pavement mix and are even more important to SFRC.

Thin placements of about 4 inches (102 mm) or less can be consolidated by a properly operating surface screed vibrator with enough energy to fluidify the mass beneath it. Double screed systems with air-actuated closely spaced vibrators have proven to work very well.

PERFORMANCE CONCERNS AND CORRECTIVE RECOMMENDATIONS

Three major concerns with SFRC airfield pavements that developed over recent years warrant attention:

(a) Exposed or loose fibers at the pavement surface.

(b) Corner and edge curling and related cracking.

(c) Load transfer devices and jointing procedures.

Surface Texture and Exposed Fibers

When SFRC is placed against forms or when it is flat troweled with proper techniques, a smooth surface results with no visible or exposed fibers. When SFRC is placed by vibrating screed or slipform paving techniques without texturing, very few exposed surface fibers can result. Power or hand troweling can totally embed all fibers in a smooth hard surface. However, most of the common techniques used in the past for texturing a fresh concrete pavement to provide skid resistance result in many partially exposed or loose fibers. These techniques include coarse brooming, tining, burlap drag, and carpet drag.

SFRC subjected to very high traffic volumes and severe weather may expose additional fibers at the surface as the cement paste begins to wear away. This normally has not been a significant problem, but it has occurred at Denver's Stapleton Airport, the Las Vegas McCarran Airport, and on part of the work at Fallon.

The exposed or loose fibers represent two concerns which have been perceived by some to be serious potential safety hazards. Experience to date has not substantiated these concerns as being real problems of significance, but the "concern" is real and the potential is conceivable. These concerns are:

(a) Injury to personnel being scraped or cut by an exposed fiber while working on the concrete surface.

(b) Ingestion of a loose fiber into an aircraft engine which ultimately results in damage to it or an unsafe operating condition.

It is common practice to use a large magnet mounted to a truck and/or to use a street sweeper to clean up loose fibers on the pavement as part of final construction cleanup operations. At least two airports (Denver's Stapleton and Las Vegas' McCarran) report continuing this type of simple cleanup after construction on a semi-routine basis with maintenance crews. The actual necessity of these cleanup operations has not, and probably cannot, be clearly established. However, good judgement has indicated that it is prudent to at least clean up the bulk of the exposed loose fibers after construction and at reasonable intervals thereafter.

Efforts have been unable to locate any documented lost-time accident or serious injury to any individual from an exposed fiber while working on an SFRC airfield pavement. However, there have been rumors concerning baggage handlers that have been "pinpricked" or cut by the fibers at commercial airports, and there reportedly have been comments made on behalf of the unions in at least one case (Salt Lake City). It is reasonable to assume that with enough hours of personnel exposure while working on an SFRC surface with partly embedded fibers, some type of abrasion, cut, or other minor injury could occur. On the other hand, many, many thousands of passengers unaware that the pavement was SFRC have walked across it at both Reno and Salt Lake City with no known incident or comment. If personnel such as mechanics or equipment handlers are exposed by kneeling or lying on an SFRC exposed surface, knee pads or a protective mat are appropriate. In extreme cases of fiber exposure and concern, a thin asphalt overlay can and has been used for permanent protection. Current technology for applying an asphalt friction course would be appropriate and serve other useful purposes.

Ingestion of a loose fiber into an aircraft engine and resulting damage is virtually impossible to identify. In some opinions, the concern is significant, but after years of use at all of the SFRC airfield pavements and with a variety of aircraft, there have been no known incidents of this occurring. Whether this is because the engines do not ingest the fibers, it occurs but causes no damage, or it occurs and causes unidentifiable damage that is cumulative and could eventually represent a problem cannot be categorically established. The evidence indicates that there is no problem, but if there were, it is reasonable to assume that it is more likely to occur with helicopters and small military aircraft close to the ground.

The combined concerns of engine ingestion and personnel exposure to loose or exposed fibers, whether real or imagined, are noteworthy enough to require improved finishing, texturing, or a protective surface treatment such as an asphalt friction course in future work. This can be accomplished by demonstrations and contract prequalification requirements in conjunction with contractors using their equipment and material suppliers.

Labor-intensive finishing techniques with magnesium floats have already demonstrated the ability to effectively cover the fibers while still providing a skid-resistant texture. This was done on an SFRC overlay for slabs on the visitor facility at Libby Dam. The concrete there is in an unsheltered area subjected to extremes in weather in the northern mountains of Montana. The work was done by contractor in 1972 (2). A close inspection 11 years later identified only one fiber on the entire surface which was not fully embedded. The facility has recorded 416,757 visitors, most of which visit during the summer and many of which walk on the surface barefoot. There never has been a complaint about exposed fibers. Although this finishing technique was successful, the area was relatively small and the finishing method was too labor intensive for production airfield pavements. It is possible that the technique could be simulated by a mechanized process such as a delayed follow-up vibrating screed bar. This could be easily tried in a pilot or demonstration project.

A "pan finish" used behind a standard Bidwell paving machine for a portion of the SFRC work done at the Salt Lake City Airport demonstrated an ability to provide a skid-resistant texture with essentially all the fibers buried. The "pan" was about 1 square yard in size with curved up edges. By trial and error, the right weight, timing, and method of mechanically pulling the pan back and forth over the fresh surface were found that resulted in this texture. The same project also showed that dragging a carpet or piece of "astro-turf" in lieu of the pan tore the surface badly and exposed extreme numbers of fibers.

Discussions have recently occurred between fiber suppliers, the manufacturer of a "vibra float," and an SFRC consultant concerning the modification of the vibra float to provide a coarse texture while burying the fibers. The vibra float is a simple hand-held magnesium "bull" float with a small air-actuated vibrator attached to it. The idea for SFRC pavements is to score the surface as it is floated so that the fibers are pressed down while ridges or streaks of mortar are left behind.

A procedure that has been tried with limited success to eliminate exposed fibers from existing pavements has been to accelerate corrosion of them by the application of de-icing salts (calcium chloride). Because the fibers have a very small cross sectional area, they should oxidize or corrode relatively rapidly in an aggressive environment. Unfortunately, some of the better quality fibers are manufactured from a high-strength wire that has good corrosion resistance, and the alkali environment due to the cement tends to slow the corrosion process. The idea may still provide a simple and economical solution to existing pavements if an acceptable fiber corroding material can be identified which does not penetrate or damage the concrete.

Curling and Related Cracking

Curl or warping of concrete slabs and related cracking is not unique to SFRC pavements. It is a relatively common problem or concern in all types of concrete pavements and overlays and is essentially independent of the fibers in SFRC. However, the high cement factors and reduced thickness typical of SFRC airfield pavements make them very prone to curling. This basically relates to the properties of the concrete itself, the environment, the construction controls, and the slab dimensions.

The vast majority of SFRC airfield pavements placed to date have exhibited some degree of curling. Table 1 summarizes actual measurements taken at each project and the degree to which the corners have developed cracks as a result of curling. An analysis of this table in conjunction with detailed notes of the inspections shows no obvious common denominator to the degree of cracking or curling. It is reasonable to assume that continued curling and some corner cracking can be expected in future SFRC airfield pavements unless an improved approach to design and/or construction is developed and followed.

Although curling with some resulting cracking has occurred in SFRC projects, it has required very little or no maintenance to date for both commercial and military installations. Some routine maintenance of corners is expected in the near future at Stapleton Airport and may develop at other projects subjected to severe curling. An area of very high traffic at Stapleton could also require reconstruction as the permanent repair solution. However, curl and corner cracking of SFRC airfield pavements should not suddently affect the military readiness or commercial capability of an installation. It is worth noting that fibers bridge the cracks and tend to hold the pieces together, whereas in conventional concrete, a corner crack is more likely to spall and require repairs sooner. This resistance to spalling and improved performance of a cracked section have been generally observed in the field and were very evident during the C5A trials by the U. S. Army (8).

Visible cracking in SFRC should not be as serious as it is when observed in conventional concrete.

The "full report" version of Reference (1) contains an extremely thorough and objective study of the problem of curling in SFRC. Typically, the amount of curl is 1/8 inch (3 mm) with a range of 0 to 5/8 inch (0 to 16 mm). Corner cracking generally begins to occur after about 1 year of service and breaks in an arc of about a 1 to 4-foot (0.3 to 1.2 m) radius around the corners where the longitudinal and transverse joints meet.

Some edge curling and its related cracking has also occurred. It occurs for the same reasons as corner curling, but to a lesser degree. It has not been a significant problem and will be improved by most of the same techniques developed for the more serious problem or corner curl.

The normal reaction to minimizing curl in concrete pavements is to reduce the joint spacing, and this has been suggested in at least one U. S. Navy document. A technical analysis (1) shows that this standard approach is not practical with typical SFRC pavements. The joint spacing would have to be decreased to such short distances of 10 to 15 feet (3.0 to 4.6 m) that it is not a reasonable solution. On the other hand, a new and opposite approach of increasing the joint spacing to say 100 to 200 feet (30 to 61 m) should be investigated. This decreases the number of corner cracks by simply decreasing the number of corners without increasing the severity of curl. It should also result in reduced over-all pavement cost.

Another standard approach to curling is to increase the thickness. This should be investigated for SFRC for several reasons. Although an increased thickness increases the amount of material used, studies may show that an over-all savings results if the required quality of the concrete and subgrade can also be reduced. The major advantage of a thicker section may in fact be that a significantly lower cement and water content of the can be used.

As discussed earlier, SFRC airfield pavements typically have had very high cement factors and associated total water contents even though the water-to-cement ratios have been low. By allowing a lower strength concrete (for example necessitating say 650 instead of 790 pounds of cement per cubic yard), less internal stress from shrinkage and thermal movement develops. The importance of this should be evaluated, and it should be included as part of future design procedures.

A consistency is lacking between "pavement load" technology and concrete "materials" technology. A general attitude exists in the concrete pavement industry and has been very prevalent in SFRC that the highest possible flexural strengths achievable should be used in design. From a materials standpoint, there are diminishing returns when pursuing high flexural strengths of say 1,000 psi (70 Kg/sq. cm) if it results in substantially greater cement contents, extra cost, more internal shrinkage stress, and high internal temperature rise from heat of hydration during initial cure. The two technologies should be used together for any concrete pavement, but it is especially true with SFRC. Unfortunately, standard thickness design procedures do not account for the disadvantages of higher cement factor mixes. One of the most important advantages of SFRC which is hidden in the design process when standard charts or tables are used, is its improved fatigue life. Because of improved fatigue characteristics, a thinner SFRC section is needed than for a comparable static strength conventional concrete. A continuation of the design approach in Reference (9) could blend both the "pavement" and "materials" technologies into one appropriate design approach for SFRC which takes into into account fatigue, curl, and internal stresses which now are not directly considered.

The full problem of curl stresses and movement has not been adequately explained or discussed in published literature (1). The problem has only been explored in the context of curl that develops by differential drying or temperature changes due to EXTERNAL influences, assuming that a plane state of equilibrium initially occurred. This addresses only part of the problem. Technology now exists to evaluate the other part of the problem - i.e., internal stresses that develop prior to the long-term effects of drying and environment. These can be significant for high cement factor mixes and may be strongly influenced by such things as the time of placement, placing temperature, climatic conditions, cement chemistry and physical properties, subgrade conditions, etc. The principal cause of these stresses is heat generated internally by hydration of the cement as it sets and hardens. If the temperature near the top of a pavement is higher than the bottom at the time it is setting and developing its early strength and modulus, the pavement will curl up at the corners and edges when the slab later reaches a uniform temperature and the upper portion cools. This phenomenon is not recorded or discussed in published literature, but it has been observed. It is very apparent in photographs that were taken by an inspector at Salt Lake City where edge curl approached 1/2 inch (12 mm) on an SFRC pavement that was less than 1 day old. It was not subjected to any drying or major external temperature changes.

Technology from mass concrete dam design can be applied to pavement technology to evaluate these initial curl stresses. The high heat generation and potential stresses of massive placements have forced that industry to develop methods of evaluating and controlling the problem. A computer program developed for this purpose has been used by Schrader to determine the initial temperatures in slabs. It can continue the evaluation throughout the long-term life of the pavement, including daily and seasonal temperature fluctuations. This type of evaluation can justify reduced cement factors and/or special conditions for placing times and temperatures which should minimize or nearly eliminate curl.

The capability exists to account for:

. Different adiabatic heat rises: cement type, quantity, chemistry, and physical properties.

. Variable temperatures throughout the day.

. Effect of wind and curing.

. Differing placing temperatures.

. Different times of placement within the day.

. Specific heat, conductivity, and density of the concrete.

. Different times of year for placing.

. Different seasonal conditions.

High-range water reducers for concrete can also be used to reduce curl in SFRC. These commercially available additives were not "state-of-the-art" when previous SFRC paving was done. They are well understood now and should be required in SFRC to reduce cement factors and their disadvantages, including heat generation and shrinkage, both of which contribute to curl.

A unique approach to design and construction which would utilize the benefits of SFRC while minimizing its disadvantages has been proposed (5). This technique sandwiches an economical lean "roller compacted concrete" (RCC) mix between thin layers of SFRC. The benefits of fatigue, strength, and impact of SFRC are efficiently used while the disadvantages of curl, high heat of hydration, etc., are eliminated or minimized. Basically, the expensive SFRC is used only in the outside portion of the pavement section where it will do the most good, and essentially non-shrink inexpensive filler is used for the slab interior. Tests confirm that curl is not a problem with this system.

Even with pre-established controls to minimize curling, some curl stress is unavoidable in virtually all concrete pavements. A list of possible ways to minimize or control it by slab layout and construction details is included in Reference (1). They vary from tiedowns to staggered jointing and pre-cut circular corner joints.

It should be noted that some methods of curl control may eliminate the curl DEFLECTION but they will not eliminate the STRESS. Solutions of this type have little value. It is the added STRESS due to curl that must be controlled. Predeflecting a curled corner only eliminates the visual aspect, but the internal stress that ultimately leads to cracking is still present.

LOAD TRANSFER DEVICES AND JOINTING

The whole issue of load transfer devices (sawcuts, dowels, tie bars, keyways, etc.), their necessity, and construction technique is one of continuing study, development, and controversy regardless of whether or not the concrete contains steel fibers. The relatively thin sections of SFRC pavements and wider joint spacings make proper jointing design and construction more critical. The issue of joint design has been argued many times. It should be carefully considered in the evaluation of any SFRC pavement. Inappropriate load transfer methods have contributed to or caused cracking in SFRC pavements. Unfortunately, some observers have been too quick to blame SFRC for cracking caused by inappropriate load transfer and jointing methods without evaluating the joint system.

Some standard design details and approaches developed many years ago and copied routinely in the more current issues of military and civilian design manuals or texts are inappropriate for today's materials (such as SFRC) and construction operations (wide slipform pavers and extended joint dimensions). For example, dowels in the longitudinal joints of paving lanes may provide load transfer, but in the process may also cause excessive restraint and cracking from drying or thermal shrinkage in the adjacent lanes.

In an example, bars used in transverse joints at the Salt Lake City Airport followed FAA and textbook guidelines. Two problems became very evident within 1 year. First, some of the bars did not slide and subsequently prevented movement of the joint. When the contraction force of the slab exceeded its tensile capacity, a full-depth tension crack appeared about 1 to 2 feet (0.3 to 0.6 m) from the joint and parallel to it. The crack location coincided with the ends of the bars. In essence, the same thing would have happened if no joint and no bars were used in construction. A partial depth sawcut joint with no bars would have been more appropriate and economical. Many of these tie bars also have cracks directly above them which reappear even after epoxy injection grouting. It appears that the bars did not slip, were not necessary, and were in fact detrimental.

Dowels or tie bars used in longitudinal joints of slipformed paving lanes can provide enough restraint against shrinkage in the adjacent lane to develop transverse cracking and/or longitudinal cracking some distance in from the end of the bars. This can be controlled or minimized by providing some other method of load transfer or by including considerations for shrinkage, strain capacity, placing temperatures, and the timing of adjacent lane placements in design. If a paving lane containing dowels has been placed, cured, and has contracted sufficiently for thermal and drying shrinkage to open the transverse contraction joints, when the adjacent lane is placed, cured, and later tries to contract the same amount, it is prevented from doing so by anchorage to the dowels. The developed internal stress can either create an overload condition when added to the applied aircraft load, develop transverse cracks, or cause a shear crack along a line parallel to the end of the dowels. Cracks resulting from this would be similar to those observed at the Reno Airport.

The idea of using tie bars to keep slabs in the outside few paving lanes from gradually floating away to the outside boundary was developed many years ago when slab sizes were on the order of 15 x 15 feet (4.6 x 4.6 m). The large slabs used today do not have the same problem or severity, yet the detail continues. As explained above, it can do more harm than good.

Keyways can be an effective load transfer method, but special precautions not currently discussed in design guides should be considered - specifically, the shear strength of the concrete (which is significantly higher for SFRC), the planeness of the key so that it can slide, and especially the angle of the key's bevel. A flat angle will transfer more load sooner than a steep angle after the joint contracts.

Partially sawcut joints which rely on aggregate-to-aggregate interlock have an added benefit in SFRC from the fibers which act as thousands of mini-dowels, but which are not considered in current design guides.

The issues discussed above provide examples of what is not specifically addressed in current standard design guides, but the experience and technology now exists so that better SFRC pavements can be designed and constructed in the future.

THE FUTURE OF SFRC PAVEMENTS

Primarily because of concerns over curling, its related cracking, and exposed surface fibers, there has been a recent slowdown in enthusiasm and the use of SFRC in pavements. However, based on experience, a study of prior work, and implementation of the current state of technology, these concerns can be overcome. SFRC uses in airfields are expected to again become a viable alternate to conventional pavements. Recent interest by highway departments and the availability of adequate quantities of fiber indicate that SFRC will probably begin to be used in heavy traffic areas. The use of SFRC for heavy-duty warehouse floors and where minimal slab thickness for new or overlay work is required will continue.

Lower water contents through the use of superplasticizers, better proportioning of aggregates, and larger aggregates are important to better field performance and can be expected. Water demand, shrinkage, and costs will probably be reduced in future work by lower cement contents. Static flexural strength requirements can remain higher than for conventional concrete, but they probably will be somewhat lower than previously required so that mixes with less potential for shrinkage and curl in the field can be built. The benefits of improved toughness, impact resistance, and fatigue performance will still provide adequate advantages to make SFRC pavements desirable even at lower flexural strengths.

A realization of the importance of placing temperature and seasonal effects will result in closer controls of placing times and conditions. This will be done primarily to achieve lower water contents with cooler mixes and reduced curl from temperature rises developed from internal and and external heat sources as the mix sets and cures.

Joint spacings and techniques, including the necessity (or lack of it) for load transfer devices will be given more attention in future work - especially with regard to the restraint of slab shrinkage and horizontal movement. Greater, rather than smaller, joint spacings can be anticipated with technical evaluations that demonstrate better economics and fewer corners/edges for potential problems.

Shrinkage compensating concrete in SFRC pavements has been used in several field trials with encouraging results. The fibers provide the internal elastic restraining mechanism necessary for the system to effectively eliminate or minimize shrinkage stresses. This combination of technologies (SFRC and shrinkage compensating cement) is expected to gradually become more common. It requires knowledgeable design with attention given to internal elastic restraint, no external nonplastic restraint from load transfer devices, properly selected mix proportions, and control during placing and curing.

The concept of using SFRC to surface a thick and economical "no slump" roller compacted concrete slab with little or no shrinkage and curl should rapidly gain acceptance. It is a most efficient design approach that minimizes shrinkage and curl and allows deep pavement sections of high strength and fatigue life without the technical and economical disadvan-disadvantages of having SFRC for the full depth.

Table 1. Summary of Construction Records and SFRC Pavement Inspection Observations

(1982-1983 CONDITION SURVEY)

Job	Thickness (Inches)	Joint Spacing (Feet)	Joint System (1)	Age (Years)	Slab Support (2)	Maintenance Performed	Fiber (3)	Corner Cracks (4)	Cement/Fly Ash (lbs/cy)/(lbs/cy)	Thermal Expansion (inx10^{-6}/°F)	Drying Shrinkage (in/in)x10^{-6}	Cure(5)	Curl (6) Amount (inch)	Curl (6) Radius (feet)
Fallon-1	5	40	L = B T = SB	2	2 AC PCC	No	B-82	C7	788/0	6.5	800	CMPD	1/8 to 1/4	1 1/2 to 4
Fallon-2	5	40	L = B T = SB	1	2 AC PCC	No	B-81	L3	766/0	5.5	770	CMPD	1/8 to 3/16	1 to 2
L.Vegas-1	6	50	L = B T = SB(?)	6	AC 18 AGG	No	F-160	C	600/250	4.8	1100	CMPD	3/8 to 3/8	4 to 5
L.Vegas-2	8	50	L = B T = SB	3	2 AC 12 AGG	No	B-85	C7	650/252	6.5	900	CMPD	0 to 1/8	2 to 3
Reno-1	4	25	L = B(?) T = SB(?)	6	GROUT PCC	No	F-200	C	658/216	5.0	800	?	NO	1 to 3
Reno-2	8	40	L = K Ti T = SB	2	6 AGG 30 ROCK	No	B-87	VL	815/0	5.0	880	CMPD	NO	0 to 2
S. Lake-1	8	30	L = BTi T = BD	2	8 CTB 24 AGG	No	B-83	NO	583/203	6.0	800	CMPD	NO	0
S. Lake-2	7 & 8	35*	L = BD&SBD T = BD&SBD	1	1 AC AC/PCC	No**	B-85	L	620/215	6.0	720	CMPD	0 to 1/8	1 to 3
Denver	7 & 8	30	L = BD T = SBD	1	FABRIC PCC	No	B-83	C	525/250	5.0	560	CMPD	1/4 to 1/2	2 to 3
Newark	3	125	L = KD T = SBC	2	EPXY 6 FRC	No	B-85	L	750/0	5.0	—	WET 7	0 to 1/8	2 to 3
Tampa-1	4	75	L = B T = None	10	12 PCC AGG	No**	F-200	NO	517/250	5.0	—	CMPD	1/8 to 1/4	0
Tampa-2	6	175	L = B T = None	10	12 PCC AGG	No**	F-200	NO	517/250	5.0	—	CMPD	0 to 1/8	0
Nrflk-1&2	5	25	L = B T = SB	3 & 5	2 AC 7 PCC	Yes	F-160	C	600/250	5.2	—	WET 7	Yes	2 to 3
Nrflk-3	5	25	L = B T = SB	2	2 AC 7 PCC	No	B-85	VL	600/250	5.2	—	WET 7	Yes	2 to 3
Nrflk-4	5	25	L = B T = SB	1	2 AC 7 PCC	No	F-160	C	600/250	5.2	—	WET 7	1/2	2 to 3
Nrflk-5	5	25	L = B T = SB	1/2	2 AC 7 PCC	No	F-160	NO	600/250	5.2	—	WET 7	1/8 to 1/4	0
JFK-1A	5 1/2	120	L = KTi&KD T = None	8	POLY AC+PCC	No**	A-175	NO	752/0	5.0	820	WET 7	NO	0
JFK-1B	8	120	L = KTi&KD T = None	8	? GRADE	No	A-175	NO	752/0	5.0	820	WET 7	NO	0
JFK-2	7	100	L = KTi T = SBD	6	POLY 12 PCC	No	A-175	NO	752/0	5.0	820	WET 7	NO	0
JFK-3	7	100	L = KTi T = SBD	2	POLY 12 PCC	No	B-110	VL	752/0	5.0	820	CMPD	NO	2 to 3
JFK-4	9	100	L = KTi T = SBD	1	AC PEN AGG	No	B-85	VL	752/0	4.8	680	CMPD	NO	0 to 2
JFK-5	7 1/2	100	L = KTi T = SBD	1	POLY 3AC+12PCC	No	B-85	L	752/0	4.8	680	WET 7	NO	1 to 3

STEEL FIBER CONCRETE
US-SWEDEN joint seminar
(NSF-STU)
Stockholm 3-5 June, 1985
S P Shah and Å Skarendahl,
Editors

STEEL FIBER REINFORCED SANDWICH PANELS

by Charles W Josifek, Vice President, Ribbon Technology Corporation, Gahanna, OH, USA

ABSTRACT

Tilt up sandwich panels made using steel fibers as the sole reinforcing material has been demonstrated to be an efficient, viable, cost savings method of building construction. The fiber reinforcement was 3" long x .035" in diameter deformed fibers applied 100 lbs per cubic yard. The fiber reinforcement additional cost was $1.72 per square foot of floor space or 1/3 the total cost of the panel. Normal procedures were followed in casting panels and tilting in place.

C. W. Josifek is Vice President of the carbon steel division of Ribbon Technology Corporation in Gahanna, OH. He was employed with Bekaert Steel Wire Corp. (20 years), Goodyear Tire & Rubber Co. (15 years), and the Army Corps of Engineers assigned to the Manhattan Project. He has been a leading promoter and advocate for reinforcing concrete with steel fibers. He is a member of the ACI Committee 544, ASTM Committee A 01.05, the TRB Committee A2E03, SAME, and NCPA.

Background History

Tilt up panel construction is the fastest, most efficient, on site construction currently available today.

Tilt up construction becomes a more important factor in the building industry every day. With cost for materials and labor constantly rising, tilt wall construction can probably be considered an evolutionary development. Tilt wall construction requires less forming and placing material than conventional building methods and in many ways creates a better structure. It permits strong contempory architectual expression and is becoming more versatile in this area all the time. Savings in time and man power are significant. With tilt wall the most important hours go into planning, close coordination, and timing in advance. Actual on site construction proceeds with remarkable speed and efficiency. The efficiency extends to the site itself. For example, the floor slab in most projects becomes the bottom form for all walls.

Tilt wall is actually precast concrete construction in its simplest form with the manufacturing moved to the site where the forms will be used. The method requires building into slabs and walls all connections including stifeners and hardwares. Sections are often moved into final position before the concrete has cured to its greatest strength.

The advantages of tilt wall construction is in the economics that can be realized. No scaffling is required since all concrete work can be done at ground level for the placing, finishing, and curing of the concrete. The project can be completed quicker and result in faster move in. This results in savings in construction and cost, such as interest rate, etc.

It requires less material and less man power than conventional construction and it meets most fire resistance codes and therefore reduces fire and earth quake insurance rates.

The tilt wall construction offers miniumum maintenance since the concrete can be left unpainted and therefore there is no cost for paint or preservatives. Color can be incorporated into the concrete. The tilt wall construction buildings resist moisture, weathering, rot, and rust. The walls are also impervious to insects and rodents. The surfaces can be easily cleaned and are very rugged. Also inherent and of more and more significance is the fact that the concrete walls discourage theft and vandalisim.

Tilt wall construction offers beauty. The walls can be smooth, textured, or exposed aggregate surfaces. They can be of natural stone design. They can use plastic form liners to simulate an almost endless variety of natural effects in concrete surfaces.

Another significant factor is that the walls can be designed to be easily removed and relocated for future expansion of the building.

Design of the Steel Fiber Reinforced Concrete (SFRC)

The Portland Cement Association of Skokie, Illinois in their report "Special Considerations for the Selection of Tilt Up Concrete Sandwich Panels" has indicated that the criteria for sandwich panels in addition to the usual need for strength and serviceability, thickness, composition, and connection between the wythes of the sandwich panel should consider the following four points.

1. Effect of relative movements of the wythes. The exterior wythes may have a tendency to expand or contract due to the temperature differentials between the two wythes. Differential creep and shrinkage due to unequal stress and humidity may also be experienced.

2. Resistance to lateral and vertical forces. One or both wythes may resist bending moments due to lateral forces such as wind, earth quakes, or soil pressure. Lateral shear to resist cracking and to transmit roof loads for the foundation can be taken by one or both wythes. If the panel is used to support a vertical load, one wythe, not both, should be designed as a load bearing wall.

3. Allowable thermal transmission. Building codes or standards may require low values for certain types of occupancy. The thickness of the insulation effects the total thickness of the panel.

4. Over all weight of the panel. The size of the panel may be limited by the availability of lifting equipment or the flexibility of the panel during lifting. Generally, sandwich panels will be heavier than solid panels since one wythe is only a non-loading-bearing facing. The use of strong backs may be considered to reduce lifting deformations and increase panel size without effecting wall thickness.

Panel Reinforcement

Ribbon Technology manufacturers steel fibers. Therefore, it was decided that the reinforcement would be 100 percent steel fiber reinforced concrete. The Xorex 3" fiber was selected which at 100 pounds per cubic yard will give 1,440 psi flexerual strength in 90 days and a flexural toughness index of 19.2. (See Figure 1)

Polystyrene bead board was selected as the insulation and it was decided that the outside wythes should be 2" and the insulation 2" thick with a "U" value of 0.0888. (See Figure 2)

The concrete mix design for the steel fiber reinforced panel was as follows per cubic yard:

Concrete Mix Design

Cement	658
Fine Aggregate	1547
Coarse Aggregate 3/8" max.	1560
Water	322
Steel Fibers (3" long)	100
Water Reducer	4 oz./100#/cement

After the selection of the reinforcing steel and the insulation, test panels were fabricated. These test panels were 35" long x 24" wide x the 6" thickness previously determined. (See Figure 3)

The flexural strength was calculated from the formula $\frac{BH^3}{12} + 2AY^2$ the modulus of Rapture (MOR) = $\frac{MC}{I}$ where

I = Moment of inertia

B = Width of the element (cap or web)

H = Height of the element (cap or web)

A = Area

Y = Distance from the center of the beam to the center of the cap.

M = Moment $\frac{(\text{load x spand})}{4}$

C = 1/2 depth of the beam

The panels were manufactured and sent to Lankard Laboratories in Columbus, Ohio for testing and evaluation.

(See Figure 4)

The load at which a crack was first observed was reported as well as the ultimate load carried by the speciman. These results are shown in Fugure 5. Typical crack patterns developed in the speciman is shown in Figure 6. From this data the design criteria was confirmed.

The panels were designed to be 12′ wide x 24′ high. The dimensions of the structure to be erected was 100′ x 200′. Forty eight panels would be required. It was therefore decided that eight casting beds would be utilized.

One of the remaining questions to be determined was the type of shear ties that would be used. Since the prime considerations was the use of our steel fibers in this building it was decided that the wythes would be connected by the fiber reinforced concrete. The use of polystyrene bead board in 2′ widths aided us in the design since we could now form the fiberous concrete ties 3" wide with the use of 6" on the perimeter of each panel.

It was noted that the use of the steel fiber reinforced concrete solid ribs at the panel edges and the ties increases the stifness of the panel considerable but also impairs the insulating value of the wall. The cold transfer through the panel for this small amount of area was not considered a problem.

Pick Up Points and Lifting Behavior

It was our belief from the calculations of the strength that there would be no problem in lifting these panels in the conventional method. We contacted the Dayton Sure Grip and Shore Company to advise us on the proper use of rigging equipment and lift up points. The report from Dayton Sure Grip is attached to this paper.

Construction of the Panels

As mentioned before the decision was made to use eight locations for casting beds and cast six panels on top of each other. It was also decided that the outside finish would be exposed aggregate and this could then be used as the bond breaker. Eight areas were selected and the exposed aggregate was placed in a thickness of approximately 1". In order to control the movement of the aggregate after it had been placed it was sprayed with grout to hold it into position while the steel fiberous concrete was placed on top. The steel fiberous concrete was poured on top and leveled. The next step was to place across the 12′ width the fiber polystyrene bead boards on the wet steel fiberous concrete and then place the remaining 2" of steel fiberous concrete on top of the polystyrene bead board. As you will see from the slides the bead board was held in place by the use of 2 x 8 board on which the workman walked.

The eight casting beds were chosen also since the workman could make four casting in one day quite easily alternating bed pours. The second casting was made upon the first on alternate days.

In those panels that were necessary for doors and windows to be placed the blockouts were provided in the panels during the pouring. For small door frames, the frame itself was used as a form.

All the full panels weighed approximately 7 1/2 tons each. All were lifted by mobile crane without incident. The crane operator had placed many other tilt wall panels and commented that these were the first tilt wall panels that he had lifted that did not show signs of cracking during the lift.

Cost Savings

The cost savings were realized from the use of steel fibers which were uniformally distributed throughout the concrete over the cost of placing welded wire mesh or rebar and holding in position.

The manufacture of the panels at the site resulted in savings due to transportation and handling cost.

The reduced weights of the 6" panels allowed smaller less costly equipment to be used.

The actual panel cost was $5.26 per square foot of which $1.72 per square foot was for the steel fiber reinforcement. Cost measured on a square foot of the floor area of the building amounted to $3.53. If measured on the the basis of a cubic foot of enclosed volume it was $.17 1/2 per cubic foot.

Conclusion

In conclusion we may say that steel fibers can:

1. Replace welded wire mesh and rebar in certain applications.

2. Can be used in conjunction with present concrete technology.

3. Can lower cost significantly

I would now like to show slides on tilt wall panels and construction.

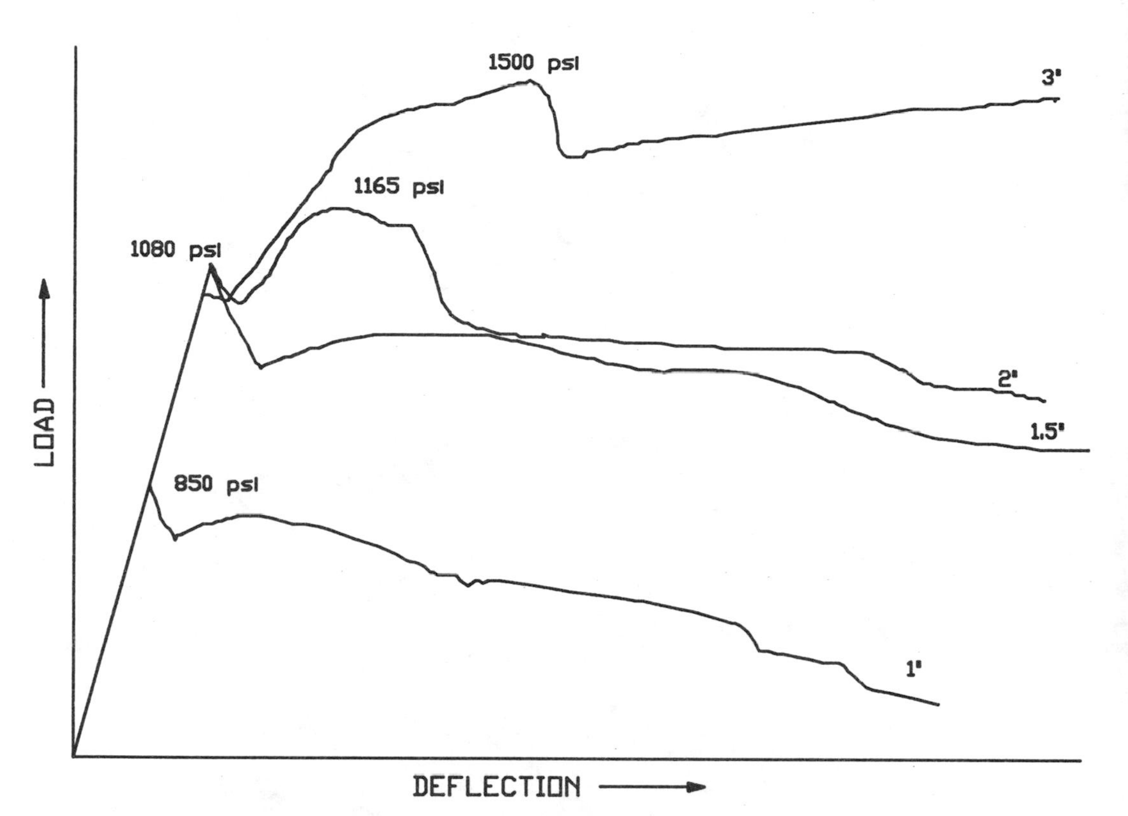

FIGURE 1. LOAD DEFLECTION BEHAVIOR FOR XOREX STEEL FIBERS. 4 X 4 X 14 BEAMS - 3rd POINT LOADING - 100 LBS/CU.YD.

FIGURE 2 U-Values* for Concrete Sandwich Panels

Thickness, in.			U-value	
Outside wythe	Polystyrene insulation**	Inside wythe	Normal-weight concrete (145 pcf)	Lightweight concrete (110 pcf)
3	1	3	0.158	0.142
3	1-1/2	3	0.120	0.105
3	2	3	0.088	0.083
2-1/2	1	4	0.158	0.142
2-1/2	1-1/2	4	0.120	0.105
2-1/2	2	4	0.088	0.083
2-1/2	1	5	0.156	0.138
2-1/2	1-1/2	5	0.110	0.102
2-1/2	2	5	0.088	0.082
2-1/2	1	6	0.154	0.134
2-1/2	1-1/2	6	0.111	0.100
2-1/2	2	6	0.087	0.080
2-1/2	1	7	0.152	0.131
2-1/2	1-1/2	7	0.110	0.098
2-1/2	2	7	0.086	0.078

*A measure of heat transmission--British thermal units per hour per square foot per degree Fahrenheit of temperature difference for a combination of materials. The lower the U-value, the higher the insulating value.

**With other types of insulation, the U-value for a wall assembly can be adjusted. Use information given in Ref. 1 (see list at end of report).

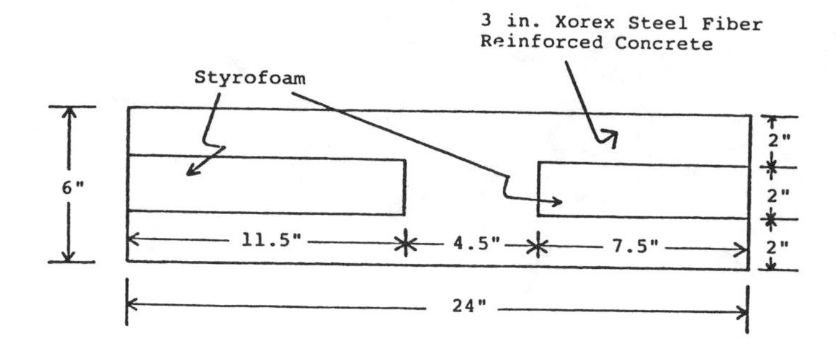

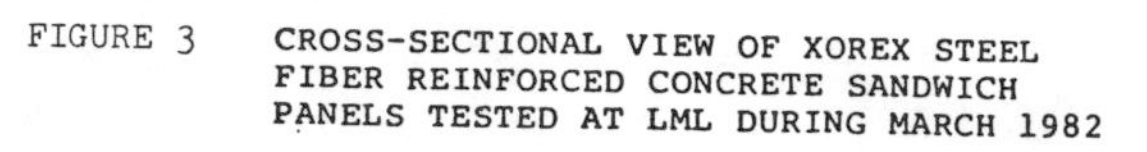
FIGURE 3 CROSS-SECTIONAL VIEW OF XOREX STEEL FIBER REINFORCED CONCRETE SANDWICH PANELS TESTED AT LML DURING MARCH 1982

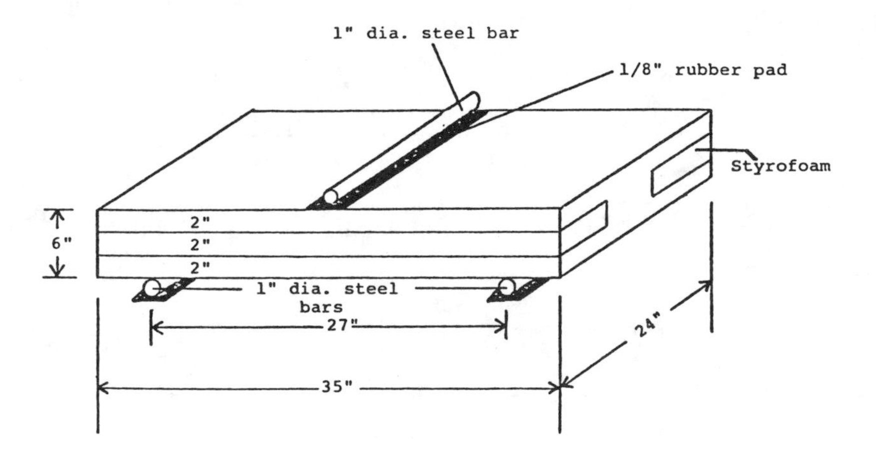

FIGURE 4 SET-UP USED AT LML FOR MEASURING FLEXURAL LOAD-BEARING CAPACITY OF RIBTEC SANDWICH PANELS TESTED AT LML DURING MARCH 1982

FIGURE 5 Results of Flexural Load Testing of Xorex Fiber Reinforced Concrete Sandwich Panels Tested at LML During March 1982

Panel Specimen Number[a]	Load at Which First Crack Observed, lb.	Ultimate Load Carried By Specimen, lb.	Flexural[b] Strength of Specimen, psi
1	10,620	13,750	680
2	12,000	13,000	640
3	9,750	13,600	670
4[c]	10,500	13,670	675

(a) Geometry of panel specimens and testing set-up shown in Figures 1 and 2

(b) Calculated according to the equation shown on page 1

(c) This specimen was filled in at one end to a depth of about 6 in. with concrete. LML chipped away about 6 in. of styrofoam from the other end and filled the area with a non-shrink grout formulation. This was done to determine whether or not having the ends filled with concrete affected the flexural strength of the panels.

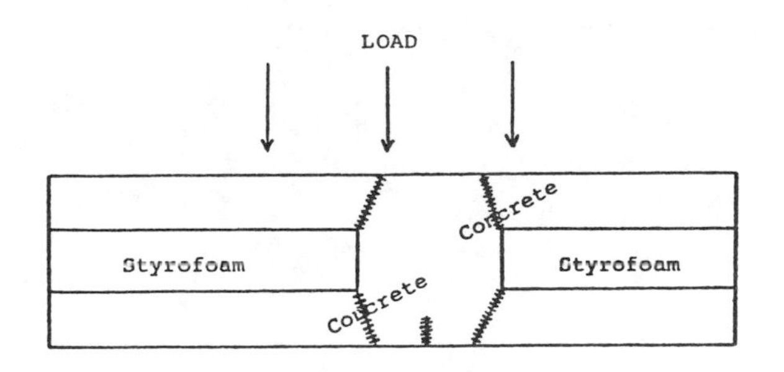

FIGURE 6 TYPICAL CRACK PATTERNS DEVELOPED IN XOREX STEEL FIBER REINFORCED CONCRETE SANDWICH PANELS FOLLOWING FLEXURAL LOAD APPLICATION TO FAILURE

References

1. "Measurement of Flexural Strength of Xorex Steel Fiber Reinforced Sandwich Panels at LML", Lankard Laboratories Report, June 8, 1982.

2. "Tilt Up Construction, Design Considerations-An Overview", By Ralph E. Spears, Concrete International, April 1980.

3. "Special Consideration for the Selection of Tilt Up Concrete Sandwich Panels", PCA Seminar E44, World of Concrete, 1981.

4. "Burke on Tilt Up", 1985

5. "Tilt Up Construction Handbook" Rev. 1-84A, Dayton Superior.

STEEL FIBER CONCRETE
US-SWEDEN joint seminar
(NSF-STU)
Stockholm 3-5 June, 1985
S P Shah and Å Skarendahl,
Editors

STEEL FIBRE CONCRETE - DEVELOPMENT AND USE BY EKEBRO AB

by Arne Nordin, Man Dir, Ekebro AB, Örebro, Sweden and Åke Skarendahl, DSc(Eng), Åke Skarendahl AB, Djursholm, Sweden

---

ABSTRACT

The paper describes experience from manufacturing of precast products in steel fibre concrete, development and use of a new production technique and discusses some important aspects on steel fibre concrete.

Ekebro AB has over a number of years made use of the idéa of steel fibre reinforcement for manufacturing of thin precast products. The first production line was based on a slightly modified conventional mixing and casting technique, and that production is still in operation. Due to restraints encountered in the production process as well as in the properties of the material and the products, Ekebro has developed a fibre shotcrete technique characterized primarely by the cutting of wires to fibres in the spray nozzle.

Among the aspects to consider in work with steel fibre concrete, the following are of major importance:

- improvement of the fiber efficiency
- reduced risk for corrosion
- more versatile production techniques
- improved steering and control of fibre inclusion

Arne Nordin is managing director of Ekebro AB, one of Sweden's major producers of ready-mixed concrete. Nordin is building engineer with experience from various concrete technology specialities as well as from foundation engineering. Arne Nordin is leading the development and use of fibre concrete within Ekebro AB.

Åke Skarendahl is working with research and development on concrete technology, especially fibre concrete, within Åke Skarendahl AB. He recieved his D Sc degree from the Royal Institute of Technology in Stockholm, and worked at the Swedish Cement and Concrete Research Institute between 1969 and 1978. He has been working with RILEM Coordinating Group for six years.

---

1. INTRODUCTION

The obvious advantages of being able to produce thin, light and durable products in concrete using the concept of fibre reinforcement has been made use of by Ekebro AB since 1976. A routine production was then started on a pilote scale, and the production process used was based on a rather conventional mixing and casting technique. Soon the production was increased to a full scale operation following the same principles, and that production is still continuing.

In order to improve the quality and performance of the products, as well as to gain in productivity, Ekebro has over a number of years been working of developments of the production process. The prime outcome has been a technique based on a spray process in which steel fibres are cut to fibres in the nozzle. The technique offers several advantages in comparison with conventional mixing and casting, in particular the possibility of using very long fibres. The technique, here called Ekebro fibre shotcrete, has been used in commercial production by Ekebro since 1983, and in 1985 licensing of the technique was commenced.

2. CONVENTIONAL MIXING AND CASTING

The first product made by Ekebro in 1976 was a box beam consisting of a 7 mm thick steel fibre concrete covering a core of expanded polystyren, see FIG 1. The product, which still is made on a regular basis, is used as edge insulation for slabs on ground and offers, apart from heat insulation, a permanent

FIG 1. Installation of edge insulating box beam prior to casting a slab on ground.

formwork for the the slab, a finished outer surface, support when levelling the slab as well as support for the facade wall.

The dimensions of the beam are 1 800 mm length, 400 - 800 mm height and 60mm width. On the back of the beam are two lightweight aggregate concrete blocks attached, acting as supports when the beam is installed. The fibre concrete contains 400 kg/m^3 cement, fly ash, sand 0 - 2 mm and 1 vol% of steel fibres 0.35 - 35 mm with hooked ends.

Other products, e g balcony panels, permanent formwork and decorative elements, are also produced using the conventional mixing and casting technique. For products with limited load bearing requirements but with high requirements on aesthetical appearance, as for example outdoor ornamental elements, polypropylene fibres of the Krenit type are used in stead of steel fibres.

Cast steel fibre products have limitations regarding material and product properties as well as production technology. The main deficiencies encountered by Ekebro are:

- Limitations in mechanical properties, mainly concerning ductility, due to short fibre lengths.
- Risk of corrosion due to the very thin cover on fibres close to the surface.

- Limited possibilities to produce curved surfaces as well as large vertical surfaces.
- Variations in fibre content and difficulties in steering and controlling the fibre content in various sections.

In order to overcome these deficiencies for products where they constitute major restrictions, the Ekebro fibre shotcrete technique was developed.

3. EKEBRO FIBRE SHOTCRETE

3.1 Development and use

The Ekebro fibre shotcrete technique is based on a spray process in which steel fibres are cut to fibres in the nozzle. The matrix is normally sprayed wet, and the nozzle is so designed as to give a matrix flow of cylindrical shape with the cutted fibres ejected concentrically with the ejected matrix. The method makes possible the use of fibres of lengths up to 200 mm, the products can be built up in separate layers, curved as well as large vertical surfaces can be made, and the fibre content can be varyed locally and be kept under a good control. By the use of long fibres, this type of steel fibre concrete is approaching the concept of ferrocement but making the production technique rational, see Skarendahl /1985/.

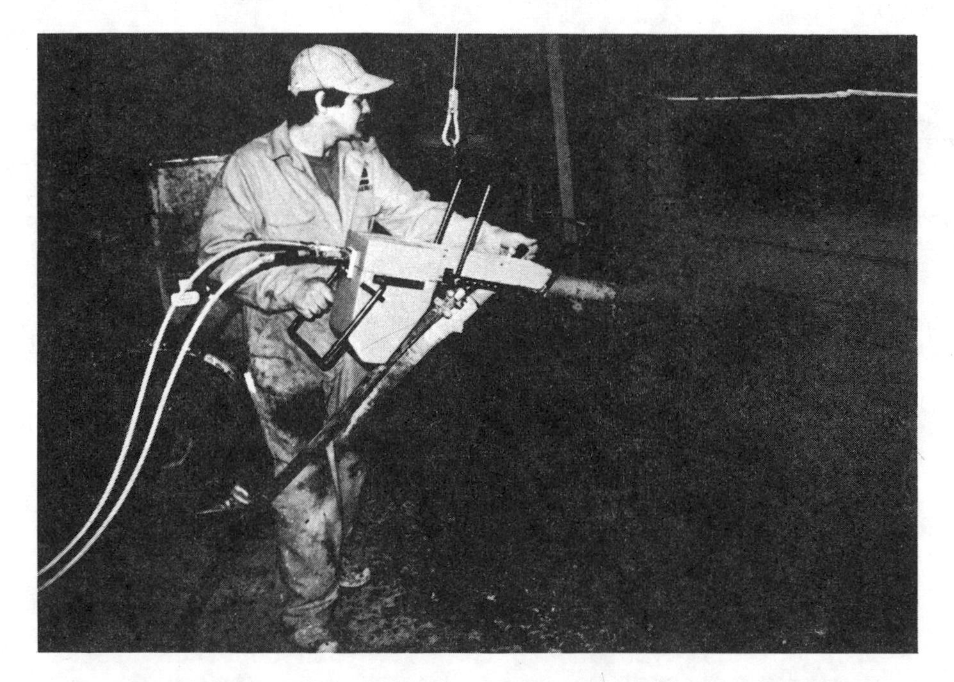

FIG 2. An Ekebro fibre shotcrete gun in operation.

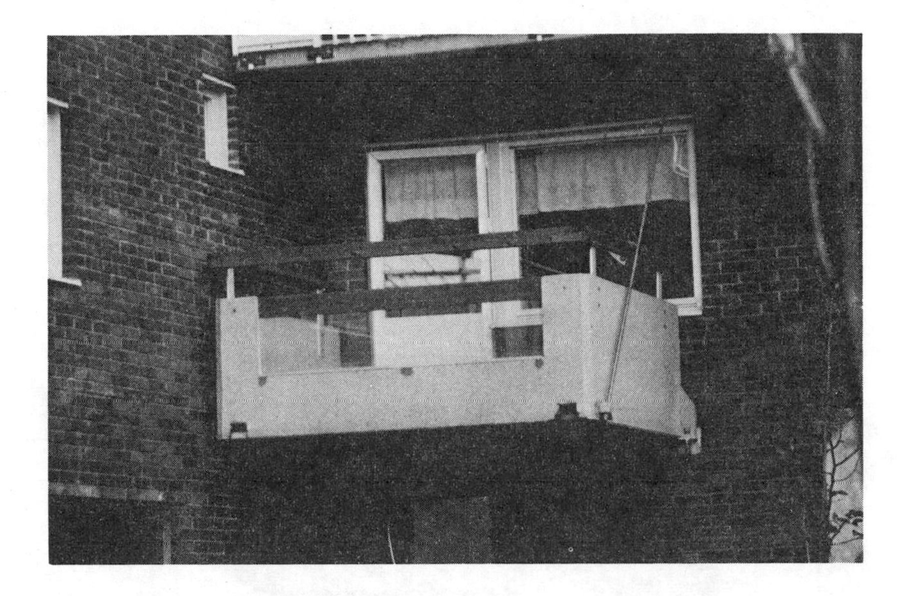

FIG 3. Ekebro fibre shotcrete balcony.

The technique has been used by Ekebro in commercial production for two years, and the first main product has been a balcony measuring 2.3 x 2.3 m with a wall height of 0.80 m, see FIG 3. The balcony is attached to the facade through support on small cantilevers and hanging in tension bars. The product is thus well suited for renovation and rehabilitation applications. The first project, into which 475 balconies are delivered, is however for a new residential housing area in Stockholm.

FIG 4. Examples of prototypes made in Ekebro fibre shotcrete (boat hull and shell).

Ekebro fibre shotcrete has tentatively also been used for other products than balconies, e g water tank, boat hull, various types of shells etc, see examples in FIG 4.The technique has also been used for in situ repair of concrete structures and for tunnel strengthening.

3.2 Production equipment

The Ekebro fibre shotcrete production equipment consists of a "gun" and a "wheel rack" of a special design, see FIG 5. In addition a screw pump, a mixer and an air compressor of rather conventional types are used. The gun incorporates a wire cutter with a fibre ejector pipe and a wet concrete spray nozzle. The

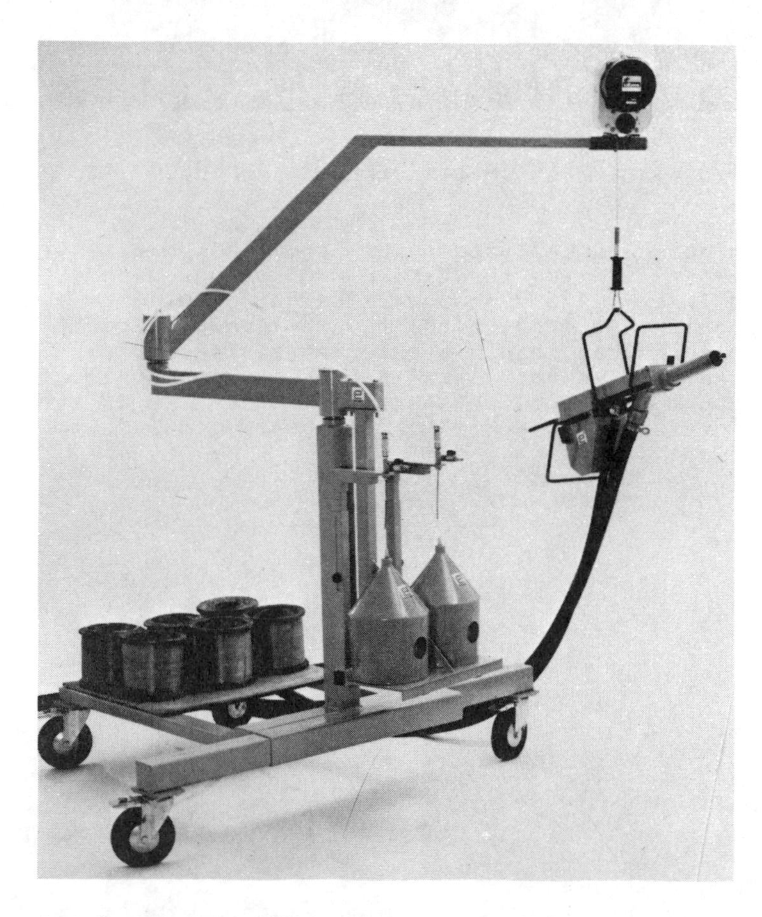

FIG 5. Ekebro fibre shotcrete equipment. The gun (right) is hanging in a balanced block. The block is attached to a boom supported by a wheel rack. On the wheel rack, two wire spools are mounted under bobbin caps.

concrete matrix is sprayed with compressed air led through a special channel arrangement in the pipe.

The gun is attached to a boom mounted on a wheel rack. The boom gives a wide working range without moving the wheel rack, and a balanced block gives simple steering of the gun as well as low physical loading on the operator. The wheel rack also supports the wire bobbins, which normally carries 30 kg of steel wire each. Two wires are simultaneously fed into the gun. The wire bobbing and the bobbin caps are designed to give an undisturbed wire supply.

The concrete matrix is mixed batchwise in a pan mixer and pumped with a screw pump to the gun. The capacity of the complete equipment is 0.5 - 5 m^3 per hour, depending mainly on the choice of various fibre parameters.

3.3 Production technique

The spraying of Ekebro fibre shotcrete is carried out on one sided moulds. The mould material is normally steel, wood board or glass reinforced plastics. The choice of mould material will depend on the size of the serie, the mould configuration etc.

The gun operator controls the flux of fibres as well as of matrix through the gun and can thus spray matrix with or without fibres as well as fibres without matrix. This enables the use of e g unreinforced outer layers for corrosion protection, increased fibre concentrations in highly stressed sections etc.

In most productions, an unreinforced layer is first sprayed on the mould surface.Then follows simultaneous spraying of fibres and matrix until the desired thickness of the fibre reinforced layer is achieved. The amount of fibres is typically between 0.5 and 1.5 vol% (between 1.8 and 5.5 weight%) depending on desired performance. In cases when fibres close to the surface should be avoided, a final unreinforced layer is sprayed. The surface layer can easily be given various textures and the whole section, or a part of it, can be pigmented for colouring. The products can of course also be painted with appropriate (e g acrylic) paints.

The needed thicknesses of the various layers of a product are determined through structural design as well as from life length analysis, mainly considering corrosion of the fibres. Typical thicknesses will be 20 - 40 mm, but a total thickness of 100 mm can if necessary be sprayed on vertical surfaces without the use of accelerators.

When product design, raw material quality etc so require, extraordinary measures are taken to secure various properties. If e g the tightness has to be improved, either a surface treatment or a modification of the matrix is applied. The surface sealing compound is either a cement or a polymer based product. For matrix modification, either an inorganic filler or a polymer additive is used.

4. IMPORTANT ASPECTS ON STEEL FIBRE CONCRETE

Experience from work with steel fibre concrete has indicated some aspects very important to consider in development. Among these aspects are 1/ improvement of the mechanical properties, mainly crack control and ductility,together with methods for their evaluation, 2/ service life estimation techniques concerning corrosion, 3/ improved and more versatile production methods including formwork technology and 4/ improved steering and control methods of the fibre inclusion.

4.1 Mechanical properties

Fracture in steel fibre concrete is normally caused by fibre debonding followed by fibre pull-out. Very seldom the full potential of the steel can be utilized as the production processes limit the fibre lengths possible to incorporate. In conventional mixing and casting, as well as in most shotcrete techniques, the length over diameter ratio (aspect ratio) of the fibres used seldom exceeds 100, and sometimes is as low as below 40. The short lengths, coupled with the rather low fibre volumes used which also is a restriction from the production processes, give a post cracking behaviour normally characterized by a significant drop in the load bearing capacity after first crack. To a varying extent the fibre reinforcement limits further growth of the cracks and thereby induce ductility.

Strengthwise, steel fibres normally attribute to only limited improvements. The effect of the fibres come from their ability to hinder crack growth, to cause multiple cracking and to give the cracked composite a load bearing capacity. The increased ductility, as well as the way in which the cracking develops, are properties of great importance for steel fibre concrete. The lack of testing and evaluation methods for these properties, as well as the lack of established relevant failure criteria for different applications,are at present serious drawbacks for the efficient use of steel fibre concrete.

The approach used by Ekebro in improving the mechanical properties has been to use a high strength matrix together with long fibres. The Ekebro fibre shotcrete equipment allows the use of fibre lengths of up to 200 mm. With a fibre diameter of 0.4 mm the aspect ratio thus is a maximum of 500. The steel used has a yield strength of 1 200 N/mm^2, is indented and has a thin zinc coating. The quality of the matrix is very high with compressive strength measured on 100 mm cubes of 80 - 100 MN/m^2. The maximum aggregate size is 4 mm, and silica fume as well as superplasticisers are used.

Before initial cracking, the strength mechanism is steel fibre concrete is virtually the same regardless of fibre volume and fibre length. The stiffness and the first crack strength are primarely depending on the matrix. The basic experience with short fibre concrete has been that the post cracking strength is direct proportional to fibre volume and fibre length. In a study of long fibre concrete, Skarendahl and Mwamila /1984/, the direct proportional effect of the volume was observed also for long fibres, while the effect of increases in fibre length was progressively reduced the longer the fibres over a certain length. The matrix strength and the bond between matrix and fibres are strongly influencing the properties in addition.

Examples of stress-strain diagrams from measurements on sawn beams 100 x 25 mm under third point loading with a support span of 300 mm,are shown in FIG 6. The fibre length varies from 48 to 192 mm.

The study of long fibre concrete also indicates that long fibres give an increased tendency for multiple cracking,and even very low volumes of long fibres have a substancial effect on the impact strength.

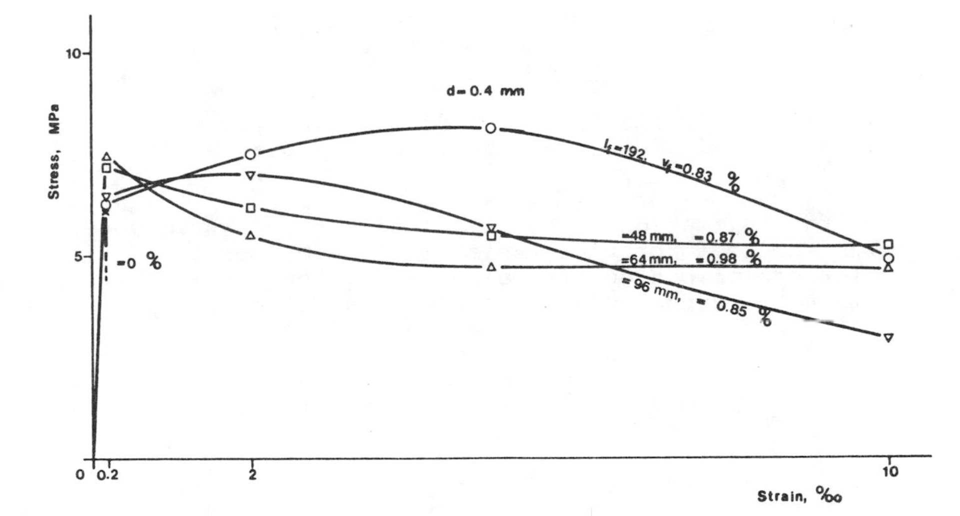

FIG 6. Mean stress-strain curves (4 specimens) for fibre diameter 0.4 mm, fibre volume 0.85 to 0.95 vol% and varying fibre lengths. Third-point loading, beam section 100 x 25 mm and support span of 300 mm.

4.2 Corrosion

When using steel fibres in concrete, attention has to be given the question of corrosion of the fibres. As the steel volume locally is very small when fibres are used, only limited expansion forces develop due to the corrosion and normally no spalling occurs. The corrosion might however give aesthetical defects in the form of staining, and the loss of the contribution of a corroded fibre has also to be considered.

The corrosion process can,according to Tuutti /1982/, be separated into an initiation phase and a propagation phase, see FIG 7

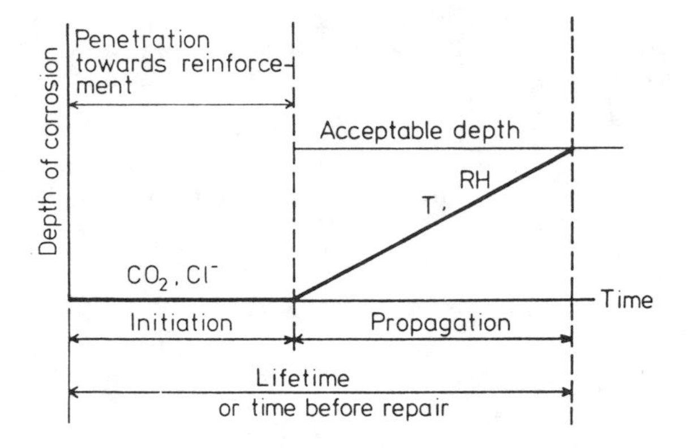

FIG 7. Schematic sketch of steel corrosion sequence in concrete according to Tuutti /1982/.

During the initiation phase the alkalinity of the pore water is high enough to protect the steel from corrosion. Furthermore, the cloride ion content is below a dangerous limit where corrosion is not induced. Due to carbon dioxide penetration and the resulting carbonation, and/or an increase in cloride ion concentration, conditions making corrosion possible will develop. At that stage the propagation phase is entered. How far the corrosion might propagate before the serviceability is exceeded is very much depending on the application in question.

When producing steel fibre concrete in a conventional mixing and casting process, some of the fibres will come very close to the surface and be covered only with a very thin layer of the concrete matrix. As the carbonation of the matrix proceeds, corrosion of the fibres close to the surface starts which might cause staining on the surface. The experience with the Ekebro insulating beams produced in conventional casting, see 2 above, is that a limited pit staining might occur, but that is very seldom a matter of controversy for that particular application. The susceptibility of rust staining for conventionally cast steel fibre concrete products is however to be considered as a serious drawback in many applications.

In the Ekebro fibre shotcrete technique it is possible to operate the fibre flow and the matrix flow separately. Unreinforced matrix layers can be sprayed whenever needed, and a common sequence when building up a product is to first spray a matrix layer followed by a composite fibre and matrix layer followed finally by an unreinforced matrix layer. Sufficient cover for the steel can thus be secured.

The design of the cover thickness can be based on an evaluation of the influencing factors considering mix design and environmental factors. For carbonation induced corrosion, the factors of importance are available carbon hydroxide, diffusitivity of carbon dioxide and carbon dioxide content in the air. When cloride induced corrosion is likely to take place, also other factors such as cloride ion threshold value, cloride ion diffusion rate etc are of importance.

The design of the cover for products in Ekebro fibre concrete is normally based on the end of the initiation period for the fibres adjacent to the cover. In ordinary concrete a certain degree of corrosion propagation is normally accepted as a criterion for serviceability, why the criterion here chosen gives results on the safe side. Another positive effect not taken into account is the zinc coating applied on the fibres. The degree of improvement is however difficult to quantify.

The product at present regularly produced in Ekebro fibre shotcrete, the balcony, is evaluated to have a service life, considering the time of initiation, of 50 years with 5 mm cover. Only carbonation induced corrosion is relevant for that type of application. The matrix has a very low water-cement ratio and is very dens due partly to silica fume addition. The service life estimate is, among other things, based on generally established relations between composition and gas diffusion. A more precise estimate can be made if the diffusion is measured on the matrix in use.

With a production process that enables easy application of unreinforced layers, the use of high quality matrices, and with a service life estimation approach, it is possible to consider in design, and avoid, corrosion problems in steel fibre concrete.

4.3 Versatile production technique

Fibre concrete materials are most effectively made use of in thin walled and curved components. When using the normal mixing and casting technique, the thin sections most often have to be cast horisontally as vertical casting is difficult, at least when the vertical portion is very high. Horisontal production of thin components is very inefficient when it comes to floor space utilization, a condition that is negatively effecting the productivity and economy. Furthermore, casting of curved surfaces is difficult no matter what direction is used. In a horisontal position the concrete tands to slide on the mould, and in a vertical position the casting is difficult and require extremely good workability, and the moulds become very expensive.

In contrast to the difficulties encountered in the normal mixing and casting technique in producing large vertical and curved components, the Ekebro fibre shotcrete technique is best used on vertical surfaces, and curved surfaces are as easy to produce as plain surfaces. The possibility of steering the matrix flow and fibre flow individually makes possible easy variation in thickness, variation in reinforcing effect etc. One condition, which sometimes might be considered as a restriction, is that the two surfaces of a section not will have the same finish. One surface will have the texture and smoothness of the mould, while the other will have either the rough surface obtained by the spraying or the texture and smoothness of a hand or machine finish of the sprayed surface.

When introducing thin and curved products into concrete production, a more complicated formwork technique has to be accomplished. Especially dubbel curvature products require a new approach. Experience and practice from glass fibre plastics production can be very valuable to apply also for fibre concrete production.

The rational production process possible to obtain with the Ekebro fibre shotcrete technique can be illustrated e g by the production of a cylindrical water tank, see FIG 8. First the bottom of the tank is sprayed on a rotary mould table. After placing an inner mould directly on the sprayed bottom, the connection between bottom and wall is sprayed. The wall is then sprayed under rotation of the tank. After spraying, the tank is covered with a plastic sheeting for curing. The whole production of a 10 m^3 tank can be made in less then two hours and is made continuously with no hardening or presetting needed during production.

4.4 Control and acceptance

Production of thin, load bearing, high performance products in steel fibre concrete requires efficient control systems. The quality, the amount and the dispersion of the constituent materials have to be under close control to guarantee an even and high quality, and in order to gain acceptance on the market for the products, a profound concept concerning quality assurance has to be applied. One of the problems encountered is that some of the properties relevant for steel fibre concrete products are not looked for in normal reinforced concrete production.

FIG 8. Prototype production of a 10 m^3 water tank. The whole production is made in one sequence.

The lack of agreed testing and evaluation methods is another restraint.

The testing done in the production by Ekebro is basically a control of the compressive strength of the matrix and a measurement of the flexural behaviour on beams sawn from separately made specimens. In addition, normally also the water-cement ratio is limited and controlled, the fibre content and distribution is measured and the cover thickness is controlled during the production sequence.

The fibre content is registrated during production through negative weighing of the wire bobbins and is measured on the products by use of a calibrated electromagnetic meter, following principles discussed in Malmberg and Skarendahl /1978/. A lay-out of a calibration diagram is shown in FIG 9.

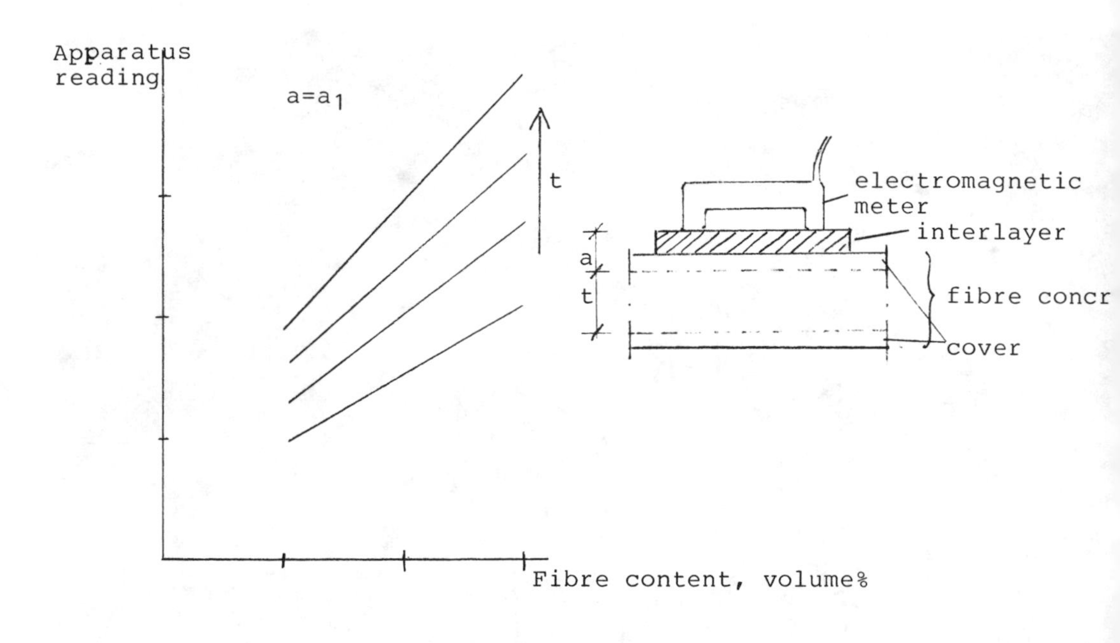

FIG 9. Lay-out of calibration diagram for electromagnetic measurement of fibre content.

In an effort of steering and controlling the fibre inclusion further, a measurement device for continuous measurement of the fibre flow and the matrix flow can optionally be installed on the Ekebro fibre shotcrete equipments. With this device the operator will continuously get information on the fibre volume sprayed.

5. REFERENCES

Malmberg Bo and Skarendahl Åke /1978/: Determination of fibre content, distribution and orientation in steel fibre concrete by electromagnetic technique. International symposium on testing and test methods of fibre cement composites, Sheffield, 1978.

Skarendahl Åke and Mwamila B L M /1984/: Sprayed fibre concrete using long steel fibres. Report March 1984 from Åke Skarendahl AB.

Skarendahl Åke /1985/: Econ fibre shotcrete system - A rational production technique. Paper presented at the Second International Symposium on Ferrocement, AIT, Bangkok, 1985.

Tuutti Kyösti /1982/: Corrosion of steel in concrete. CBI Research 4-82, Swedish Cement and Concrete Research Institute.

STEEL FIBER CONCRETE
US-SWEDEN joint seminar
(NSF-STU)
Stockholm 3-5 June, 1985
S P Shah and Å Skarendahl,
Editors

PRODUCTS MADE OF STEEL FIBRE REINFORCED CONCRETE ON THE SCANDINAVIAN MARKET

by Ingvar Sällström, MSc, Manager R&D div at Precon AB, Falkenberg, Sweden

---

SUMMARY

Precon AB has since 1978 been producing following products in steel fibre reinforced concrete:

1 Flat panels of magnet oriented steel fibre concrete

2 Partition walls for pig boxes in animal stables

3 Urine draining slabs for manure duets in pig stables

4 Overhead line poles of spun steel fibre concrete

5 Thin-walled cofferdams and form shells

6 Sandwich panels for bathroom modules

7 Heat accumulating steel fibre concrete slabs

The products are more closely described on the following pages.

Ingvar Sällström, is the head for Precon AB's R & D division at the head office in Falkenberg. Mr Sällström has a civil engineering degree and he has a long experience from concrete product design and production.

Mr Sällström has since 1975 worked with the introduction and industrial production development of steel fibre reinforced concrete. In this field he has designed and produced some very unique products.

INTRODUCTION

PRECON AB - a company in the Incentive-Group - has been developing the production technique for steel fibre reinforced concrete since 1975. Together with the Institute for Innovation technic in Stockholm (IIT) experimental work was the main activity from 1975 to 1979 with different kinds of products. In this period of time the very efficient magnetic orientation of steel fibres in concrete was brought into the production (Fig. 1 and 2) as well as additional equipment to the batch plant to get an appropriate fibre dispersion.

Fig.1

Fig.2

The possibility to use steel fibre concrete in already existing traditional products was considered, however, as the higher material costs couldn't be compensated with lower labor costs fully out, only a few traditional concrete products have come into production. A close study of the prospective use of steel fibres as replacement for traditional bar and mesh reinforcement indicated that the best profitability was to be expected from extreme thin wall panels, double curved panels or products with a complex design.

AGRICULTURAL PRODUCTS

Since Precon AB had an extensive production of agricultural concrete products, the advantages of steel fibre concrete in this production programme was studied. There was a vast demand particularly for partition walls and other components to stable interiors. The first partition walls left the production line in 1978. (Fig. 3).

Fig.3

Partition walls for pig boxes are made of 25 mm thick unoriented steel fibre concrete and are glued directly to the floor slab.

In an other application, where higher strenght was required, the panel was made out of a 20 mm thick steel fibre (Fig. 4) concrete panel in which the fibres had been magnetically oriented according to calculated stresses.

Fig.4

Another new developed article in this product range (Fig. 5) is a urine draining slab made of 20 to 30 mm thick unoriented steel fibre concrete with conical drainage holes.

Fig.5

These three products have been highly appreciated by the farmers, who in these products have got a material that isn't attached by the very damp and corrosive environments that are found in animal stables. Precon AB has during these years got a very good experience of the use of steel fibre concrete and found its performance excellent in the difficult environments.

POLES FOR OVERHEAD LINES

As Precon AB during many years has manufactured products for the power industry - mostly substation enclosures and other ready-to-use electrical buildings in co-operation with ASEA, the interest arouse to produce poles for overhead lines and lightning poles in concrete. In many countries in Europe and elsewhere the use of concrete poles is well established since many years back, but in Scandinavia, where the supply of good suitable timber is rich, the wood pole has been dominating. Considering the standardized handling equipment, fittings and work methods it was necessary to develop a concrete pole that as far as possible resembled the wood pole. After studying concrete poles in other countries we decided to more closely study the spinning technique. Precon AB has developed all machinery in house for this production. This has resulted in a machine, which can produce poles of 8 to 13 meters of length, with a thickness of material of 30 to 60 mm and a fibre content of 1,1 to 2,0 per cent by volume. The production capacity will be 27 to 40 poles per dayshift.

The machinery (Fig. 6) is built with an axial moving trolley on which the pole form is spinning. The concrete is pumped into spinning and axial moving form simultaneously with steel fibres. The fibres have a diameter of approximately 1,2 mm and a length of about 500 mm (Fig. 7) and are made of very high tensile steel. The fibres and concrete are fed into the form in several layers and by a special arrangement the fibres can be given a certain axial angle.

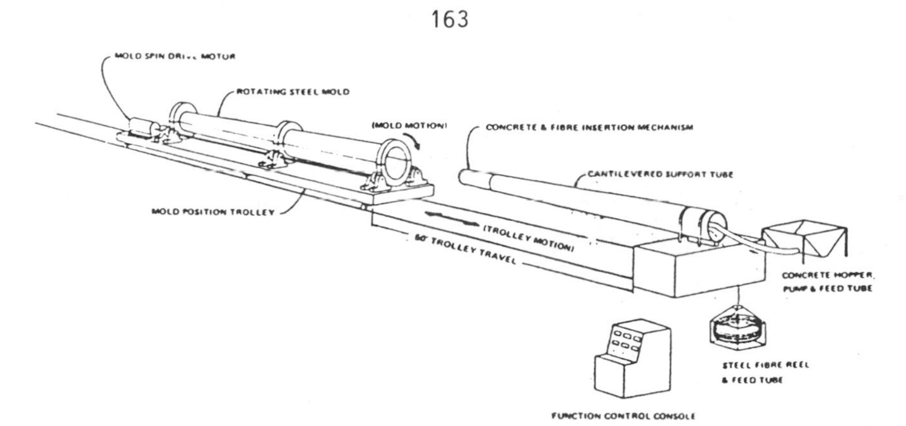

Production machine for fiber-reinforced concrete manufacture, feeds steel fibers and concrete continuously into a rotating mold.

Fig.6

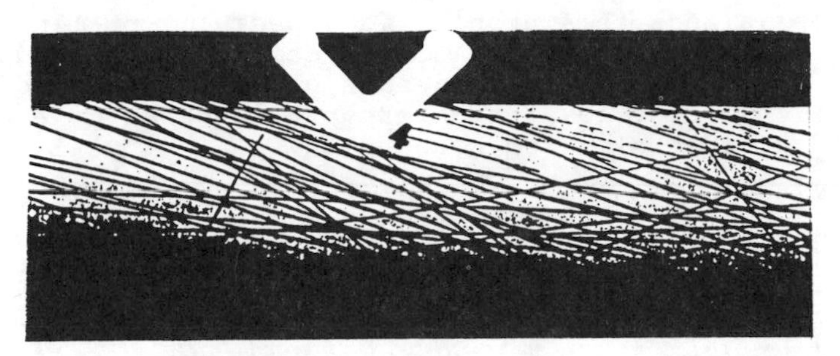

Fig. 7: X-ray photograph of a transmission pole made of steel fibre concrete.

The poles are autoclave cured in their forms and have after the programmed curing reached approximately 45 % of full strenght when stripped. The production process is controlled by a computer in which pole type, its data for concrete- and fibre quantity, number of layers and rotation- and other motion speeds is predetermied. In the quality control a bending test is included (Fig. 8). The plant is still in its running-in stage, but delivery comittments have already been undertaken to some customers.

Fig.8

Fig.9

The product is considered a great future as a replacement for wood poles in Scandinavia. Foreign concrete pole manufacturers have shown great interest to learn more about the Precon spinning process as being the first efficient automatic production method.

COFFERDAMS, HARBOUR GATES

In countries where the difference between the tides is big and strong winds creates strong waves, there are serious problems in the mooring areas with the water turbulence. From time to time in small yacht harbours the yachts are destroyed by storms. You therefore try to close the harbour entrance with gates to get a calm harbour basin. This has sofar been very expensive.

On the Faroe Islands this problem is especially prominent. In co-operation with the Harbour Authoroties on the Islands, Precon AB has designed and constructed a gate consisting of 3 box shaped sections (Fig. 10), which are put together to a gate articulated against a bottom foundation.

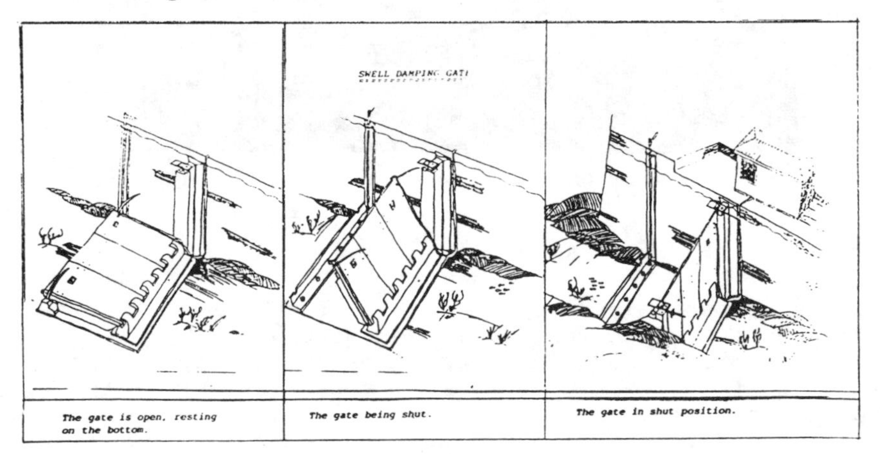

Fig.10

The gate is in open position filled with water and resting on the bottom. When the gate is to close the harbour entrance, the water is pressed out by compressed air. Now the gate rises and locks against columns on each side of the entrance.

To make this function the gate must have low weight, why steel fibre concrete was choosen. The box shaped gate is one monolittice unit (Fig. 11). The wall thickness varies between 30 and 50 mm and, in order to get the gate airtight from the inside, we put a laminate of plastic on a foam core, creating the cavity in the gate. To protect the steel fibre against ferrous corrosion we polymer impregnate the gate form the outside.

When the gate is in upright position and locked, an air valve

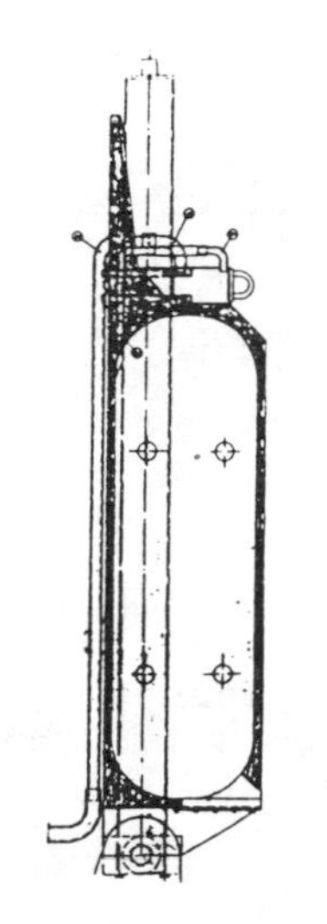

Fig. 11: Gate section.

is opened and the gate is filled with water. When the gate is to be lowered air is pressed in, in order to get a balance in the gate and water is filled in until the gate slowly is lowered to bottom position.

Wind and waves also affects the movements of the gate and heavy shock absorbers must be adapted to damp the up to 50 tons strong shock forces and at extremely low water level the gate is assisted with electric motors.

This method is patent applied.

The gate has been in operation since the autumn of 1984 (Fig. 12) and functions well even in hard weather. The harbour basin is calm and the yacht owners don't need to feel anxious about their yachts.

Fig.12

FORM SHELLS

On working sites where if of different reasons is difficult to build forms for concrete pouring or in countries where it is difficult to make salt resistant concrete form panels or form shells of steel fibre concrete are proved to be excellent.

On the Faroe Islands they have big problems to build harbour piers strong enough for the powers of the sea. Recently Precon has delivered cofferdams of different shapes for this purpose. We have manufactured thin-walled cloumn shells (Fig. 13) where the section is trapezoidal to fit into a circular arc. The shells have a length of 7 meters and weigh about 6 tons. The column shells forming the pierhead (Fig. 14) are installed beside each other in a semi circle. They are placed on a bottom foundation and are filled with stones up to the water level and thereafter with concrete.

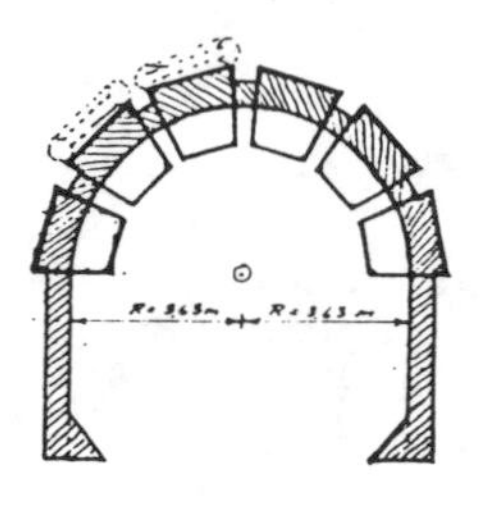

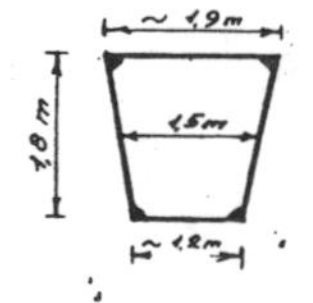

Fig. 13: Column moulds grouped for a harbour pier.

Fig.14

The columns are manufactured horizontally in one piece of unoriented steel fibre concrete. The columns are designed for wave forces up to 2 tons/m^2 and to an inner form pressure of 1 Mp/m^2.

For the same purpose we also manufacture lower cofferdams, one H-shaped shell approximately 2 x 2 meters and with a height of 1,5 meters (Fig. 15) and square shells (Fig. 16) 4 x 4 meters and 4 x 2 meters with a height of 2,0 meters. The shells are made of 30 to 40 mm thick unoriented steel fibre concrete. Both types are installed after one another, either at the side of a paved pier to reinforce it or on top of the pier to make it higher. The cofferdams are filled with stone and covered with a concrete or asphalt top.

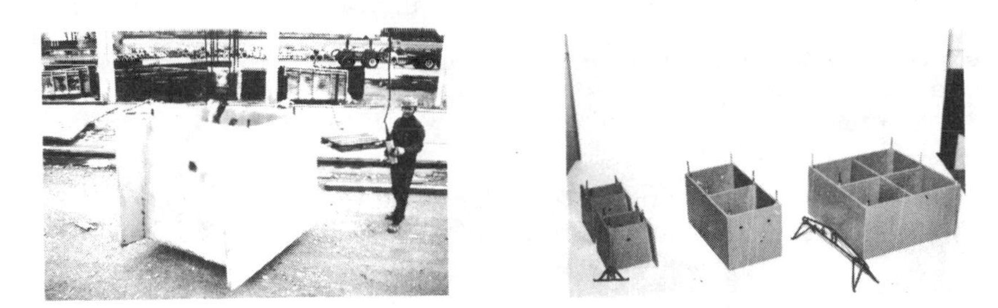

Fig.15 Fig.16

The Harbour Authoroties on the Faroe Islands have established the fact that this method to reinforce and build up inadequate piers reduces the overall cost with approximately 50 % compared to earlier methods.

SANDWICH PANELS

Through a Nordic co-operation initiated by Nordforsk, a number of producers, building contractors and architects together have started an applied fibre concrete product development under the name SCANDINAVIAN FIBRE CONCRETE DESIGN. As introduction product we have chosen turn-key bathroom modules (Fig. 17), bath- and shower rooms and restrooms, where the walls and the roof are made of sandwich panels.

Fig.17

Fig.18

The panels are made with a core of 40 mm styrofoam and 10 mm fibre concrete on each side (steel fibre in Sweden, plastic fibre in Denmark). The floor - which must be very thin - is made of 20 to 40 mm steel fibre concrete with a 50 mm high frame. The panels are poured in plastic forms and the inner coating is made according to the customer's choice. Standard design is made with a gelcoat surface, which is laid in the form before casting. The gelcoat can be chosen in a variety of colors (Fig. 18) or in a textured finish.

The production capacity will be 6 units per day. The weight of the bathroom including fitting out details is approximately 1500 to 2000 kg and is delivered completely installed ready to use to the working site. Only remaining work is to hook up power sewage and water (Fig 19).

Fig 19

The individual members of the co-operation team manages their respectively home market theirselves. The export market is managed mutually.

The bathroom modules can also be shipped desassembled in order to avoid expensive volume transports at long distance transports.

A further development for adaption of bathroom modules for repair- and rebuilding works as well as a combination bathroom-kitchen modules is next on the programme.

HEAT ACCUMULATING STEEL FIBRE CONCRETE

One of the companies in the Incentive-Group is producing food handling trollies for canteens. In order to keep the food hot during the transport they have sofar had the canteen placed on a pot stone slab in which electric wires are laid in to keep the slab heated. This slab is very expensive.

Therefore Precon has developed a heating slab made of steel fibre reinforced iron-ore cement in the size of 530 x 645 x 30 mm. This slab has shown very good heat accumulating qualities (Fig 20).

Fig 20

Compared to a pot stone slab with the same size tests have shown (Fig 21) that the slab made of iron-ore cement takes longer time to heat up to 50° (40 minutes compared to 27 minutes), but the cooling off to 40° is much slower (125 minutes compared to 88 minutes).

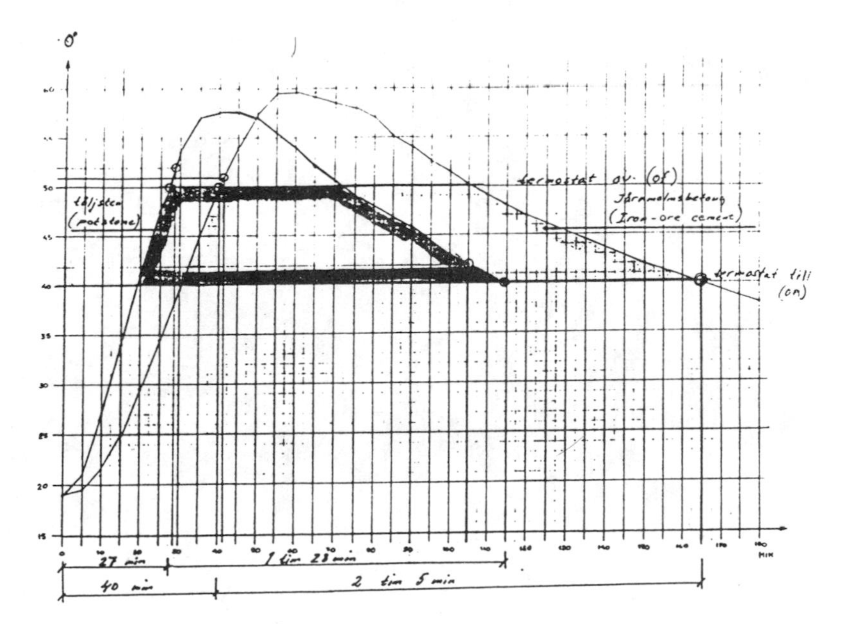

Fig 21. Heat accumulation in slab of iron-ore cement compared to slab of pot stone.

We have now installed 50 slabs in food handling trollies all over Sweden for a long testing period. This in order to see if any changes take place in the material and if the heat accumulating qualities remains during a long period of use.

STEEL FIBER CONCRETE
US-SWEDEN joint seminar
(NSF-STU)
Stockholm 3-5 June, 1985
S P Shah and Å Skarendahl,
Editors

AN OVERVIEW OF PROGRESS IN APPLICATIONS OF STEEL FIBER REINFORCED CONCRETE

by Ronald F Zollo, Prof, Dep of Civil Engineering, University of Miami, Coral Gables, Florida, USA

ABSTRACT

The past two decades of research worldwide involving steel fiber reinforced concrete (SFRC) is referenced and used to provide insight into future development. An overview of current applications and production practices in the United States is classified and presented for comparison with the experience being reported in other parts of the world.

Special applications technologies, including extrusion, high temperature (refractory) and lightweight matrix SFRC, are also discussed with emphasis on the developments in extrusion research.

Ronald F. Zollo is Professor of Civil and Architectural Engineering at the University of Miami in Coral Gables, Florida. He is a member of the American Concrete Institute, ACI Committee 544 Fiber Reinforced Concrete, since 1972 and currently serves the committee as chairperson of the subcommittee on synthetic fibers. His experience with steel fibrous concrete has been continuous over the past two decades of its development.

INTRODUCTION

It is a generally accepted principle that to gain insight into where one is going, one should pause to look back to rediscover and assess where one has been. In this way, the state of the art can be established in the heart as well as in the mind of the reviewer. As much of the history of the development of steel fibrous concrete has already been well documented (1) it will not be repeated here. However, new information or insight needs to be inserted into the historical record as it becomes available and this is what the first part of this paper attempts to do.

Still in keeping with the theme of this paper, as expressed by the title, a second goal will be the discussion of special application technologies, namely, extrusion, high temperature and lightweight concrete. The discussion is restricted to applications of relatively low, less than two percent by volume, fiber loadings. The use of fibers at four to ten times this fiber loading is a most interesting relatively new development which will be more ably discussed by another author who will be heard at this seminar.

AN UPDATE OF U.S. EXPERIENCE WITH SFRC

Fiber Type and Volume History in the U.S.

Steel fiber reinforced concrete (SFRC) in 1984 is available worldwide using fibers manufactured on at least four continents. Research and field experience has pushed yearly fiber volume production statistics upward with recent worldwide estimates of production in the 20,000 metric ton range. Figure 1 contains yearly production figures for fibers produced in the United States and in Europe (2). These data are separated according to a generic classification relating to the chemical composition of the parent metal into two principle categories: carbon steel fibers and alloy steel fibers.

The essential requirements with regard to the chemical composition of fibers for SFRC are the maintenance of some degree of fiber ductility and the assurance of yield strengths for individual fibers which exceed service stresses and which exceed fiber pull-out strengths for the particular brittle matrix containing the fibrous reinforcement. The generic classifications carbon steel and alloy steel cover a broad range of specific fiber material compositions with corresponding functional differences that should be evaluated for any particular application contemplated. Realizing this limitation to such generic classification one might place carbon steel fibers in the category of those fibers used mainly with portland cement concrete matrices and stainless steel alloy fibers in the category of those used in the production of refractory materials. Figure 1 shows the history of fiber production for fibers of both categories. The curves of the lower portion of the graph, those for

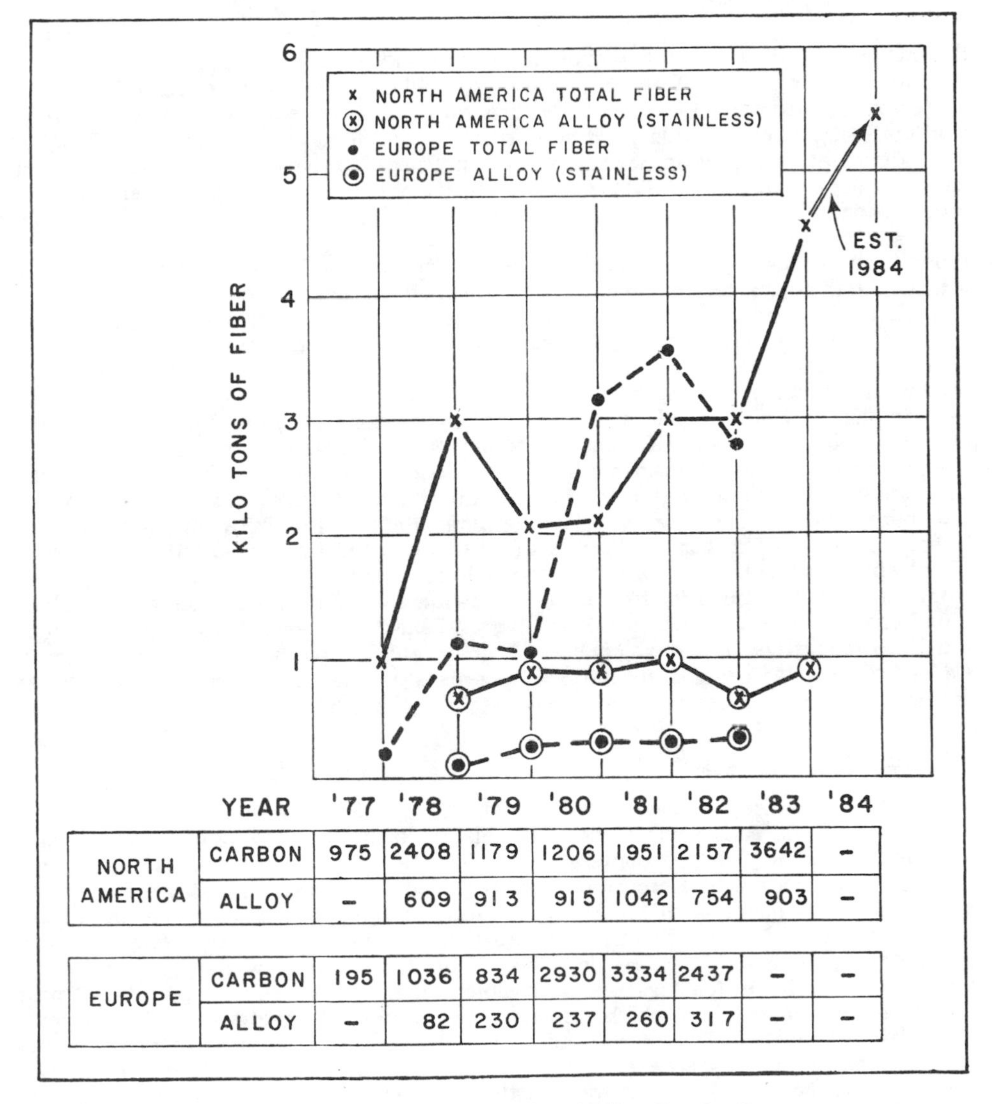

	YEAR	'77	'78	'79	'80	'81	'82	'83	'84
NORTH AMERICA	CARBON	975	2408	1179	1206	1951	2157	3642	-
	ALLOY	-	609	913	915	1042	754	903	-
EUROPE	CARBON	195	1036	834	2930	3334	2437	-	-
	ALLOY	-	82	230	237	260	317	-	-

Fig. 1 Recent History of Steel Fiber Production

alloy fiber volume, are indicative of the pace of production of fibrous concrete in the refractory materials industry. The two upper curves are for total steel fiber production. While there are special applications for carbon fiber SFRC within the refractory industry these are less often used. Fluctuations apparent from year to year on the figure are in part a reflection of worldwide free market economic considerations. The 1982 slide in alloy fiber production, for example, is a result of decreased demand for refined petroleum products which occurred during that period.

In addition to chemical composition, fibers are often classfied according to their means of production. There are four basic means of fiber production. They are cut from drawn wire, cut from sheet stock, produced directly from molten metal, and cut in a machining process from bulk steel blocks. The last process, machined fiber, is not used in the production of fibers which are commercially available in the U.S.

Depending upon the chemical composition and method of fiber production, i.e., degree of cold working, ample fiber yield strength and fiber ductility needed for SFRC performance in portland cement and refractory concrete is assured. Commercially available U.S. fibers provide yield strength of from 50 to 250 ksi (345 to 1725 MPa) and sufficient ductility to satisfy the requirements of strength and ductility for mixing, placing SFRC. ASTM A820 which is a specification for steel fibers is preparation. (3)

Fiber form and shape, both cross-sectional and longitudinal, is also properly separated into two categories termed smooth and deformed. A breakthrough in thinking with respect to the benefits derived from the use of SFRC came with the realization that for many applications post cracking, and even post ultimate, strength and deformation performance was paramount to earlier notions of possible extension of the more nearly elastic range of SFRC behavior. It is now widely understood that post cracking performance is limited by interfacial fiber/matrix bond strength and methods have been devised to improve mechanical interaction at the fiber/matrix interface. For the past several years deformed fibers have taken virtually exclusive control of the U.S. and worldwide markets. Deformed fibers affect fiber/matrix bond initially either by greater mechanical interlocking along their length or by end anchorage of fibers. Subsequent to bond failure at the interface, the work required to draw such fibers through the brittle matrix in the pull-out process is substantially increased from that which is found for smooth surfaced fibers of minimum surface area, such as round cross-sections, and for straight fibers. Table 1 contains a brief description of the deformed fiber types which are commercially available in the U.S. in 1984. Added to the table for informational purposes is the machined fiber which, as previously mentioned, is not currently available in the U.S. (4)

Current Uses Classified

The first several years in the early development of SFRC were hindered by the strict association of material performance with elastic composite strength characteristics and by the difficulties encountered in applications procedures. Relatively high volume high aspect ratio fibrous concrete required considerably greater effort in the preparation and handling of plastic state material compared to that required with plain concrete. The introduction of deformed fibers significantly contributed to development by favorably affecting the system economics and handling problems. Lower fiber volume loadings were possible which still produced beneficial post cracking performance and which also provided beneficial effects on material costs and mix workability. The use of deformed fibers simplified the use of conventional applications procedures such as pumping, gunning, slip forming, pressure or ram packing, and batch placement for on-site or precast production. The improved fiber/matrix interfacial bond characteristics of deformed fibers at low fiber volumes provided cost effective toughness and ductility durability factors for many applications while, for some fiber types, sacrificing pre-cracking strength gains.

TABLE I
DEFORMED FIBERS AVAILABLE
TO U.S. SFRC PRODUCTION

PRODUCER / LOCATION	EQUIVALENT DIAMETER * mm.	INVENTORY LENGTHS mm.	SCHEMATIC CROSS SECTION PROFILE
BEKAERT STEEL WIRE CORPORATION PITTSBURGH, PA. (1)	0.5, 0.8,	30, 50, 60	
NATIONAL STANDARD COMPANY, INC. NILES, MICH. (1)	0.4	19, 25	
	(3)		
MITCHELL FIBERCON, INC. PITTSBURGH, PA. (2)	0.4	19, 25	
RIBBON TECHNOLOGY CORPORATION COLUMBUS, OH.	0.8, 1.0	25, to 76	
	(3) 0.3 to 0.5	19 to 36	
AIDA ENGINEERING JAPAN (4)	0.4, 0.6	30	

(1) BY WIRE - DEFORMED ENDS
(2) BY SLIT SHEET - SHEAR LIP ALONG LENGTH
(3) BY MELT EXTRACTION - ROUGH IRREGULAR SHAPE, STAINLESS STEEL
(4) NOT AVAILABLE IN THE U.S.

* BASED ON DIAMETER OF CIRCULAR CROSS-SECTION OF SAME AREA (1mm. = 0.039 IN.)

As with any construction material the use of SFRC is dependent on economics, material performance, experience and availability. Satisfaction of these criteria has provided numerous opportunities for SFRC field applications in the U.S. Table II provides a listing of SFRC applications with indication of the relative position each application occupies in the spectrum which spans from experimental to production applications. It is the result of a current survey of U.S. producers. A number of applications listed in Table II are discussed in detail by other authors contributing to this seminar.

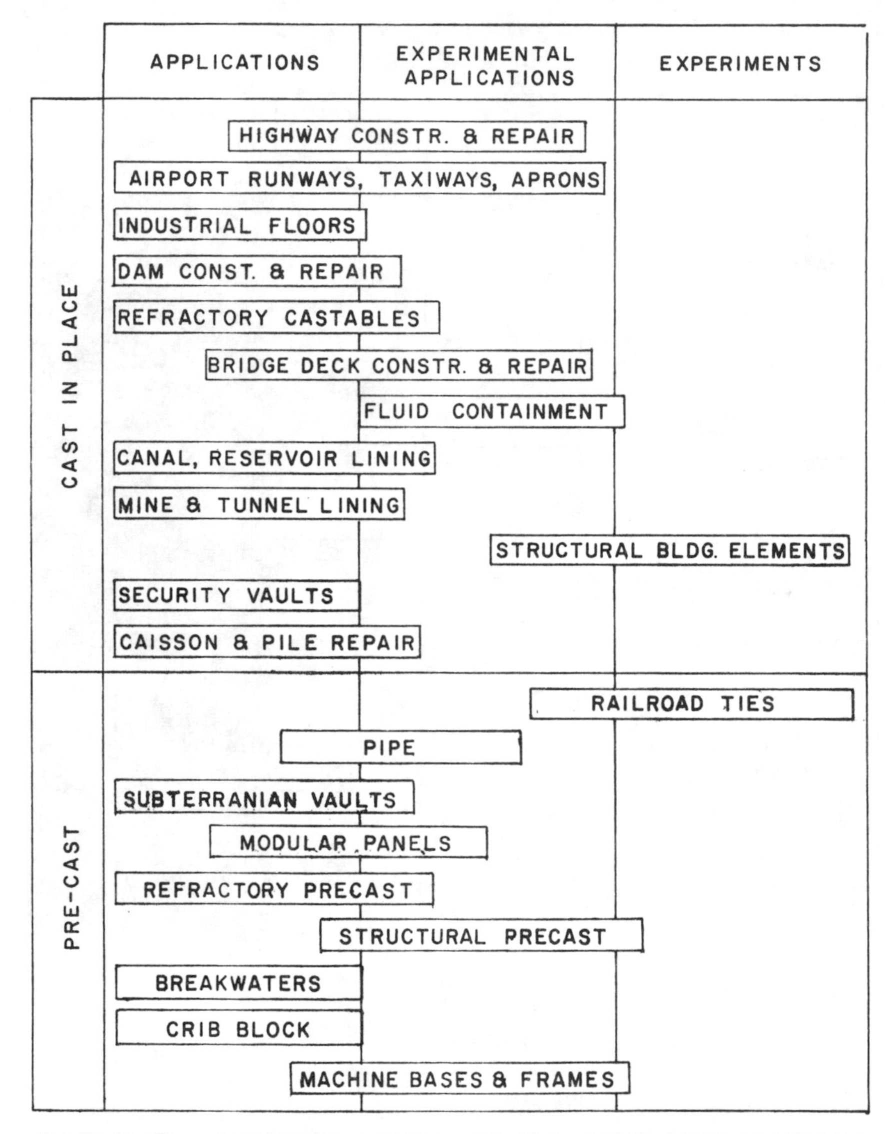

TABLE II CURRENT STATUS OF U.S. SFRC APPLICATIONS

The majority of applications appear to be classifiable as secondary structural. These best exploit those characteristics commonly associated with increased toughness and material ductility such as increased fracture strain, crack control, impact resistance, thermal shock resistance, fatigue resistance and spall resistance. Extensive research, again reported elsewhere during this seminar, is ongoing which will provide opportunity for applications of SFRC with primary structural elements although it must be expected that overall structural ductility will still be supplied by continuous reinforcement in the form of re-bars or fabric.

Some of the applications listed in Table II compete with conventional concrete construction and other construction materials strictly on the traditional basis of in place cost. Others compete on the basis of improved performance by virtue of enhanced composite material properties and increased service life even though initial in place cost may be higher than alternative material choices.

The concept of zoning or prescription casting involves the use of fibrous concrete in only certain areas or zones of either cast in place or precast products. In this way refractories, for example, have been produced which afford protection to otherwise vulnerable fibers in hot zones by casting these fibers away from the hot zones or by protection with layers of ablative or heat damage resistant materials. In structural applications joints or sections of structural elements can be protected as necessary from destructive seismic forces, or even static shear forces by placement of zones of fibrous concrete to, for example, replace or augment more conventional reinforcement.

This discussion of applications has been limited to those which require 0.5 to 2.0 as a maximum volume percentage of fiber. In fact, a practical upper limit on fiber volume when considering material costs and materials handling problems is 1.0 to 1.5 volume percent. New developments in SFRC involving much higher fiber volume loadings requires a separate classification terminology. Fiber volumes of from 2 to 10 percent are possible with creative production technologies. (Details of the research and applications with SFRC using high fiber volume is discussed by another author contributing to this seminar.) The range of fiber volumes from 0.5 to 2.0 percent may now properly be considered 'low' fiber volume loadings. A third category, further down the scale, for fiber volume loading of less than 0.5 percent may be termed 'very low' fiber volume loading. While no significant benefit has been reported for currently commerically available steel fibers introduced at these very low concentrations in portland cement concrete, the very low fiber volume loading has been applied using synthetic fibers of high aspect ratio. These fibers are being tested at 0.08 to 0.5 volume percent. Economic and material performance differences for these three system regimes -- high, low and very low fiber volume loadings -- assure that there will be very little, if any, overlapping applications.

Current Practice for Specifying, Mixing and Placing

The American Concrete Institute (ACI) Committee 544, Fiber Reinforced Concrete, has published ACI 544.3R-84 (5) which contains guidelines for the preparation and placement of SFRC. This document details the solution to practical problems commonly associated with SFRC use. There has been enough experience with SFRC worldwide to have encountered practically all of the potential problems associated with the material for commercially available fibers and the reference is a compilation of that experience for batch casting methods, including pumping. Similar experience has been gained regarding shotcrete applications and this is further discussed by yet another contributor at this seminar and is the subject of ACI 506.1R-84. (6)

SPECIAL APPLICATIONS TECHNOLOGIES

Extrusion

Enough experience has been gained with batch casting and shotcreting SFRC so that these may be considered conventional applications techniques. Batch casting, in this sense, is meant to include pumping and slip forming. Extrusion in the true sense of the word is the process of shaping by forcing material through a stationary die. A research program initiated a decade ago provided evidence that SFRC could be produced by this process. (7,8)

Extrusion through a stationary die of any portland cement based matrix is made difficult with or without the presence of steel fibers due to the problem of dewatering in the high pressure forming process. If water is permitted to be squeezed from the mix then high intergranular stresses accrue which restrict required plastic flow. Flow conditions deteriorate rapidly and the mix solids 'freeze up' into a solid immovable mass in the extrusion chamber. The device is thus clogged and must be disassembled before hydration of the portland cement prevents convenient cleanout of the device. Maintenance of the plasticity of the mix through the die by preventing undesirable dewatering is accomplished by appropriate mix design and the use of mineral and chemical admixtures. Unfortunately, the mineral which best facilitates extrusion of the mix is also a hazardous material, asbestos.

Extrusion with asbestos cement mixtures had been successfully accomplished prior to 1950. The control required on the process, however, discouraged all but a few from developing commerical enterprise utilizing the process. In 1965, a U.S. patent was issued (9) which described how certain chemical admixtures would alter the liquid phase in the mix to substantially reduce dewatering problems. Among the chemicals which performed best in this regard is methyl cellulose which is commerically available as Methocel 60HG and Methocel 65HG from the Dow Chemical Company. The patent also described other chemical admixture alternatives. Figure 2 shows the regions of extrudable mixtures as a function of the amount of chemical admixture by weight percent of mortar and of the total water content. The figure shows that the penalty for achieving extrudable mixes without asbestos is that relatively high amounts of chemical admixtures need to be added with the liquid phase and that a very narrow range of permissible water content allows extrusion. The range of water content is actually so small as to represent almost insurmountable practical production problems. The high pressure extrusion process, therefore, and at least as of this writing, does require the use of asbestos in the mix. The use of 10 to 20 percent by weight of asbestos in the mortar substantially broadens the range of water content and reduces the amount of chemical admixture required in the mix water. Having obtained an extrudable mix little or no difficulty is encountered by the addition of as much as 1.5 volume percentage of steel fibers.

A typical extrudable mix of portland cement mortar contains 11.6% asbestos, 40.6% portland cement, 24.5% sand (dry), 23% water and 0.3% methyl cellulose, by weight. Reference (9) describes a number of other extrudable mix formulas. Dry ingredients are first mixed, then water, with chemical admixtures, if any, is added in solution, and finally fibers are broadcast into the mix. The mix with asbestos is dry in appearance, with the asbestos absorbing and holding the free mix water, and thus a pan mixer with kneading rollers is required to achieve thorough mixing of the ingredients.

The extruding machine may be a piston or auger type. The one pictured in Figure 3 is an auger type machine with a 15 horsepower electric motor driving the 6-inch diameter auger. This is a scaled down version of larger extrusion equipment, with 12-inch diameter augers, which have been used. The smaller extruder is better suited to laboratory scale testing. The auger feed extruder is a continuous process production device in that as long as material is fed into the hopper, continuous, prismatic extrusions will exit the die.

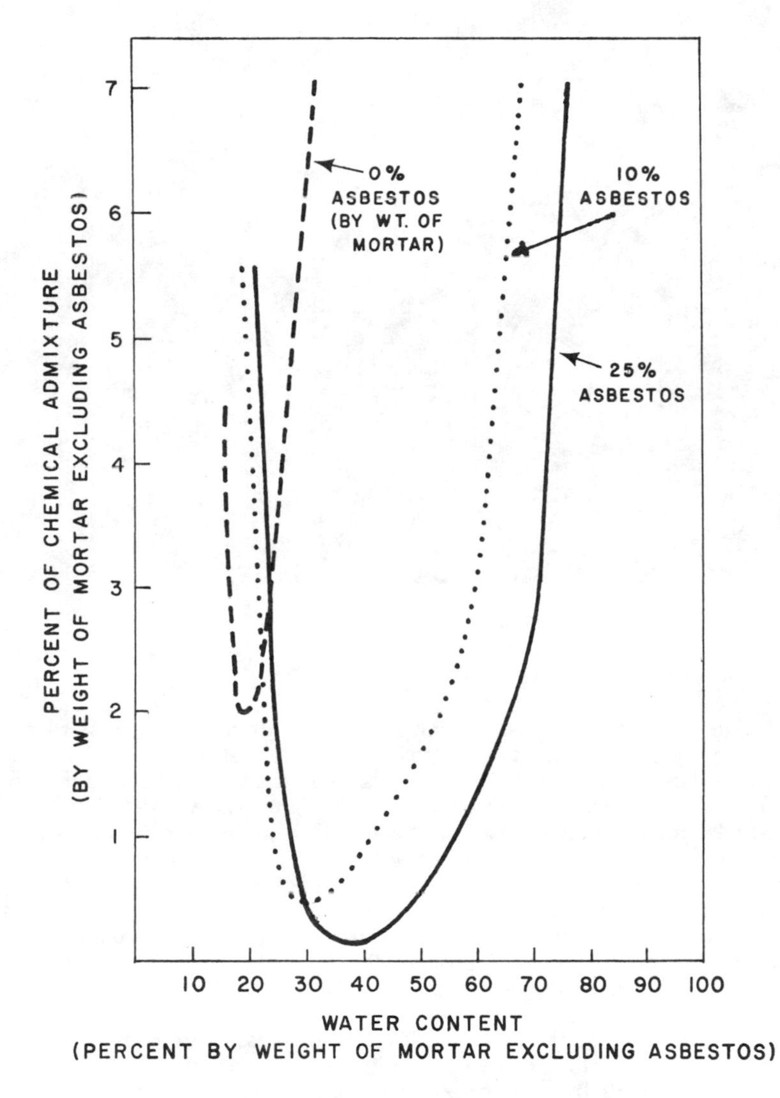

FIG. 2 EXTRUDABLE MIX DESIGNS

Fig. 3 Extrusion Machinery

Extrusion Results

Dies of high cross-sectional aspect ratio are difficult to use because of the tendency to tear off projecting elements or edges as the material exits the die. This, however, is still somewhat a function of the mix design and the size of the auger or piston feed, and can also be addressed by improved die design. Aspect ratios of six have been achieved with high quality production. Shape retention is excellent and although sharp edges may best be avoided, they can be achieved and will remain sharp.

Steel fibrous extrusions provide enhanced modulus of rupture and post cracking ductility material properties when compared to nonfibrous product but there are penalties with regard to surface properties. SFRC extrusions are rough due to fiber popout. This makes handling of hardened product somewhat hazardous.

Comparisons for strength and ductility (toughness) between plain and fibrous extrusions can be obtained by producing extrusions from the same batch by withholding the fibers from the mix until plain extrusions have been obtained. Like cast SFRC, increases in the modulus of rupture found in this way range from 5 to 40 percent but the post cracking performance is more significantly enhanced.

Post cracking performance can be quantified by the toughness index which is the ratio of the area under the load deflection curve of the SFRC samples determined to an arbitrary deformation limit to the area similarly determined for the otherwise identical specimen without fibers. Values thus obtained are between five and six for

SFRC extrusions and indicate directly the benefit gained with the fiber addition, i.e., 500 to 600 percent increases in material toughness.

The low water/cement ratio, high cement content mortar containing asbestos provides a high quality matrix by itself. Incorporation of steel fibers favorably affects resulting extrusions as the pressure forming process provides significant improvement in fiber/matrix bond characteristics and provides a favorable fiber alignment along the axis of extrusions. Figure 4 is a typical autographic trace of load-deflection response for SFRC extrusion. The post cracking sawtooth curve which results corresponds, at sawtooth peaks, to audible sounds of fiber pull-out. The curve shown is for one half inch (12.7 mm) smooth surfaced wire fiber of aspect ratio

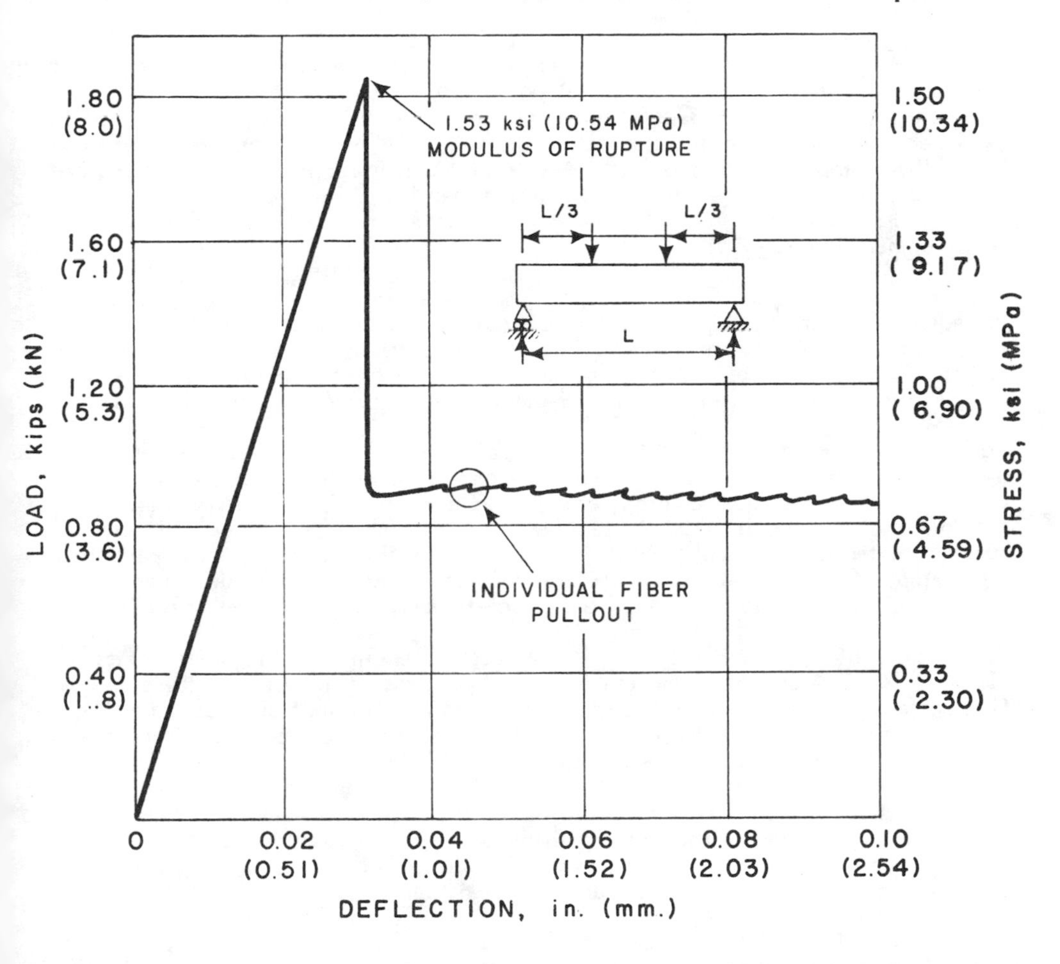

FIG. 4 TYPICAL SFRC EXTRUSION LOAD-DEFLECTION

50. Similar, although admittedly less effective, benefit in terms of the toughness index measure was found for one quarter inch fiber (6.4 mm) of aspect ratio 40. In this latter case, a much finer sawtooth load deflection response was recorded. This behavior is known to be strongly dependent on the pressure forming process as no such behavior is observed when the material is cast into forms. Similar behavior would be

expected in a pressure ramming or forging type production process although this has not been attempted. The achievement of significant post cracking ductility and pull-out behavior indicative of individual fiber bond failure for the short fiber is a noteworthy response. It indicates that low aspect ratio fiber, 40 in this case, is achieving significant fiber/matrix bond strength even though smooth surfaced fibers were used in the trials discussed. This reaffirms the position of dominant importance that interfacial fiber/matrix bond strength has on SFRC performance.

Research has continued with the extrusion process in an attempt to eliminate the asbestos component of the mix. Fibers other than steel have also been used and many of these eliminate the surface popout and handling problems but at the expense of post cracking ductility performance.

Research began at Canada Cement Lafarge of Montreal, Canada, with the support of the Canadian National Research Council in 1983. Not much is known relative to their progress but they have produced nonsteel fibrous extrusions using the same basic extrusion machinery as in the research in the U.S. The dies have been redesigned as have other components of the machine so that much lower extrusion pressures are achieved and no asbestos is needed in the mix design.

High Temperature SFRC

High temperature environments produce thermal gradients and thermal shock loading in addition to sustained high temperature loadings. Under these conditions material toughness and ductility remain the important material performance criteria. In addition, however, the effect of temperature on fiber strength and the effect of both temperature and corrosive gases on fiber durability are important considerations.

For sustained and severe high temperature applications stainless steel alloy fibers are usually most effective. (For certain special applications where short term exposure or short service lives are required carbon steel fibers have been used.) Lankard (10) cautions that the use of stainless steel ought not be considered in a generic sense as specific material alloys will yield different performance characteristics in particular high temperature applications. Therefore, refractory batch mixture design should be based on prototype or sample testing at service temperatures whenever possible.

The results of tests for refractory composite strength and toughness tested at elevated temperatures are reported for the two forms of alloy steel fibers used in U.S. refractory production; melt extraction fibers and slit sheet fibers. (10) These data and subsequent investigation verify that expected reductions in strength and toughness of composites are a function of the particular fiber alloy type. Fiber alloy type also affects individual fiber tensile strength at elevated temperatures which may lead to a fiber tensile strength failure mode of composite behavior instead of the preferred or expected fiber pull-out behavior. Fiber rupture behavior reduces performance in a number of categories of durability in which fibrous refractories are expected to endure such as abrasion and spall resistance.

Applications of SFRC in refractories at 'low' fiber volumes (0.5 to 2.0 percent by weight) are found in virtually all high temperature heavy industries including ferrous metal production, metal processing furnaces, petroleum refining, cement production (kiln) facilities and experimental coal gasification facilities. SFRC has proven to be a cost effective choice and the volume production of such materials has slowly but steadily grown yearly, worldwide.

Lightweight SFRC

Where materials handling problems might exist due to bulk weight of precast elements, lightweight aggregate concrete SFRC exists as a viable alternative. For example, SFRC has found application as mine crib blocks. Tests by the U.S. Bureau of Mines (11) in 1980 led to the conclusion that normal weight SFRC cribbing systems were effective based on system economics at equivalent support capacity when compared to timber cribbing. Further, the volume of cribbing materials of SFRC was very much lower than conventional wood cribbing for equal support capacities, thus producing shipping and handling economies and causing far less blockage of work and ventilation spaces in application.

Research at the University of Alabama (12) used expanded clay lightweight course and fine aggregate to produce lightweight SRFC (LSFRC) for standard testing and for tests involving scaled model cribbing systems. LSFRC in these tests used only 0.7 percent of hooked end steel fiber by volume.

The tests on various crib configurations produced load-deflection response which is geometrically similar to the bell shaped curve which is characteristic to that found by the Bureau of Mines for normal weight SFRC cribbing. Projections to full-scale applications for equivalent crib bearing areas indicate an approximate capacity for LSFRC cribs of 70% of that of normal weight cribs can be achieved. LSFRC cribs in these tests had a unit weight of approximately 90 pcf (1440 kg/m^3) which indicates a reduction in crib bearing capacity which is approximately equivalent to the decrease in unit weight of the SFRC. Reduced unit costs with regard to transportation and handling associated with mine installation make LSFRC a viable alternative in this application.

SUMMARY AND CONCLUSIONS

Growth in experience and application of SFRC in the U.S. has climbed upward in its relatively short two decade history. It has not, however as of this point in its history, achieved the status of anything more than a specialty building material. It does provide a cost effective alternative in a number of applications and it is the only viable materials alternative for some applications. For these reasons, therefore, it is a material that should continue to develop and to stimulate research into the next two decades.

REFERENCES

(1) "State of the Art Report on Fiber Reinforced Concrete," ACI 544.1R-82, American Concrete Institute, Detroit, Michigan, 1982.

(2) Personal Correspondence. Steve Dickerson, Battelle Development Corporation, Columbus, Ohio.

(3) "Standard Specification for Steel Fibers for Fiber Reinforced Concrete," American Society for Testing and Materials (ASTM), A820 (to be published).

(4) Kobayashi, K., "Development of Fiber Reinforced Concrete in Japan," International Journal of Cement Composites and Lightweight Concrete, Vol. 5, No. 1, February 1983.

(5) "Guide for Specifying, Mixing, Placing and Finishing Steel Fiber Reinforced Concrete," ACI 544.3R-84, American Concrete Institute, Detroit, Michigan, 1984.

(6) "State of the Art Report on Fiber Reinforced Shotcrete," ACI 506.1R-84, American Concrete Institute, Detroit, Michigan, 1984.

(7) Zollo, R.F., "Extrusion of Steel Fiber Reinforced Concrete," Journal of the American Concrete Institute, Vol. 101, No. ST12, December 1975.

(8) Zollo, R.F., "Fiber Reinforced Concrete Extrusion," Journal of the Structural Div., American Society of Civil Engineers, December 1975.

(9) U.S. Patent, No. 3,219,467, November 23, 1965.

(10) Lankard, D.R., "Factors Affecting Selection and Performance of Steel Fiber Reinforced Monolithic Refractories," Ceramic Bulletin, Vol. 63, No. 7, 1984.

(11) Anderson, G.L. and T. W. Smelser, "Development, Testing and Analysis of Steel Fiber Reinforced Concrete Mine Support Members," Report R.I.-8412, U.S. Bureau of Mines, Washington, D.C., 1980.

(12) Corbitt, C.L., "An Investigation of Lightweight Steel Fiber Reinforced Concrete Mine Support Cribbing," A Masters Thesis, Department Civil Engineering, University of Alabama, 1983.

STEEL FIBER CONCRETE
US-SWEDEN joint seminar
(NSF-STU)
Stockholm 3-5 June, 1985
S P Shah and Å Skarendahl,
Editors

PRODUCING AND PROMOTING OF SWEDISH STEELFIBERS TO THE MARKET

by Gösta Odelberg , MSc, Spec Products Div, SSAB, Borlänge, Sweden

---

SUMMARY

Some aspects on different fiber-making processes are shown and the fibers' suitability as reinforcing elements in ordinar mortars is discussed. Especially is discussed the behaviour of SFRC when exposed to dynamic loads of repeated drop-weight type, and comparisons of cleavage surfaces for hit beams versus those which are bent to break are made. The dropped weight causes more fiber ruptures than the bending.

Results from own drop-weight tests are presented.

Economics of fiber addition are presented as well as comparisons of disposal reinforcement areas for different fiber sizes.

A presentation of our actual hopes and plans for raising Scandinavian fiber use and our own definition of six different users and their applications will end this paper.

Gösta Odelberg, KTH Bergsingenjör - metallurgy

- responsible for the marketing and promoting of steel fibers within SSAB* since 1983
- responsible for the reuse of slags, waste products and surplus equipment from the Domnarvet steel mill. Production of road materials, fertilizers and powder. 1978-1982
- technical service engineer for heavy plates within Scandia Plate 1969-1978

*SSAB Svenskt Stål AB, Special Products Division makers of wire rod, reinforcement, rail and since 1981 distributors of mechanical and ME steel fibers

---

INTRODUCTION

If, as in our case, we already make conventional reinforcement it is natural to watch all signs from the cement consuming industry in development of the products.

A primary observation is that the addition of fibers to a mortar gives disappointing or small changes to its properties as compression strength and wear. Fibers should not substitute or replace the conventional reinforcement. On the other hand some other properties which normally are not considered for mortars could be obtained and improved by adding fibers into the mix.

The best results are noted for steel fiber reinforced concrete. The market should demand steel fibers for thin wall structures, shotcrete, refractories and specific purposes.

The research for optimal steelmaking methods has led into processes where solidification of molten steel will occur very close to the final shape of product, wire rod for instance.

A virgin wire rod could be spun from the melt in a dynamic process directly in dimensions corresponding to the desired properties in the finished wire. A result of this development is the melt extracted fiber. From the steelmakers' point of view the melt extracting of fibers will promote the development of continuous shape rod making, while the fibers can also be made in the conventional mechanical way from wire or strips.

MARKETING OF STEEL FIBERS

Since the concrete unit price will be raised three times or more by addition of sufficient lots of steel fibers, it is of great interest to produce a useful fiber at lower costs. Melt-extracted fibers of carbon steel could be produced at 2/3 of costs of mechanical fibers. They are still not available in market.

The necessary annual lot for a production plant could be in the range 3-4000 tonnes, but the oxide-protection difficulties in producing them and also embarrassing results compared to mechanical fibers in SFRC-tests have retarded the process.

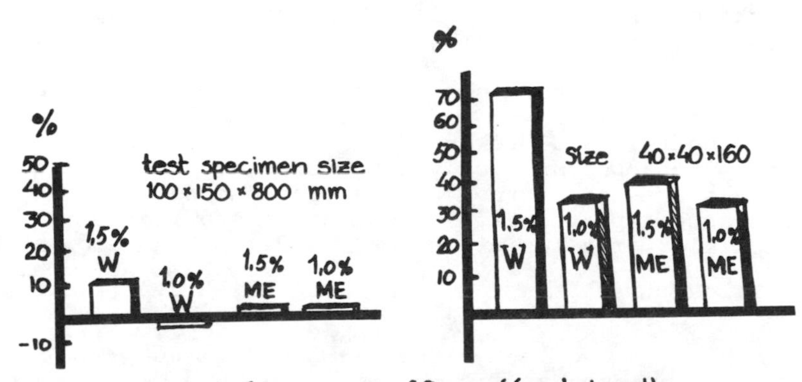

Fig 1. Improvment of flexural bend strength for two types of fiber in two contents in large specimen (left) and small specimen (right).

The marketing of steel fibers differs from the marketing of conventional reinforcement in almost all ways. In our attempt to market steel fibers we must encourage any cementite consumer in his developing of new types of elements, overlays and precastings. In the specific fiber technology the potential customer is not fed with SFRC knowledge from his own branch, where much knowledge is instead hidden to protect own expensive experience. Especially do we think that much work has been wasted on comparing tests of compression, where the concrete itself already has its best properties. The good of a representative and repeatable toughness measuring method is quite obvious. This method must correspond to the advantages and properties of SFRC which can be studied in practice.

DIFFERENT FIBERS

Elements of SFRC can be hit, struck, vibrated and fatigued in practice and this leads to the conclusion that SFRC allows any designer or user to utilize concrete instead of wood, plates or PVC.

As mentioned we have a serious interest in comparison between melt extracted fibers and mechanically produced fibers, in our case cut wire. This interest is combined with the ambition to serve the customers with measured toughness or impact resistance figures for correct calculation of sufficient additions of fibers and necessary surface bond in the mortar.

IMPACT RESISTANCE - VISIBLE CRACKS

Almost all examinations of properties consider the matrix behaviour till the point of total break or collapse. The improvement of impact resistance in the fiber reinforced concrete element during tension, where no visible or measurable cracks occur, is not easily expressed by destructive or even non-destructive testing.

The balcony front (1" thickness shot or cast) resists hits with a hammer and mechanical wearing is good, but still it will be judged dangerous and broken if it is supposed to serve also after the appearance of visible cracks. Its improved function is present in the "before crack" condition. An impact resistance repeated drop weight test, showing the resistance till "first visible crack" is described below.

TEST EQUIPMENT

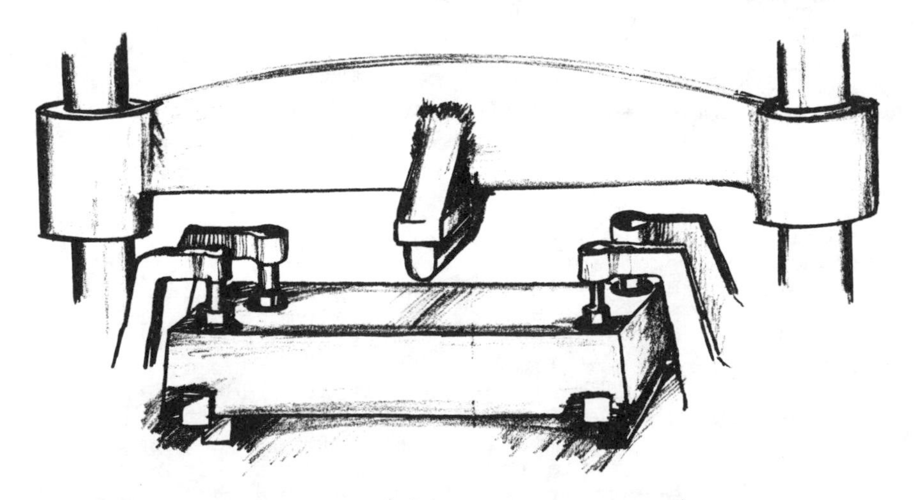

Fig 2 Drop weight test equipment

We simply built a suitable impact/toughness drop weight hammer in which the energy should correspond to the strength of the beams or specimen tested. Ideas to this hammer were published by Clifton J.R. and Knab L.I. (Cement & Concrete Research vol 14 1983). They indicate a hammer weight of 5 kg and fall heights of 1000 mm or more for slab-like specimens of 610 x 610 x 76 supported by frame around all four edges.

We have tested smaller specimens of 40 x 40 x 300 and 60 x 60 x 300 as well as 80 x 60 x 300 in a hammer, where the specimens have been supported only transverse the ends, fastened by clamps and hit from 30-50 mm with a 8 or 6 kg heavy drop-weight. The specimens are cast to size or cut from larger pieces, mainly from large shotcrete samples. The tests, i.e. the height of the repeated drops, are adjusted so that the specimens will withstand at least 25 drops before the first visible crack appears. Finally the specimen falls apart and the two halves have been subject to flexure bending test.

Especially for concrete mix suitable for shotting we have tested longer series of sawn specimens, where two types of fiber have been mixed in. In order to see if the flexural tensile corresponds to the "impact", number of drops to first crack, a diagram is drawn. It shows two levels of flexural strength depending on choice of fiber and also probability of drop-weight resistance if one type is preferred. It is obvious that the impact properties vary a lot even within one piece of concrete divided into a number of specimens, where all visible and measurable factors are similar.

Our drop-weight testing can, as practiced or modified, give useful information on specific differences in SFRC due to type and amount of fiber. The program will continue drop-weight testing, cast specimen with variation of fiber length and fiber content. The equipment has been used in a scholar job, where ME-fibers have been compared to mechanical fibers in two fractions, 1.0 and 1.5 volume %. The mechanical fibers are better than the ME and the higher addition is much better than the lower one.

A conclusion may be that the reinforcing of a mortar by fibers can be as effective with the more easy-mixed plain fibers provided they exceed a certain length. For fibers with a structured surface this length may be even shorter.

Studies of broken or separated test specimens give information of the fiber behaviour in the cleavage.

By the specimens we have tested we found that the number of fibers visible in the cleavage per square unit was indicating homogeneous fiber presence in matrix.

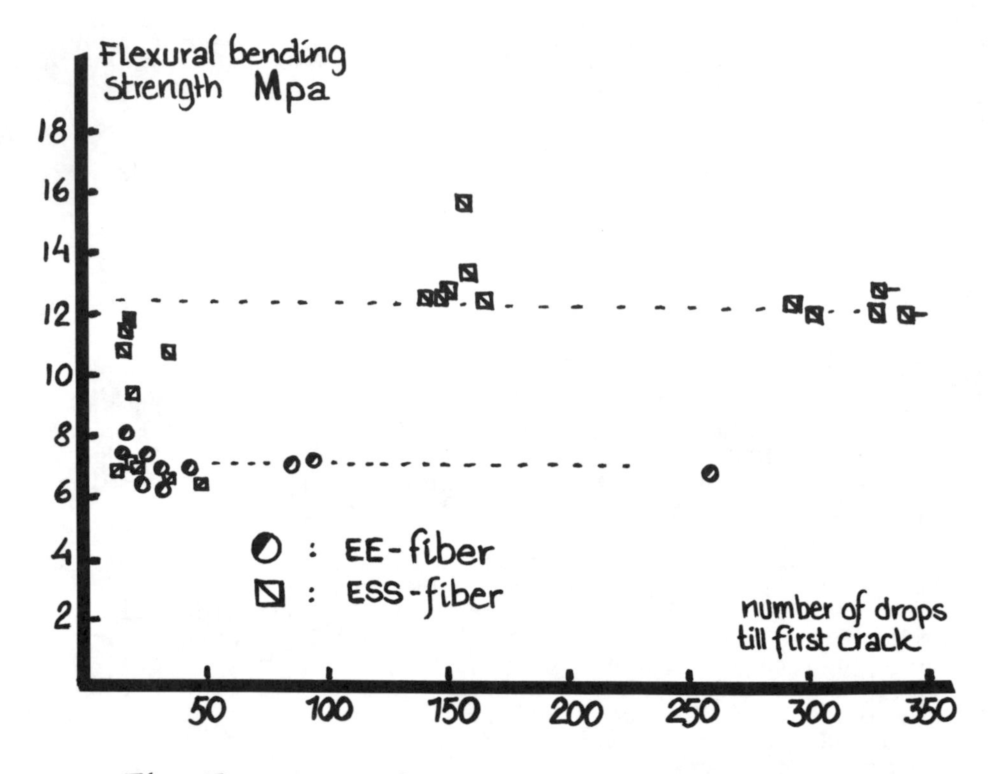

Fig. 3 dropweight and flexural testing of shotcrete

For the drop-weight tested specimens the fibers were broken with the matrix cleavage in the tensioned or stressed zone, while they were unbroken in neutral or compressed zone, where they later - by specimen separation - were drawn out. When these separated halves were tested by flexural bending test the major part of the fibers were drawn out unbroken also in the stressed zone. This observation is done for the series we drop-weight tested containing 18 mm fibers abt 1 % volume.

SIX FIELDS OF USE

Our interest in SFRC has led us into knowledge which the different fields of fiber consumption are and also ideas of where a raised consumption can be expected. We also understand that pavements in SFRC are well-established outside Scandinavia.

A SFRC made in turbine mixers for element making, where the element is cast, vibrated and treated carefully.

B SFRC safe deposit equipment, also turbine mixed with rising content of fibres and special cement or paste for optimal destruction resistance.

C Refractory industry. Higher content of heat-resistant fibers in alloyed grades. Turbine-mixers are used and the valuating of process is somewhat different than for SFRC. This field is well-established and gives good improvements. ME-fibers are commercially used.

D SFRC shotcrete instead of conventional vibrated casting. Shotcrete with addition of fibers can be sprayed directly to a single mould wall to form an element. This type of shotcrete overlaps the field under A.

E SFRC shotcrete where the presence of steel fibers eliminates the necessity of reinforcement nets and a two-layer spraying. Steel fiber reinforced shotcrete gives a lower price per m^2, reinforced rock as well as speeded up capacity per shift or day. This field is so far the major fiber consumer in Scandinavia. The fibers will not exceed 1" of length.

F SFRC mixed in concrete for capacitive production for pavements, overlays, road surfaces and aircraft fields etc. The concrete can be reinforced with larger fibers and can even be added into a rotating truck mixer.

F (F_1) A certain field is to stabilize bitumenous masses with steel fibers. This has been subject for several tests, where the results will soon verify the good of fiber addition to the mix.

Generally it can be assumed that the track deformation in the pavement can be reduced by 50 to 100 per cent if 1.5 % (weight!) of steel fibers is mixed in the bitumenous skeleton.

How do results obtained in practice correspond to the optimum laboratory results and which type and amount of fibers are the best for real improvement of actual properties?

At first it is noted that a length/diameter relation (aspect ratio) of abt 100 is desired. In practice this relation is not realistic but for some specific cases of type F, and for the cases where fibers are cut or injected immediately in the spraying or casting moment.

In Scandinavia there has been a development on such systems for a long time, but no system has yet a capacity that can compete with especially the wet method with mixing of prepared fibers together with the concrete.

Fig. 4 Shotcrete - sprayed (left) and cast specimen

FIBERCOST (SEK) PER REINFORCMENT AREA AT A CERTAIN ADDITION LEVEL			
fiber	PRICE SEK/kg	m²/kg	SEK/m² area
Ø 0,4	7,50	12,5	0,60
Ø 0,5	7,00	10,0	0,70
Ø 0,6	6,50	8,3	0,80
Ø 1,0	5,00	5,0	1,00
□ 0,3× 0,4	8,00	14,5	0,55

In practice the chosen length of fiber is limited by risk of inhomogeneous mix. The turbine mixing allows fibers about 40 mm of length, related to certain levels of fiber content. Concrete mixed in a station and/or in a drum-mixer is limited to shorter fibers in the mixing sequences and way of wetting and plastising by additors.

For longer fibers, used for pavements, there seems to be more liberal limitations by station-mixed concrete. In figure **5** I have tried to exemplify how these fields of use (A-F) can be placed in areas due to fiber length and fiber volume.

From many users of SFRC we have learnt that the ease of mixing and the shape and sharpness of fiber ends are important. A disadvantage by the end-hooked fibers is the difficulty to avoid balls, and the sharp ends of almost all cut wire fibers are against them. The fibers (cut from band) with enlarged ends have nicer shape of end, but are bound to certain length since the fibers are cut transversely.

See photo below showing examples of differently shaped ends.

The advantages of the enlarged hooked ends on steel fibers are obvious in all studies where required tension energy is calculated for a matrix. In practice, though, the effort is somewhat debatable since the fibers adhere along the entire length and the considering with ideal sliding fiber with a fiber-long elasticity between its hooked ends is not applicable.

For fibers exceeding a certain length the adhesion between fiber surface and paste is the major improving factor for SFRC, implying a not so elastic but impact-resistant behavior in the tension area before first visible crack. This certain length is not empirically found by property examination, but is well-defined in practice, especially for shotcrete applications where the length 3/4" or 18-20 mm has been widely used in Scandinavia and elsewhere because of ease in mixing and ejecting.

The adhesion between fiber surface and hardened cement paste is sufficient already for plain cut drawn wire, and SFRC using such fibers in elements and pavings are serving its purpose quite well, even if flexural bending tests are not very impressing compared to plain concrete beams. In order to improve the adhesion the plain fiber can be surface-structured or profiled. Flexural bend tests show the best results for such profiled fibers and end-hooked fibers.

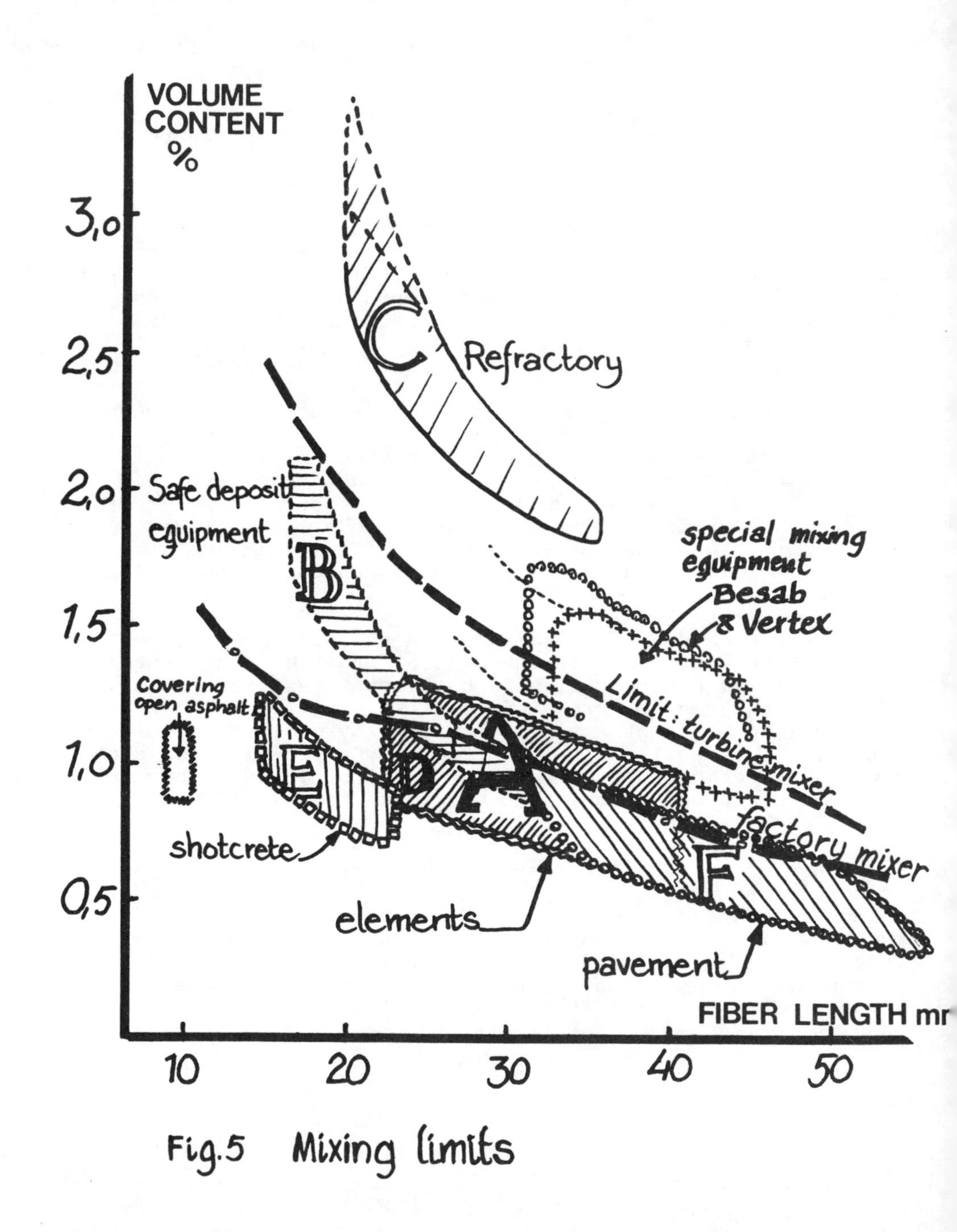

Fig.5 Mixing limits

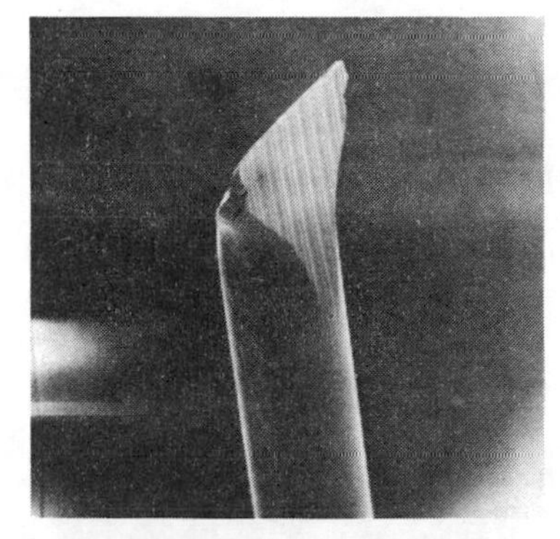

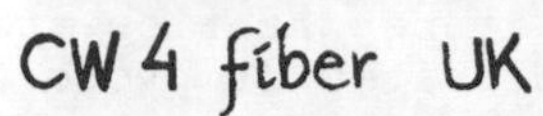

CW 4 fiber UK

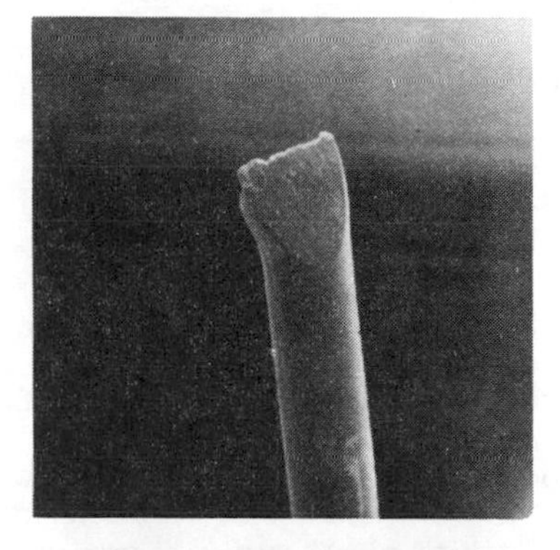

unknown Japan

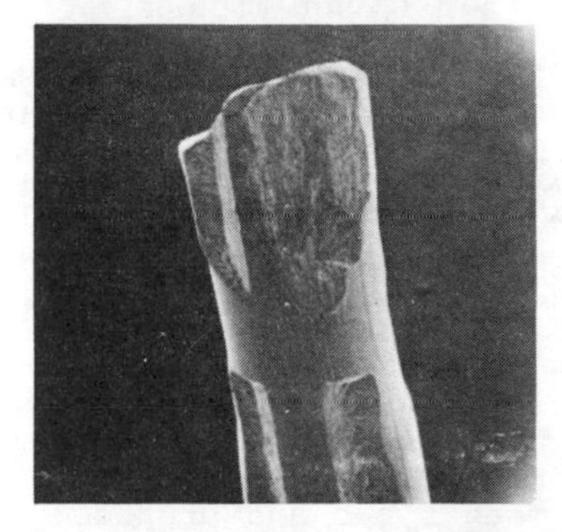

Ess-fiber Sweden

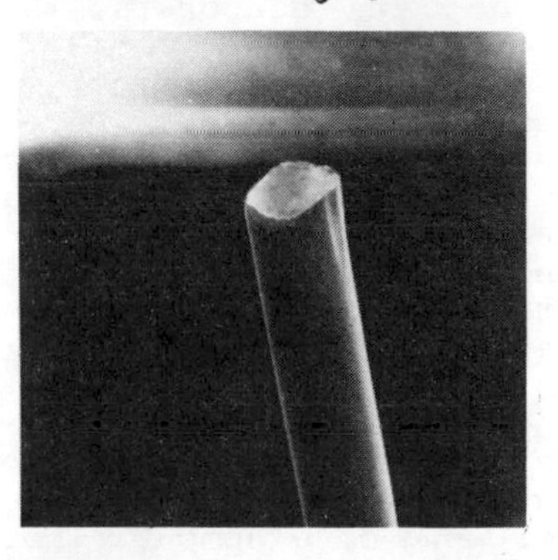

Trefill-Arbed-Belgum

EE-fiber Australia

Fig. 6 Various sharpness on fibers

SURFACE REACTIONS

When using steel fibers it is important to put attention to the fact that a wide metal area is exposed to a alkaline reactive solution. This wide area per unit of weight is, of course, the condition for a good reinforcement effect, but the bond between fiber surface and paste can easily be disturbed or even ruined if reaction conditions are present. A clean and fresh steel surface is binding well in the paste and is probably not decreasing with time in aging or even carbonated mortars, but a coated surface can react if the coating contains e.g. zinc or other reactive metals. This is a contradiction, since the handling, transportation and storing of fibers is easier if they are coated or protected, but the best reinforcing is obtained by clean fibers.

The fiber making from cut wire is faster if the wire reduction can be lubricated by a pure Zn-coating or brass coating (cord wires). This coating must, though, be treated with a alkaline-resistant chromate or similar. Some cements do naturally contain sufficient lots of such protective chromate, but some do not. The effects can be easily measured on water-pressure tests of beams. The penetration in a water-tight mortar is clear if the Zn-Ca(OH)-reaction has occured.

Our experiences after five years of researching and marketing steel fibers show a safely but rather slowly growing consumption. The steel fiber reinforced products are established after long development. There are many types and producers of steel fibers and the field is carefully covered by patents. We have had a study performed by CBI last year and this study is concluding the same observations.

The use of fibers for reinforcing pavements in Scandinavia is still on an undecided level. The cement-producing industry is anyhow very urgent to market solutions, where cement-based pavements and layers are competing bitumenous masses, which so far have been widely used in our regions. This is a fine occasion for SFRC to show its advantages in this field and there is a lot to learn from other experiences in Europe as well as in USA.

REFERENCES

(1) Nordforsk, Fiberbetong (fiber concrete), project committee for FRC material, "Part reports and main report", Stockholm 1977

(2) Kobayashi, K, "Development of fibre reinforced concrete in Japan", Construction Press 1983

(3) Clifton, J.R. % Knab, L.I. "Impact testing of concrete, Cement & Concrete Research vol. 14, 1983

(4) Lind, H & Broström, H, "Impact strength of SFRC comparison of two fibers in two contents, Chalmers Institute of Technology 1985

STEEL FIBER CONCRETE
US-SWEDEN joint seminar
(NSF-STU)
Stockholm 3-5 June, 1985
S P Shah and Å Skarendahl,
Editors

PREPARATION, PROPERTIES AND APPLICATIONS OF CEMENT-BASED COMPOSITES CONTAINING 5 TO 20 PERCENT STEEL FIBER

by David R Lankard, Dr, Lankard Materials Laboratory, INC, Columbus, Ohio, USA

ABSTRACT

Slurry-infiltrated-fiber-reinforced concrete (SIFCON) is preplaced-fiber concrete (analagous to preplaced aggregate concrete) with the placement of steel fibers in a form or mold, or on a substrate, as the initial construction step. The preplaced fibers are then infiltrated with a fine-grained, hydraulic cement-based slurry.

Laboratory and field experience has shown SIFCON to be a unique concrete material, combining high strength with high ductility. Its ability to resist cracking and spalling in many static and dynamic loading situations is far superior to conventional SFRC and to conventionally reinforced portland cement concrete. Its ability to be constructed as massive sections as well as in very thin sections, permits consideration of a broad range of potential applications.

SIFCON technology has now gone beyond the laboratory stage with experimentation in pavement and bridge repair and rehabilitation, security concrete, refractory concrete, precast concrete products, and explosive-resistant structure applications.

David R. Lankard is President of Lankard Materials Laboratory, Inc., Columbus, Ohio, and currently serves as chairman of American Concrete Institute Committee 547, Refractory Concrete and is a member of Committees 227, Radioactive Waste Management, and 544, Fiber Reinforced Concrete. He received his B.S. in geology from Indiana University in 1959, his M.S. in ceramics from the University of Illinois in 1965 and his Ph.D. in ceramic engineering from Ohio State University in 1970. Du ng 1965-77 he was with Battelle, Columbus Labs. Dr. Lankard is an adjunct assistant professor in the Dept. of Ceramic Engineering at Ohio State. He has been involved in the research, testing, and failure analysis of refractory and portland cement concretes since 1965.

BACKGROUND

In preparing conventional steel fiber reinforced concretes, the steel fibers are mixed along with the other concrete ingredients (cement, aggregates, and water) using conventional mixing equipment. As the fiber concentration is increased along with fiber aspect ratio, it becomes difficult to mix and place these materials. In practice, it has been found that the amount of fiber must be kept under 2 volume percent and aspect ratio must be kept under 100. This situation places bounds on the improvements in the engineering properties of concrete (flexural strength, flexural toughness index, impact resistance, fatigue resistance) that can be gained through the use of steel fibers.

In 1978, Lankard Materials Laboratory began an investigation of means to incorporate larger amounts of steel fibers in steel fiber reinforced cement-based composites. The result of this investigation was the development of slurry infiltrated fiber reinforced concrete (SIFCON) in which steel fiber contents up to 20 volume percent are attained.

Laboratory and field investigations of SIFCON over the past 7 years have demonstrated the constructability and the outstanding engineering properties of this unique cement-based composite material [1-8].

PREPARATION OF SIFCON

SIFCON is preplaced-fiber concrete (analagous to preplaced aggregate concrete) with the placement of steel fibers in a form or mold, or on a substrate, as the initial construction step. Fiber placement is accomplished by hand or through the use of commercial fiber dispensing units. The steel fiber content achieved is dictated by the packing density of the fiber which, in turn, is controlled by fiber aspect ratio and length. For a given fiber, slightly higher loadings are achieved if the mold or form is lightly vibrated during the fiber filling step. Using commercially available steel fibers, fiber loadings of 5 to 20 volume percent have been achieved.

The placement of steel fibers in a 2 ft. x 2 ft. x 2 in. slab mold is shown in Figure 1. As might be expected, when steel fibers are "rained" onto a substrate or into a mold, a preferred fiber orientation occurs. The orientation is essentionally two-dimensional, perpendicular to the gravity vector. The orientation effect is more exaggerated with some fibers than with others. In general, there is a trend toward a three-dimensional fiber orientation that accompanies decreases in fiber diameter and aspect ratio.

FIGURE 1. PLACEMENT OF STEEL FIBERS IN A MOLD - THE FIRST STEP IN THE PREPARATION OF SLURRY INFILTRATED FIBER REINFORCED CONCRETE (SIFCON).

The fiber orientation phenomenon must be taken into account when designing field installations of SIFCON or in preparing laboratory specimens.

The preparation of test specimens of SIFCON requires special considerations relating mainly to the desirability of avoiding nonuniform fiber distributions and of avoiding unfavorable fiber orientations.

In preparing molded cylindrical specimens of SIFCON, edge effects can occur (as illustrated in Figure 2) in which fiber density near the surface of the specimen is lower than that in the interior. Additionally, a disproportionate number of fibers may align vertically (parallel to the long cylinder axis) along the outer surface.

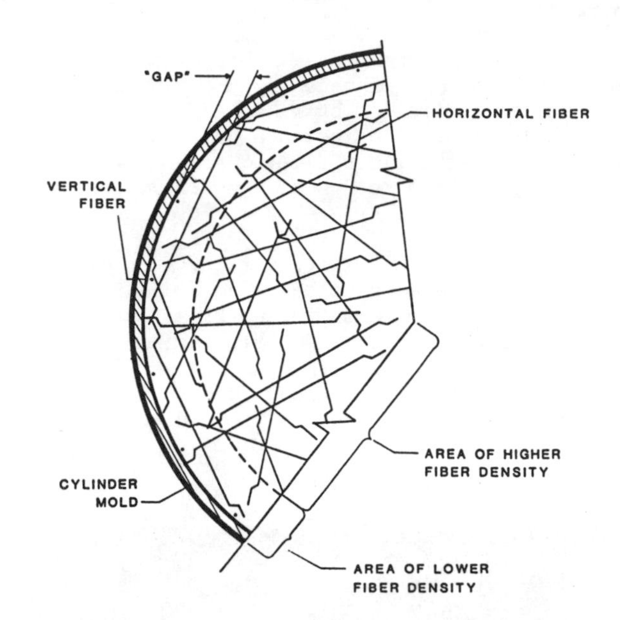

FIGURE 2. FIBER ORIENTATION "EDGE EFFECTS" IN A MOLDED SIFCON CYLINDER SPECIMEN

A preferred, but more expensive method of preparing cylindrical SIFCON specimens is shown in Figure 3.

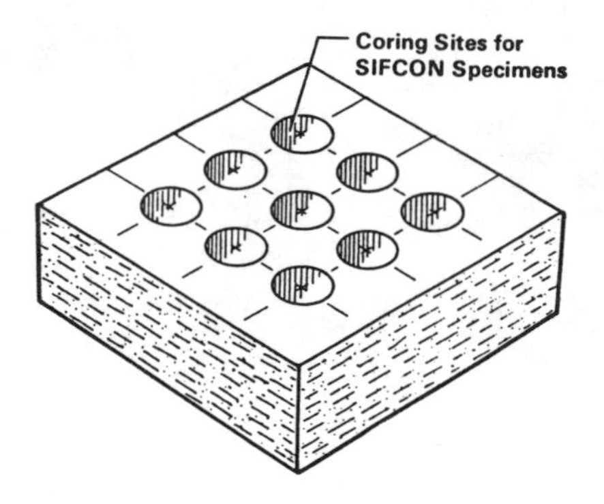

FIGURE 3. ILLUSTRATION OF MANNER IN WHICH CYLINDRICAL CORE SPECIMENS CAN BE TAKEN FROM A CAST SIFCON SLAB

Core specimens, drilled from precast slabs, eliminate both of the problems just discussed for the molded cylindrical specimens. However, even with core specimens, consideration must be given to coring direction with respect to fiber placement direction as this can dramatically affect the orientation of fibers in the core (as shown in Figure 4). Core specimens molded and loaded in the same direction (i.e., the gravity vector) may exhibit twice the compressive strength of specimens molded in one direction with the load being applied perpendicular to that axis.

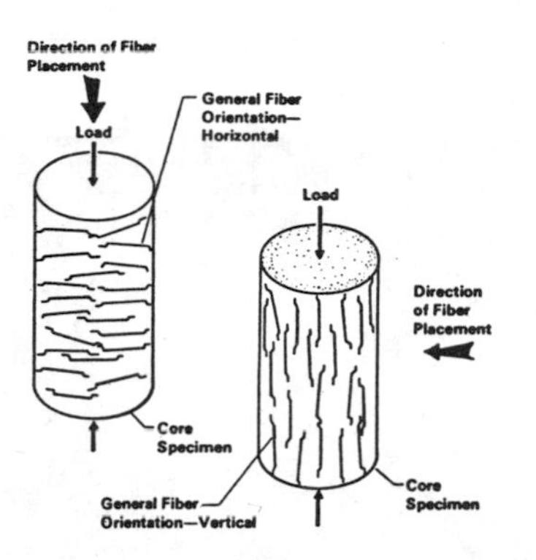

FIGURE 4. ORIENTATION OF FIBERS IN CORED SIFCON SPECIMENS AS INFLUENCED BY CORING DIRECTION WITH RESPECT TO FIBER PLACEMENT DIRECTION

A similar situation with respect to fiber orientation occurs for SIFCON beams tested in flexure. Figure 5 shows fiber orientations resulting from preparation of SIFCON specimens as beams in one instance and as columns in another. The flexural strength of SIFCON specimens molded as beams can be two to three times higher than that of specimens molded as columns.

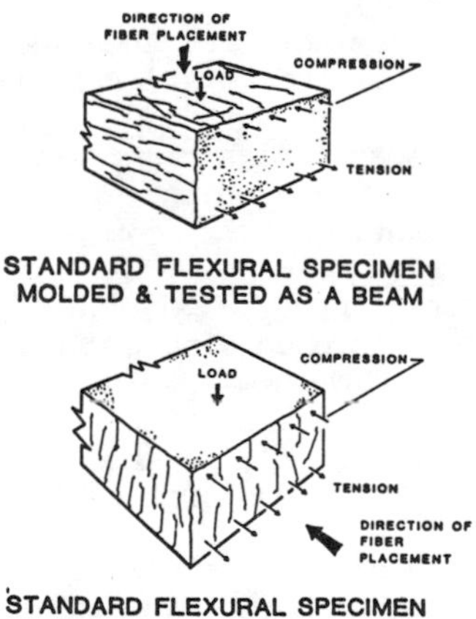

FIGURE 5. EFFECT OF FIBER PLACEMENT DIRECTION ON FIBER ORIENTATION IN TEST BEAMS MOLDED AS COLUMNS AND AS BEAMS

Once the steel fibers have been placed on a substrate or in a mold, then are infiltrated with a fine-grained, cement-based slurry. The infiltration step is accomplished by (1) simple gravity-induced flow, (2) gravity flow aided by external vibration, or by (3) pressure grouting. Slurry infiltration of a 3.81 cm (1-1/2 in.) thick tilt-up panel, by gravity-induced flow, is shown in Figure 6.

FIGURE 6. CEMENT-BASED SLURRY INFILTRATION OF PREPLACED STEEL FIBER USING SIMPLE GRAVITY-INDUCED FLOW

The choice of infiltration technique is dictated largely by the ease with which the slurry moves through the packed fiber bed.

Both cement pastes (cement, flyash, water) and mortars (cement, flyash, fine sand, water) have been successfully infiltrated. When mortars are used, grain sizing of the aggregate phase in the slurry must be such that a minimum of particles exceed the smallest opening in the packed fiber bed. Obviously, if this condition is not met, the fiber bed becomes clogged with aggregate particles and further infiltration is impossible.

In most of the work done to date, high-range water-reducers (superplasticizers) have been used to provide a suitable slurry viscosity while maintaining a low water-cement ratio.

During the slurry infiltration step, a portion of the exposed packed fiber surface is intentionally not covered with slurry. This procedure permits the escape of air contained below the casting surface. Additionally, it is possible to verify that slurry infiltration has been complete as slurry rising from the bottom to the top of the packed fiber bed begins to show in this uncovered fiber area. This condition is illustrated in Figure 7.

Once the slurry infiltration step is complete, the curing of SIFCON is the same as for any hydraulic cement-based material.

ENGINEERING PROPERTIES OF SIFCON

Property measurements have included,

1. Flexural strength and load-deflection behavior in flexure.
2. Compressive strength and load-deflection behavior in compression.

FIGURE 7. SLURRY INFILTRATION STEP OF A SIFCON BEAM SHOWING PORTION OF PACKED FIBER BED THAT IS INTENTIONALLY KEPT FREE OF SLURRY COVERAGE (to permit expulsion of entrapped air and to permit verification of complete slurry infiltration)

3. Modulus of elasticity (in compression).
4. Drying shrinkage strains.
5. Impact resistance.
6. Fatigue resistance.

Unit Weight

The unit weight of SIFCON depends both on slurry unit weight and on steel fiber loading. For a slurry unit weight of 1.92 gm/cc (120 lb/ft^3), SIFCON unit weights varied from 2.16 to 3.13 gm/cc (135 to 195 lb/ft^3) for steel fiber contents ranging from 5 to 20 volume percent (see Figure 8).

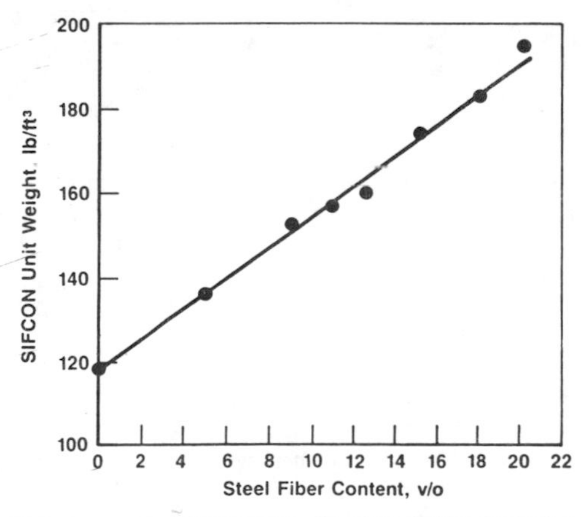

FIGURE 8. EFFECT OF STEEL FIBER CONTENT ON THE UNIT WEIGHT OF SIFCON (Cement/Flyash Slurry)

Flexural Strength Properties

Flexural strength measurements have been made on moist-cured 10.2 x 10.2 x 35.6 cm (4 x 4 x 14 in.) beams using third-point loading and a 30.5 cm (12 in.) span.

One day flexural strength values for SIFCON typically have ranged from 14.0 MPa (2000 psi) to 42.1 MPa (6000 psi). Twenty-eight day values typically have ranged from 21.0 MPa (3000 psi) to 63.1 MPa (9000 psi). Ultimate flexural strengths as high as 84.2 MPa (12,000 psi) have been achieved.

Flexural load-deflection curves for SIFCON containing 5 to 14 volume percent steel fiber are shown in Figure 9 where a comparison is made with conventional SFRC. In addition to high values of ultimate flexural strength, SIFCON demonstrates a very large strain to failure.

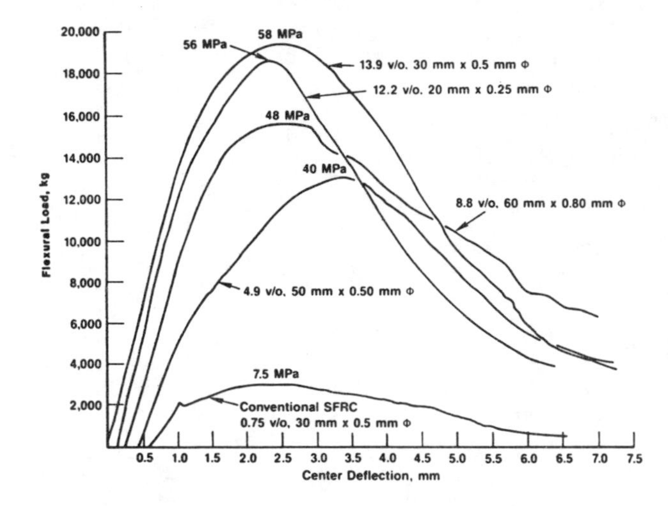

FIGURE 9. TYPICAL LOAD-DEFLECTION BEHAVIOR IN FLEXURE FOR SIFCON CONTAINING 4.9 TO 13.9 VOLUME PERCENT BEKAERT STEEL WIRE FIBER (deformed ends) AND FOR CONVENTIONAL SFRC. 10.2 x 10.2 x 35.6 cm (4 x 4 x 14 in.) BEAM SPECIMENS (28 day fog cure), THIRD POINT LOADING, 30.5 cm (12 in.) SPAN

Figure 10 illustrates the mode of failure for a SIFCON beam specimen at various points in its loading history. Cracking is observed with the unaided eye at a load considerably below the ultimate load (Figure 10A). At a deflection corresponding to ultimate load (Figure 10B), about five cracks have occurred in the specimen, but none of the cracks have yet opened to any great extent. On the descending portion of the load-deflection curve (Figure 10C), the cracks have advanced upward into the specimen with the

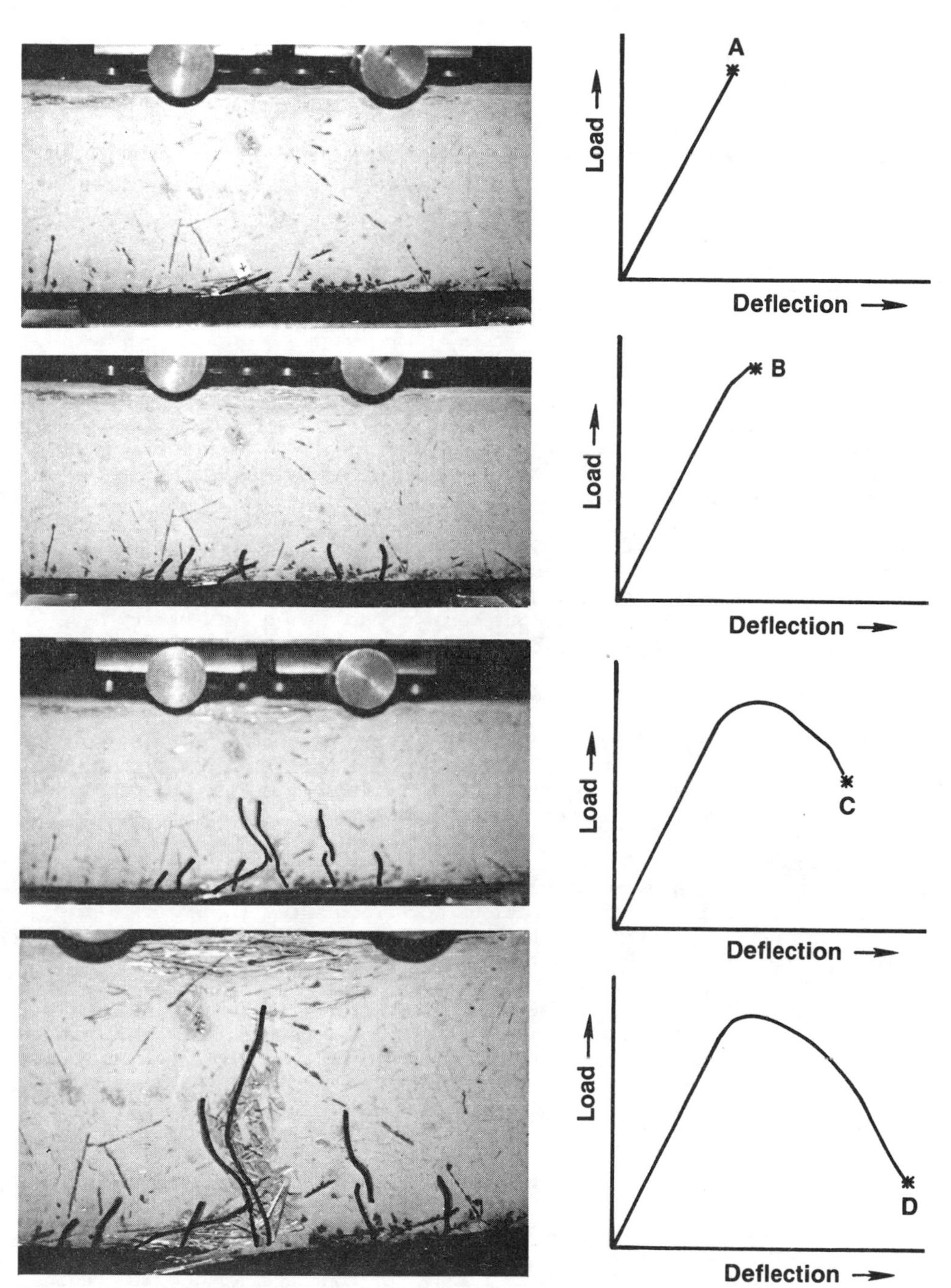

FIGURE 10. APPEARANCE OF SIFCON (10.2 x 10.2 x 35.6 cm (4 x 4 x 14 in.) FLEXURAL BEAM SPECIMEN CONTAINING 13.4 VOLUME PERCENT, 30 mm x 0.50 mm DIAMETER STEEL WIRE FIBER (Bekaert Steel Wire Corp.) AT VARIOUS POINTS IN ITS LOADING HISTORY. CRACKING HIGHLIGHTED.

failure occurring principally in only one of the cracks. At the end of the test, at a deflection of around 0.76 cm (0.3 in.) - Figure 10B), the failure crack has opened considerably.

In general, as with conventional SFRC, the ultimate flexural strength and flexural toughness index of SIFCON increase with increases in fiber quantity and fiber aspect ratio.

Compressive Strength Properties

Compressive strength measurements have been made on both molded and cored SIFCON specimens.

For conventional SFRC with fiber contents not exceeding 2 volume percent, the compressive strength of the fiber-containing concrete typically does not exceed that of the same concrete without fibers. For slurry infiltrated fiber concrete (SIFCON), dramatic improvements in compressive strength are achieved with some fiber types.

The portland cement/flyash slurry used in much of the laboratory and field work to date has a 1 day unconfined compressive strength of 35.1 to 42.1 MPa (5000 to 6000 psi) and a 28 day compressive strength of 77.1 to 91.7 MPa (11,000 to 13,000 psi). SIFCON specimens prepared with this slurry have exhibited unconfined 28 day compressive strengths of 63.1 to 189.3 MPa (9000 to 27,000 psi). Overall, unconfined compressive strengths greater than 210.4 MPa (30,000 psi) have been attained.

Figure 11 shows load-deflection behavior in compression for a number of SIF-CON composites. It is worth emphasizing the fact that the deflection scale in Figure 11 represents a 2.54 cm (1.0 in.) total deflection full-scale. The failure mode of SIFCON under an unconfined uniaxial compressive load is a gradual radial deformation prior to failure across a shear plane. Figure 12 shows examples of SIFCON specimens which were loaded in unconfined compressions to very high deflections. In one case, the specimen has remained nominally intact with large deflections occurring along the shear plains. In the other case, a catastrophic failure has occurred with breakage of all fibers in the shear plane.

In general, the highest values of compressive strength are attained in SIFCON when relatively small diameter fibers (<0.051 cm - 0.020 in.) are used at relatively high concentrations (>10 volume percent).

Modulus of Elasticity

As is obvious from the compressive load-deflection curves shown in Figure 11, it is very difficult to define the elastic range for SIFCON composites. Under unconfined compressive loadings, significant deviations from linear behavior occur at stress levels as low as 15 to 20% that of the ultimate stress. Additionally, considerable flaking of the specimen surface begins to occur at low stress levels.

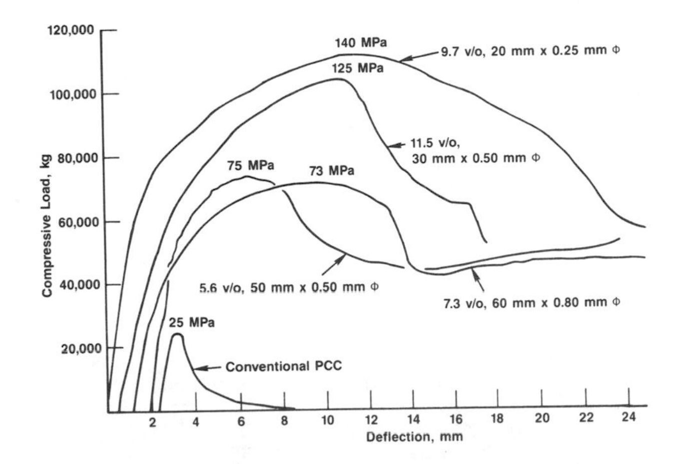

FIGURE 11. TYPICAL LOAD-DEFLECTION BEHAVIOR IN COMPRESSION FOR 10.2 x 17.8 CM (4 x 7 IN.) CYLINDRICAL SIFCON SPECIMENS (28 day fog cure) CONTAINING 5.6 TO 11.5 VOLUME PERCENT BEKAERT STEEL WIRE FIBERS (deformed ends).

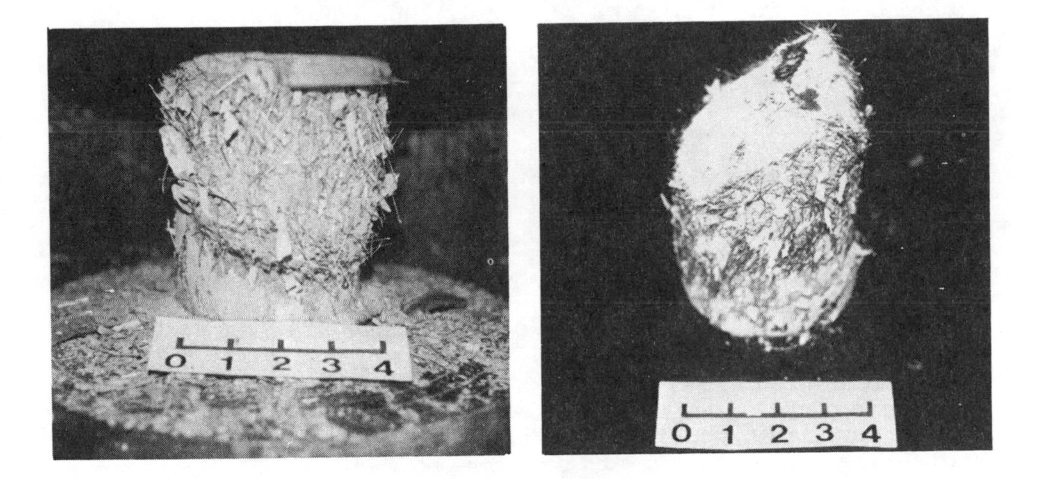

FIGURE 12. EXAMPLES OF 10.2 x 17.8 CM (4 x 7 IN.) CYLINDRICAL SIFCON SPECIMENS FOLLOWING UNCONFINED COMPRESSIVE LOADING. SPECIMEN ON LEFT EXHIBITED DUCTILE BEHAVIOR THROUGHOUT LOADING RANGE. SPECIMEN ON RIGHT FAILED CATASTROPHICALLY WHEN FIBERS BROKE IN SHEAR PLANE.

Despite this situation, attempts have been made to measure the elastic modulus of SIFCON composites. For these measurements, the compressive stress-strain relationship between zero stress and 20% of the ultimate stress was used. Values obtained (28 day) have typically ranged from 14,025 to 24,545 MPa (2.0 to 3.5 x 10^6 psi).

Drying Shrinkage Strain

The drying shrinkage behavior of SIFCON is illustrated in Figure 13 along with that of the unreinforced slurry. The data shown in Figure 13 were obtained on 7.6 x 7.6 x 28.6 cm (3 x 3 x 11-1/4 in.) beam specimens that were fog room cured for 28 days before being placed in a room at 23C (74F) and 50 percent relative humidity.

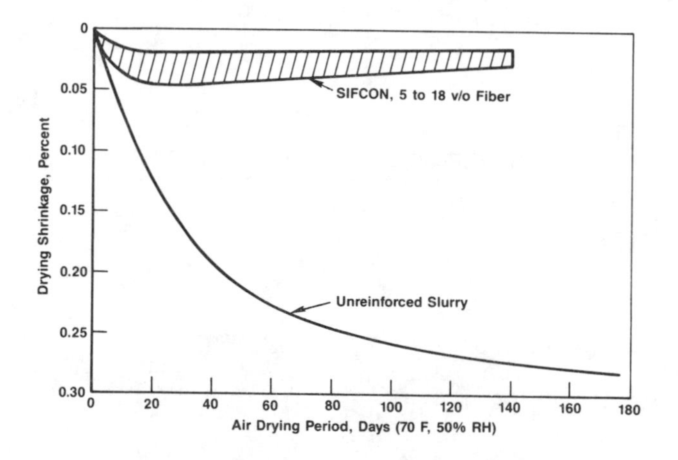

FIGURE 13. DRYING SHRINKAGE OF SIFCON AND PLAIN, UNREINFORCED SLURRY. 7.6 x 7.6 x 28.6 cm (3 x 3 x 11-1/4 in.) BEAM SPECIMENS FOG CURED 28 DAYS AND THEN EXPOSED TO 50 PERCENT RELATIVE HUMIDITY AND 23C (74F).

The unreinforced slurry has exhibited a continual and large drying shrinkage strain over the 180 day exposure period. The SIFCON composites subjected to air drying conditions exhibited a relatively small amount of shrinkage strain (0.02 to 0.05%) that peaked somewhere between 7 and 28 days. For exposure periods beyond 28 days, the SIFCON specimens have shown no significant further shrinkage despite continued shrinkage strain in the plain matrix material. Note that the magnitude of drying shrinkage strain for SIFCON is within the range exhibited by conventional portland cement concrete(PCC).

Other Engineering Properties of SIFCON

Investigations have been made of the resistance of SIFCON to impact loads, to explosive loads, and to stress cycling (fatigue). Although quantitative results cannot be presented here, it can be reported that order of magnitude improvements in impact and explosive loading resistance have been demonstrated over conventional SFRC and conventional PCC. The beneficial effect of steel fibers on fatigue resistance observed in conventional SFRC also exists for SIFCON.

APPLICATIONS FOR SIFCON

Since its introduction in 1979, SIFCON has been prepared as both precast shapes and as cast-in-place construction. Trials have now been undertaken with SIFCON in the areas of (1) pavement rehabilitation, (2) bridge deck rehabilitation, (3) structural restoration of concrete, (4) security concrete applications, (5) precast concrete products, (6) explosive-resistant structures, and (7) refractory applications.

Pavement Rehabilitation

SIFCON has been considered as a thin (1.9 to 5.1 cm - 0.75 to 2 in.), bonded overlay for distressed PCC pavements[1,4,7]. The only field trial conucted to date was a 2.54 cm (1 in.) bonded SIFCON overlay placed on a deteriorating 5.5 m x 15.2 m (18 ft. x 50 ft.) concrete parking lot slab in 1979. On this project, the fibers were placed directly on the prepared concrete slab surface which had been primed with a bonding layer (cement paste). The slurry was poured through a steel grate to which air-actuated vibrators were attached to facilitate infiltration (as shown in Figure 14).

FIGURE 14. INFILTRATION STEP OF SIFCON PREPARED AS A 2.54 CM (1.0 INCH) THICK OVERLAY OF A PARKING LOT SLAB IN AUGUST 1979. THE SLURRY IS BEING POURED THROUGH A VIBRATING STEEL GRATE WHICH SITS ON THE PACKED STEEL FIBER BED

Currently (1985), the overlay remains fully bonded to the old concrete and is in good condition.

It is believed that SIFCON can be a cost/effective rehabilitation material for concrete pavements. However, additional research is needed in the areas of (1) automated construction procedures, and (2) placement techniques for the wearing surface.

Bridge Deck Rehabilitation

In 1984, SIFCON was used to repair spalled areas on the decks of three Interstate highway bridges near Albuquerque, New Mexico. On this project, the slurry infiltration step was accomplished by gravity flow alone (no external vibration). Aggregate was applied to the surface of the still fresh SIFCON as a dry shake and trowelled in to form the wearing surface (as shown in Figure 15).

FIGURE 15. CONSTRUCTION OF A SIFCON REPAIR OF A BRIDGE DECK ON AN INTERSTATE HIGHWAY NEAR ALBUQUERQUE, NEW MEXICO (June 1984). INFILTRATION OF THE FIBER BED WAS GRAVITY-INDUCED FLOW (no vibration). PHOTOGRAPH SHOWS DRY SHAKE APPLICATION OF COARSE AGGREGATE TO FORM THE WEARING SURFACE.

Structural performance of the SIFCON bridge deck repairs have been satisfactory to date (1985). However, information is needed concerning the effect of the relatively high levels of steel in SIFCON as related to the corrosion of reinforcing steel in bridge decks.

Structural Restoration of Concrete

SIFCON has been used in one instance by the New Mexico Department of Transportation for repair of damage in a prestressed concrete beam. This repair was made by forming the spalled area on the beam, placing the fibers in the form, and infiltrating the packed fiber bed. The results to date have been satisfactory.

Security Concrete Applications

A modified version of SIFCON incorporating preplaced fiber and aggregate prior to slurry infiltration has demonstrated superior performance as a security concrete (safes, vault doors).

Precast Concrete Products

A number of precast concrete products have been produced of SIFCON including cast pipe sections, planks, flat plates, and what is believed to be the world's first concrete basketball backboard[1,4,6].

Thin (2.54 to 5.08 cm - 1 to 2 in.) precast SIFCON slabs have been constructed and used as an impact and wear-resistant surfacing over conventional PCC. Slabs as large as 2.4 m x 3.0 m (8 ft. x 10 ft.) were prepared which contain anchors cast into the bottom surface. Figure 16 shows the placement (in October 1983) of one of four 2.4 m x 3.0 m x 5.1 cm (8 ft. x 10 ft. x 2 in.) thick SIFCON slabs into an excavated area containing fresh PCC (at an airfield gate area). The slabs (which supported the main wheel carriage of Boeing 727 aircract showed no cracking through a one year service period.

FIGURE 16. INSTALLATION OF A PRECAST SIFCON SLAB 2.4 m x 3.0 m x 5.1 cm (8 ft. x 10 ft. x 2 in.) INTO AN EXCAVATED AREA CONTAINING FRESH PORTLAND CEMENT CONCRETE. SLAB INSTALLED AT A COMMERCIAL AIRPORT GATE AREA IN OCTOBER 1983.

The 1.9 cm (0.75 in.) thick SIFCON basketball backboard, installed in June, 1984, is shown in Figure 17. The backboard shows promise of lasting forever.

FIGURE 17. THE WORLD'S FIRST CONCRETE BASKETBALL BACKBOARD - CONSTRUCTED OF SIFCON IN 1984.

Explosive-Resistant Structures

Because of its high flexural and compressive strength, combined with high ductility, SIFCON is being considered for use in structures to resist the effects of explosive loadings[5].

In 1984, a 1/8-scale model of a hardened silo structure was constructed of SIFCON by the New Mexico Engineering Research Institute in Albuquerque, New Mexico. The silo [6.1 m (20 ft.) long having 15.2 cm (6 in.) thick SIFCON walls] is shown in Figure 18. In practice, the inner steel liner was completely erected and the outer liner installed in 1.5 m (5 ft.) long sections. SIFCON placement was done within the liners in 1.5 m (5 ft.) lifts with slurry infiltration aided by external vibration.

Following construction, the fully instrumented structure was buried underground and subjected to a surface-mounted explosive loading. The performance of the SIFCON structure in this test exceeded expectations. Currently, larger scale models of SIFCON silo structures are under construction.

SIFCON has also demonstrated excellent resistance to the penetration of projectiles.

Refractory Applications

Precast SIFCON shapes have been used in a number of high temperature applications, where refractory concretes are typically used[2,3,8]. For these applications, the SIFCON composites are prepared with stainless steel fibers and slurries based on calcium aluminate cement.

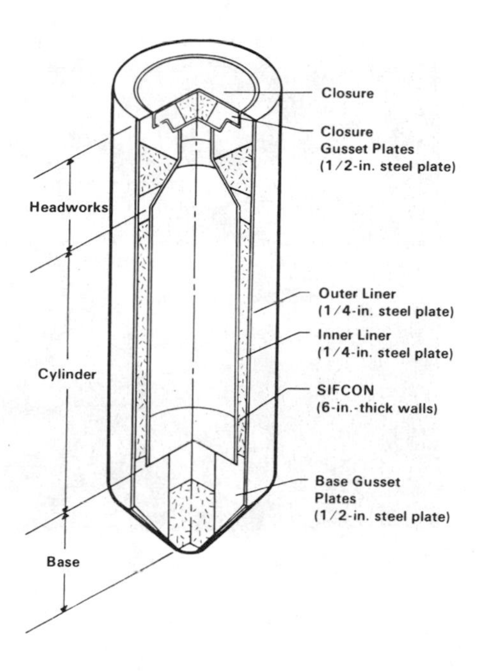

FIGURE 18. SCHEMATIC DIAGRAM OF 1/8 SCALE MODEL [6.1 m (20 ft.) long] OF A HARDENED SILO STRUCTURE (containing SIFCON) CONSTRUCTED BY THE NEW MEXICO ENGINEERING RESEARCH INSTITUTE, ALBUQUERQUE, NEW MEXICO

At these high temperature applications, the SIFCON composites provide excellent resistance to spalling under thermal shock conditions and under conditions of high mechanical abuse. Figure 19 shows a refractory SIFCON seal plate used on the perimeter of a steel soaking pit cover. In this application, the SIFCON is subjected to both thermal shock and mechanical abuse. Refractory SIFCON has worked far better than any other material for this application.

A number of other refractory applications are currently being considered for SIFCON.

THE FUTURE FOR SIFCON

SIFCON is a unique concrete material, combining high strength with high ductility. Its ability to resist cracking and spalling in many static and dynamic loading situations is far superior to conventional SFRC and to conventionally reinforced portland cement concrete. Its ability to be constructed as massive sections as well as in very thin sections, permits consideration of a broad range of potential applications.

FIGURE 19. A STEEL INGOT SOAKING PIT COVER SEAL PLATE - CONSTRUCTED OF SIFCON CONTAINING STAINLESS STEEL FIBERS AND CALCIUM ALUMINATE CEMENT-BASED SLURRY.

SIFCON technology has now gone beyond the laboratory stage with experimentation in pavement and bridge repair and rehabilitation, security concrete, refractory concrete, precast concrete products, and explosive-resistant structure applications. The degree of success of SIFCON in these and other as yet unexplored applications (such as seismic-resistant structures) will depend upon the development of a design philosophy and the development of equipment and procedures that will permit construction with SIFCON on a large scale. Some efforts in these areas have already begun in the U.S. by agencies such as the National Bureau of Standards, various branches of the Defense Department, and the Portland Cement Association. Only time will tell just how SIFCON will fit into the construction and building scene.

REFERENCES

(1) Lankard, D.R. and Newell, J.K., "Preparation of Highly Reinforced Steel Fiber Reinforced Concrete Composites", American Concrete Institute Special Publication SP-81, Proceedings from the 1982 Fall Convention, American Concrete Institute, Detroit, Michigan, September 19-24, 1982, pp. 286-306.

(2) Lankard, D.R. and Lease, D.H., "Highly Reinforced Precast Monolithic Refractories", Bulletin American Ceramic Society, Volume 61, No. 7, 1982, pp. 728-732.

(3) Lankard, D.R., "Factors Affecting the Selection and Performance of Steel Fiber Reinforced Monolithic Refractories", Bulletin American Ceramic Society, Vol. 63, No. 7, 1984, pp. 919-925.

(4) Lankard, D.R., "Slurry Infiltrated Fiber Concrete (SIFCON)", Concrete International, Vol. 6, No. 12, December 1984, pp. 44-47

(5) Schneider, B., Mondragon, R., and Kirst, J., New Mexico Engineering Research Institute, Task Report NMERI, TA8-69 (8.36/01), pp. 1-83, June, 1984.

(6) Lankard, D.R., "Slurry Infiltrated Fiber Concrete (SIFCON): Properties and Applications", Proceedings of Materials Research Society 1984 Fall Meeting, Volume 42, "Potential For Very High Strength Cement-Based Materials", Edited by J.F. Young and S. Mindess.

(7) "Concrete Overlay Construction", U.S. Patent 4,339,289, July 13, 1982.

(8) "Highly Reinforced Refractory Concrete With 4-20 Volume Percent Steel Fibers", U.S. Patent 4,336,255, December 28, 1982.

STEEL FIBER CONCRETE
US-SWEDEN joint seminar
(NSF-STU)
Stockholm 3-5 June, 1985
S P Shah and Å Skarendahl,
Editors

MICROMECHANICAL MODELING OF STEEL FIBER REINFORCED CEMENTITIOUS MATERIALS

by James A Mandel, Prof, Dep of Civil Engineering, University of Syracuse, Syracuse, New York, USA

ABSTRACT

To increase the fracture resistance of cementitious materials, steel fibers are frequently added, thus forming a composite material. To predict the properties of the composite, many types of mathematical models have been formulated, ranging from law of mixture type models, which consider only the mechanical properties of the matrix and fibers and the volume percentage of the fibers, to more complicated micromechanical models which consider the internal structure of the material.

Researchers agree that the behavior of the fiber-matrix interface has a significant influence on the fracture resistance of steel fiber reinforced cementitious materials. As a result, several recent models have considered the influence of fiber debonding and pullout. Predictions by these models of the cracking resistance of steel fiber reinforced mortar are in agreement with experimental data.

A micromechanical multiplane finite element model was recently developed to simulate the behavior of a stressed region in a fiber reinforced material from the onset of loading to failure. The model uses separate finite elements to represent the matrix, fibers, and fiber matrix interface. The failure load and sequential failure mechanism predicted by the model for edge cracked epoxy tension specimens, reinforced with carefully placed steel fibers, was in agreement with experimental results.

Dr. James A. Mandel is a Professor in the Department of Civil Engineering at Syracuse University. He is member of the American Society of Civil Engineers, The American Concrete Institute and the American Academy of Mechanics and is a registered Professional Engineer. Before joining the faculty at Syracuse University, he was employed by Richardson, Gordon, and Associates in Pittsburgh Pennslyvania and Goodyear Aerospace Corporation in Akron, Ohio. He is married to Rita (Sommer) Mandel and has two children Belinda and Robert.

INTRODUCTION

Cementitious materials such as concrete, mortar, and cement paste have a primary deficiency in their weakness in resisting tensile, impact and other energy loadings. This weakness is due to the materials' inability to prevent small cracks, which can originate at a flaw or void in the materials, from becoming unstable under small tensile stresses. To overcome this weakness, steel, glass, polymer, and other type fibers have been added as reinforcement, thus producing a composite material. Even though the percentage by volume of the fibers is small, the mechanical behavior of the resulting composite material is substantially different from that of the matrix material, the composite generally possessing higher tensile strength, ductility, and fracture toughness. For both design and research purposes, mathematical models have been formulated to predict the mechanical behavior of the composite from the mechanical properties of the matrix and fibers and the percent by volume, location, and size of the fibers.

Many types of models have been formulated ranging from law of mixture type models, which consider only the mechanical properties of the matrix and fibers and the percent by volume of the fibers, to far more complicated models which attempt to consider the internal structure of the composite. As the model becomes more complicated it requires more input data and generally yields a more accurate description of the interactions between the components comprising the composite.

The type of model that is pertinent depends upon the objective. For example, if the purpose is to predict the tensile strength of a cementitous material reinforced with a specific type of steel fiber, a simple law of mixtures or strength of material type model, which computes average stresses, is sufficient. However if the objective is to determine "why" the composite behaves as it does or "how" the mechanical properties can be improved, a more sophisticated model which must also consider the properties of the fiber-matrix interface is required. In addition the model must consider both the magnitude and distribution of the stresses in the composite and the influence of cracks in the matrix and along the fiber-matrix interface on the stress conditions. If one views models as microscopes to critically examine the mechanical behavior of the composite, a law of mixtures or strength of materials type model would be a microscope with a low magnification which can therefore see only the average stresses in the fiber, matrix, and interface. A micromechanical finite element model, using finite elements to model the matrix material, fibers, and the fiber-matrix interface would in turn be a microscope with a much higher magnification capable of computing both the magnitude and distribution of the stresses. These computed stresses along with failure criteria for the matrix, fibers and fiber matrix interface can then be used to model the progressive failure process in the composite. Understanding of the influence of the mechanical properties of the matrix material, fibers, and fiber-matrix interface and the geometry of the composite on the failure mechanism and failure load can lead to more efficient design with existing composites. A model, with a higher magnification than that of a micromechanical finite element model, such as model on an atomic scale, is probably not necessary. The reason for this is that a crack in the matrix material or along the fiber-matrix interface constitutes a flaw of a much larger size than atomic size flaws. This is not to say that electron microscopy work is not necessary, as these studies are essential to gain the basic understanding of the materials' structure and behavior required to formulate the model.

To formulate a micromechanical finite element model to predict the mechanical behavior of a steel fiber reinforced cementitious material, several basic questions must be answered. Among these are:

1. Can fracture mechanics be used to describe the stress conditions and crack stability of a crack in the matrix material?
2. What is a crack front in the matrix? Can it be considered as a single crack front or should it be modeled as a process zone?
3. How can we define debonding on the fiber-matrix interface? Is it simply a bond failure or should it be modeled as an interface crack?
4. What are the critical values defining failure along the fiber-matrix interface? How can they be measured? Note these critical values are functions of many variables including fiber diameter, length and surface characteristic as well as the structure and properties of the matrix.

The opinions of researchers concerning these and other questions will be discussed later in this paper.

BACKGROUND

From extensive experimental and analytical work over more than the last thirty years, several facts have been demonstrated concerning steel fiber reinforced cementitious materials. Steel fiber reinforced concrete has been shown to have a higher tensile strength, fracture toughness, and ductility than that of the matrix alone. When fiber reinforced concrete specimens are loaded in flexure, a load deformation curve such as the one shown in Figure 1 is generally observed. The curve is linear up to point A and reaches a maximum at point B. The stresses corresponding to points A and B are designated as the first crack strength and the ultimate strength respectively.

In the 1960's two types of theories to predict the first crack and ultimate strengths of the composites were proposed, one relating the strength of the composite to the average fiber spacing and the other relating it to the fiber volume, spacing, and aspect ratio. The former mechanism is based on linear elastic fracture mechanics while the latter mechanism is based on the law of mixtures of composite materials. Discussion of these two early theories, as applied to fiber reinforced cementitious materials, may be found in Hannant's book, "Fiber Cements and Fiber Concretes" (1).

In 1963 a fracture mechanics model was published by Romualdi and Batson (2) to theoretically predict the fiber cracking stress. They experiment verified their model through testing of concrete beams reinforced with closely spaced continuous steel reinforcements. In 1964, Romualdi and Mandel (3) applied this concept to matrices reinforced with uniformly distributed short steel fiber reinforcements. Test results from beam and split cylinder specimens indicated that the tensile cracking stress of the fiber reinforced concrete is inversely proportional to the square root of the fiber spacing.

In 1968, Romualdi (4) derived an expression for the linear elastic fracture mechanics parameter, G_c, in terms of the bond strength and the length, diameter, and volume percentage of the fibers. In 1974, Parimi and Rao (5) published a model, based on the fiber spacing to compute G_c and the cracking stress of discontinuous fiber composities. As part of their development, an equation was derived for the effective fiber spacing. Their equation ($11.1\ d/\sqrt{p}$) was different from the expression derived by Romualdi and Mandel ($13.8\ d/\sqrt{p}$), where d and p are the diameter and volume percentage of the fibers. Their equation for G_c was also different from the equation developed by Romualdi. The percentage increases in cracking stress, obtained by Parimi and Rao from the testing of fiber reinforced mortar beams, were generally not in agreement with values computed for G_c, using either the equation derived by Parimi and Rao or the equation developed by Romualdi.

There is additional experimental evidence both to support (Snyder and Lankard (6), and to dispute Shah and Rangan (7), Johnson and Coleman (8) a fiber spacing model. This is not surprising as these early models (based on a fracture mechanics concept) consider only the average fiber spacing when calculating the tensile strength of the composite. There is even disagreement as how the fiber spacing is to be calculated (3) (5) (9). The models stipulate that the percent by volume and fiber aspect ratio must be such that there is fiber overlap, however except for this, the fiber aspect ratio is not a parameter in the model, as the average spacing is calculated directly from the percent by volume and diameter of the fibers.

The "law of mixtures" approach (10) has been used extensively in the mathematical modeling of fiber-resin and other composites. The assumptions of this approach are:

1. The fibers are aligned in the direction of the load
2. The fibers are bonded to the matrix
3. Poisson's ratio of the matrix and fibers is zero.

Using statistics and compatibility of strain at the fiber matrix interface the average stresses in the matrix and fibers (σ_m and σ_f) and the effective modulus of the composite (Ec) are calculated as:

$$\sigma_m = \frac{\sigma_c}{1 + V_f\ (E_f/E_m - 1)} \qquad (1)$$

$$\sigma_f = \sigma_m\ E_f/E_m \qquad (2)$$

$$E_c = E_f\ V_f + E_m\ (1 - V_f) \qquad (3)$$

where V_f is fiber fraction of the volume, E_f and E_m are Young's modulus of the fibers and matrix, and σ_c is the applied stress on the composite. To account for fiber orientation, correction factors for the 1-D aligned fibers and the 2-D and 3-D random fiber orientation cases have been proposed by Cox (11) and Krenchel (12). A second correction factor for the fiber length: $L/2L_c$ for $L < L_c$ and $1-L_c/2L$ for $L < L_c$ was proposed by Allen (13). L_c, the critical fiber length, is twice the fiber length required to cause fiber failure in a pullout test and L is the actual fiber length. Unfortunately the critical fiber length, L_c is difficult to estimate. The fiber volume fraction, V_f, is multiplied by the two correction factors to account for fiber orientation and length.

Law of mixture type models have been used to explain the mechanical properties of many type composites. They have been most successfully used in composites with high percentages by volume of continuous fibers. Shah and Rangan (7) showed experimentally that for fiber aspect ratios up to 150, the following simple linear model could be used to relate the ultimate strength of fiber reinforced concrete to the percent by volume of the fibers.

$$S_c = A\ S_m\ (1 - V_f) + B\ V_f L/d \qquad (4)$$

S_c and S_m are the strength of the composite and the matrix material respectively. L and d are length and diameter of the fibers. A and B are experimentally determined constants. Swamy, Mangat, and Rao (14) report values for A and B of 0.97 and 3.41 for the ultimate composite strength and 0.843 and 2.93 for first crack composite strength. Note constant A represents the contribution of the matrix and constant B depends on the bond strength between the matrix and fibers.

Swamy, Mangat, and Rao also pointed out that the above mentioned strength equations were based solely on geometrical considerations and did not consider the bond transfer length. Note, the fiber's effectiveness in transfer of load through bond is influenced by the length of the fiber. They derived an equation for the "effective" spacing, which considered the influence of the fiber's length and diameter on the bond efficiency of the fiber. Values for the ratio of the first crack strength and the ultimate flexural strength of steel fiber reinforced cementitious materials to those of the matrix material alone, computed with, relationships based on the "effective" spacing, were shown to agree with experimental results.

Based on the law of mixtures Aveston et al (15) developed a model to calculate the complete stress strain curve for a composite with a brittle matrix and long fibers parallel to the load. Aveston assumes that the matrix will crack before failure of the composite if the fibers can carry an additional load, equal to the load carried by the matrix material when it reaches its ultimate strain. This additional load is then transferred linearly back into the matrix over a transfer length x'. Eventually, the matrix is broken down into blocks of length between x' and 2x'. Aveston related the resulting crack spacing, crack width, and the complete stress strain curve of the composite (including the ultimate stress) to the elastic moduli of the matrix and the fibers, the volume percentage of the fibers, the ultimate stresses of the matrix and the fibers and the limiting value for the maximum shear stress between the matrix and the fibers. Aveston then modified his theory for continuous fibers to predict the properties of brittle matrices with discontinuous fibers.

Naaman, Argon, and Moavenzadeh (16) developed a statistical fracture model to predict the tensile properties of discontinuous fiber reinforced cementitious materials. The model is based on the hypothesis of the weakest link chain. There are two parts to this model. Part 1, for ductile behavior, based on mechanics (law of mixtures) and the statistics of failure, investigate the behavior of one characteristic fiber to predict the macroscopic response of the composite. Part 2 for brittle fracture includes a fracture mechanics criterian in the analysis.

The work of these early investigators is excellent for estimating the effect of certain parameters such as fiber volume, fiber strength, and matrix strength on the performance of composites. Use of these models, along with experimental data, can provide a basis for design. A disadvantage of both the "law of mixtures" and fiber spacing based models is that they are based primarily on average stresses. Since the distribution of stresses is not considered the influence of changes in the geometry of the composite on peak stresses and the resulting sequential failure process in the composite cannot be evaluated. To accurately model the region near a crack in the matrix or near the fiber matrix interface, the stress gradient must be considered. Without consideration of the stress gradients in these regions, a model cannot logically explain "why" a certain failure mode takes place and thus is lacking as a research tool.

A primary factor in determining the mechanical behavior of a fiber reinforced material is the properties of the fiber-matrix interface. In applications where high tensile strength is required, failure must not occur at the fiber-matrix interface prior to the failure of the fibers. However in applications where toughness is important, slip along the fiber matrix interface is essential for the material to absorb energy. In composites with brittle matrix materials, such as steel fiber reinforced cementitious materials, understanding of the behavior of the fiber-matrix interface is especially critical, since both the tensile strength and toughness of the composite must be significantly higher than those of the matrix material. This is difficult to achieve in fiber reinforced cementitious materials. A combination of a matrix with low tensile strength and the difficulty of achieving fiber volumes greater than 10 percent results in matrix cracking at relatively low stress levels. In addition the resistance to crack extension from the strain energy due to plastic flow at the crack tip is small. Hence resistance to crack extension must be provided by work done in breaking the bond between the fiber and the matrix and work done against friction during fiber pullout.

Kelly (17) discussed, in detail, both his work and the work of other investigators on the mechanics of fiber pullout. He reports (with experimental verification) that the load-extension curve during a fiber pullout test for a composite with a brittle matrix is of the form shown in Figure 2. The straight line portion of the curve (O-A) is characterized by debonding. The load then drops and the area under the remainder of the curve is a measure of the energy required to overcome friction during fiber pullout. For short fibers, the debonding may all occur at the maximum load (point A), and the energy per unit area necessary for debonding may be estimated by dividing the area O AB (Figure 2) by the surface area of the fiber. Using this value as a design criterion for other than a pure fiber pullout geometry with a short fiber length is incorrect because:

1. The peak load required for fiber pullout is dependant on the pressure between the fiber and matrix.
2. For long fibers the O-A portion of the curve (Figure 2) has a sawtooth form indicating that complete debonding does not take place at the peak load, partial debonding taking place at the peak of each of the sawtooths. For this case, the energy for debonding can also be estimated (in composites with transparent or translucent matrix materials) using portions of the area under the curve associated with the drops in load and the observed length of debonding that has occurred at each drop in load.

Proper application of the results of a fiber pullout test to the fracture analysis of a composite is dependent on the crack length. If the crack length is shorter than the fiber spacing, as shown in Figure 3a, only the energy required for debonding provides resistance to crack growth. For crack length larger than the crack spacing with fiber bridging the crack (Figure 3b), both the work of debonding and the work of pullout contribute to the resistance to crack growth. Keep in mind, however, that the energy at the crack tip available for crack extension is also larger for the latter case as it is generally proportional to the square root of the crack length. An added complication is that the energies required for both debonding and pullout are dependent on the normal pressure between the fiber and the matrix, and for the cases illustrated in Figures 3a and 3b, this pressure is not the same as it is in a fiber pullout test.

To estimate the tensile strength and resistance to crack growth in a matrix reinforced with discontinuous fibers parallel to the direction of the loading, Kelly develops equations to estimate the average stress carried by a fiber at the ultimate tensile strength of the composite and the average work of pullout per fiber. These equations are based on the assumption (verified experimentally) that the value of the shear stress, , associated with the critical fiber length is the same as the shear stress developed during sliding friction during fiber pullout. The critical fiber length L_c (equation 5) is the longest fiber length permitting fiber pullout before fiber rupture.

$$L_c = \frac{d\ \sigma_f}{2} \tag{5}$$

where d and σ_f are the diameter and ultimate strength of the fiber. This approach assumes, that the shear stress distribution is uniform over the fiber length. If the direction of the fiber is inclined to the load or if there are any flaws along the fiber-matrix interface or in the matrix material near the fiber, a more detailed analysis may be required to model the failure sequence in the composite. A technique to model debonding, which is based on the stress conditions associated with an interface crack between two dissimilar materials, will be presented later in this paper.

Pinchin and Tabor (18) studied the pull-out behavior of steel wires embedded in a cement paste or mortar matrix. They report that prior to debonding, microcracking due to shrinkage reduced the stress transfer at the fiber-matrix interface. This leads to scatter in the results of pullout tests. They also report that an increase in the bond strength, as measured in pull-out tests, does not necessarily result in a corresponding increase in composite tensile strength. This is not surprising since both the stress distribution in the matrix and fibers and the cracking pattern in a tensile test is different from those in a pull-out-test. This fact indicates the need to consider the stress distributions when defining a failure criteria for debonding and when computing the energy required for pullout. The large sudden release of energy associated with debonding suggests consideration of a fracture mechanics approach when modeling debonding.

Pinchin and Tabor also examined the fibers, pulled from cement paste and mortar, in a scanning electron microscope. The fibers pulled from cement paste had cement paste adhered to the surface of the fiber indicating that the failure was primarily in the matrix. When steel wires were pulled from a mortar matrix, the failure occurred both in the matrix and at the wire surface. Pinchin and Tabor also report that, for fibers with surface roughness, the debonding load is higher. This appears to be due to mechanical interaction between the matrix and the fiber. However after debonding, the stress transfer is still entirely frictional and of the same order as a smooth fiber.

In 1976 Naaman and Shah (19) reported the results of an extensive testing program to investigate the effect of fiber diameter, spacing (groups of fibers being pulled out simultaneously), and orientation on the fiber pullout load. One inch long fibers with diameters of 0.016 inch, 0.010 inch, and 0.006 inch (aspect ratios of 62.5, 100, and, 167) were pulled from one half of a standard tensile briquette specimen. They concluded that the peak pullout loads for fibers inclined to the loading direction is almost as high as those for fibers parallel to the loading direction, and that the work to pull out an inclined fiber is higher than that of a straight fiber. They also noted that there are differences in the pullout mechanisms of inclined and straight fibers, as evidenced by the differences in the shapes of the pullout load verses pullout distance curves (Figure 4). The investigators attribute this to the fact that the straight and inclined fiber pullout tests have different stress distributions and cracking patterns.

The effect of fiber spacing (2, 4, 9, 16, and 36 fibers per square inch) was found not to have a significant effect on the peak pullout load per fiber in tests with 0.016 inch diameter fibers, parallel to the direction of the load. Note this is a wide range of spacings, 0.50 inch (31 fiber diameters) for 4 fibers per square inch to 0.166 inch (10.4 fiber diameters) for 36 fibers per square inch. Naaman and Shah conclude that pullout behavior cannot be predicted from elastic continuum considerations.

Pullout load versus pullout displacement curves, similar to those reported by Kelly and Naaman and Shah, were observed in fiber pullout tests conducted by Mandel, Said, and Sun-Wei (Figure 5). In these tests, groups of fibers* (length 2 inches, diameter 0.02 inches (aspect ratio = 100)) manufactured with end hooks and groups of the same fibers with the hooked portion removed (length 1.75 inches) were pulled from a mortar matrix. The initial portion of the pullout load versus pullout displacement curves (zero to peak fiber load), for both the straight and hooked ended fibers, had a sawtooth form, indicating that a portion of the debonding takes place prior to the peak loading.

There have been many studies of the transfer of stress between a conventional reinforcing bar and concrete. Among these are Jiang, Shah, and Andonian (20), Somayaji and Shah (21), Nilson (22), and Ingraffea et al (23). As in the case for transfer between a fiber and mortar or cement paste, all investigators agree that the shear stress distribution between the reinforcing bar and the concrete is not linear and the cracking patterns influence the stress distribution.

The above discussion, concerning the fiber pullout indicates that during this process two types of mechanism are present, debonding and frictional slip. To properly model fiber reinforced cementitious materials, both of these mechanisms must be considered and the stress distribution and cracking patterns must be considered. A micromechanical finite element model, recently developed at Syracuse University (24), (25), treats fiber debonding as a mixed mode (opening and sliding modes) fracture mechanics problem involving an interface crack between two dissimilar materials. This approach is based on analytical expressions developed by Rice and Sih (26), (27), Cherepanov (28), and others. The procedure uses the critical value of the crack tip energy field, based on the opening and sliding mode stress intensity factors at the load level causing debonding, as the criterion for fiber-matrix debonding. This procedure will be discussed in more detail later in this paper.

---

*DRAMIX ZL 50/50 fibers manufactured by the Bekaert Wire Corporation

In modeling crack growth in the matrix material, one must decide whether or not a criterian based on linear elastic fracture mechanics can be used. There has been much controversy concerning this question. A recent study by Saouma, Ingraffea, and Catalano (29) reevaluated the results of a significant experimental investigation by Kesler, Naus, and Lott (30), which previously concluded that linear elastic fracture mechanics was not applicable to cement, mortar, and concrete. For five of the eight mixes tested, use of the correct stress intensity factor relationship to compute toughness from the test results, resulted in a value of toughness independent of the specimen geometry. Reevaluation of the load records produced improvement in the results from the remaining three mixes tested. They concluded that linear elastic fracture mechanics was applicable to cementitious materials.

To model crack growth in concrete in a finite element model, Hillerborg, Modeer, and Peterson (31) developed another criterion based on fracture mechanics. An approach similar to the Barenblatt model, was used to describe the microcracked (process) zone near the crack tip. Their analysis led to the first criterian for crack growth for concrete which is based on a stress-crack opening displacement relationship. Their method is very well suited for finite element analysis as it eliminates the need for dealing with a stress singularity at the crack front. In a recent publication, Hillerborg (32) describes techniques for modeling the region near a crack front in a cementitious material. He uses a "fictitious" crack zone to model the fracture process zone near the crack front where the crack opening displacement is below a critical value. The magnitude of the stress transferred in this zone is related to the crack opening displacement. For concrete, there is reason to believe that the stress-crack opening displacement relationships obtained from tension tests can be used.

MODELS FOR FRACTURE RESISTANCE OF FIBER REINFORCED CEMENTITIOUS MATERIALS

A model to predict the fracture resistance or ultimate tensile strength of a fiber reinforced materials, must consider the mechanics of the fiber-matrix interface. Opposition to unstable crack growth in the matrix material is provided primarily by the resistance of the fiber-matrix interface to fiber debonding and fiber pullout. In addition, it is necessary to identify and measure the interface parameters that will be used as failure criteria in the model. For the model to be applicable to all loading conditions, it must be capable of computing both the magnitude and distribution of the stresses in the composite, a difficult task given the nonlinear mechanical behavior and cracking associated with cementitious materials. In this section, descriptions of some of the recent models will be presented.

Bazant and Cedolin (33), (34), (35) developed techniques for finite element modeling of cracks in reinforced concrete. The cracks were considered to be distributed (smeared) and were represented by finite width bands of elements. Three methods were proposed for use as a criterian for crack growth: an energy-release rate calculated from change in the strain energy of the element just ahead of the crack band, the stress intensity factor computed from the stresses in the element just ahead of the crack band, and consideration of the nodal displacements near the crack front, with the assumption that the crack propagation direction is such that the crack is one of pure opening mode. In the models proposed, the effect of bond slip of the reinforcement is also considered.

Velasco, Visalvanich, and Shah (36) studied the applicability of fracture mechanics approaches to fiber reinforced concrete. K_{cr}, J integral, critical crack opening displacement, compliance technique to determine slow crack growth, and R-curve analysis were considered. The R-curve method was the most promising, as the relationship between stress intensity factor and measured crack extension was independent of the initial crack length.

The research of Wecharatana and Shah (37), (38), (39), (40), (41) has resulted in the formulation of an excellent theoretical model, based on the concepts of nonlinear fracture mechanics, to predict the crack propagation resistance of fiber reinforced concrete. In their initial studies, a method was developed to calculate R curves for mortar, concrete, and fiber reinforced concrete. The concept of strain energy release rate, based on linear elastic fracture mechanics, was modified to include the observed nonlinear effect of the process zone length in cementitious materials. In fiber reinforced mortar, R curves values were significantly influenced by the crack openings, the larger the crack opening, the higher the energy release rate. Double cantilever and double torsion specimens were used, with the results independant of the specimen size and type.

In the theoretical model developed by Wecharatana and Shah (41), a crack in the matrix material (Figure 6) is divided into three zones:

1. A traction free zone where the fibers bridging the crack have been pulled out or broken and thus provide no resistance to crack opening.
2. A fiber bridging zone, of length L_f, where the closing pressure due to the fibers is a function of the opening mode crack opening displacement.
3. A matrix process zone, of length L_p, where closing pressure is provided by aggregate interlock and microcracking.

In the traction free zone, the crack opening displacements are larger than η^f_{max}. This value is assumed to be one half the fiber length (L/2). Note it cannot exceed this value, since any larger value of crack opening displacement must result in fiber pullout. In the fiber bridging zone, the crack opening displacements are less than η^f_{max}, and greater than η^m_{max} the critical crack opening displacement of the matrix. η^m_{max} can be obtained from the descending portion of the tensile stress-displacement relationship for the matrix at the point where the tensile stress is close to zero. Note, in this model, a given crack in the matrix is assumed to propagate when the crack opening displacement reaches a value of η^m_{max}.

The theoretical model requires two material parameters, the critical crack opening displacement of the matrix (η^m_{max}) and the relationship between the fiber bridging (crack closing) pressure (σ) and the cracking opening displacement (η). The σ-η relationship can be determined experimentally from data obtained during the post cracking stage in a uniaxial tensile test of the fiber reinforced material. The following unique σ-η relationship was obtained for cement composites reinforced with smooth steel fibers regardless of their length, diameter, or volume percentage.

$$\frac{\sigma}{\sigma_{max}} = [1 - \frac{\eta}{\eta^f_{max}}]^2 \qquad (6)$$

σ_{max} is the maximum post cracking stress and η^f_{max} is taken as half the fiber length. Note there are analytical and empirical expressions to predict σ_{max} (14). One expression relates σ_{max} to the fiber reinforcing index (V_f L/d) and the average bond strength of the fibers ( ).

$$\sigma_{max} = \beta\tau \ (V_f \ L/d) \qquad (7)$$

V_f, L, and d are the volume fraction, length, and diameter of the fibers. β is a constant which depends on the length and spacial distribution of the fibers. A value of $\beta\tau$ can be determined from the results of fiber pullout tests.

Since the closing pressure depends on the crack opening displacement, which in turn depends on the specimen geometry, external loading, and the closing pressure itself, an iterative procedure is required to solve for a crack profile and crack closing pressures that are consistant with the crack initiation criteria. Starting with a given crack length and an external loading, the initial step in the iterative procedure is to assume the crack profile and the length of the process zone (L_p). From the assumed crack profile and the $\sigma - \eta$ relationship, L_f and the fiber bridging closing pressures are computed. Next, using an elastic analysis, the crack opening displacements caused by the applied loading and the crack closing pressures (σ) are calculated. If the crack opening displacement at the end of the matrix process zone is equal to η^m_{max}, the initiation criteria is satisfied and the assumed value of L_p is correct. If not, a new value for L_p is assumed and the procedure is repeated.

As shown in Figures 7 and 8, theoretical predictions, by this model, of crack mouth opening displacements are in agreement with experimental data from notched beam and double cantilever beam specimens. Note in the theoretical analysis, a straight line crack profile was assumed.

An expression for the fracture energy (strain energy release rate at crack initiation) of fiber reinforced cementitious materials, based on the compliance method, was also developed by Wecharatana and Shah (40). This expression considers the inelastic energy absorbed during crack growth. In this case, the energy is primarily due to fiber pullout. Using this expression and the results from repeated application of this model with increasing crack lengths and load intensity, the entire R-curve can be constructed. As shown in Figure 9, results predicted by the theoretical model agreed well with experimental data from double cantilever beam specimens. Note in the analysis, a straight line crack profile was assumed.

The theoretical model developed by Wecharatana and Shah has several advantages over previous models. Among these are:

1. The influence of the process zone on the effective crack length is considered.

2. The pullout resistance of the fibers is considered.
3. Only two material parameters are required for the model (σ-η relationship and η_{max}^{m}) and these parameters can be measured experimentally.
4. The model is general in that it can be applied to virtually all fiber reinforced cementitious materials.
5. Predictions, by the model, of the cracking resistance (strain energy release rate) of steel fiber reinforced mortar are in agreement with experimental data.

Another theoretical model to predict the fracture resistance of fibers reinforced cementitious materials was developed by Visalvanich and Naaman (42). In their model, two parameters are required to compute the R-curve of the composite, the relationship between crack closing pressure and crack opening displacements (σ-η) and the shape of the crack when propagation takes place.

Although the following (σ-η) relationship, used by Visalvanich and Naaman (eqn. 8) was different from the expression used by Wecharatana and Shah (eqn. 6), both yield approximately the same results and both are shown to be in agreement the results of tests of steel fiber reinforced mortar.

$$\frac{\sigma}{\sigma_{max}} = [0.1(\frac{2\eta}{L})+1][(\frac{2\eta}{L})-1]^{2\eta} \qquad (8)$$

Both models recommend use of eqn (7) to estimate the maximum post cracking stress (σ_{max}).

The crack shape was assumed to have a straight profile with an opening angle equal to the observed critical crack opening angle (the angle, independent of crack length, at which the crack starts propagating). Experimental results (43) indicate that there is a relationship between the critical crack opening angle and the fiber reinforced index. As the loading increases in this model, the crack grows slowly and the opening between the crack faces increases. Using an energy balance concept along with the crack geometry shown in Figure 10, the σ - η and σ_{max} relationships (eqns. 7 and 8), and the assumed crack profile as described above, an expression (eqn. 9) was developed for the energy per unit area of crack extension of steel fiber reinforced mortar during slow (stable) crack growth.

$$G_a = \beta\tau\,(V_f\,L/d)\,\gamma c\,[1-1.9(\frac{\gamma c}{L})+ 1.067\,(\frac{\gamma c}{L})^2 + 0.2(\frac{\gamma c}{L})^3] \qquad (9)$$

γ is the tangent of the critical crack opening angle and c is the crack extension beyond the original crack length a_o (Figure 10). When δ_t reaches a value of one half the fiber length (L/2), the fiber pullout zone is fully developed and the energy per unit area of crack extention (Fracture energy) reaches a steady state (critical) value, G_c (equation 10).

$$G_c = 0.171\;\beta\tau\;(V_f\;L^2/d) \qquad (10)$$

Comparison of equations (9) and (10) with the results of double cantilever beam tests are shown in Figure 11.

This model is similar in many respects to the model developed by Wecharatana and Shah. In both models, fiber pullout energy is considered to be the major contributor to the fracture properties of the composite and hence both models depend on the relationship between the crack opening displacements and the resistance provided by fiber pullout. Both models use the same relationship to estimate the maximum post cracking stress of the composite and assume a straight crack profile.

MICROMECHANICAL MODELING OF FIBER REINFORCED MATERIALS

The models developed by Wecharantama and Shah (41) and Visalvanich and Naaman (42) to predict the fracture resistance of fiber reinforced material are excellent models for design purposes. They are accurate and require relatively little material input data. The required material input data, however, includes the relationship between the fiber bridging (crack closing) pressure and the crack opening displacement, thus requiring testing of the composite material. These models are therefore, to some degree, semiempirical.

A micromechanical model is primarily a research tool, an objective of the model to determine "why" the material behaves as it does, with the hope that this knowledge will result in the design of composites with improved properties. In this type of model, frequently the matrix and fibers are treated as separate elements and the mechanical properties of the composite are predicted from the mechanical properties of its constituent materials and the geometry of the composite.

Paramasivam, Curiskis, and Vallioppan (44) recently used a three dimensional finite element micromechanics analysis to study the linear elastic and nonlinear properties of fiber reinforced cement composites. In this study, two idealized unidirectional composites, with short fibers arranged in rows, (Figure 12) were investigated, with the assumption that the behavior of this idealized materials can be related to the behavior of a composite with randomly located fibers through fiber orientation factors. Because of symmetry, typical repeat units, indicated by the dashed lines in Figure 12, were considered to be representative regions of the composites and were analyzed using 3-D finite element analysis. A perfect bond was assumed along the interface between finite elements representing fiber material and matrix material. The model was used to investigate the effect of the volume fraction of the fibers on the stress-strain curve, Poisson's ratio and internal stress distribution in steel fiber reinforced cement paste. Predictions of the yield and ultimate stresses of the composite, by the model, were in reasonable agreement with the test results from randomly oriented steel fiber reinforced cement specimens (Table 1). Note, the fiber volume fractions used in the finite element analyses were obtained by dividing the fiber volume fractions of the actual test specimens by a fiber orientation factor of 0.47.

The micromechanical model developed in this study appears adequate for modeling the stress strain behavior of steel fiber reinforced cementitious materials. But since fiber delamination and pullout are not considered, it cannot be applied to predict the fracture properties of the composite.

Tolf (45) also developed a micromechanical model for static analysis of fiber reinforced materials which investigated a small representative region of the composite. He found that long fibers result in a brittle composite while short fibers result in a more ductile material. Tolf also investigated the feasibility of using micromechanical models to describe the dynamic behavior of fiber composites.

The model used by Tolf considered a two dimensional representative region (Figure 13) in the composite. The constituent materials were considered to be linearly elastic and the bond between the matrix and fibers was assumed to be perfect (no slip). To simplify the boundary conditions in the analysis, a representative domain (Figure 14) was selected for analysis. The elasticity equations, representing this domain, were solved using the finite difference method. Parametric investigations, applying this model to composites with a high volume percentage of fibers, were shown to yield reasonable results for stresses in the matrix, fibers, and fiber-matrix interface and for the effective elastic modulus in the fiber direction. The calculations demonstrated that a better understanding of the behavior of composite materials can be achieved using micromechanical analysis.

Recent studies have been conducted at Syracuse University, the major objective being to develop a micromechanical finite element model to simulate the behavior of a stressed region in a fiber reinforced material from the onset of loading to failure. Of particular interest in these studies is the fracture behavior of short fiber composites. For this type of material there are several modes of failure which include:

Crack growth in the matrix between fibers
Fiber delamination and pullout
Fiber yielding
Crack growth in the matrix around fibers

The problem is further complicated because the mechanism of failure is sequential and can include some or all of the above mentioned modes. Furthermore the mechanism of failure can change with changes in the mechanical properties of the constituent materials, changes in the geometry of the composite, or changes in the type of loading.

A micromechanical finite element model was developed which uses three element types: a triangular or rectangular quarter point crack tip element, developed by Barsoum (46), (47), to represent a region of the matrix material near a crack tip, isoparametric quadratic quadrilateral (or triangular) elements to model the remaining matrix material and the fibers, and a linear zero thickness bond interface element (48) to model the fiber matrix interface. Note Henshell and Shaw (49) and Barsoum showed that the stresses in the crack tip element had an inverse square root singularity at the corner of the element adjacent to quarter point nodes, thus making it possible to obtain accurate values of the stress intensity factor at the crack tip.

The input quantities to the model consist of Young's modulus and Poisson's ratio for the matrix material and for the fibers, the shear and normal stiffnesses of the fiber matrix interface, and the boundary conditions and applied loading to the portion of the composite being analyzed.

Since the failure mechanism is sequential, the load is applied in increments. After each increment of loading, the finite element mesh and element material properties are modified to reflect changes due to crack growth or a material failure. This procedure is continued until the portion of the composite being analyzed can no longer carry the applied load. Nodal displacements and element stresses are computed by the model at each load level.

The failure criteria used for crack growth in the matrix and along the fiber-matrix interface are based on linear elastic fracture mechanics. When the stress intensity factor (computed in finite elements near the crack tip from nodal displacements) is equal to or greater than the critical stress intensity factor of the matrix material (a material property), growth of a crack in the matrix material is assumed to occur.

The primary objective of the initial portion of this research was to establish the validity of this approach. In these studies (50), (51) the micromechanical model was applied to edge cracked tensile specimens (Figure 15) to predict the effect of fibers near a crack front in the matrix material on the load required for crack growth.

The test specimens were cast, using methyl-methracrylate for the matrix material with a row of closely spaced steel fibers near the crack front. Note Young's modulus of the fibers (30,000 ksi) was much larger than the Young's modulus of the matrix material (300 ksi). This enhances the effect of the fibers on the stresses in the matrix material near the crack front. In addition methyl methacrylate is transparent, thus making it easier to observe crack growth. To determine the critical stress intensity factor of the matrix material, edge cracked tensile specimens, without fibers, were tested. Note this value is required in the criterian for crack growth in the matrix material.

Since the spacing of the fibers is very small (two fiber diameters), the fiber areas were assumed, in the analysis, to be uniformly, distributed through the thickness direction of the specimen, thus reducing the problem to a two dimensional one. Finite element analyses were run for two different assumptions regarding the bond properties between the fibers and matrix material. The loads required for the initiation of crack growth calculated using the finite element analyses were in close agreement with experimental values (see Table 2). From these studies, it was concluded that:

1. Higher modulus fibers near a crack can increase the magnitude of the loading required for the initiation of crack growth.
2. Micromechanical finite element modeling can be a useful tool in studying fracture mechanics of fiber reinforced materials.

To model the complete fracture process in a fiber reinforced material, crack growth around the fibers, as well as fiber delamination and pullout must be considered. For these portions of the failure mechanisms, the model must be extended to three dimensions, as the assumption of distributing the fibers areas uniformly through the thickness of the specimen is not valid. For these cases however a three dimensional finite element analysis is not feasible for this application, because of limitations on computer storage and central processing unit time. To rectify this dilemma, an efficient finite element modeling technique (multiplane analysis) (52), (53), (25) was developed. This method can be used to obtain close approximate solutions to problems belonging to a class of three dimensional elasticity problems.

In this technique, the continuum is considered to be divided into two or more laminae parallel to an x-y plane FIgure 16. Each lamina is modeled using a plane of two dimensional finite elements. The z direction thickness associated with the 2D finite elements in each x-y plane of elements is equal to the thickness of the lamina being modeled by that plane of elements. Each of these laminae, if isolated, must be suitable for 2D analysis, that is all x-y cross sections are identical within a lamina, and the loading and boundary conditions within each lamina are also independent of z. Corresponding points in different laminae (same x and y coordinates) may be of different materials or differ in boundary restraints or surface tractions, provided such tractions act only in an x-y plane. Adjacent planes of 2-D finite elements are connected by special finite elements ("connector elements"), whose purpose is to model the variations of the x and y direction displacements in the z direction.

To investigate the accuracy of a multiplane finite element model, it was used to predict the sequential failure mechanism and failure loads leading to crack growth past the row of fibers in an edge cracked tension specimen (Figure 17). The test specimens were cast using epoxy for the matrix material, with a row of steel fibers near the crack front. Young's modulus of the fibers (30,000 ksi) was much longer than that of the matrix material (530 ksi), thus emphasizing the effect of the fibers. In addition, epoxy is translucent and doubly refractive, allowing the observation of changing stress patterns in a circular polariscope and making observations of crack growth easier.

The multiplane finite element analyses of the test specimens modeled a typical portion of the thickness of the specimen (Figure 17) which includes one half of a fiber and one half of the matrix material between adjacent fibers. In Figure 18, part of the modeled portion of the cross-section of the edge cracked tension specimen is shown. The fibers shown are two in a row of fibers of constant z direction spacing. This region is suitable for a multiplane analysis with two x-y planes of elements (dual plane analysis). One plane contained matrix material only, while the second plane included both matrix material and fiber elements. In the second plane, zero thickness bond interface elements were placed between matrix material and fiber elements to model the fiber-matrix interface. The two planes of elements were joined using 16 noded connector elements to account for the z direction variation of the x and y direction displacements. Since the specimen is symmetric about the crack axis, only one half of the specimen was modeled. The mesh had 954 nodes.

In the dual plane finite element analyses, the round fiber was replaced with a fiber of rectangular cross-section. The dimensions of the rectangular cross-section fiber were selected to yield the same cross-section area and in plane moment of inertia (resistance to bending of fiber in x-y plane (Figure 18)) as the round cross-section fiber. Since the perimeters of the round and rectangular fiber cross-sections are not equal, finite element results for shear stresses along the fiber-matrix material interface must be corrected to obtain shear stresses for specimens with round fibers.

Complete failure of a member fabricated from a fiber reinforced material is generally preceeded by local failures. Therefore sequential dual plane finite element analyses were performed to model the progressive failure mechanism of the side cracked specimens. After a local failure was determined from a finite element analysis, material properties and/or the finite element mesh was changed to reflect the local failure and another finite element analysis was performed. The procedure was continued until the dual plane analysis indicated a complete failure of the specimen. Local failure modes considered in the analysis are yield in either the matrix material or fiber, crack extension in the matrix material, and failure of the matrix to fiber bond.

Criteria for crack growth in the matrix material and along the fiber-matrix interface was based on the principles of linear elastic fracture mechanics. The crack in the matrix material was assumed to grow when the opening mode stress intensity factor (computed form the x and y direction displacements at nodes in elements near the crack tip) reached the critical stress intensity factor of the matrix material (a material property determined from tensile tests of edge cracked epoxy specimens). Fiber debonding, breaking of the matrix to fiber bond around the perimeter of the fiber, was assumed to occur in two steps. That is, debonding was considered to develop on a half circumference of the fiber when the stresses in the matrix adjacent to the fiber indicates a yield condition in the matrix or when a crack in the matrix reaches the fiber (see Figure 19). Continuation of debonding along the length of the fiber is treated as growth of an interface crack between two dissimilar materials. It is considered to occur when the energy release rate at the crack tip (computed from the opening and sliding mode stress intensity factors) reaches its critical value (determined from the results of fiber pullout tests).

The sequential failure mechanism predicted by successive applications of the multiplane finite element model was the same as the failure mechanism observed during the experimental program with the failure occurring in three major stages:

1. Initiation of rapid crack growth
2. Crack arrest at the row of fibers
3. Slow crack growth past the fibers, followed by unstable rapid growth of the crack front across the specimen.

During the third major stage in the failure mechanism, the multiplane finite element model indicated the initiation and continuation of fiber debonding, first on the half circumference of the fiber adjacent to the crack (left half circumference of fiber (Figure 19)) and later on the other half circumference of the fiber. The debonded lengths reached 7.5 and 2.5 fiber diameters on the left and right half circumferences of the fiber. Due to the small diameter and reflective surface of the fibers, debonding could not be observed during the experimental program.

The ultimate load in the analysis (367 lbf) overestimates the average experimental ultimate load (300 lbf) by 22%. The average load for first observed crack growth past the fibers is 279 lbf. The analytical estimate of this value, 311 lbf, exceeds the experimental value by 11%.

The agreement between multiplane analytical results and laboratory test results show that the multiplane method provides a useful tool for micromechanical study of fiber reinforced composite materials. The effect of fiber spacing normal to, as well as in the plane of loading, can be modeled. In addition, a bond failure over a part of the fiber's length, and over a portion of the fiber's circumference can also be modeled. The fiber's ability to inhibit crack growth is substantially influenced by the extent of fiber debonding. In the dual plane finite element analyses of the edge cracked specimens, there was no increase in loading after debonding occurred around the complete perimeter of the fiber cross-section. Because of the importance of modeling fiber debonding, the following research to develop improved finite element modeling techniques for representing the fiber-matrix interface was undertaken.

To help accomplish this a new, five noded, zero width interface element (54) (see Figure 20) was developed to model the interface between two dissimilar materials. The stiffness matrix of this element was shown to have the theoretical $1/\sqrt{r}$ singularity in its stress field near the crack tip that is present in closed form solutions of interface crack problems. This finite element is compatible with both singular quarter point finite elements and nonsingular interface elements. When used in a mesh, the finite length sides of this five noded singular element are connected to quadratic singular quarter point elements and the zero width side is joined to a six noded quadratic interface element (55) (see Figure 21). The shear and normal stiffnesses of the interface are input values, so that the new finite element can be used to model interfaces between many materials. To verify the accuracy of the element, it was used in a finite element mesh to represent a bimaterial cracked plate (Figure 22) subjected to a biaxial tension loading. The crack length was much smaller than the plate dimensions so that the finite element solution could be compared to a theoretical solution (26). Because of symmetry about the y axis only one half the plate was analyzed. The finite element mesh had 1109 nodes. As shown in Table 3, values of the opening and sliding mode stress intensity factors (K_I and K_{II}) from the finite element analysis are in close agreement with theoretical values.

The most widely accepted parameter used as a criterion for growth of an interface crack between two different elastic media is a critical value of the energy release rate at the crack tip, Γ_{cr} (a material property of the interface). Interface crack propagation begins when:

$$\Gamma \geq \Gamma_{cr} \tag{11}$$

where Γ, the energy release rate at the crack tip is related to the opening mode and sliding mode stess intensity factors (equations developed in reference (28)).

For applications where the mode of failure is predominantly the sliding (shear) mode and there is adhesion between the surfaces of the two materials, Cherepanov (28) formulates and solves a different type of two material interface crack problem (different boundary conditions), which he refers to as a slip crack problem. The slip crack is defined as the interface length along which the adhesion between the two materials is broken. This formulation allows for the development of friction forces between the crack surfaces. These friction forces contribute to the resistance to movement along the length of the slip crack even though the adhesion between the two materials is broken. The form of the solution for stresses near a slip crack and the criterion for slip crack growth are different than that of an open crack problem. However, the criterion for slip crack growth is also based on a critical value for the flow of energy (Γ_{cr}) at the tip of the crack. For a slip crack $_{cr}$ is related to the critical value of the sliding mode stress intensity factor (K_{IIc}), which in this case is a constant of the adhesion between the two materials (slip toughness).

For the case of a cylindrical shaped fiber being pulled from an infinite elastic media in which a fiber length of at least several fiber diameters has broken free to form an interface crack, Cherepanov derived the following equation relating Γ_{cr} to the force (P_{cr}) causing the crack to propagate:

$$P_{cr} = \frac{\pi}{\sqrt{2}} \; d^{3/2} \; \sqrt{\Gamma_{cr} E} \qquad (12)$$

where d and E are the diameter and Young's modulus of the fiber. This force is independent of crack length. Therefore once interface crack propagation begins, it will continue, with an increase of force required only to overcome lateral friction. P_{cr} can be estimated from the load-displacement curve of a fiber pullout test (Figure 23) and the above equation can then be used to compute Γ_{cr}.

To determine if an interface crack will propagate in a loaded fiber reinforced material, the energy release rate at the crack tip (Γ) must be calculated for comparison with Γ_{cr}. Using the nodal displacements from a multiplane finite element analysis of the stressed region in the composite, the opening mode and sliding mode stress intensity factors can be calculated (26). These in turn can be used to calculate Γ (28).

To verify the accuracy of the multiplane finite element model, the case, just discussed, of a fiber being pulled from an infinite elastic media was analyzed using a two plane (duo plane) analysis. Isoparametric, quadratic, quadrilateral and triangular elements; triangular, quarter point, singular crack tip elements (46); Zero width, quarter point, singular interface crack tip elements (54); and quadratic interface elements (55) were used in the analysis. The mesh used had 882 nodes. Nodal displacements from the multiplane analysis were used to compute the opening mode and sliding mode stress intensity factors which in turn were used to calculate the energy release rate at the crack tip (Γ). This value of Γ was compared to a theoretical value of Γ_{cr} computed from equation 12 using a value for P_{cr} equal to the load applied to the fiber in the duo plane finite element analysis.

The value of Γ computed with the duo plane finite element analysis was within six percent of the values of Γ_{cr} computed from equation 12, thus demonstrating the capability of the multiplane finite element model for use in predicting fiber debonding.

From the results of the studies conducted using this micromechanical finite element model, several conclusions can be made.

1. The model is a useful tool for study of both the mechanism of failure and the failure loads of fiber reinforced composites. Since the model gives information as to the cause of the failure, it can be used as an aid in the design of new materials as well as in studies to improve the properties of existing materials.
2. The material input data includes only the mechanical properties of the matrix material, fibers, and the fiber-matrix interface. Techniques have been developed to measure these properties.

Studies are presently underway to investigate crack growth in steel fiber reinforced mortar using this model. Testing of edge cracked mortar specimens, with carefully placed steel fibers, is included to both verify the model and as an aid in improving the model.

In summary there are several valid approaches to the modeling of fiber reinforced cementitious materials. The type of the model that is pertinent depends on the objective. For design applications involving the materials' resistance to cracking, the models developed by Wecharatana and Shah (41) and Visalvanich and Naaman (42) are excellent. For applications involving the development of new composite materials, the micromechanical finite element model is useful as it focuses on the causes of failure, thus indicating how the properties of the composite can be improved.

REFERENCES

(1) Hannant, D.J. "Fibre Cements and Fiber Composites", John Wiley and Sons, Chichester, 1978.

(2) Romualdi, J.P. and Batson, G.B., "Mechanics of Crack Arrest in Concrete", Proc. ASCE, Vol. 89, EM3, June 1963, pp. 147-168.

(3) Romualdi, J.P. and Mandel, J.A. "Tensile Strength of Concrete Affected by Uniformly Distributed and Closely Spaced Short Lengths of Wire Reinforcement" ACI Journal Proc. V61, No. 6, June 1964, pp. 657-671.

(4) Romualdi, J.P., Proc. Int. Conf., The Structure of Concrete, Brooks, A.E. and Newman K. (Eds.) , London, 1968, pp. 190-201.

(5) Parimi, S.R. and Rao, J.K.S., "On the Fracture Toughness of Fiber Reinforced Concrete", Fiber Reinforced Concrete, SP-44, American Concrete Institute, Detroit, 1974, pp. 79-92.

(6) Snyder, M.J. and Lankard, D.R., "Factors Affecting Cracking Strength of Steel Fibrous Concrete, ACI Journal, Proc. V69, No. 2, Feb. 1972.

(7) Shah, S.P. and Ranjan, B.V., "Fiber Reinforced Concrete Properties" ACI Journal, Proc. Vol. 68, No. 2, Feb. 1971, pp. 126-135.

(8) Johnson, C.D. and Coleman, R.A., "Strength and Deformation of Steel Fiber Reinforced Mortar in Uniaxial Tension", Fiber Reinforced Concrete, SP-44, American Concrete Institute, Detroit, 1974, pp. 177-207.

(9) McKee, D.C., "The Properties of an Expansive Cement Mortar Reinforced with Random Wire Fibers", Ph.D. Thesis, University of Illinois, Urbana, 1969.

(10) Holiday, L., "Composite Materials", Elsevier, Amsterdam, 1966.

(11) Cox, H.L., "The Elasticity and Strength of Paper and other Fibrous Materials", British Journal of Applied Physics, 3, 1952, pp. 72-79.

(12) Krenchel, H., "Fiber Reinforcement" Akademish Forlag, Copenhagen, 1964.

(13) Allen, H.G., "Glass-Fibre Reinforced Cement, Strength and Stiffness", CIRIA Report 55, Sept. 1975.

(14) Swamy, R.N., Mangat, P.S., and Rao, CV.S.K., "The Mechanics of Fiber Reinforcement of Cement Matrices", Fiber Reinforced Concrete, SP-44, American Concrete Institute, Detroit, 1974, pp. 1-28.

(15) Aveston, J., Mercer, R.A., and Sillwood, J.M., "The Mechanism of Fibre-Reinforcement of Cement and Concrete", National Physical Laboratory Report (Great Britain) No. Sl, No. 90/11/98 Part I, Jan. 1975, Part II, DMA 228, Feb. 1976.

(16) Naaman, A.E., Argon, A.S. and Moavenzadeh, "A Fracture Model For Fiber Reinforced Cementitious Materials", Cement and Concrete Research, Vol. 13, pp. 397-411, 1973.

(17) Kelly, A., "Interface Effects and the Work of Fracture of a Fibrous Composite", Proc. Royal Soc. of London, A319, 1970, pp. 95-116.

(18) Pinchin, D.J. and Tabor, D., "Interfacial Phenomena in Steel Fiber Reinforced Cement II: Pull-Out Behavior of Steel Wires", Cement and Concrete Research, Vol. 8, 1978, pp. 139-150.

(19) Naaman, A.E. and Shah, S.P., "Pull-Out Mechanism in Steel Fiber-Reinforced Concrete", J. of the Struct. Div., Amer. Soc. of Civil Engr., Aug. 1976, pp. 1537-1549.

(20) Jiang, D.H., Shah, S.P., and Andonian, A.T., "Study of the Transfer of Tensile Forces By Bond", ACI Journal, May, June 1984, pp 251-259.

(21) Somayaji, S. and Shah, S.P., "Bond Slip Relationship and Cracking Response of Tension Members", ACI Journal, May, June 1981, pp. 217-225.

(22) Nilson, A.H., "Internal Measurement of Bond Slip", ACI Journal, Proc. V69, No. 7, July 1972, pp. 439-441.

(23) Ingraffea, A.R., Gerstle, W.H., Gergely, P., and Saouma, V., "Fracture Mechanics of Bond in Reinforced Concrete", Journal of Structural Engineering, Vol. 110, No. 4, April 1984.

(24) Pack, S.C., and Mandel, J.A., "Micromechanical Multiplane Finite Element Modeling of Crack Growth in Fiber Reinforced Materials", Engng. Frac. Mech. 20, 335-349, 1984.

(25) Pack, S.C., "Multiplane Finite Element Method, Application to Fiber Reinforced Material", Ph.D. Dissertation, Syracuse University, Sept. 1982.

(26) Rice, J.R. and Sih, G.C., "Plane Problems of Cracks in Dissimilar Media", Transactions of ASME, Journal of Applied Mechanics, June 1965.

(27) Sih, G.C. and Rice, J.R., "The Bending of Plates of Dissimilar Materials with Cracks", Transactions of ASME, Journal of Applied Mechanics, Sept. 1964.

(28) Cherepanov, G.P., "Mechanics of Brittle Fracture", McGraw Hill International Book Co., 1979, (Translated from Russian).

(29) Saouma, V.E., Ingraffea, A.R., and Catalano, D.M., "Fracture Toughness of Concrete Revisited", J. of Engr. Mechanics Div., Amer. Soc. of Civil Eng., Dec. 1982, pp. 1152-1165.

(30) Kesler, C., Naus, D. and Lott, J., "Fracture Mechanics - Its Applicability to Concrete", Proc. of the 1971 International Conference on Mechanical Behavior of Materials, Vol. IV, Japan, 1972, pp. 113-124.

(31) Hillerborg, A., Modeer, M., and Petersson, P.E., "Analysis of Crack Formation and Crack Growth in Concrete by Means of Fracture Mechanics and Finite Elements", Cement and Concrete Research, Vol. 6, 1976, pp. 773-782.

(32) Hillerborg, A., "Analysis of One Single Crack", Fracture Mechanics of Concrete (edited by F.W. Wittmann), Elsevier Science Publishers, 1983, pp. 223-249.

(33) Bazant, Z.P. and Cedolin, L., "Blunt Crack Band Propagation in Finite Element Analysis", J. Engr. Mech. Div., Am. Soc. of Civil Eng., Apr. 1979, pp. 297-315.

(34) Bazant, Z.P. and Cedolin, L, "Fracture Mechanics of Reinforced Concrete", J. Engr. Mech. Div., Am. Soc. of Civil Eng., Dec. 1980, pp. 1287-1305.

(35) Bazant, Z.P. and Cedolin, L., "Fracture Mechanics of Reinforced Concrete", Fracture in Concrete, Publ. by Amer. Soc. of Civil Engr., Proceedings of a session sponsored by the Committee on Properties of Materials at the ASCE National Convention in Hollywood, Florida, Oct. 1980, pp. 28-35.

(36) Velazco, G., Visalvanich, K. and Shah, S.P., "Fracture Behavior and Analysis of Fiber Reinforced Concrete Beams", Cement and Concrete Research, Vol. 10, 1980.

(37) Wecharatana, M. and Shah, S.P., "Double Tension Test for Studying Slow Crack Growth of Portland Cement Mortar", Cement and Concrete Research, V. 10, Nov. 1980, pp. 833-844.

(38) Wecharatana, M. and Shah, S.P., "Resistance to Crack Growth in Portland Cement Composites", Fracture in Concrete, Publ. by Amer. Soc. of Civil Engr. Proceedings of a session sponsored by the Committee on Properties of Materials at the ASCE National Convention in Hollywood Florida, Oct. 1980, pp. 82-105.

(39) Wecharatana, M. and Shah, S.P., "Predictions of Nonlinear Fracture Process Zone in Concrete", J. Engr. Mech. Div., Amer. Soc. of Civil Engr., Vol. 109, No. 5, 1983.

(40) Wecharatana, M. and Shah, S.P., "Slow Crack Growth in Cement Composites", J. Struct. Div., Am. Soc. of Civil Eng., Vol. 108, No. 6, 1982, pp. 1400-1413.

(41) Wecharatana, M. and Shah, S.P., "A Model for Predicting Fracture Resistance of Fiber Reinforced Concrete", Cement and Concrete Research, Vol. 13, No. 6, Nov. 1983, pp. 819-829.

(42) Visalvanich, K. and Naaman, A.E., "Fracture Model for Fiber Reinforced Concrete", ACI Journal, March/April 1983, pp. 128-138.

(43) Visalvanich, K. and Naaman, A.E., "Fracture Methods in Cement Composites" J. of Engr. Mech. Div., Amer. Soc. of Civil Engr., Vol. 107, No. 6, Dec. 1981, pp. 1155-1171.

(44) Paramasivam, P., Curiskis, J.I., and Villiappan, S., "Micromechanics Analysis of Fiber Reinforced Cement Composites", Fibre Sci. and Tech., 20, 1984.

(45) Tolf, G., "Mechanical Behavior of a Short-Fibre Composite", Fibre Sci. and Tech., 19, 1983.

(46) Barsoum, R.S., "Triangular Quarter-Point Elements as Elastic and Perfectly Plastic Crack Tip Elements", International Journal for Numerical Methods in Engr., Vol. 11, 1977, pp. 85-98.

(47) Barsoum, R.S., "On the Use of Isoparametric Finite Elements in Linear Fracture Mechanics", International Journal for Numerical Methods in Engr., Vol. 10, 1976, pp. 25-37.

(48) Goodman, R.E., Taylor, R.L., and Brekke, T.L., "A Model for the Mechanics of Jointed Rock", Transactions of the ASCE Journal of the Soil Mechanics and Foundations Division, Vol. 94, No. SM3, May 1968, pp. 640-643.

(49) Henshell, R.D. and Shaw, K.G., "Crack Tip Elements are Unnecessary", International Journal for Numerical Methods in Engr., Vol. 9, 1975, pp. 495-507.

(50) Mandel, J.A., Pack, S.C., and Tarazi, S., "Micromechanical Studies of Crack Growth in Fiber Reinforced Materials", Engng. Frac. Mech. 16, 741-754, 1982.

(51) Mandel, J.A. and Pack, S.C., "Crack Growth in Fiber Reinforced Materials", J. of the Engr. Mech. Div., Amer. Soc. of Civil Eng., Vol. 108, No. 3 & 4, 1982.

(52) Pack, S.C. and Mandel, J.A., "2-D Multiplane Finite Element Technique For Solving a Class of 3-D Problems", Int. J. for Numerical Methods in Engineering, Vol. 19, 1983.

(53) Pack, S.C. and Mandel, J.A., "An Improved Finite Element for Connecting Adjacent Laminae of 2D Elements" accepted for publication in the Int. J. for Numerical Methods in Engineering.

(54) Tarazi, S. and Mandel, J.A., "Zero Thickness Quarter Point Crack Tip Finite Element for Modeling an Interface Between Two Materials", being prepared for submission to a journal.

(55) Ngo, D., "A Network-Topological Approach to a Finite Element Analysis of Progressive Crack Growth in Concrete Members", Ph.D. Thesis, University of California, Berkeley, California, June 1975.

(56) Wecharatana, M. and Shah, S.P., "Nonlinear Fracture Mechanics Properties", Fracture Mechanics of Concrete (edited by F.H. Wittmann), Elsevier Science Publishers, 1983, pp. 463-480.

(57) Mindess, S., "The Fracture of Fiber Reinforced and Polymer Impregnated Concrete: A Review", Fracture Mechanics of Concrete (edited by F.W. Wittmann), Elsevier Science Publishers, 1983, pp. 481-501.

(58) Visalvanich, K., and Naaman, A.E., "Evaluation of Fracture Techniques in Cementitious Composites", Fracture in Concrete, published by Amer. Soc. of Civil Engr., Proc. of a Session sponsored by the Committee on Properties of Materials at ASCE National Convention in Hollywood, Florida, Oct. 1980.

(59) "State-of-the-Art Report on Fiber Reinforced Concrete", ACI Committee 544, Concrete International, May 1982.

TABLE 1*

Comparison Between Numerical and Experimental Values of Composite Properties

Volume fraction adopted for FEM	Volume fraction modified due to orientation factor	l_1 (mm)	Numerical results				Actual volume fraction	Experimental results	
			Row-column arrangement		Staggered arrangement			Stress at first crack	Ultimate stress
			Yield stress	Ultimate stress	Yield stress	Ultimate stress			
			($N mm^{-2}$)		($N mm^{-2}$)			($N mm^{-2}$)	
2·67	5·68	30	1·70	4·60	1·75	5·20	5	2·20	5·25
1·42	3·02	30	1·55	2·60	1·65	3·00	3	1·83	2·61

* (Table 5 from reference 44)

TABLE 2

COMPARISON OF ANALYTICAL AND EXPERIMENTAL RESULTS

METHYL METHACRYLATE TENSION SPECIMENS

Specimen	a Inches	Applied Load for Initiation of Crack Growth Pounds — Finite Element Analyses: no Fibers	Finite Element Analyses: Fibers Bonded At Ends Only	Finite Element Analyses: Fibers Bonded Along Entire Length	Experiment: Fiber Cast Into Matrix Material	% Deviation Between Experimental & Finite Element With Fibers Bonded Along Entire Length
1	0.875	457	649	726	735	1.2
2	0.910	437	626	715	690	-3.5
3	0.930	426	615	711	720	1.3
4	0.875	457	649	726	670	-7.7

Table 3 Comparison of the Finite Element and Closed Form Solutions at $\theta = \pi$

E_2/E_1	K_I $K/in^{3/2}$ Theoretical Value	K_I Finite Value $r/a=0.125$	K_{II} $K/in^{3/2}$ Theoretical Value	K_{II} Finite Value $r/a=0.125$
1.5	27.96	27.44	0.31	0.17
3	27.70	28.0	0.66	0.53
10	27.34	29.1	0.92	0.86

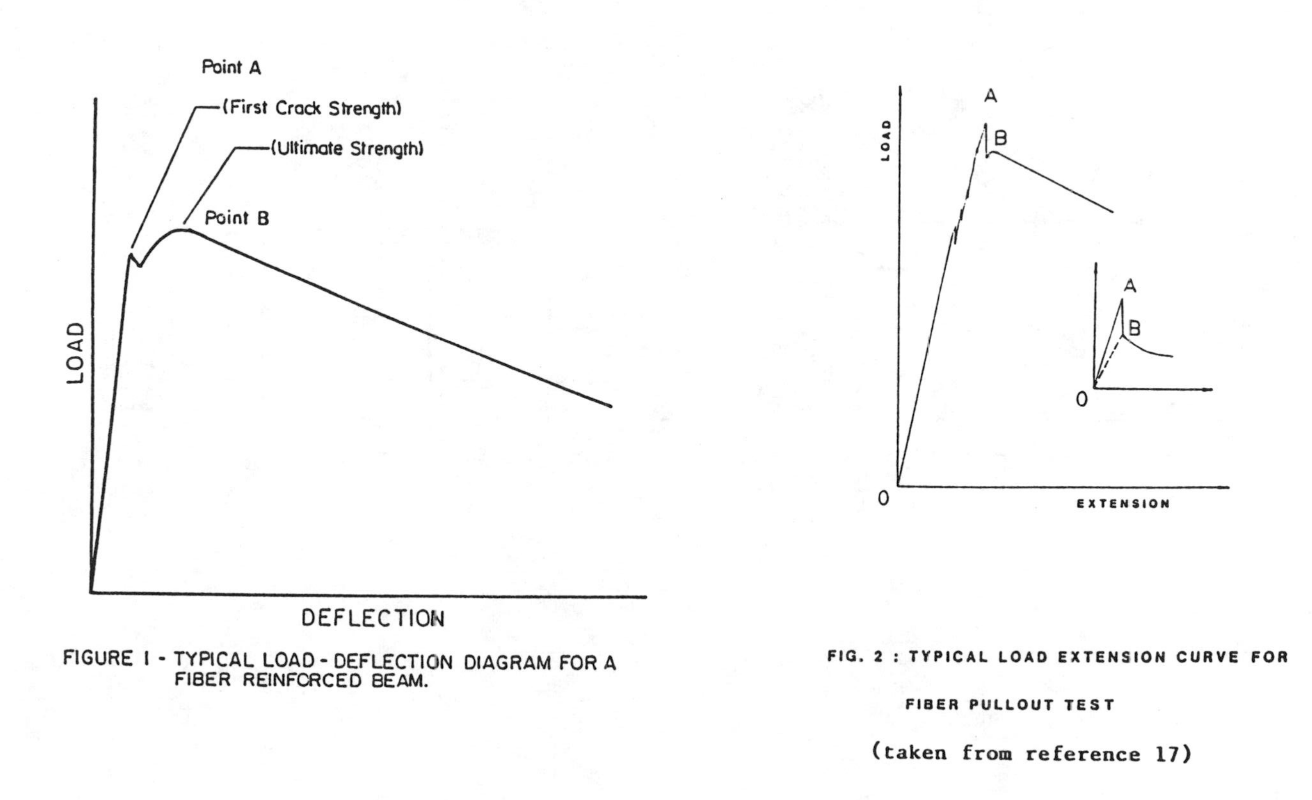

FIGURE I - TYPICAL LOAD - DEFLECTION DIAGRAM FOR A FIBER REINFORCED BEAM.

FIG. 2 : TYPICAL LOAD EXTENSION CURVE FOR FIBER PULLOUT TEST

(taken from reference 17)

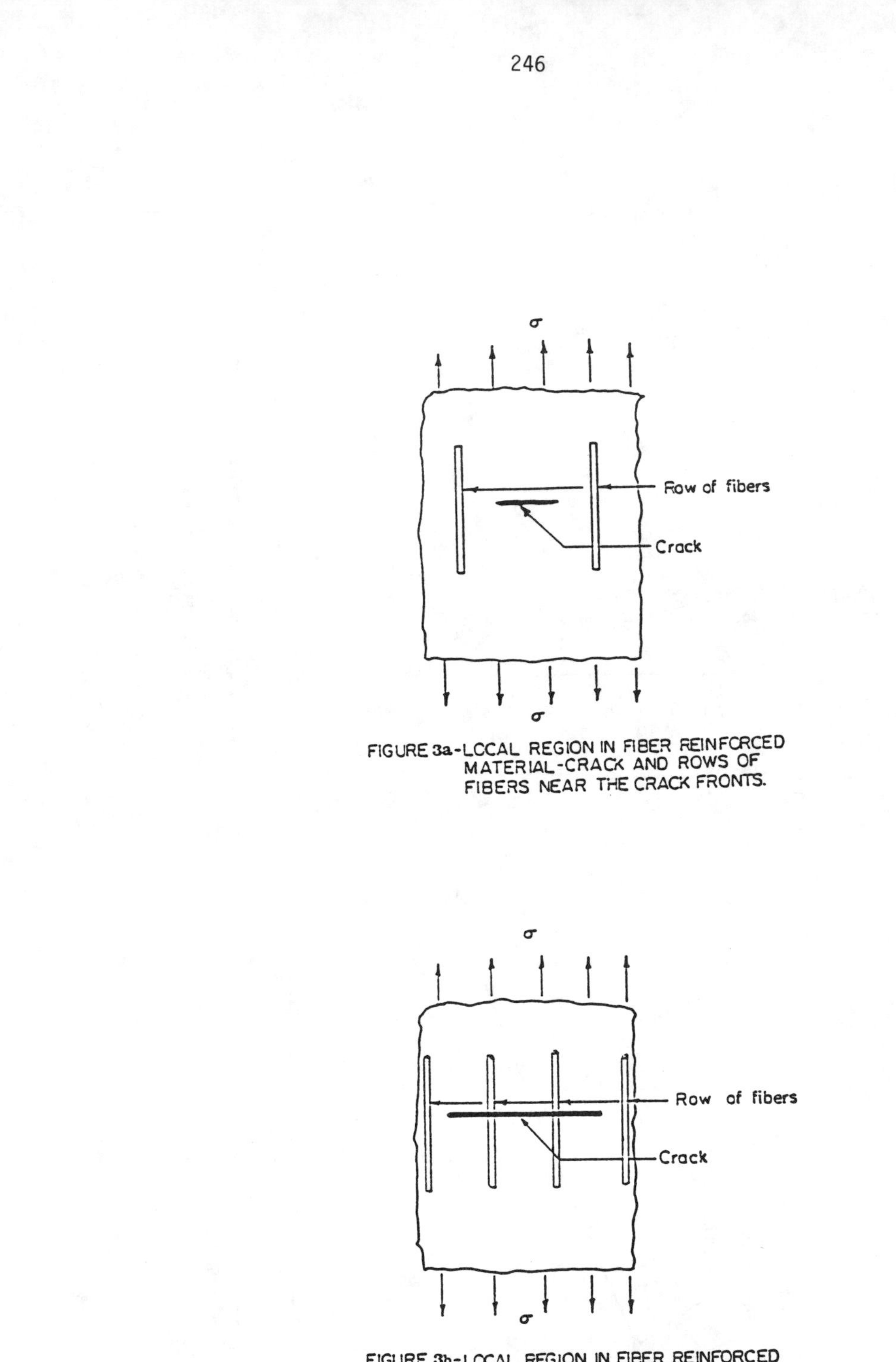

FIGURE 3a-LOCAL REGION IN FIBER REINFORCED MATERIAL-CRACK AND ROWS OF FIBERS NEAR THE CRACK FRONTS.

FIGURE 3b-LOCAL REGION IN FIBER REINFORCED MATERIAL- CRACK AND ROWS OF FIBERS SPANNING THE CRACK AND NEAR THE CRACK FRONT.

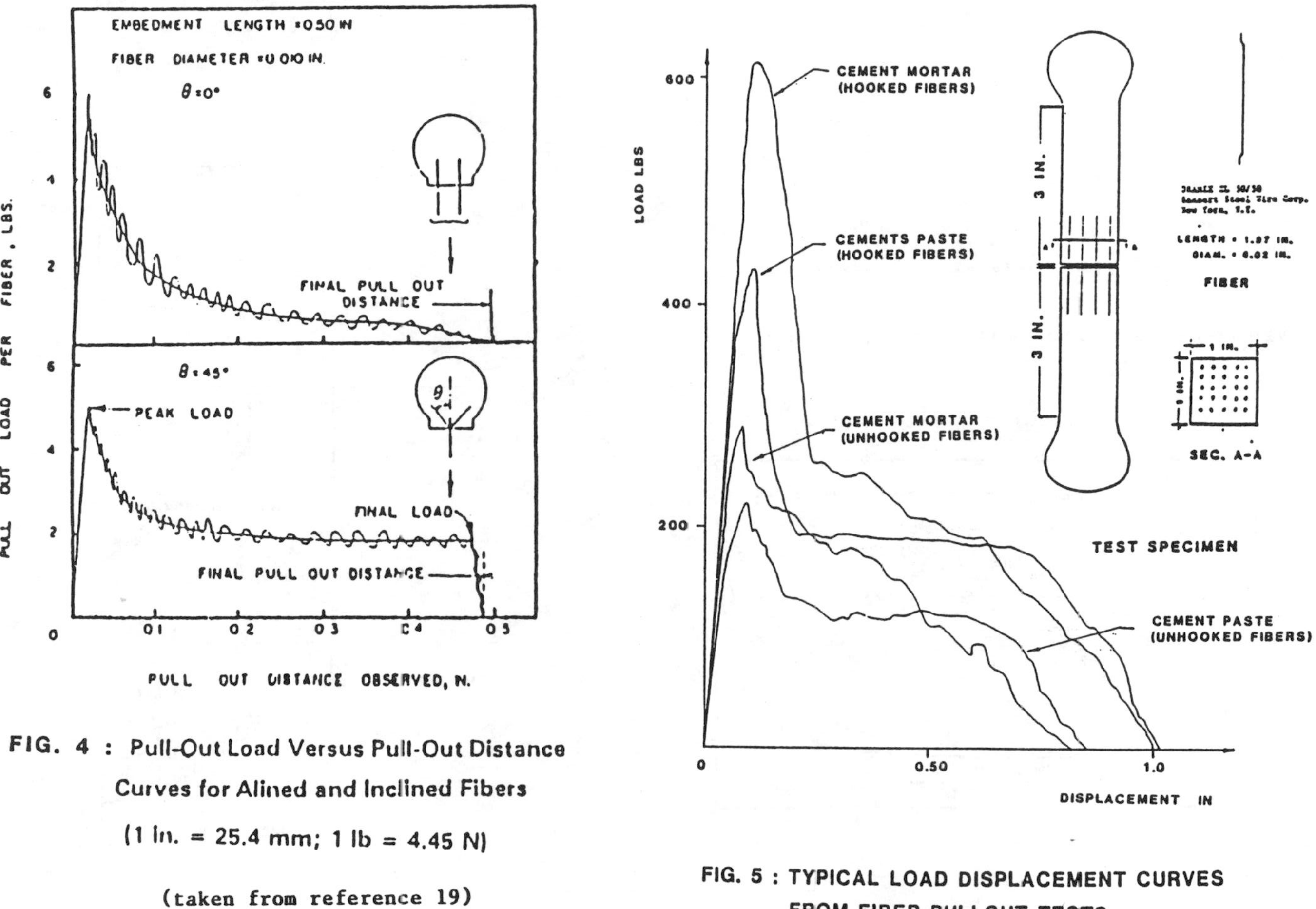

FIG. 4 : Pull-Out Load Versus Pull-Out Distance Curves for Alined and Inclined Fibers

(1 in. = 25.4 mm; 1 lb = 4.45 N)

(taken from reference 19)

FIG. 5 : TYPICAL LOAD DISPLACEMENT CURVES FROM FIBER PULLOUT TESTS

FIG.6 - An Idealized Representation of a Crack

(taken from reference 41)

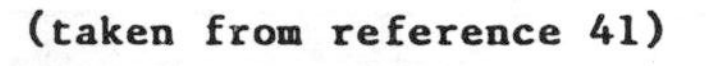

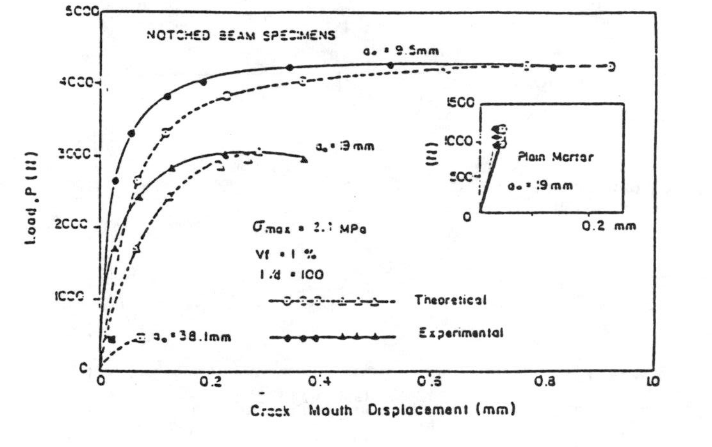

FIG 7 : A COMPARISON OF LOAD-CMD CURVES FOR BEAMS

(taken from reference 41)

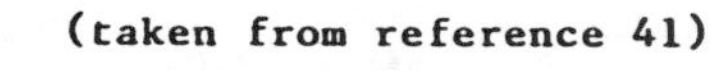

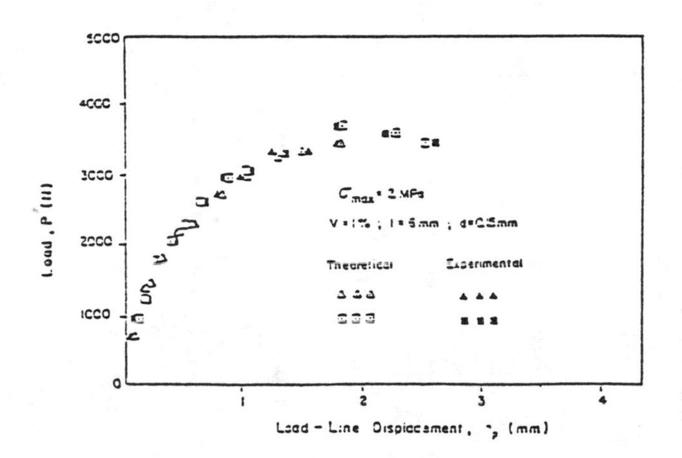

FIG 8 : A COMPARISON OF LOAD-CMD CURVES FOR DCB

(taken from reference 41)

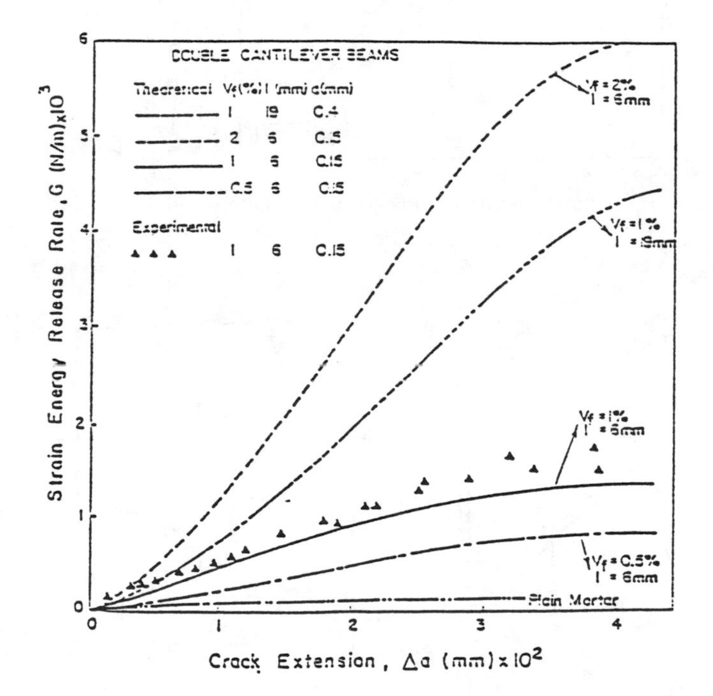

FIG 9 : RESISTANCE (R) CURVES

(taken from reference 41)

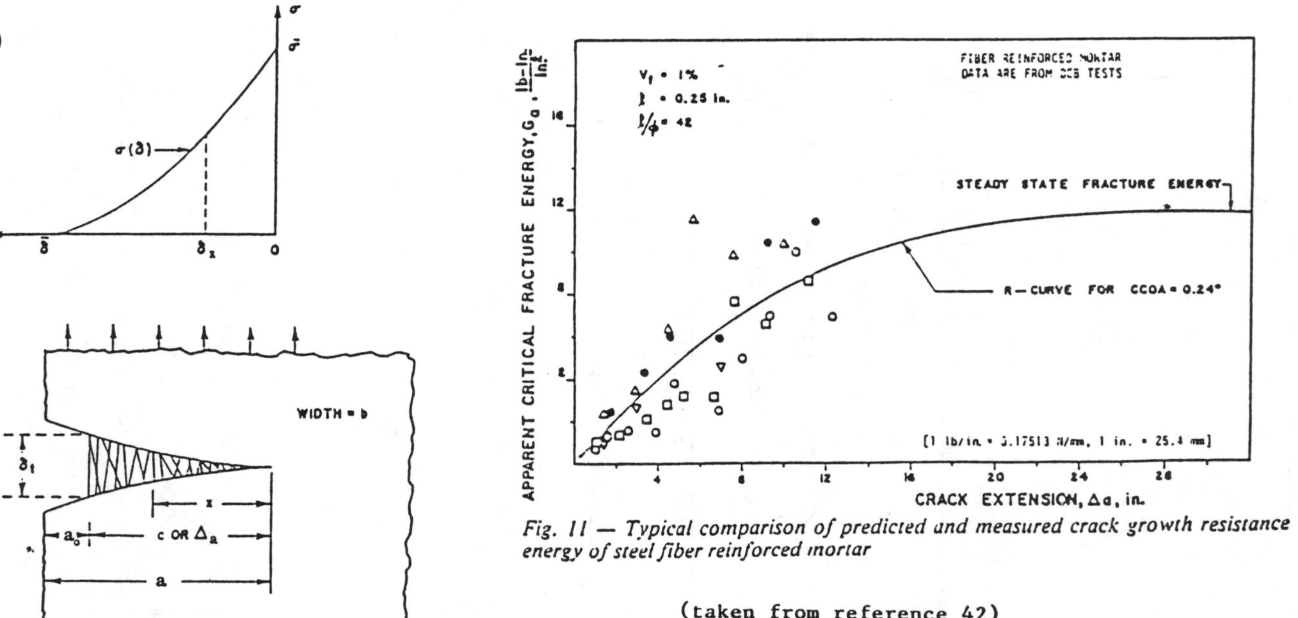

Fig. 11 — Typical comparison of predicted and measured crack growth resistance energy of steel fiber reinforced mortar

(taken from reference 42)

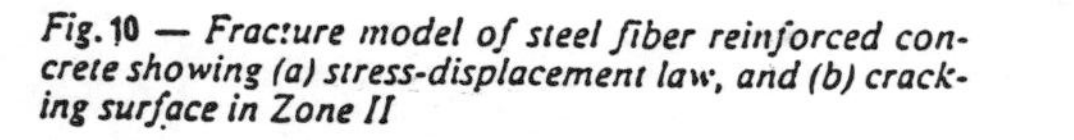

Fig. 10 — Fracture model of steel fiber reinforced concrete showing (a) stress-displacement law, and (b) cracking surface in Zone II

(taken from reference 42)

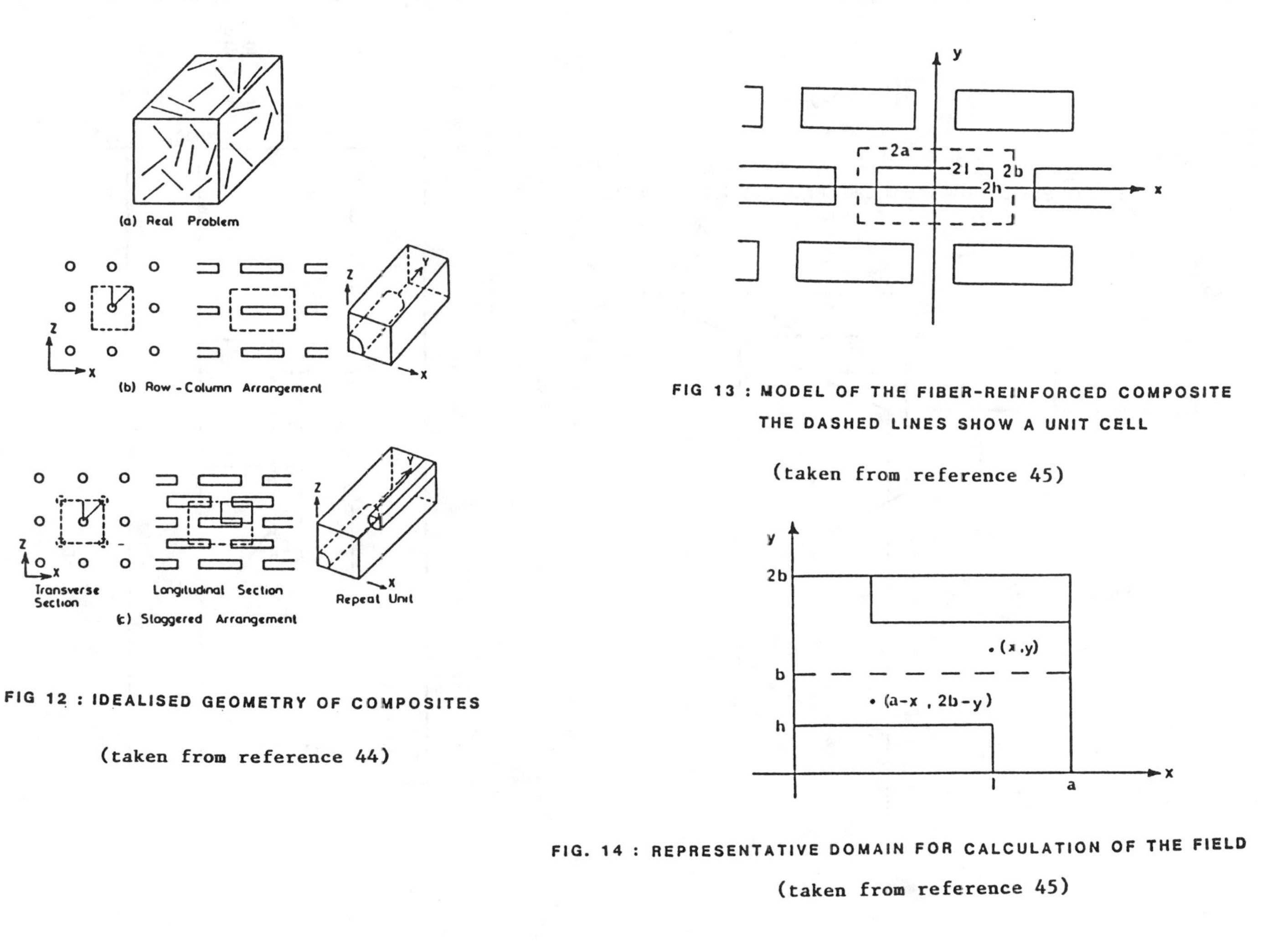

FIG 12 : IDEALISED GEOMETRY OF COMPOSITES

(taken from reference 44)

FIG 13 : MODEL OF THE FIBER-REINFORCED COMPOSITE
THE DASHED LINES SHOW A UNIT CELL

(taken from reference 45)

FIG. 14 : REPRESENTATIVE DOMAIN FOR CALCULATION OF THE FIELD

(taken from reference 45)

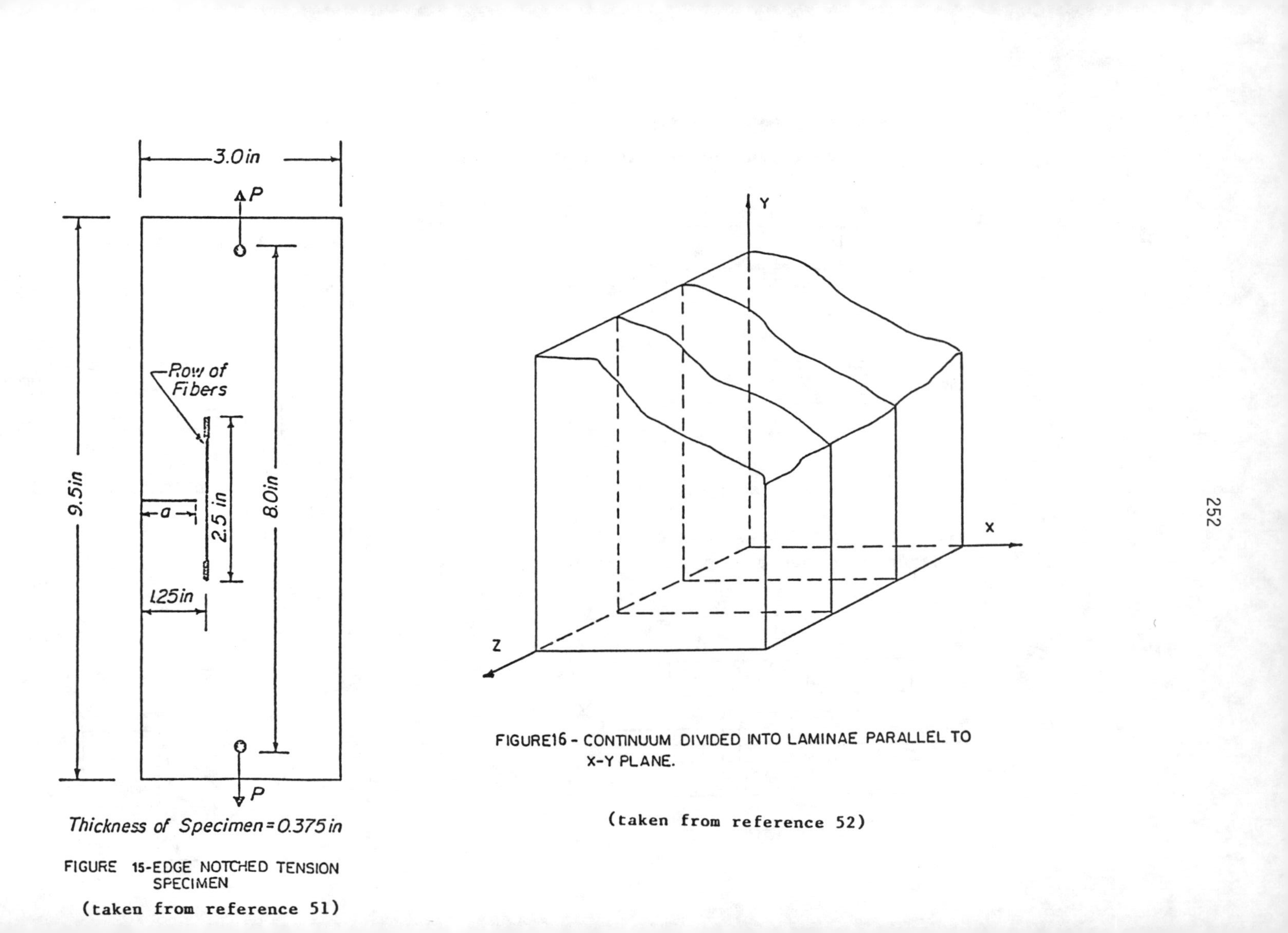

FIGURE 15-EDGE NOTCHED TENSION SPECIMEN

(taken from reference 51)

FIGURE 16 - CONTINUUM DIVIDED INTO LAMINAE PARALLEL TO X-Y PLANE.

(taken from reference 52)

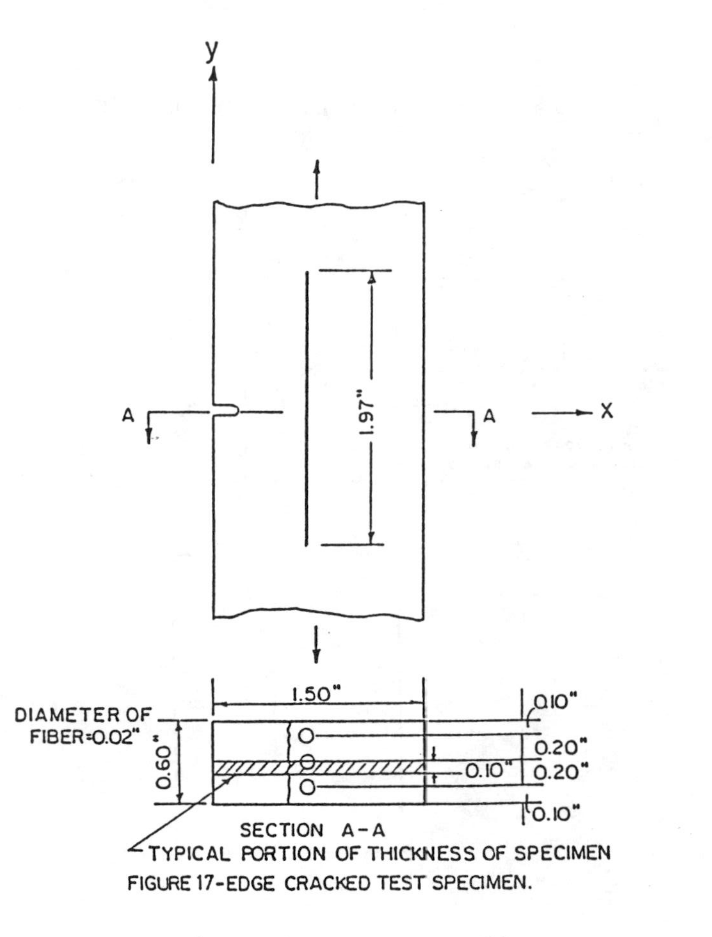

FIGURE 17-EDGE CRACKED TEST SPECIMEN.

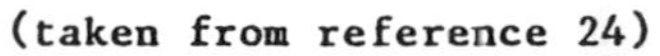

(taken from reference 24)

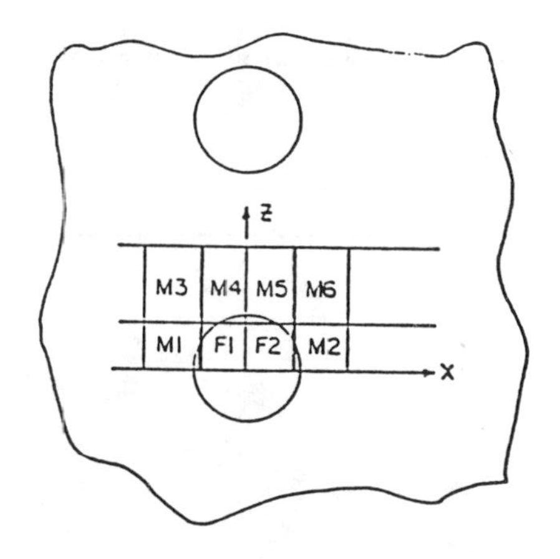

FIG. 18 - LAMINA REGION MODELED IN DUAL PLANE ANALYSIS

(taken from reference 24)

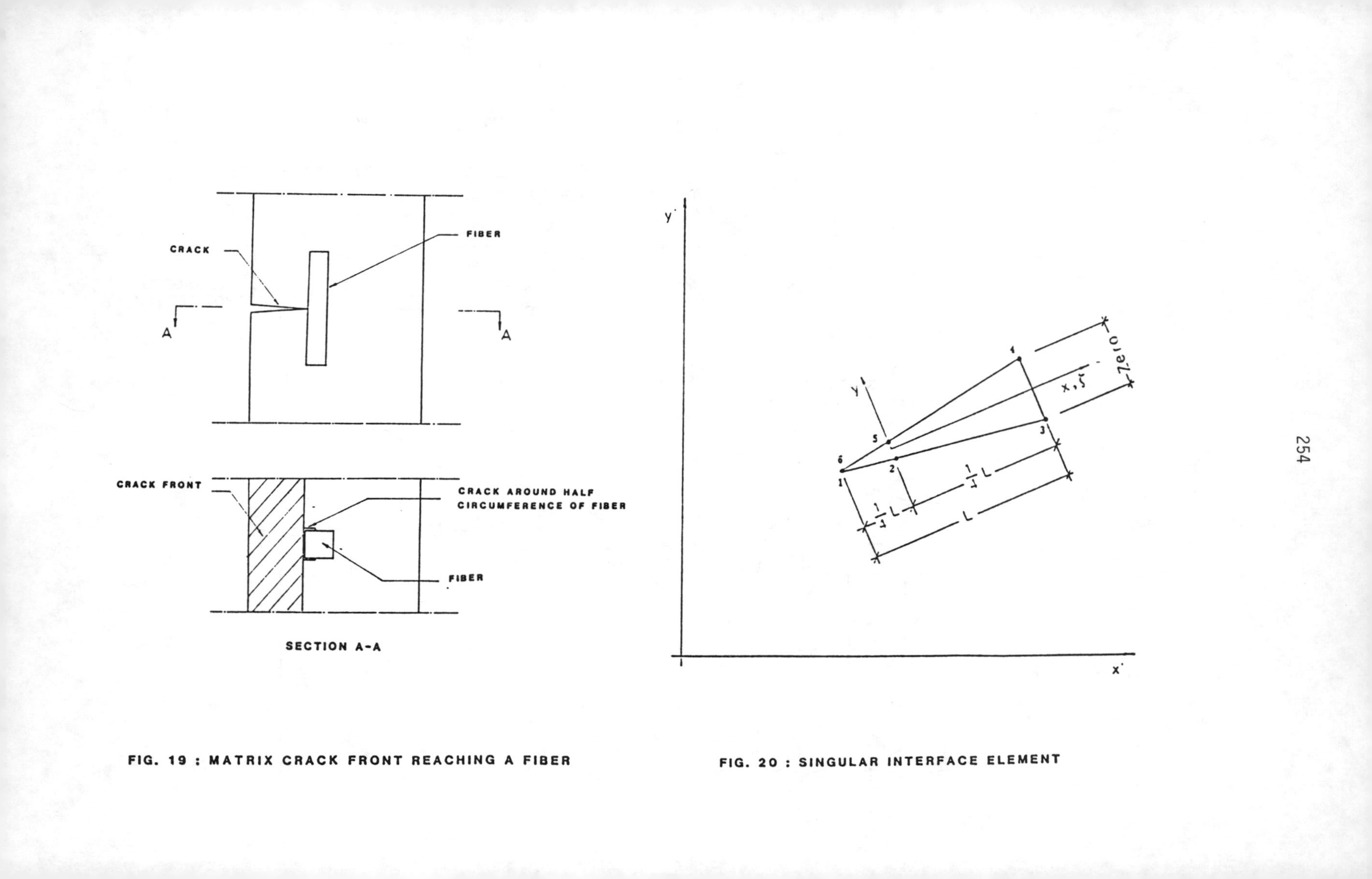

FIG. 19 : MATRIX CRACK FRONT REACHING A FIBER

FIG. 20 : SINGULAR INTERFACE ELEMENT

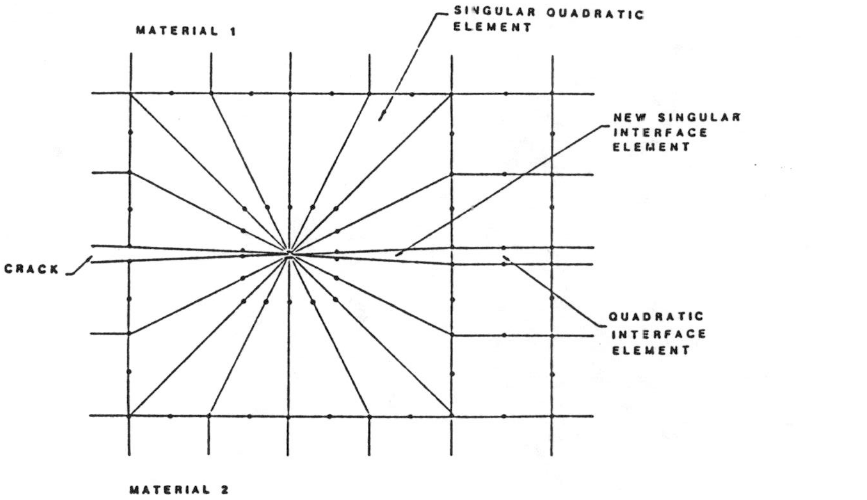

FIG. 21 : CRACK TIP REGION IN A FINITE ELEMENT MESH

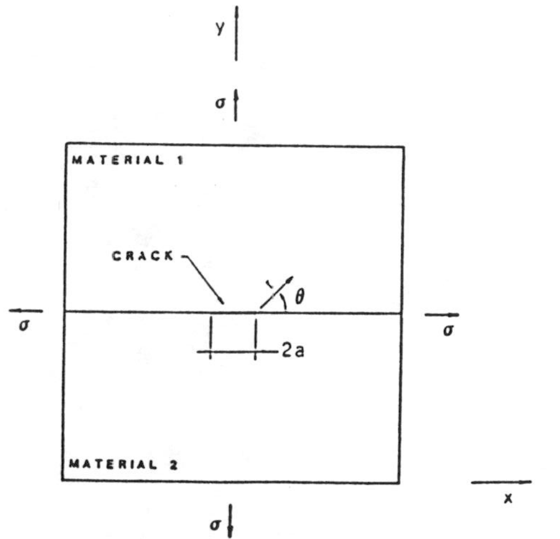

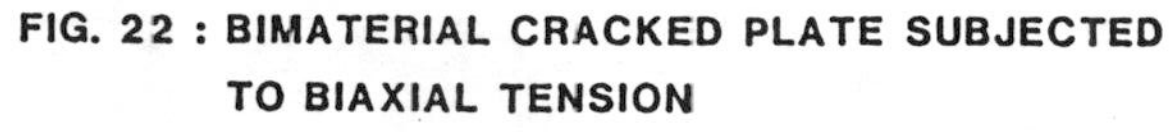

FIG. 22 : BIMATERIAL CRACKED PLATE SUBJECTED TO BIAXIAL TENSION

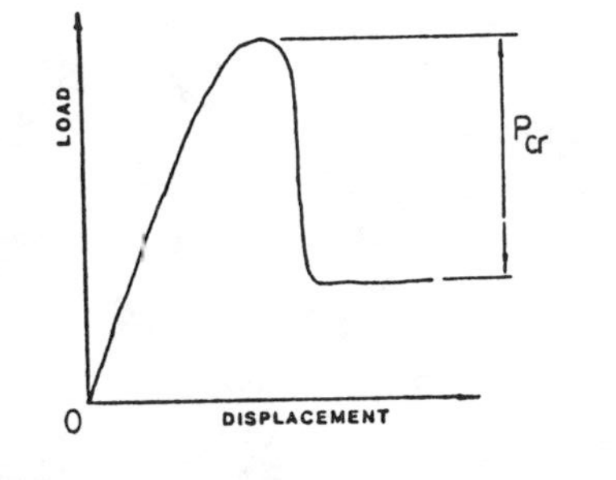

FIG. 23 : LOAD DISPLACEMENT CURVE OF A FIBER PULLOUT TEST

STEEL FIBER CONCRETE
US-SWEDEN joint seminar
(NSF-STU)
Stockholm 3-5 June, 1985
S P Shah and Å Skarendahl,
Editors

DETERMINATION AND SIGNIFICANCE OF THE FRACTURE TOUGHNESS OF STEEL FIBRE CONCRETE

by Arne Hillerborg, Prof, Div of Building Materials, Lund Inst of Technology, Lund, Sweden

---

ABSTRACT

The tensile fracture behaviour of steel fibre concrete is discussed and compared to that of plain concrete. Whereas plain concrete ceases to transfer stresses already after a damage zone deformation of a few tenths of a millimeter, steel fibres continue to transfer appreciable stresses after wide cracks have appeared in the concrete. The effect of this difference on the behaviour of notched and unnotched beams has been analysed by means of finite element calculations, using the fictitious crack model and a simple approximation for the stress-deformation properties of the damage zone. It is concluded that linear elastic fracture mechanics is never applicable to steel fibre concrete and that steel fibre concrete structures can often be analysed by means of the theory of plasticity.

Arne Hillerborg is Professor of Building Materials at Lund Institute of Technology, University of Lund, Sweden. He is a member of RILEM TC 50 - Fracture Mechanics of Concrete. He has earlier been active within the field of structural engineering, where he developed the Strip Method for design of reinforced concrete slabs.

INTRODUCTION

Steel fibres are used in concrete in order to change the tensile fracture behaviour. It is therefore natural to use the tensile fracture behaviour as a starting-point in all discussions regarding the influence of steel fibres on strength and deformation properties.

TENSILE FRACTURE BEHAVIOUR

Fig 1 shows the stress-deformation behaviour in a tensile test with a bar with a constant cross section, made of an homogenous material. The test is made under conditions of deformation control. Before the maximum stress is reached the strain is the same at all points along the bar. When the deformation exceeds the point of maximum stress a damage zone with a reduced strength is formed somewhere along the bar.

As the deformation increases still further the damage in the damage zone also increases. This leads to a decrease in strength and consequently in stress in that particular zone. The other parts of the bar are unloaded. The strain at that stage is no longer the same at all points along the bar. The strain within the damage zone is greater than that within the other parts.

The complete stress-deformation behaviour can be described by means of two curves, as shown in Fig 1, viz one stress-strain curve (σ-ε), valid for all the material, and one stress-deformation curve (σ-w) for the additional deformation within the damage zone. This additional deformation is

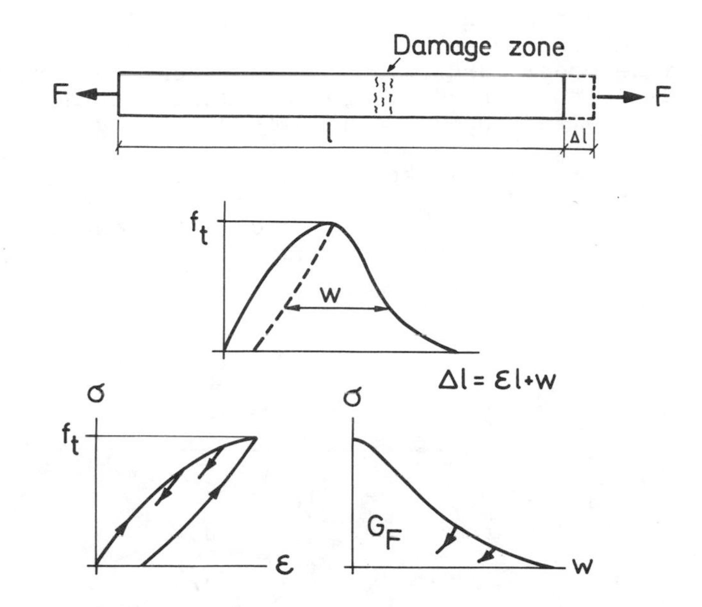

FIG 1. The total elongation is the result of a distributed strain and an additional deformation w within the damage zone.

denoted w. The deformation $\Delta \ell$ on any gauge length ℓ, containing one damage zone is

$$\Delta \ell = \ell \cdot \varepsilon + w \qquad (1)$$

The values of ε and w are taken from the curves in question for the same stress, using the unloading branches where unloading occurs.

Energy is absorbed during the deformation. This energy can be divided into two parts, viz those belonging to the two types of diagrams in Fig 1. The area enclosed by the σ-ε-curve (between the loading and unloading branches) corresponds to an energy absorbed per unit volume of the material, whereas the area below the σ-w-curve corresponds to an energy absorbed per unit cross sectional area of the damage zone.

The area below the complete σ-w-curve is denoted by G_F. It can be interpreted as the energy which is absorbed per unit (projected) area of the fracture surface.

The σ-ε-curve of a material is completely defined if the values of E (modulus of elasticity) and f_t (tensile strength) and the shape of the curve are known. The σ-w-curve is completely defined if the values of f_t and G_F and the shape of the curve are known.

As two curves are required in order to define the stress-deformation properties of the material in a structure, ordinary simple model laws are not valid when two structures are compared, neither are ordinary equations for the design of structures valid. More general model laws have to be developed.

Suppose that we have a structure made of a material with σ-ε- and σ-w-curves according to Fig 2. If model laws are to be valid, the shape of the stress-deformation curve for any part of the structure must be independent of the size of the structure.

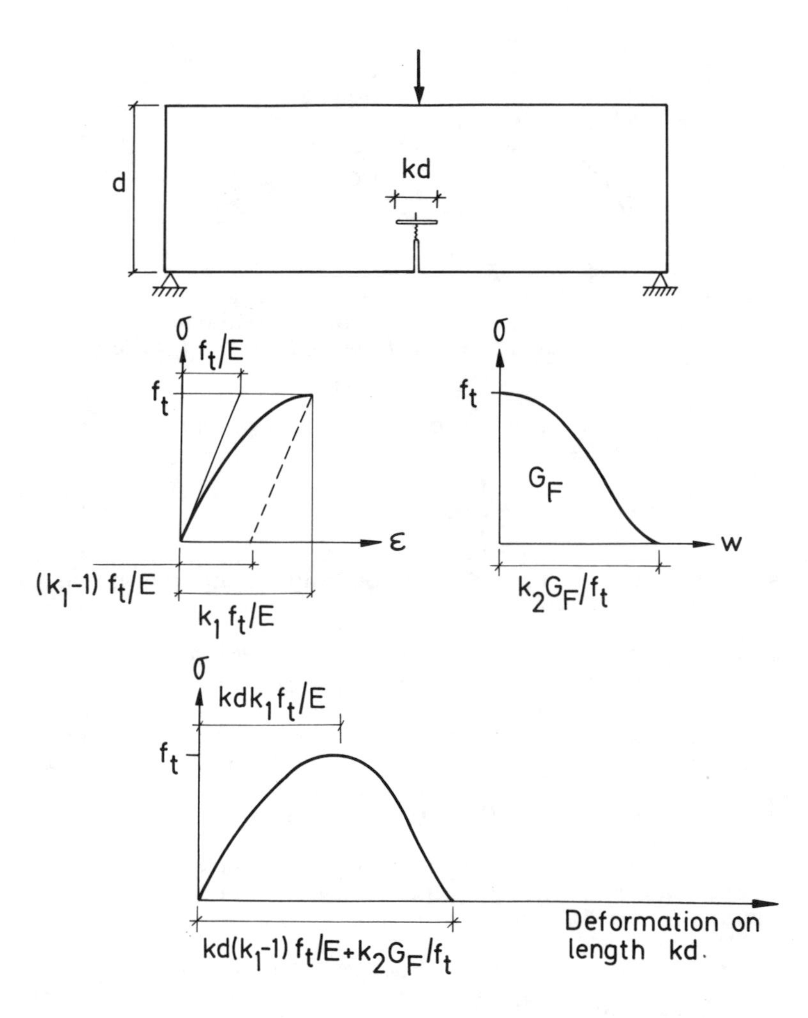

FIG 2. Stress-deformation relation for a length containing one damage zone.

Suppose that we calculate the stress-deformation curve for an arbitrary length kd, comprising a damage zone. This length is proportional to the size of the structure. The length change $\Delta \ell$ can be calculated by means of Eq (1). It consists of two parts, belonging to the two curves. The part belonging to the σ-ε-curve is proportional to f_t/E and to kd, whereas the part belonging to the σ-w-curve is proportional to G_F/f_t.

The shape of the complete stress-deformation curve is unchanged if the relation between these two parts is unchanged. Thus we obtain the following condition, which must be fulfilled if model laws are to be valid

$$\frac{kdf_t/E}{G_F/f_t} = \text{constant} \tag{2}$$

or

$$\frac{d}{\ell_{ch}} = \text{constant} \tag{3}$$

with

$$\ell_{ch} = \frac{EG_F}{f_t^{\,2}} \tag{4}$$

The value ℓ_{ch} is called the <u>characteristic length</u> of the material. It is a material property, which has no direct physical interpretation. It is however of great practical significance, as it determines the length of damage zones at fracture, the suitable element size in finite element analyses and many other things.

If the shapes of the σ-ε- and σ-w-curves and the value of the relation (3) are known for a structure, the response of the structure to a load is completely defined if the load is expressed as a calculated stress σ, divided by the tensile strength f_t. The calculated stress can be arbitrarily chosen as a flexural stress, a shear stress etc. The most important factor is the stress at fracture, i e the strength f.

Thus any formal strength f of a structure can be expressed as a relation between f/f_t and d/ℓ_{ch}. It can also be expressed as a relation between the logarithms of these values. For small changes of the values the relation can be expressed by means of a straight line (the tangent to the curve):

$$\ell n \frac{f}{f_t} = A - B \ell n \frac{d}{\ell_{ch}} \tag{5}$$

This can also be written

$$\ell nf = A - B\ell nd + B\ell nE + B\ell nG_F + (1-2B)\ell nf_t \tag{6}$$

Differentiation of this expression gives

$$\frac{df}{f} = B\,\frac{dd}{d} + B\,\frac{dE}{E} + B\,\frac{dG_F}{G_F} + (1-2B)\,\frac{df_t}{f_t} \tag{7}$$

This equation shows the <u>sensitivity</u> of the strength to changes in different properties, i e the relative change in strength due to a relative change in a property. All these sensitivities depend on the slope B of a curve showing the relation between f/f_t and d/ℓ_{ch} in a diagram with logarithmic scales. The sign of B is chosen so that a downward slope is positive, because the slope is always downward.

One extreme value of B is zero. In such a case the sensitivity to changes in G_F is 0 and the sensitivity to changes in f_t is 1. This corresponds to the ordinary theory of strength of materials.

The other extreme value of B is 0.5. In this case the sensitivity to changes in G_F is 0.5 and the sensitivity to changes in f_t is 0. This corresponds to linear elastic fracture mechanics.

It is evident from the above that the sensitivities concerning G_F and f_t are interdependent, and that one increases as the other decreases.

Diagrams showing the relation between f/f_t and d/ℓ_{ch} for some structures of plain concrete can be found e g in /1, 2, 4, 5/. For beams of an ordinary size the sensitivity with regard to G_F is 0.15-0.35 with a corresponding sensitivity with regard to f_t of 0.3-0.7.

FRACTURE ENERGY OF STEEL FIBRE CONCRETE (SFC)

The fracture energy G_F of a material is determined by the area under the σ-w-curve, see Fig 1. For concrete the stress in the damage zone decreases rather rapidly as the deformation w increases. Typical σ-w-curves for concrete are shown in Fig 3 /2/. When the deformation w has reached a few hundredths of a millimeter the stress has fallen to a small fraction of the strength.

When steel fibres are present they will also be able to transfer stresses across the damage zone when the deformation w is a matter of millimeters. Therefore there is a great difference between the σ-w-curves of plain concrete and of steel fibre concrete (SFC). In SFC the concrete matrix mainly transfers stresses just as the damage zone starts forming, whereas practically all stresses across the damage zone are transferred by the fibres as w grows from a few hundredths of a millimeter up to several millimeters.

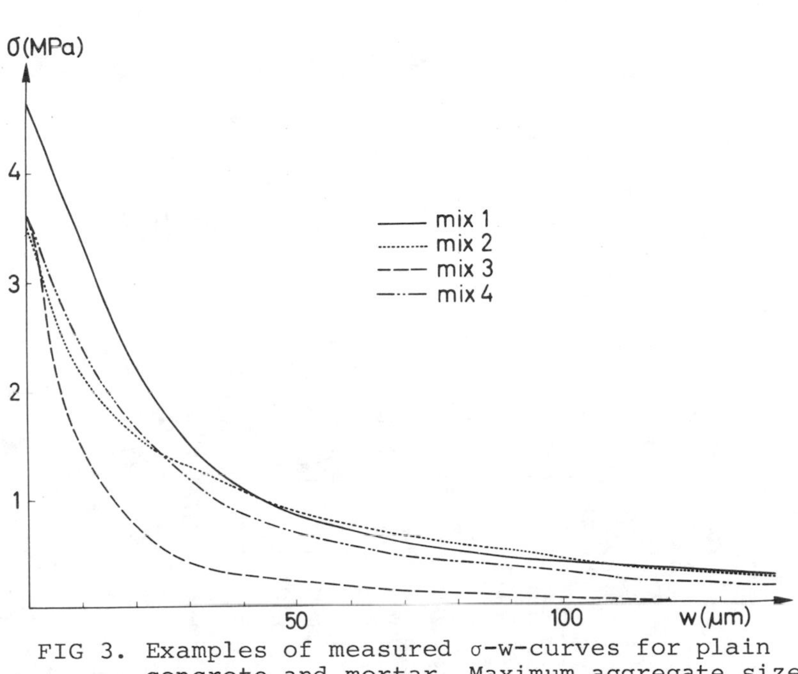

FIG 3. Examples of measured σ-w-curves for plain concrete and mortar. Maximum aggregate size is 2 mm for mix 3, and 8 mm for the other mixes /2/.

The stress transfer across the damage zone in SFC can be analysed theoretically, see e g /6/. For such an analysis it is necessary to know - or to make assumptions regarding - the laws governing the stress-deformation properties of the fibres as they are pulled out of the concrete in different directions. In practice it is hardly possible to make realistic calculations of the σ-w-curves for SFC. Therefore analyses of real structures should be based on measured σ-w-curves for the material in question.

Examples of measured σ-w-curves for steel fibre mortar are shown in Fig 4 /3/. It must be noted that the highest of these curves is not a true σ-w-curve, as it has a maximum for $w>0$. This type of curve will be discussed below.

All the curves in Fig 4 for steel fibre mortar show appreciable stresses when w reaches values of several millimeters, corresponding to cracks of the same order. In this respect SFC differs very much from plain concrete. For plain concrete the stresses fall to zero before the width of the cracks becomes greater than a few hundredths of a millimeter. Therefore the complete σ-w-curve is of interest for plain concrete, and also the area G_F below that curve.

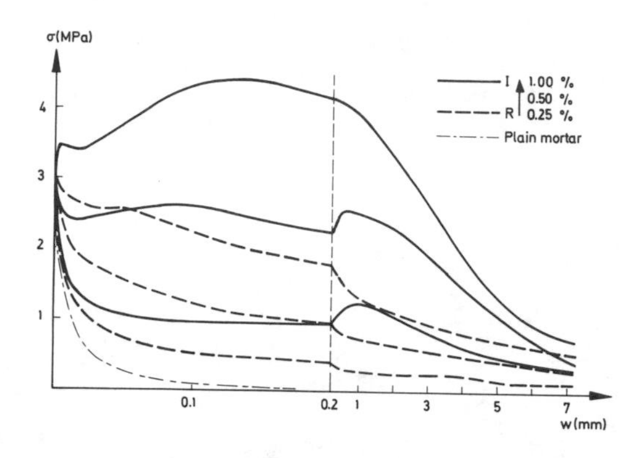

FIG 4. Examples of σ-w-curves for steel fibre mortars. R=round smooth fibre, I=indented fibre. The curves represent fibre contents of 0.25, 0.5 and 1.0 % by volume /3/.

For most real structures a behaviour with cracks wider than about 1 mm is of little interest. The complete σ-w-curve for such a material is therefore of no practical significance for an analysis, nor is the area G_F below that curve. The use of the σ-w-curve in a general analysis and discussion must therefore be different for plain concrete and for SFC.

Fig 4 indicates that the σ-w-curve for SFC can be approximated by a horizontal line for w-values between a few hundredths of a millimeter and a few tenths of a millimeter, which is the most interesting part of the curve for practical applications. The first part of the curve can be approximated by a straight line. With these approximations the shape of the σ-w-curve can be assumed to be according to Fig 5 for SFC.

As the value of G_F is undetermined (very high) for such a curve, the ordinary definition of ℓ_{ch} is impossible to use. Instead a formal value G_F' will be used, corresponding to a continuation of the sloping line, see Fig 5. A corresponding value

$$\ell'_{ch} = \frac{EG_F'}{f_t^2} \tag{8}$$

will be used in the analysis. In this way a direct comparison can be made with an unreinforced concrete, where the σ-w-curve is approximated by a straight line.

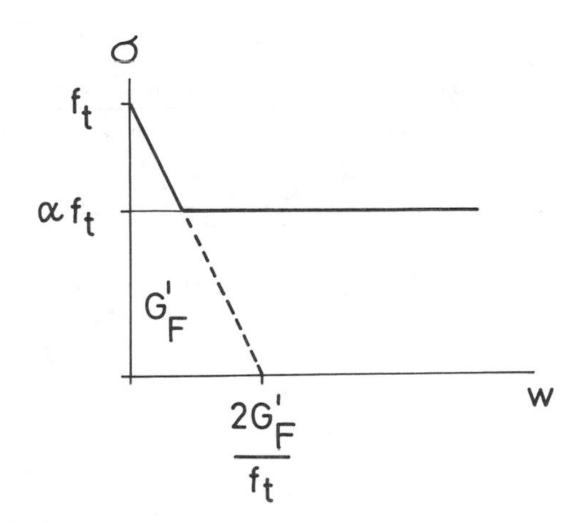

FIG 5. Assumed approximate σ-w-curve for steel fibre concrete.

The shape of the σ-ε-curve also has an influence on the fracture behaviour. The use of non-linear σ-ε-curves in finite element analyses is rather complicated and expensive. Therefore all analyses so far have been based on the assumption that the σ-ε-curve is a straight line, i e a purely elastic behaviour.

Some results of analyses performed under these assumptions are shown in Figs 6-8. The calculations are made with the Fictitious Crack Model and finite elements. A further simplifying assumption is that only one damage zone occurs. Fig 6 shows how the ratio between the net bending strength f_{net} and the tensile strength f_t varies with $(d-a)/\ell'_{ch}$ for an unnotched and for a notched beam in bending. Different α-values refer to Fig 5. A higher α-value corresponds to a higher fibre content.

The influence of the fibres on the slope of the first part of the σ-w-curve (and thus on G'_F) is not known. It can be expected that the fibres change the slope so that the value of G'_F increases. This is not taken into account in the discussion below, but it is assumed that G'_F has the same value as G_F for the matrix concrete.

From Fig 6 it can be seen that the fibres theoretically have no influence on the bending strength for beams with low values of $(d-a)/\ell'_{ch}$, particularly where the fibre content is low. On the other hand for high values of $(d-a)/\ell'_{ch}$ the bending strength is nearly constant, corresponding to a purely plastic behaviour, governed by the horizontal part of the assumed σ-w-curve.

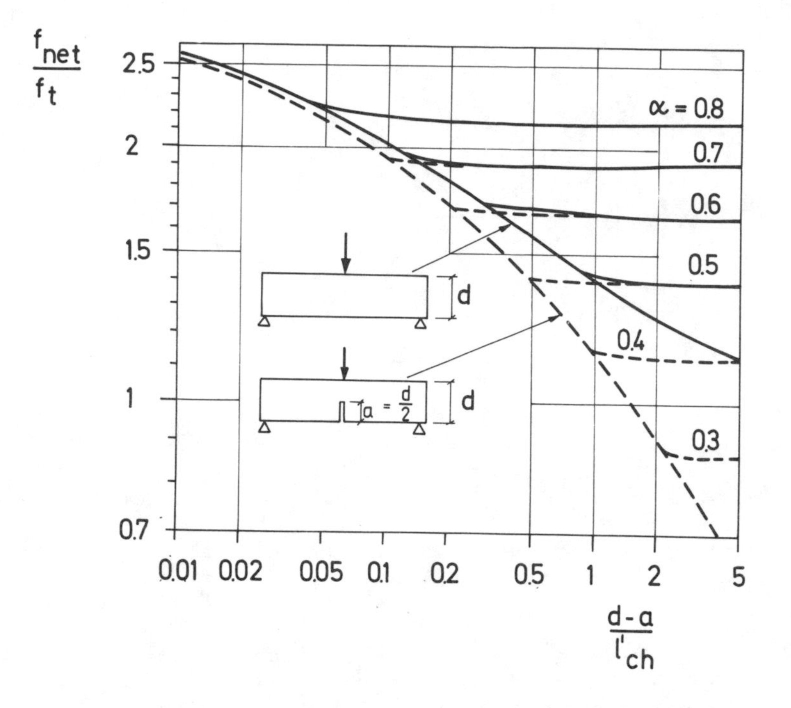

FIG 6. Theoretical flexural strength of unnotched and notched beams.

It is important to notice that the curves in Fig 6 become horizontal when the size of the beam grows, and that there is no size effect for large beams, even if they are notched (as long as the used assumptions are valid). This is in sharp contrast to materials where conventional fracture mechanics is applicable.

Where the curves in Fig 6 are horizontal the sensitivity of the structural strength f according to Eq (7) is 1 with regard to f_t and 0 with regard to G_F'. This means that the strength of the structure can be analysed by means of the ordinary theory of strength of materials - in this case the theory of plasticity - but not by means of fracture mechanics.

Figs 7-8 show theoretical load-deformation diagrams for unnotched and notched beams with d/ℓ'_{ch} =1. It can be noticed that the value of α, i e the amount of fibres, is of primary importance where toughness is concerned.

DETERMINATION OF FRACTURE TOUGHNESS

For plain concrete the fracture energy G_F can suitably be measured by means of a bending test on a notched beam /2, 5, 7/.This type of test is not suitable for SFC. The main reason for this is that the material is too tough, so that the beam cannot be completely broken in such a test. Another reason is

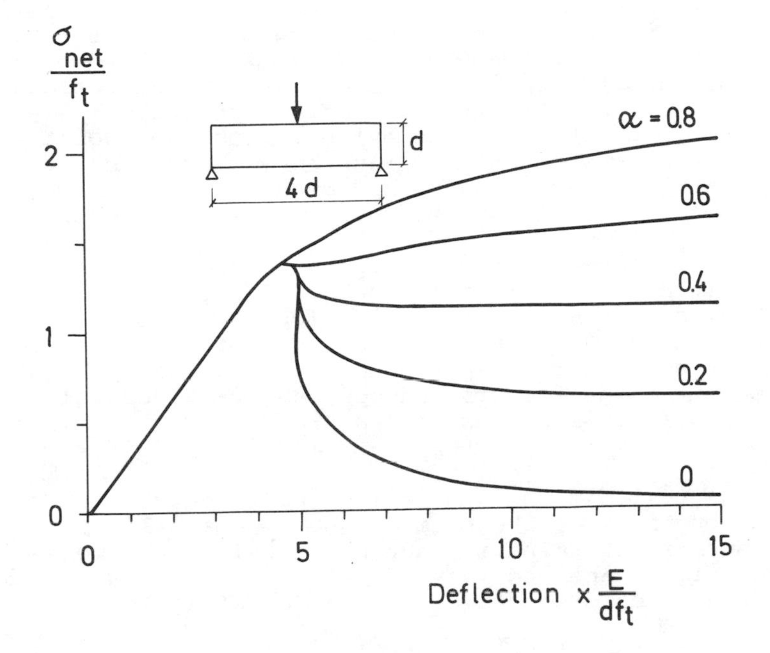

FIG 7. Theoretical load-deflection curves for unnotched beams with $d/\ell'_{ch}=1$.

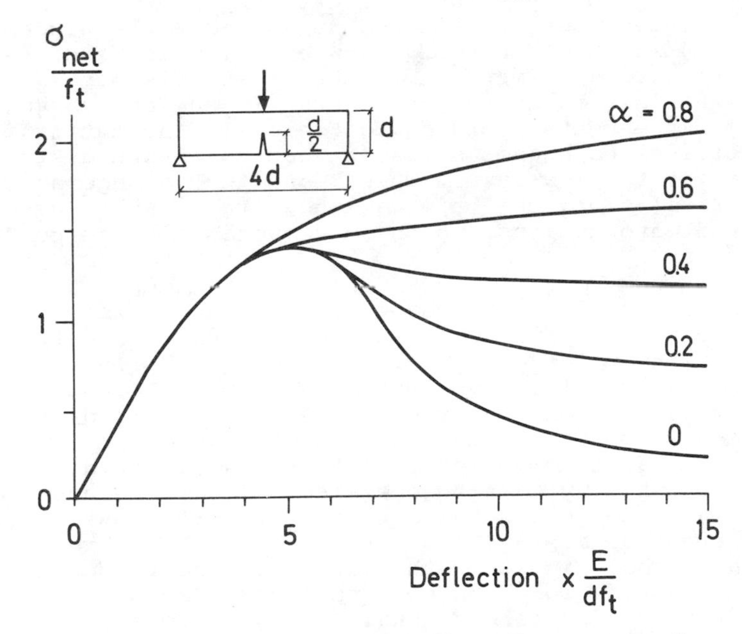

FIG 8. Theoretical load-deflection curves for notched beams with $d/\ell'_{ch}=1$, $a/d=0.5$.

that the value of G_F, i e the area below the complete σ-w-curve, is not of any practical significance for SFC.

Methods based on conventional fracture mechanics, i e basically linear elastic fracture mechanics (LEFM), are not applicable to SFC. This is evident from the fact that the real crack (a stress-free crack) never advances in SFC, as large stresses are transferred even when the crack width in the concrete matrix has reached many millimeters.

It must be noted that more sophisticated methods based on LEFM are also unapplicable to SFC. One example of such a method is the R-curve analysis. Tests on a specimen with a constant size and a variable notch depth may give R-curves which seem rather independent of the notch depth. If the size of the specimen is changed, the R-curves will also change /3/. This proves that a R-curve is not a material property.

The only valid method to define and to determine the fracture toughness of SFC seems to be by means of the σ-w-curve. For such a determination notched tensile specimens can be used, Fig 9. The best way of performing such a test in order to obtain representative results has still to be investigated.

As long as the fibre content is small ($\alpha<1$ in Fig 5) the fibres have little influence on the strength of the material or on the shape of the σ-ε-curve. The σ-ε-curve can then be approximated by means of a straight line. With a great fibre content the situation changes. Thus the highest curve in Fig 4 is an example of a σ-w-curve, determined by means of a notched specimen, where the stress rises to a maximum for $w>0$. This is no real σ-w-curve in the sense shown in Fig 1. In a test according to Fig 1 all the material will be stressed to higher stresses than the assumed starting point for the σ-w-curve. This means that damage zones will appear all along the bar, causing large, rather evenly distributed deformations, which on a macroscale can be treated as strains.

Consequently if we, by means of a notched tensile specimen, measure a σ-w-curve according to Fig 10a, the part before the maximum point on a macroscale should belong to the σ-ε-curve. The resulting σ-ε- and σ-w-curves are shown in Fig 10b. The shape of the σ-ε-curve cannot be determined from the σ-w-curve in Fig 10a, as the process of creation of many damage zones is rather complicated. It is however evident that the creation of many damage zones leads to a σ-ε-curve which is non-linear to such an extent that it more closely resembles the curve for a plastic material. When the fibre content reaches a certain level the material can be looked upon as being plastic, in which case the term fracture toughness in the ordinary fracture mechanical sense looses its meaning.

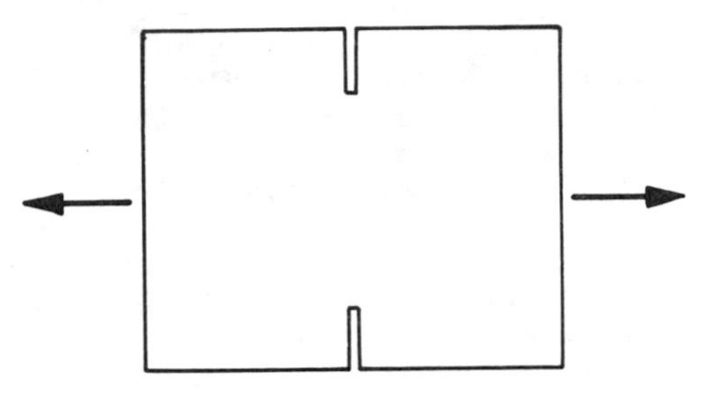

FIG 9. Example of a tensile specimen for the determination of σ-w-curves.

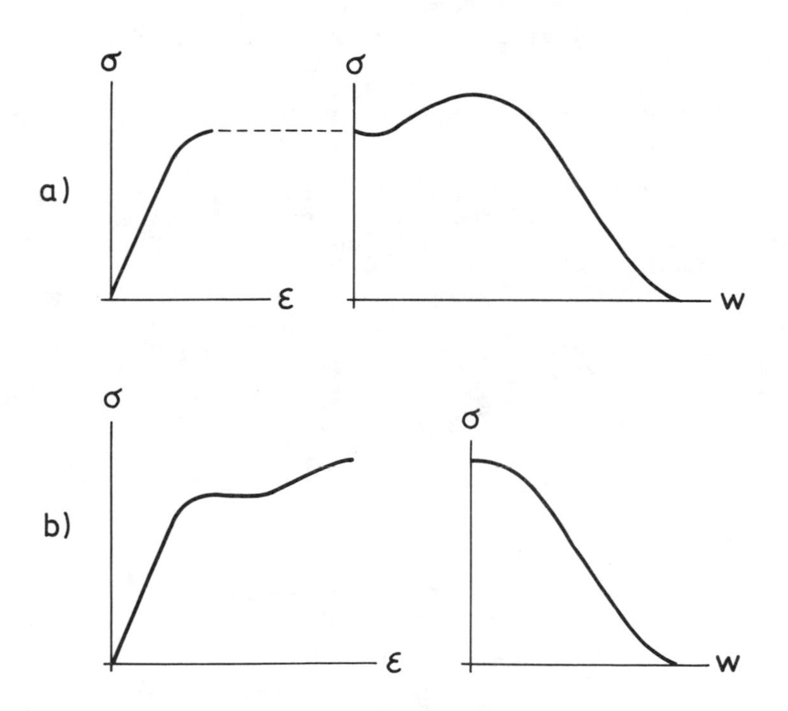

FIG 10. If the measured σ-w-curve has a higher maximum than the starting point, only the part to the right of this maximum belongs to the true σ-w-curve.

The transition from a nearly elastic to a nearly plastic σ-ε-curve can - at least theoretically - happen rather suddenly when the fibre content reaches a certain limit. The corresponding change in behaviour of a structure in bending does not take place suddenly. In bending the number of damage zones will increase as the fibre content increases, and the behaviour of the structure will change towards a more plastic behaviour. Thus the structure may be analysed as if the material were plastic even with much lower fibre contents than those which cause a plastic behaviour in pure tension. A plastic material behaviour (with a tensile strength equal to αf_t) can be assumed in all cases where the curves in Fig 6 are horizontal.

CONCLUSIONS

1. Linear elastic fracture mechanics can never be expected to give realistic predictions of the behaviour of steel fibre concrete. This statement is also valid for R-curve analysis and other methods based on LEFM.

2. In many cases the analyses can be based on the theory of plasticity.

3. A more general analysis has to be based on complete σ-w-curves.

4. The only valid way of defining and determining the fracture toughness of steel fibre concrete is by means of σ-w-curves. Such curves can be determined by means of a suitably designed tensile test.

REFERENCES

(1) Hillerborg, A., Modéer, M., and Petersson, P.-E., "Analysis of crack formation and crack growth in concrete by means of fracture mechanics and finite elements", Cement and Concrete Research, Vol 6, pp 773-782, 1976.

(2) Petersson, P.-E., "Crack growth and formation of fracture zones in plain concrete and similar materials", Lund Inst. of Tech., Div. of Building Materials, Report TVBM-1006, 1981.

(3) Petersson, P.-E., "Fracture mechanical calculations and tests for fibre reinforced cementitious materials", Advances in Cement-matrix Composites. Proceedings, Symposium L, Materials Research Society, Annual meeting, Boston November 17-18, 1980, pp 95-106.

(4) Gustafsson, P.J., and Hillerborg, A., "Improvements in concrete design achieved through the application of fracture mechanics", Application of fracture mechanics to cementitious materials, NATO Advanced Research Workshop September 4-7, 1984, Northwestern University.

(5) Hillerborg, A., "Influence of beam size on concrete fracture energy determined according to a draft RILEM recommendation", Lund Inst. of Tech., Div. of Building Materials, Report TVBM-3021, 1985.

(6) Hillerborg,A., "Analysis of fracture by means of the fictitious crack model, particularly for fibre reinforced concrete, Int. J. of Cement Composites, Vol 2, pp 177-184.

(7) "Determination of the fracture energy of mortar and concrete by means of three-point bend tests on notched beams", Proposed RILEM Recommendation, January 1982, revised version October 1984.

STEEL FIBER CONCRETE
US-SWEDEN joint seminar
(NSF-STU)
Stockholm 3-5 June, 1985
S P Shah and Å Skarendahl
Editors

FRACTURE RESISTANCE OF STEEL FIBER REINFORCED CONCRETE

by S P Shah, Prof, Northwestern University, Evanston, Illinois, USA

Y S Jenq, MS, Northwestern University, Evanston, Illinois, USA

ABSTRACT

A fracture mechanics based theoretical model is presented to predict the crack propagation resistance of steel fiber reinforced cement based composites. Mode I crack propagation and steel fibers are treated in the proposed model. The mechanism of fracture resistance for FRC can be separated as: subcritical crack growth in matrix and beginning of fiber bridging effect: post critical crack growth in matrix such that the net stress intensity factor due to the applied load and the fiber bridging closing stresses remain constant; and a final stage where the resistance to crack separation is provided exclusively by fibers. The response of FRC during all these stages was successfully predicted from the knowledge of matrix fracture properties and the pull-out load vs. slip (σ-ω) relationship of single fiber. The model was verified with the results of experiments conducted on notched-beams reported here as well as by other researchers. Beams were loaded in a closed-loop testing machine so as to maintain a constant rate of crack mouth opening displacement.

•Dr. S.P. Shah is Professor of Civil Engineering at Northwestern University. After receiving his Ph.D. from Cornell University, He has taught at the University of Illinois at Chicago and at Massachusetts Institute of Technology. He was a Guest Professor at Delft Institute of Technology in 1976 and at Denmark Technical University in 1984. His current research interests include: constitutive relations of concrete, application of non-linear fracture mechanics to rocks and concrete, impact loading, fiber reinforced concrete and hysteretic behavior of reinforced concrete structures.
Mr. Y. S. Jenq is a Ph.D. student at Northwestern University. He received his M.S. from Ohio State University. His research interests include application of fracture mechanics to cementitious composites.

INTRODUCTION

Research conducted during the last twenty years has shown that the addition of fibers significantly improves penetration, scabbing and fragmentation resistance of concrete. The possible applications of fiber reinforced concrete (FRC) include explosion and shock resistant protective structures. Even though the enhanced "cracking resistance" is the most important attribute of FRC, there are no rational methods of measuring or predicting this important material property.

For fiber reinforced cement based composites, the principal beneficial effects of fibers accrue after the cracking of matrix has occurred. For loads beyond which the matrix has initially cracked, the further crack extension and opening is resisted by bridging of fibers across the crack. Many investigators have used fracture mechanics concepts [1,3,5,6,11,13,14,-19,20] to incorporate the effects of fibers bridging.

In this paper a fracture mechanics based theoretical model is presented. To aid in development of the proposed theoretical model, experiments were conducted on unreinforced and steel fiber reinforced notched-beam specimens of various sizes. Beams were loaded in a closed-loop testing machine so as to maintain a constant rate of crack mouth opening displacement (CMOD).

PROPOSED MODEL FOR CRACK PROPAGATION

The fracture resistance of fiber reinforced concrete is dominated by the coupling effects between fibers and matrix. As cracks start propagating in matrix, fibers tend to resist further crack propagation. To accurately predict this fiber bridging mechanism, the crack propagation in unreinforced matrix is first described.

Unreinforced Matrix

Crack propagation (only Mode I is considered here) in plain matrix (portland cement paste, mortar and concrete) can be separated into four stages. (1) Prior to crack initiation, the load vs. crack mouth opening displacement (CMOD) response can be considered essentially linear (Fig. 1a). (2) The shape of the load vs. CMOD response becomes nonlinear and significant permanent displacements are observed upon unloading in the nonlinear range (Fig. 1b). The inelastic displacement during crack growth in cement based composites is primarily due to friction associated with roughness of cracks, geometrical interlock and microcracking outside the critical section. (3) The structure reaches its critical load carrying capacity when the critical stress intensity factor K_{Ic}^{S} is reached (Fig. 1c). (4) If the structure is loaded at a moderate rate using a displacement control, then cracks propagate in a steady state (in the sense of K_I) condition [9]. For a three-point bend test the load will decrease and the softening type of post-peak response is observed (Fig. 1d).

Determination of K_{Ic}^{S} and $CTOD_c$ Using Three-Point Bend Tests

A plot of load vs. crack mouth opening displacement (CMOD) of a concrete beam obtained from the experiments (Fig. 2) can be used to describe the observed inelastic displacement and slow crack growth phenomena. A significant amount of inelastic displacement can be observed when the specimens are unloaded immediately after the peak load. The total crack mouth opening displacement ($CMOD^T$) is composed of a sum of the elastic crack mouth opening displacement without slow crack growth ($CMOD_o^e$) , inelastic crack mouth opening displacement (CMOD*), and the elastic crack mouth opening displacement due to slow crack growth ($CMOD_s^e$) (as shown in Fig. 2).

It is clear that in order to apply LEFM, the elastic crack mouth opening displacement ($CMOD^e = CMOD^T - CMOD^*$) should be extracted from the total displacement ($CMOD^T$). Also, to overcome the difficulties in measuring the exact length of the crack, an effective crack length is defined in this paper. The effective crack length (a) is the sum of the initial notch length (a_o) plus an effective crack extension (ℓ_e) . For a three-point bend test of notched-beam with span-depth ratio of four, the elastic crack mouth opening displacement can be expressed by an empirical formula with accuracy of 1% error [17] as:

$$CMOD^e = \frac{6\ P\ s\ a}{b^2 t\ E'} V_1 \left(\frac{a}{b}\right) \qquad (1)$$

where

$$V_1\left(\frac{a}{b}\right) = 0.76 - 2.28A + 3.87A^2 - 2.04A^3 + \frac{0.66}{(1 - A)^2}$$

E' = Young's modulus of elasticity, $A = \left(\frac{a}{b}\right)$, $a = a_o + \ell_e$

P, s, a_o, b, t are indicated in Fig. 3.

For a given measured peak load (P_{max}), initial notch length (a_o) and the measured elastic $CMOD^e$, the effective crack length (a) is determined so that the calculated $CMOD^e$ is equal to the measured $CMOD^e$. Once the effective crack length is determined, then one can calculate using LEFM the values of K^S_{Ic} and $CTOD_c$. The K^S_{Ic} can be calculated as [17]

$$K^S_{Ic} = \frac{1.5\,P_{max}S}{b^2\,t}\,F_1\left(\frac{a}{b}\right)\sqrt{\pi a} \tag{2}$$

in which

$$F_1\left(\frac{a}{b}\right) = \frac{1}{\sqrt{\pi}}\,\frac{1.99 - A(1 - A)\,(2.15 - 3.93A + 2.7A^2)}{(1 + 2A)\,(1 - A)^{3/2}}$$

$$A = \frac{a}{b}$$

The value of $CTOD_c$ is the crack opening displacement at original notch tip and can be calculated using LEFM [9] and expressed as:

$$CTOD_c = CMOD^e\left[\left(1 - \frac{a_o}{a}\right)^2 + \left(-1.149\,\frac{a}{b} + 1.081\right)\left(\frac{a_o}{a} - \left(\frac{a_o}{a}\right)^2\right)\right] \tag{3}$$

Some values of K^S_{Ic} and $CTOD_c$ obtained from testing results of three different sizes of beams (L, M, S) (see Fig. 3) and five different mixes [8] are reported in Table 1. It can be seen that the values of K^S_{Ic} as suggested here are independent of the size of the beam. This was not true for the values of the conventionally calculated critical stress intensity factor using the initial notch length (a_o) and the maximum load (P_{max}). The critical crack tip opening displacement ($CTOD_c$) were also found to be size-independent (Table 1) for all test series except for the large beams of C1 series.

Fiber Reinforced Composites

The prediction of the response of the fiber reinforced composites (FRC) containing an initial notch of length a_o can be facilitated by dividing the response into four stages (Fig. 4).

(a) Linear Range:
Within the linear range of the unreinforced matrix (Fig. 1a) the FRC also behaves linearly. The initial Young's modulus for FRC may be different than that of matrix depending upon the amount of fibers.

(b) Nonlinear Crack Growth:
When the stress intensity factor (calculated using the conventional linear elastic fracture mechanics - LEFM) becomes greater than half the critical stress intensity factor (K^S_{Ic}) nonlinear crack growth commences and the fibers will tend to resist this growth. The fiber bridging forces will depend on the total crack opening displacement while the calculations of

stress intensity factor will depend on the effective crack growth and the elastic crack opening displacement.

The total load P acting on the composite structure can be divided into three parts (Fig. 5):

$$P = P^M + P_k^f + P_s^f \qquad (4)$$

where P^M is the contribution due to matrix and is related to K_I, P_k^f is related to K_I^f and accounts for the singularity effect due to fiber bridging, and P_s^f satisfies global equilibrium due to fiber bridging forces (Fig. 5).

(c) Steady State:
At this stage, cracks propagate in a steady state (i.e., $K_I = K_{Ic}^S$) . Note that the value of P for FRC does not necessarily attain a maximum when K_I just reaches K_{Ic}^S . Depending upon the volume of fibers, the maximum load for FRC may occur for a larger crack length than that corresponding to the peak load in the unreinforced matrix (Fig. 4).

(d) Completely cracked matrix:
When the crack opening displacement (CMOD in Fig. 4) becomes very large, the resistance offered by matrix becomes negligible and eventually the stress intensity factor (K_I) becomes zero. Further crack separation is now mainly resisted by fibers. At this stage the load (P) and the corresponding CMOD can be calculated from only the global equilibrium consideration. That is:

$$P = P_s^f \qquad (5)$$

RESISTANCE PRODUCED BY FIBERS

The resistance offered by fibers depends primarily on the interfacial bond between the fibers and the matrix. This is because of the short length (on the order of 25mm) of the fibers and rather weak zone in the matrix that is observed in surrounding fibers [4]. It is assumed that the fiber-bridging forces can be calculated from the pull-out test results of a single aligned fiber. Load-slip relationship of the two types of steel fibers observed from pull-out tests are shown in Fig. 6. For both types of fibers, the initial debonding can be approximated by a vertical straight line up to the maximum load. With this approximation, the load-slip curve can be represented by:

$$\frac{\sigma(w)}{\sigma_{max}} = \left[1 - \frac{w}{w_{max}}\right]^m , \quad 0 \leq w \leq w_{max} \qquad (6)$$

where σ_{max} is the maximum load divided by the total cracked area, w = slip or crack opening displacement, w_{max} = slip at zero load which can be assumed to be half the length of the fiber, and $\sigma(w)$ is the load at any given slip divided by the cracked area and is termed fiber pull-out stress, or the fiber bridging closing pressure and m is constant which depends on the type fibers and was assumed to be 2 for the straight fibers used in this investigation [2,10,12,16].

To account for the spatial distribution of fibers an effective volume fraction of fibers (V_{ef}) was used in this study. Based on single fiber pull-out tests of straight steel fibers, the following value for σ_{max} was used for present study:

$$\sigma_{max} = 240\ V_{ef}\ \text{psi} = 1.655\ V_{ef}\ \text{MPa} \qquad (7)$$

The determination of V_{ef} was empirical and is based on global equilibrium condition at the point when matrix is completely cracked.

CALCULATION OF LOAD-CMOD CURVES FOR FIBER REINFORCED COMPOSITES

The procedure involves calculating the applied load and the corresponding CMOD for a given value of ℓ_e and the associated value of K_I (which is a function of ℓ_e). To obtain the entire P - CMOD response the above procedure is repeated for different values of ℓ_e and K_I. For example, in the steady state, for any given ℓ_e ($\ell_e > \ell_{ec}$) and K^s_{Ic}, the associated P^M in Eq. 4 (see Fig. 5) can be calculated as:

$$P^M = \frac{K^s_{Ic}\ b^2 t}{1.5\ \sqrt{\pi a}\ F_1(\frac{a}{b})}, \quad a = a_o + \ell_e \qquad (8)$$

It was assumed that fibers are rigid inclusions and the crack profile for fiber reinforced composites remain the same as unreinforced matrix. Thus, the $CMOD^e$ can be calculated by substituting P^M for P into Eq. 1. $CMOD^e$ was transformed into $CMOD^T$ according to empirical relationship observed from loading - unloading tests of unreinforced matrix.

From the knowledge of $CMOD^T$, V_{ef} and σ-w relationship (Eq. 6), the stress intensity factor due to fiber bridging force can be determined using the solution of a crack in an infinite strip of unit thickness subjected to a unit point load (Fig. 7) as Green's function and integrated over the closing pressure zone. Similarly P^f_K and K^f_I can be related by Eq.8. The value of P^f_s is determined from the global equilibrium conditions [Fig. 8]. Thus, the applied external load can be calculated from Eq. 4. Repeating the above procedures for different values of ℓ_e and the associated K_I, the entire load-CMOD response is achieved.

CALCULATION OF LOAD-DEFLECTION RESPONSE OF FRC

To relate the loads with the load-point deflection, the concept of global energy balance was used. The total strain energy release rate (termed G_R) for the critical section can be derived as:

$$G_R = \frac{d}{da}(W = W_e + W_p) \approx C\frac{K_I^2}{E'} + \int_0^{CTOD} \sigma(w)\ dw \qquad (9)$$

where W_e and W_p are the elastic and inelastic energies consumed during the formation of new crack, K_I = net stress intensity factor due to applied load and closing pressure, CTOD = crack opening displacement at the original crack tip, and C = energy correction factor for the plain matrix which ranged from 1.5 to 2.5.

For a certain amount of increment of effective crack extension, the total energy absorbed determined from G_R vs. ℓ_e relationship is equated to that obtained from the load vs. load-point deflection curve (shaded areas in Fig. 9). From this equality and knowing the R-curve, load-point deflections can be determined. It was assumed that unloading was elastic prior to peak, whereas for load-deflection curve beyond the peak, it was assumed that the elastic deflection after unloading remained constant and was equal to that at the peak (Fig. 9). More details about the derivation of load-deflection and load-CMOD relationship are given in Ref. 21.

TEST PROGRAM

Three point bend test was used to verify the validity of the proposed model. Fiber reinforced beams with dimensions 11 in. (280mm) x 3 in. (76mm) x 0.75 in. (19.1mm) (span x depth x thickness) and different fiber volume fractions (ranging from 0% to 1.5%) were prepared. One inch long brass coated smooth steel fibers with 0.016 in. diameter were used. Four different series [Table 2] with the same mortar matrix were cast.

DISCUSSION OF TEST RESULTS

The material properties of unreinforced matrix were directly calculated from the experimental results.

To account for distribution of fibers from section to section of the beam, from beam to beam and to include spatial distribution at a given section an effective volume fraction (V_{ef}) rather than global volume fraction (V_f) was used in the theoretical analysis [Table 2].

The experimental results of load-CMOD curves for beams made with different fiber volume fractions (including unreinforced matrix) are plotted in Fig. 10 and compared with the theoretical prediction. The theoretical prediction is judged to be quite satisfactory. Good agreement was also found between the theoretical prediction and experimental results of load-deflection curves (Fig. 11). Fig. 12 shows a plot of peak load values vs. effective fiber volume fractions. The strength of FRC beams with effective fiber volume fraction of 2.5% is about twice the strength of unreinforced matrix. The G_R values at ℓ_e equals 1.8 in. (a/b = 0.933) are plotted in Fig. 13 for different vlaues of V_{ef}. It can be seen that the energy absorption ability for beams with V_{ef} = 2.5% is about 30 times that of unreinforced matrix. In comparison to the improvement of energy absorption, the strength improvement due to addition of fibers is less significant. This was also shown by Shah and Rangan [15].

Load-CMOD curves reported by Velazco, et al. [18] were also analyzed. The material properties used to predict the composite behavior were calculated from the reported data and are given in the figures. Experimental results reported by Velazco, et al. [18] are plotted in Fig. 14. The theoretical predictions are in good agreement with the experimental results for all different effective fiber volume fractions.

CONCLUSIONS

1. A Fracture Model is proposed for fiber reinforced concrete. This model is based on the properties that the crack propagation in the matrix can be described by two parameters: critical stress intensity factor and critical crack tip opening displacement. Because of the nonlinear slow crack growth, two fracture parameters are required. The stress intensity factor is calculated at the tip of the effective crack rather than the initial crack length.

2. Effect of fibers is to reduce the stress intensity factor at the tip of the effective crack and to provide additional energy due to debonding. These effects can be incorporated in the proposed model if the pull-out load-slip relationship of a single fiber is known.

3. The model predicted load vs. deflection and load vs. crack mouth opening displacement relationship compared favorably with the experiments described here as well as those reported by others.

4. In this investigation only steel fibers and mode I crack propagation are considered.

ACKNOWLEDGEMENT

The research reported here was supported by the Air Force Office of Scientific Research [Lt. Col. Lawrence D. Hokanson, Program Manager] under Grant No. AFOSR-82-0243.

REFERENCES

1. Ballarini, R., Shah, S. P., and Keer, L. M., "Crack Growth in Cement Based Composites," Engineering Fracture Mechanics, Vol. 20, No. 3, 1984, pp. 433-445.

2. Burakiewicz, A., "Testing of Fiber Bond Strength in Cement Matrix," Proceedings, International Symposium, RILEM-ACI-ASTM, Sheffield, Sept. 1978, pp. 355-365.

3. Bowling, J., and Groves, G. W., "The Propagation of Cracks in Composites Consisting of Ductile Wires in a Brittle Matrix," Journal of Materials Science, Vol. 14, 1979, pp. 443-449.

4. Diamond, S., and Bentur, A., "On the Cracking of Concrete and Fiber-Reinforced Cement," in Application of Fracture Mechanics to Cementitious Composits, ed. by S.P. Shah, to be published by Martinus Nijhoff Publishers, 1985.

5. Foote, R. M. L., Cotterell, B., and Mai, Y. W., "Crack Growth Resistance Curve for a Cement Composite," in Advances in Cement Matrix Composites, Proceedings, Symposium L., Materials Research Society, Annual Meeting, Boston, Massachusetts, Nov. 17-18, 1980, pp. 135-144.

6. Hillerborg, A., "Analysis of Fracture by Means of the Fictitious Crack Model, Particularly for Fiber Reinforced Concrete," International Journal of Cement Composites, Vol. 2, No. 4, Nov. 1980, pp. 177-184.

7. Jenq, Y. S., and Shah, S. P., "Nonlinear Fracture Parameters for Cement Based Composites: Theory and Experiments" in Application of Fracture Mechanics to Cementitious Composites, ed. by S. P. Shah, to be published by Martinus Nijhoff Publishers, 1985.

8. Jenq, Y. S., and Shah, S. P., "A Frcture Toughness Criterion For Concrete," Engineering Fracture Mechanics (to appear).

9. Jenq, Y. S., and Shah, S. P., "A Two Parameter Fracture Model for Concrete," (submitted for publication).

10. Johnston, C. D., and Gray, R. J., "Uniaxial Tensile Testing of Steel Fiber Reinforced Cementitious Composites," Proceedings, International Symposium, RILEM-ACI-ASTM, Sheffield, Sept. 1978, pp. 451-462.

11. Lenian, J. C., and Bunsell, A. R., "The Resistance to Crack Growth of Asbestos Cement," Journal of Materials Science, Vol. 14, 1979, pp. 321-332.

12. Naaman, A. E., and Shah, S. P., "Pull-out Mechanism in Steel Fiber Reinforced Concrete," Journal of ASCE-STD, August 1976, pp. 1537-1548.

13. Nishioka, k., Yamakawa, S., and Hirakawa, k.,. "Test Method for the Evaluation of the Fracture Toughness of Steel Fiber Reinforced Concrete", Proceedings, International Symposium, RILEM-ACI-ASTM, Sheffield, Sept. 1978, pp. 87-97.

14. Petersson, P. E., "Fracture Mechanics Calculations and Tests for Fiber-Reinforced Cementitious Materials," in Advances in Cement-Matrix Composites, Proceedings, Symposium L., Materials Research Society, Annual Meeting, Boston, Massachusetts, Nov. 17-18, 1980, pp. 95-106.

15. Shah, S. P. , and Rangan, B. V., "Fiber Reinforced Concrete Properties," ACI Journal, Vol. 62, No. 2, Feb. 1971, pp. 126-135.

16. Stroeven, P., Shah, S. P., deHaan,Y. M., and Bouter, C., "Pull-out Tests of Steel Fibers," Proceedings, International Symposium, RILEM-ACI-ASTM, Sheffield, SEPT. 1978, PP. 345-353.

17. Tada, H., Paris, P.C., and Irwin, G.R., The Stress Analysis of Cracks Handbook, Del Research Corporation, Hellertown, Pennsylvania, 1973.

18. Velazco, G., Visalvanich, K., Shah, S. P., and Naaman, A. E., "Fracture Behavior and Analysis of Fiber Reinforced Concrete Beams," Progress report for National Science Foundation, March 1979.

19. Wecharatana, M., and Shah, S. P., "A Model for Predicting Fracture Resistance of Fiber Reinforced Concrete," Cement and Concrete Research, November 1983, pp. 819-830.

20. Wecharatana, M., and Shah, S. P., "Prediction of Nonlinear Fracture Process Zone in Concrete," Journal of EMD, ASCE,June 1982, pp. 1100-1113.

21. Jenq, Y. S., and Shah, S. P., "Crack Propagation Resistance of Fiber Reinforced Concrete," (submitted for publication).

Series	Compressive Strength f'_c (psi)	Young's Modulus E(x 10^6 psi)	K^S_{Ic} (psi $\sqrt{in.}$)				$CTOD_c$ (x 10^{-3} in.)			
			Large	Medium	Small	Average	Large	Medium	Small	Average
C1	3650	4.87	930.8	813.5	904.5	882.9	0.3845	0.7035	0.792	0.63
M1	3942	3.68	631.1	654.0	644.1	643.1	0.324	0.463	0.302	0.363
M2	5718	4.71	918.9	879.3	816.3	871.5	0.475	0.336	0.332	0.381
M3	7950	5.41	-	-	963.5	963.5	-	-	0.394	0.394
P1	4013	3.01	595.5	544.8	547.1	562.1	0.267	0.244	0.301	0.271

Table 1 - Relative material properties of all series.

TABLE 2 - MIX-PROPORTION OF FIBER SERIES

SERIES	CEMENT	SAND*	WATER	FIBER VOLUME FRACTURE (V_f) (%)	EFFECTIVE FIBER VOLUME FRACTURE (V_{ef}) (%)
F0	1.0	2.0	0.4	0	0
F1	1.0	2.0	0.4	0.5	0.4 ~ 1.1
F2	1.0	2.0	0.4	1.0	0.7 ~ 1.4
F3	1.0	2.0	0.4	1.5	1.4 ~ 2.6

* Maximum aggregate size equals 0.1875 in.

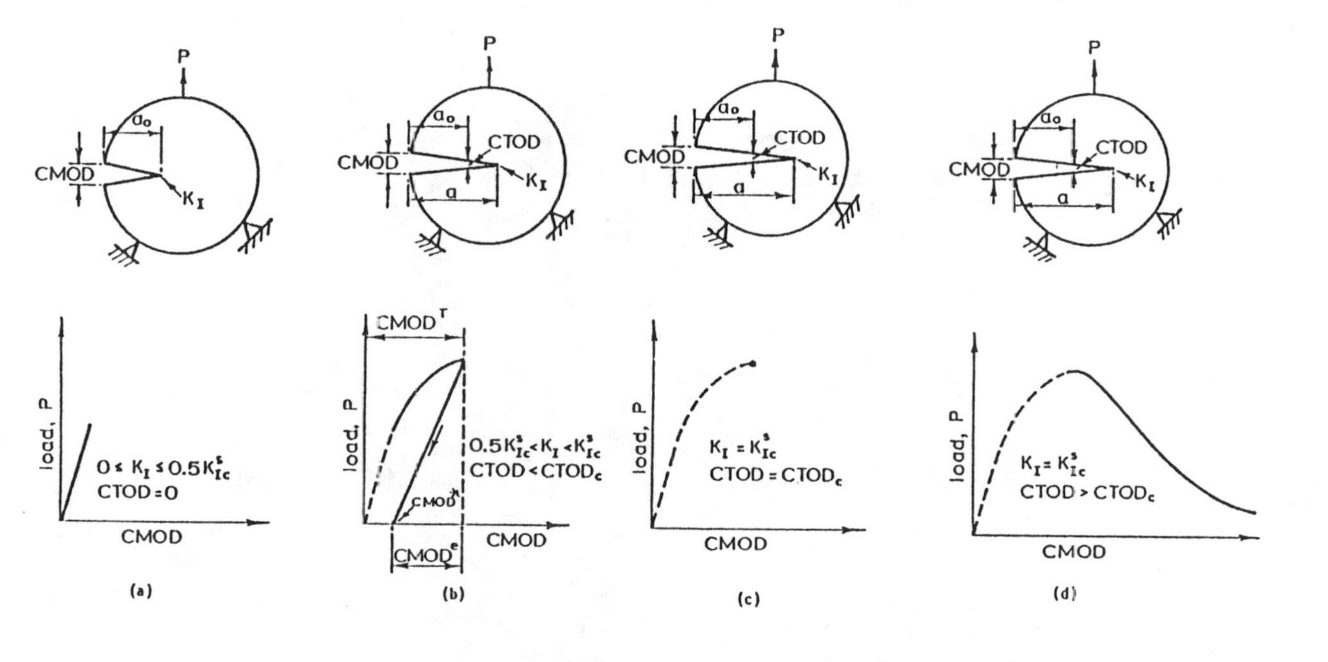

Fig. 1 - Fracture Resistance Stages of Unreinforced Matrix

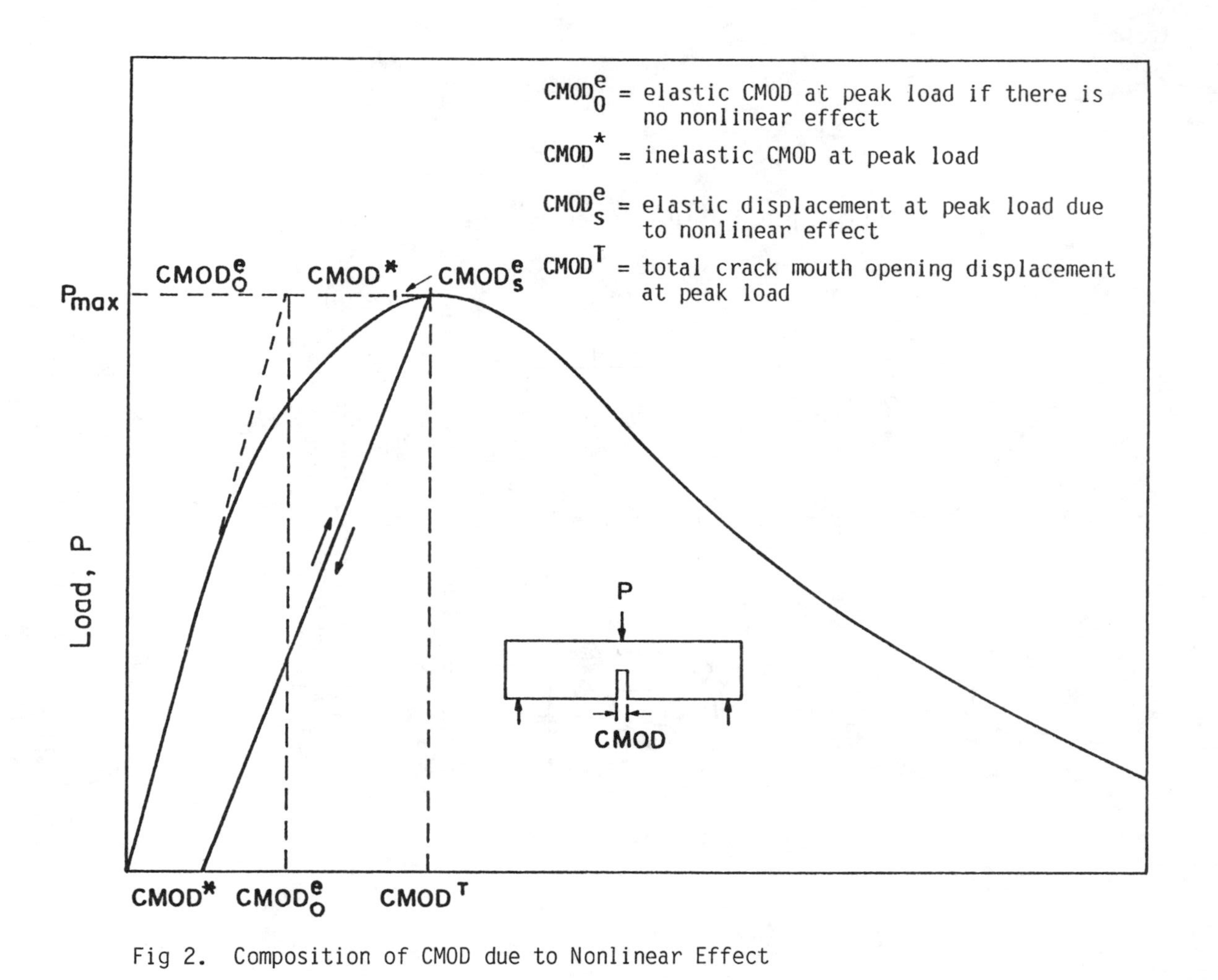

Fig 2. Composition of CMOD due to Nonlinear Effect

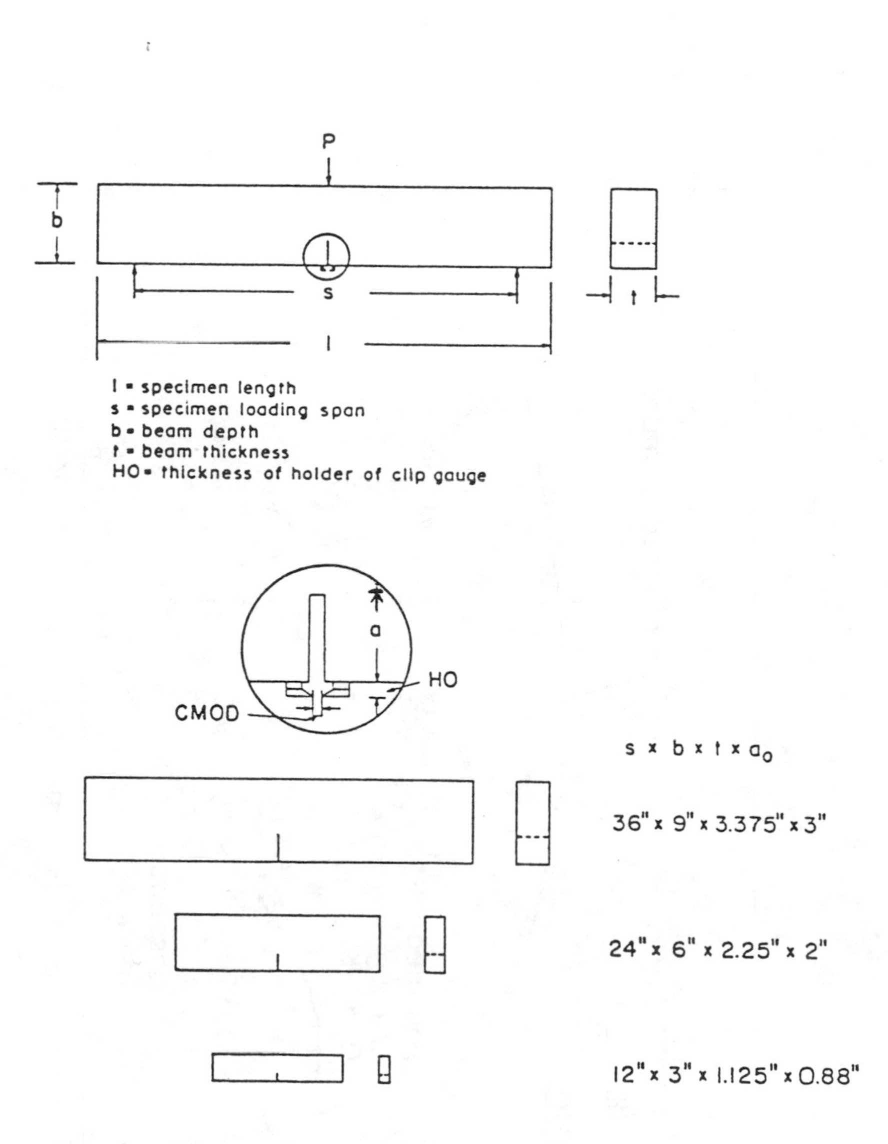

Fig. 3. Specimen Dimensions of Unreinforced Matrix

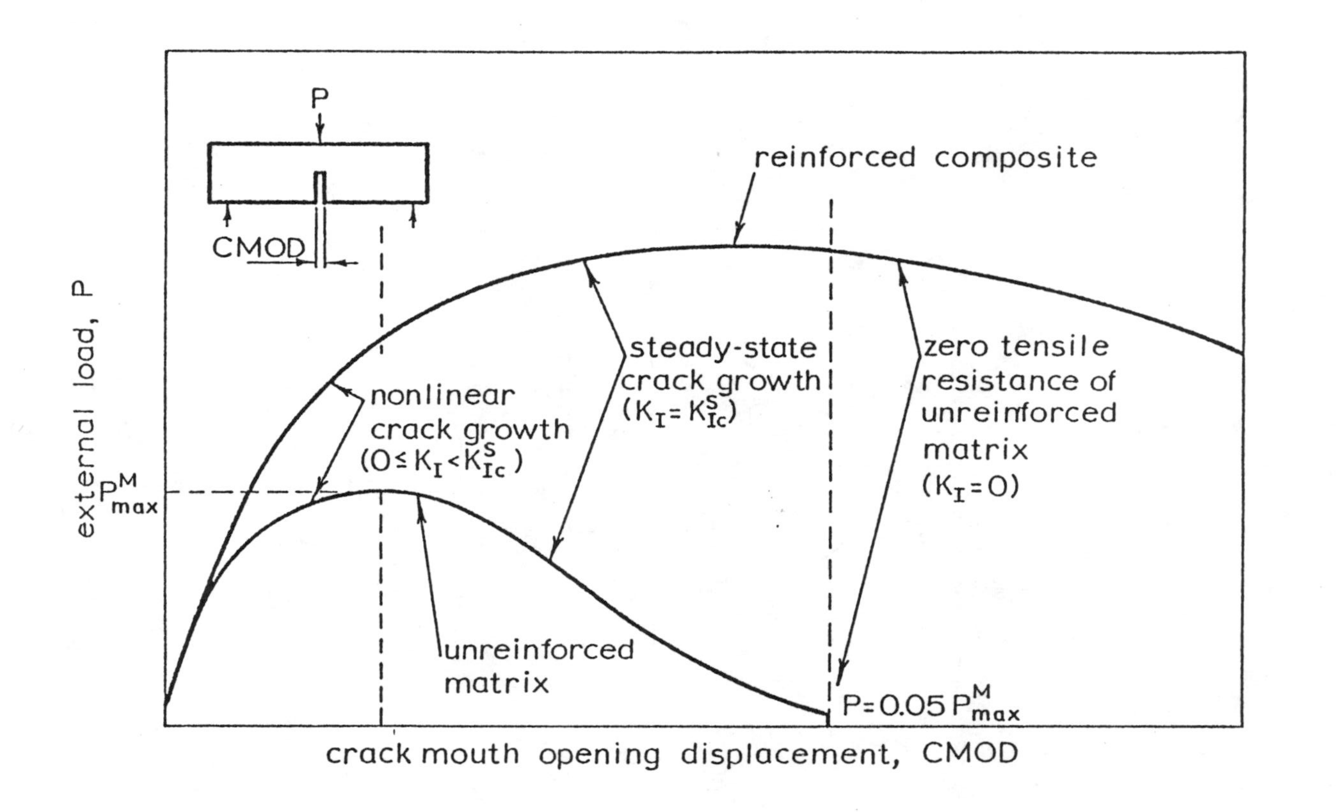

Fig. 4. Fracture Resistance Mechanisms of Fiber Reinforced Concrete

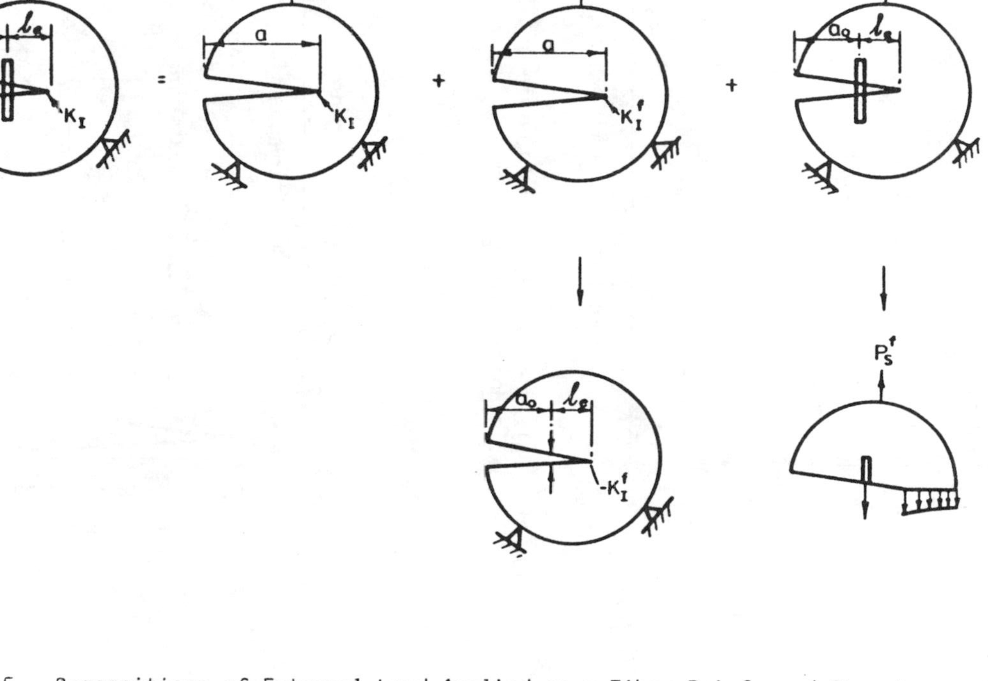

Fig. 5 - Compositions of External Load Applied on a Fiber Reinforced Structure

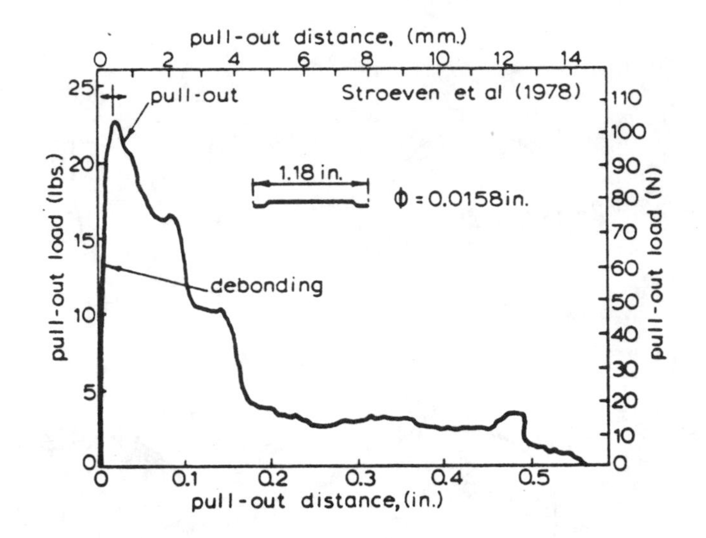

Fig. 6a. Pull-out Load-Slip Relationship of Single Fiber with Hook at the Ends

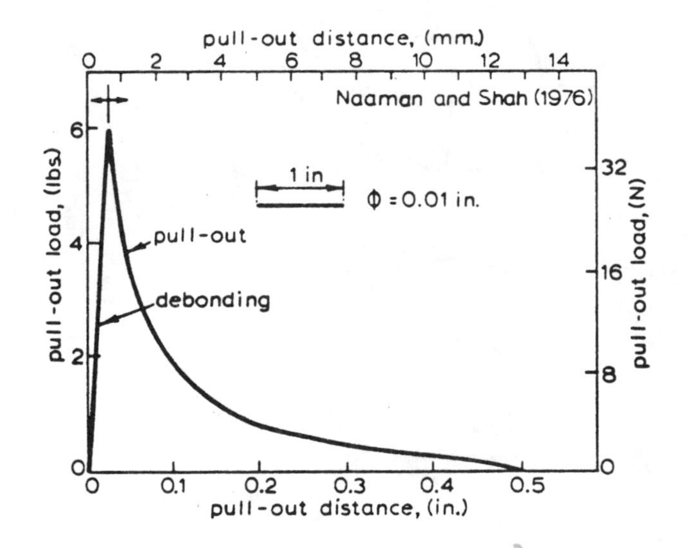

Fig. 6b. Pull-out Load-Slip Relationship of Single Straight Fiber

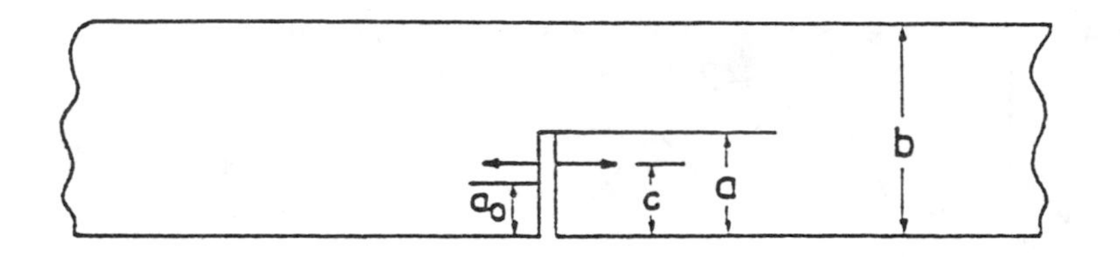

Fig. 7. Stress Intensity Factor of an Infinite Strip Subjected to a Unit Point Load

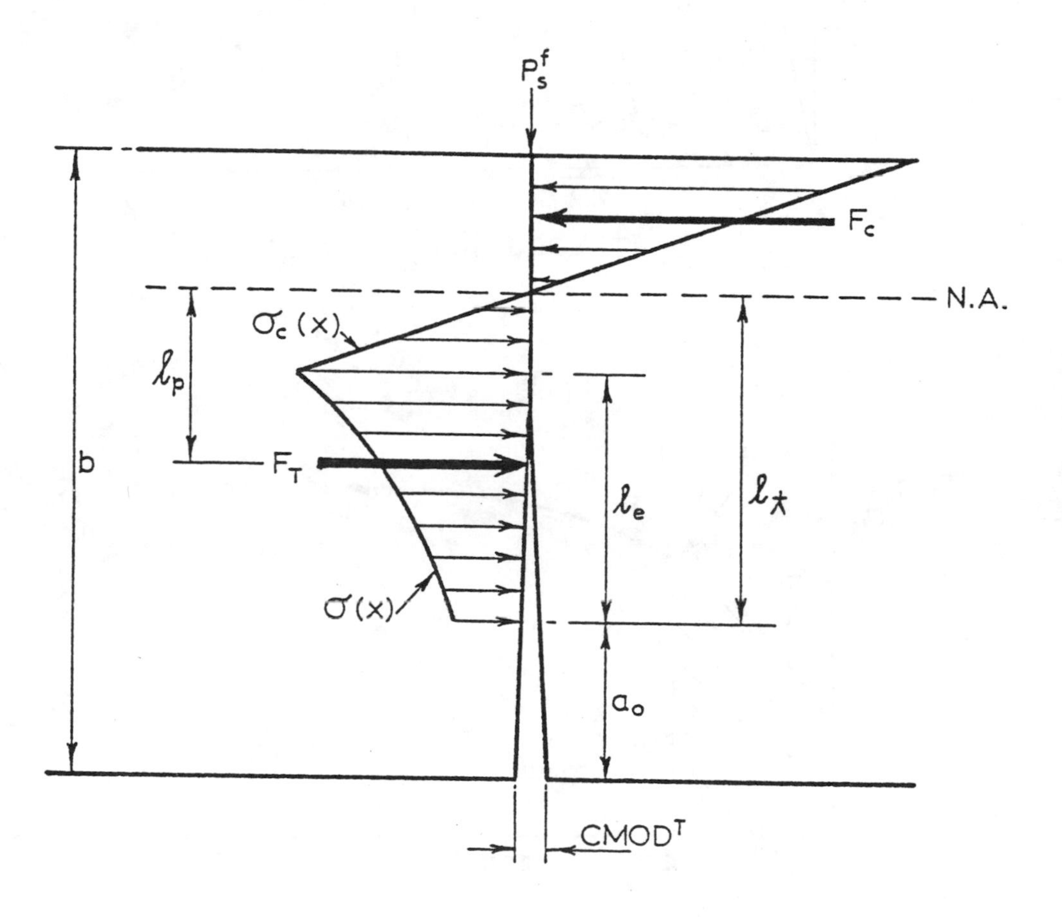

Fig. 8. Global Equilibrium Condition of 3-PB Notched Beam

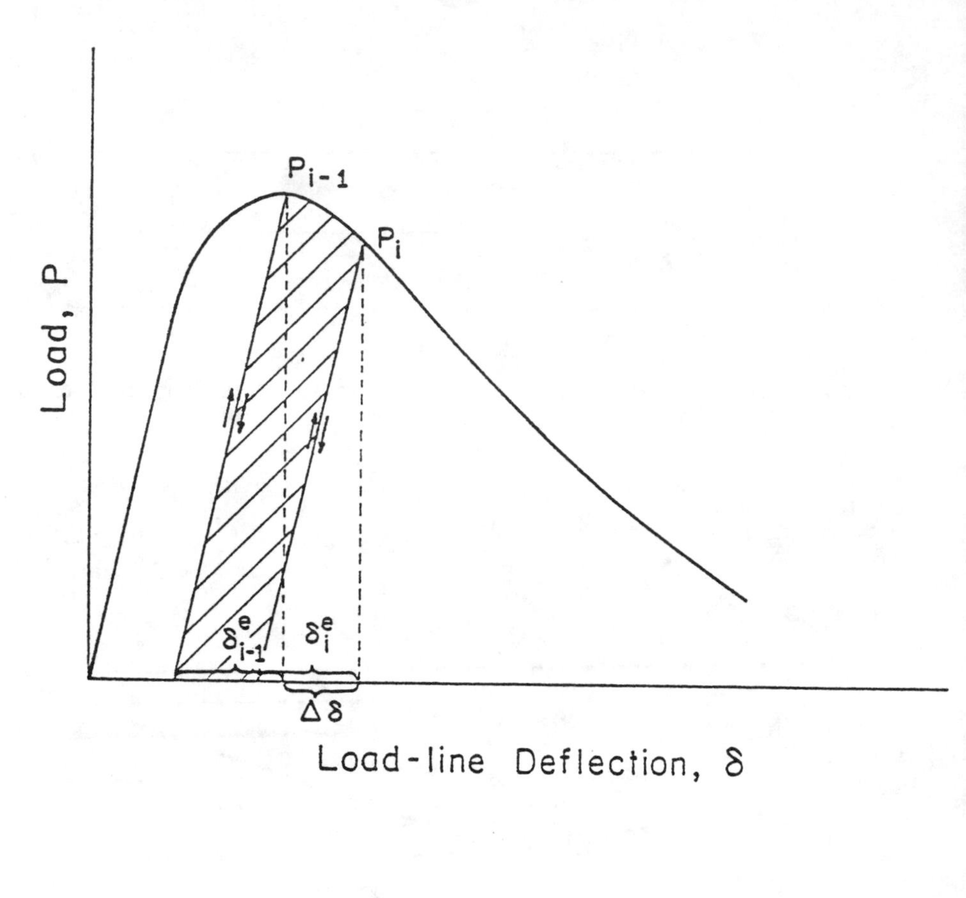

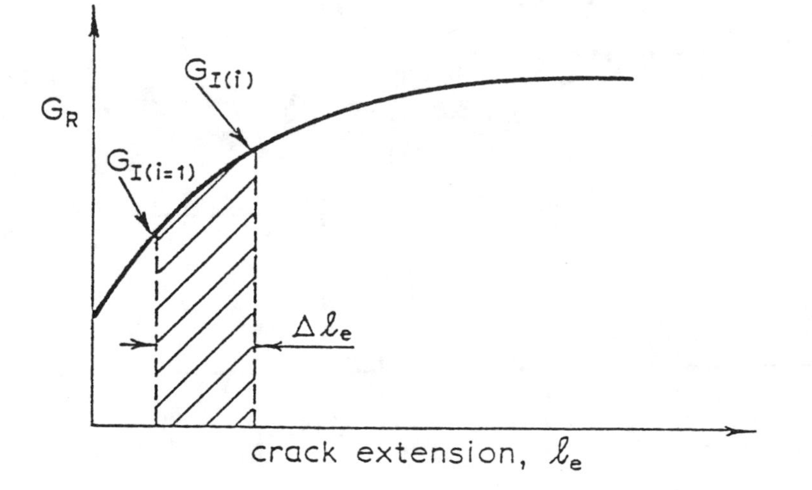

Fig. 9. Consideration of Energy from Load-Deflection and G_R curves

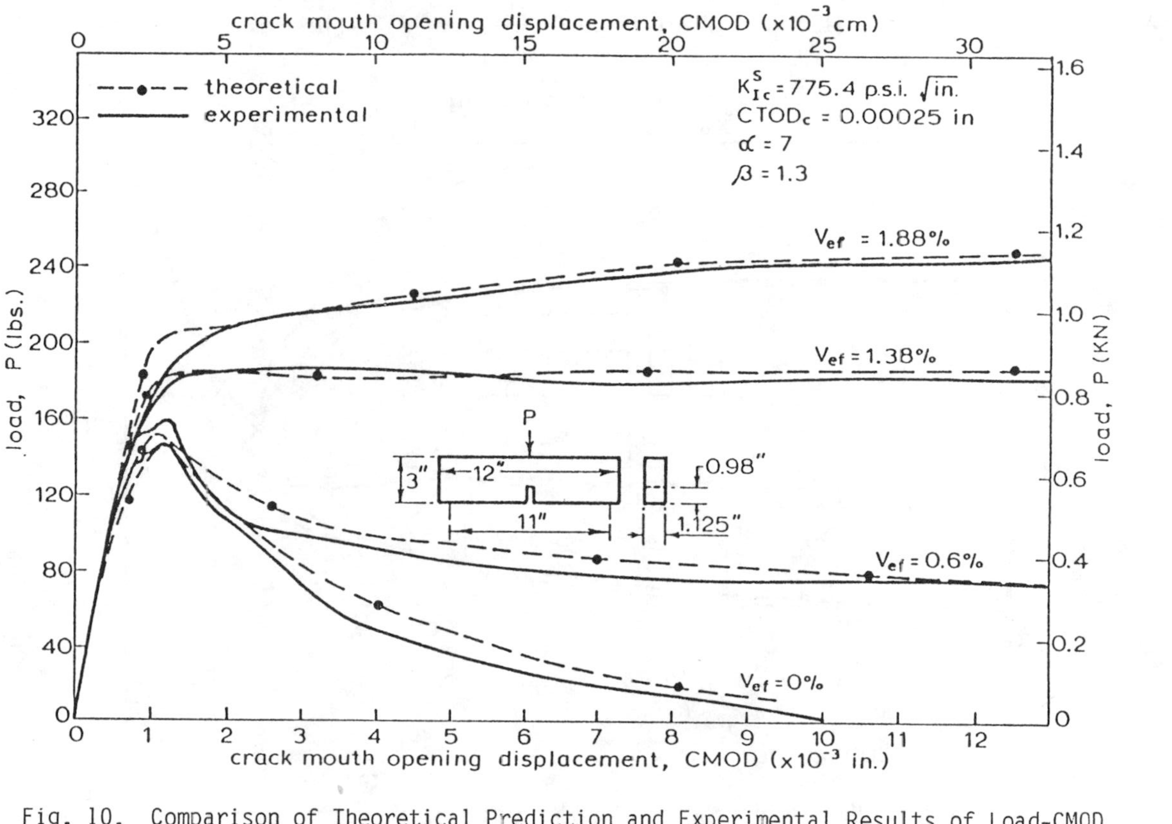

Fig. 10. Comparison of Theoretical Prediction and Experimental Results of Load-CMOD Curve

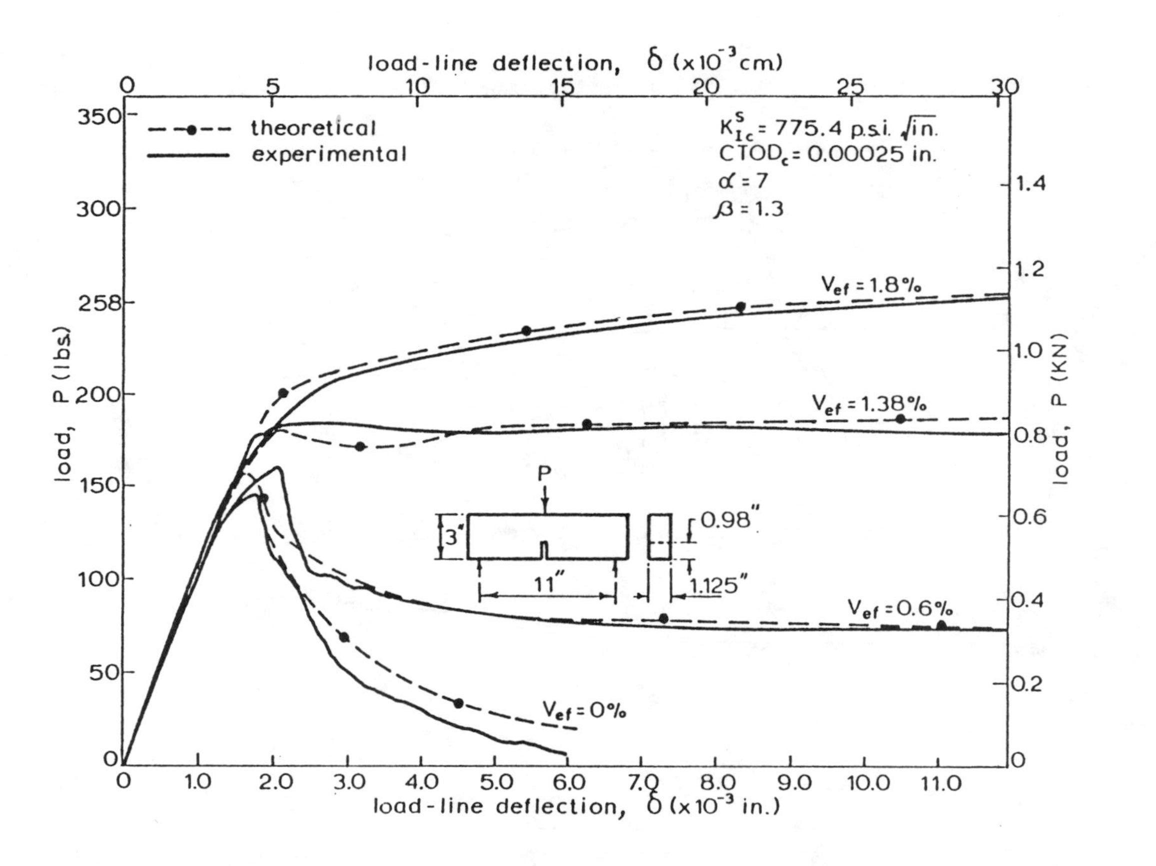

Fig. 11. Comparison of Theoretical Prediction and Experimental Results of Load-Deflection Curve

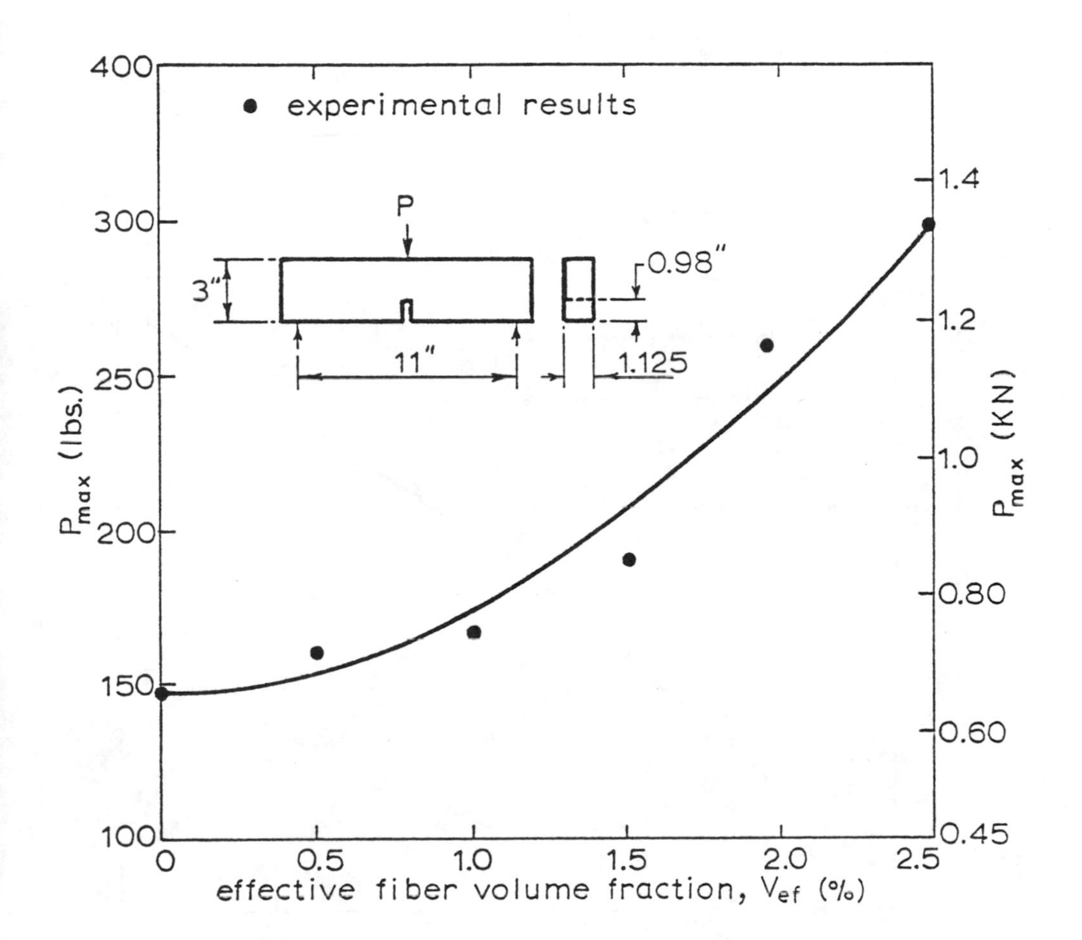

Fig. 12. Relationship Between Peak Load and Effective Fiber Volume Fraction

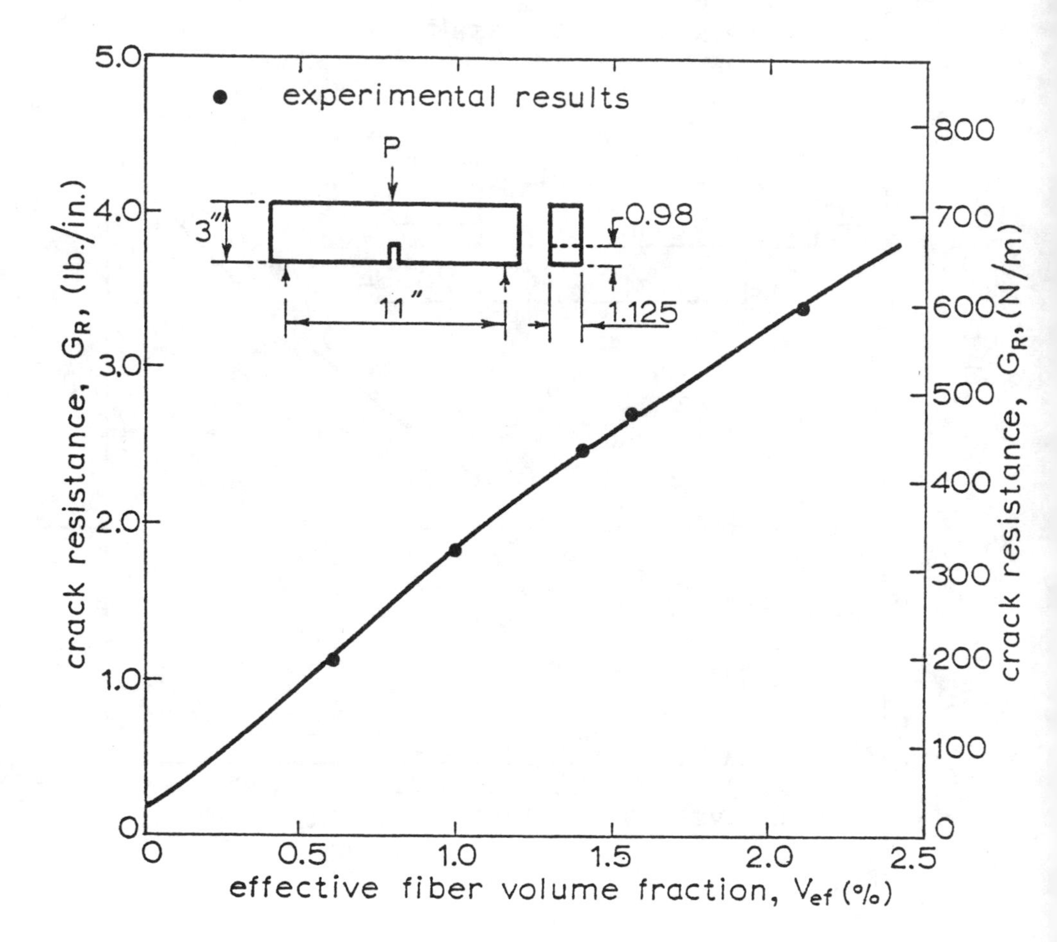

Fig. 13. Relationship Between Crack Resistance and Effective Fiber Volume Fraction

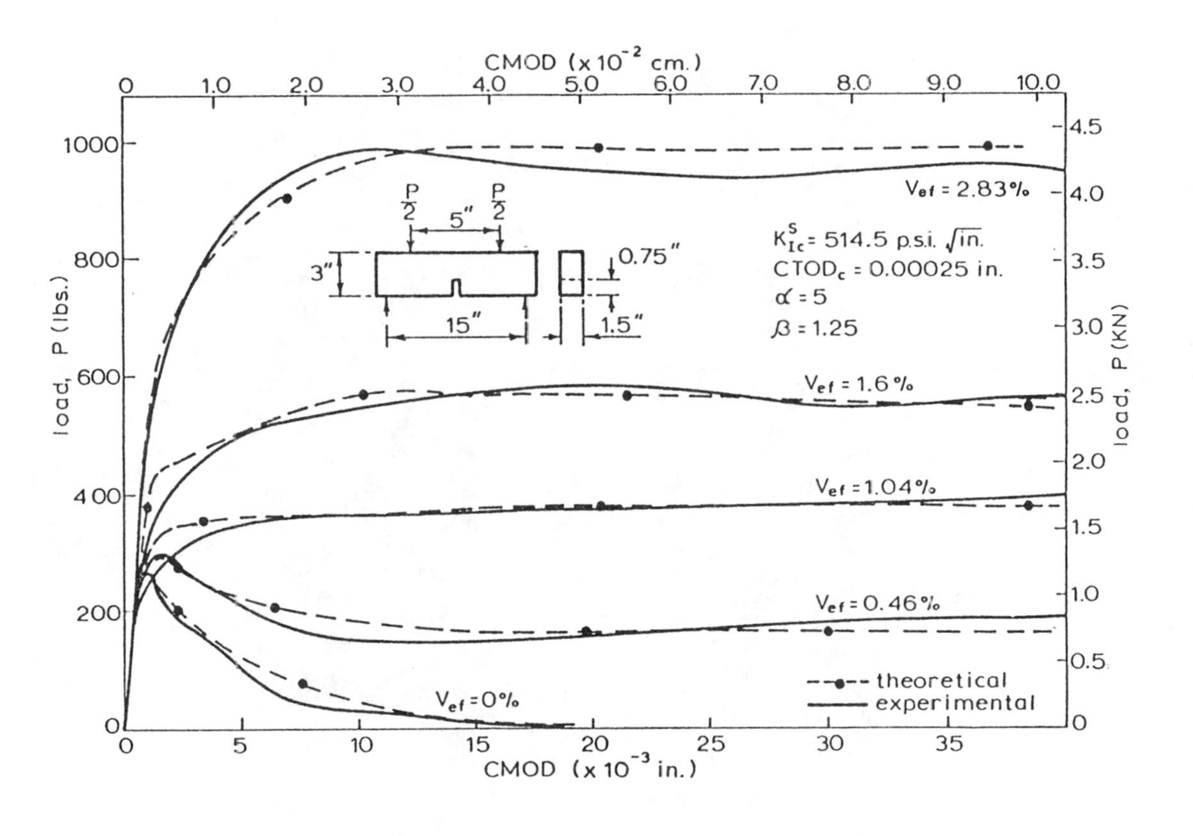

Fig. 14. Comparison of Theoretical Prediction and Experimental Results of Load-CMOD Curves Reported by Valazco et al. [18]

STEEL FIBER CONCRETE
US-SWEDEN joint seminar
(NSF-STU)
Stockholm 3-5 June, 1985
S P Shah and Å Skarendahl,
Editors

STRENGTH, DEFORMATION AND FRACTURE TOUGHNESS OF FIBER CEMENT COMPOSITES AT DIFFERENT RATES OF FLEXURAL LOADING

by V S Gopalaratnam, Asst Prof, University of Missouri - Columbia, Columbia, Missouri, USA

S P Shah, Prof, Northwestern University, Evanston, Illinois, USA

ABSTRACT

Several test methods used for impact tests of fiber reinforced concrete (FRC) are reviewed in the article with a view to evaluate the reliability of the material responses obtained therefrom. Parasitic effects of inertia observed while conducting instrumented impact tests on concrete composites are discussed at length. Based on experience gained during the course of the development of a modified instrumented Charpy test scheme, useful guidelines for selection of the various test parameters are proposed in order to minimize parasitic inertial loads.

The effect of strain-rate on the flexural behavior of unreinforced matrix and 3 different fiber reinforced concrete (FRC) mixes are discussed. Results obtained from the modified instrumented Charpy tests on cement composites compare well with results from several similar investigations that use an instrumented drop-weight set-up.

FRC mixes are more rate-sensitive than their respective unreinforced matrices, showing increases in dynamic (strain-rate of 0.3/s) strength of up to 111% and energy absorption (up to a deflection of 0.1 in.) of up to 70% (V_f = 1.5%) over comparable values at the static (strain-rate of 1 x 10^{-6}/s) rates. Composites made with weaker matrices, higher fiber contents and larger fiber aspect ratios are more rate sensitive than those made with stronger matrices, lower fiber contents and smaller fiber aspect ratios. Several observations made in the study suggest that the rate sensitivity exhibited by such composites is primarily due to a change in the cracking process at the different rates of loading.

Relative improvements in performance due to the addition of fibers as observed in the instrumented tests are also compared to those from the conventional impact and static tests. Resulting from this comparison, it is proposed that static flexural toughness tests could be used to approximately estimate the dynamic performance of FRC.

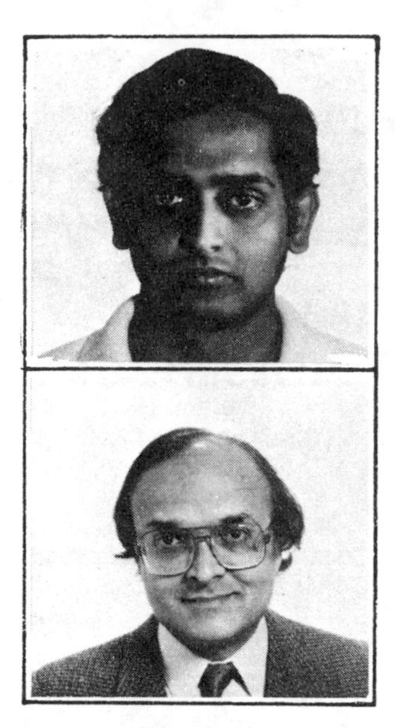

•Dr. V.S. Gopalaratnam is Assistant Professor of Civil Engineering at University of Missouri-Columbia. He received his Ph.D. from Northwestern University. His research interests include fracture mechanics and strain-rate behavior of concrete and related composites.
•Dr. S.P. Shah is Professor of Civil Engineering at Northwestern University. After receiving his Ph.D. from Cornell University, he has taught at University of Illinois at Chicago and at Massachusetts Institute of Technology. He was a Guest Professor at Delft Institute of Technology in 1976 and at Denmark Technical University in 1984. His current research interests include: constitutive relations of concrete, application of non-linear fracture mechanics to rocks and concrete, impact loading, fiber reinforced concrete and hysteretic behavior of reinforced concrete structures.

INTRODUCTION

Despite its extensive use, low tensile strength has been recognized as one of the major drawbacks of concrete. Although one has learned to avoid exposing concrete structures to adverse static tensile loads, these cannot be shielded from short duration dynamic tensile stresses. Such loads originate from sources such as impact from missiles and projectiles, wind gusts, earthquakes and machine vibrations. The need to accurately predict the structural response and reserve capacity under such loading has led to an interest in the mechanical properties of the component materials at high rates of straining.

One method to improve the fracture resistance and the resistance of concrete when subjected to impact and/or impulsive loading is by the incorporation of randomly distributed short fibers. Concrete (or Mortar) so reinforced is termed fiber reinforced concrete (FRC). Moderate increase in tensile strength and significant increases in energy absorption (toughness or impact-resistance) have been reported by several investigators [1-3] in static tests on concrete reinforced with randomly distributed short steel fibers. Studies on the dynamic behavior of FRC are rather limited in comparison. This, despite the fact that the most important property of such composites is its superior impact resistance.

As yet no standard test methods are available to quantify the impact resistance of such composites, although several investigators have employed a variety of tests including drop weights, swinging pendulums and the detonation of explosives. These tests though useful in ascertaining the relative merits of different composites do not yield basic material characteristics which can be used for design.

More recently instrumented impact tests have been developed to obtain reliable and continuous records of the characteristics of brittle materials when they are subjected to high straining rates [4-10].

Results from such tests could be used to formulate constitutive relations for these composites which would probably lead to more rational design procedures for structures subjected to impact. As in the static analysis of FRC, it is expected that the composite behavior under dynamic loading could be predicted from the knowledge of the behavior of its constituent materials as well as their interaction, at higher rates of loading.

TYPES OF IMPACT TESTS

A review of some of the test methods and results obtained therefrom, based on tests on fiber reinforced concrete specimens is presented in the following sub-sections. Drawbacks of some of these popular conventional tests are also discussed whenever relevant.

Charpy Impact Test

Charpy test is a standard impact test recommended for metals (ASTM E29). The energy consumed to totally fracture a notched beam specimen is computed from the rise angle of the pendulum after impact and is used as a measure of the impact resistance of the material. In one such test, Battelle Development Corp. [11] reported increases in Charpy impact energy from 2.2 kJ/m^2 for plain concrete (25x25x102mm beams) to 21.7 kJ/m^2 for FRC (2% by volume of 0.15mm diameter steel fibers). In similar tests with different specimen sizes Krenchel [12] and Johnston [13] observed a similar magnitude of increase in impact resistance achieved with the incorporation of steel fibers. However, Radomski [14] reports of different impact energy values from that reported in [11]. This is probably due to the different test set-up compliances in these studies. Abe et al, [15] while testing rate-insensitive silicon carbide specimens, have shown using an elaborate energy balance, that energy absorbed by the specimen is only of the order of 30% of the total energy recorded in the Charpy test. Bluhm [16] conducted Charpy impact tests on metallic specimens and observed that the stiffness of the pendulum does significantly affect both the peak load and fracture energy recorded.

The differences in the results of some of the earlier studies [11-14] can also be attributed to size dependent characteristics of inhomogeneous materials. Representative results from tests on cement composite specimens of sizes comparable to that recommended for metal specimens (10mm x 10mm x 50mm) cannot, for obvious reasons, be obtained.

In addition to the above machine stiffness and specimen size dependent characteristics of such tests, the conventional Charpy test yields only the total energy absorbed in fracturing a specimen. Properties like ultimate strength, corresponding strains and deformations, influence of the rate of loading, etc. which are invaluable to the development of rational design procedures cannot be ascertained.

Drop-Weight Test

In the drop-weight type of test a stationary specimen is struck by a falling weight. The number of blows to produce failure yields a qualitative measure of the impact resistance of the material. The number of variables involved in such tests are larger than that in conventional Charpy type tests. Examples are the specimen size, support configuration, size and shape of the hammer, drop weight and the prescribed failure criteria (first crack, perforation or total fracture, fixed extent of damage, fixed amount of deformation). All the above variables make such a test less meaningful for any purpose other than a qualitative (and/or comparative) measure of the impact resistance of the material being tested. Nanda and Hannant [17] using a "number of blows to no rebound" test found that plain concrete failed after 5 blows while concrete reinforced with 5% steel fibers withstood up to 100 blows. Dixon and Mayfield [18] also recorded an increase in the number of blows to no rebound when concrete was reinforced with 1% by volume of steel fibers.

Jamrozy and Swamy [19] have published results of tests conducted to study the behavior of FRC cubes (200mm) subjected to repeated drop-weight impact loading applied by a 50 kg hammer falling through 300mm. Three types of steel fibers were used: straight round fibers (0.25 x 15mm, 0.25 x 25mm), crimped fibers (0.25 x 25mm) and hooked fibers (0.4 x 40mm). The number of blows to produce first crack was used as a measure of impact resistance. For straight fiber (0.25 x 25mm, volume fraction 1%) reinforced FRC, first crack was found to appear after about 150 blows. Increasing fiber aspect ratio (ℓ/d) and volume fraction (v_f) was found to increase the impact resistance. They also noted that crimped and hooked fibers performed better under impact loading than smooth fibers.

Bailey et al [20] have reported results of drop weight tests conducted on FRC stair treads to access their impact behavior. Using 1% (volume fraction) of (a) fibrillated polypropylene fibers (50mm long), (b) crimped steel fibers (150mm long) and indented steel fibers (63mm long) they noted that first cracking occurred at approximately the same drop-weight irrespective of whether the tread contained fibers. However, the inclusion of fibers was found to reduce the severity of the subsequent cracking behavior. The American Concrete Institute (ACI Committee 544 on FRC, Ref. 21) recommends a drop-weight type test to evaluate the impact resistance of concrete. A 2½ in. (64mm) diameter hardened steel ball is placed on a cylindrical specimen (6 in. diameter, 2½ in. height, 152 x 64mm). A 10 lb. (4.54 kg) hammer is dropped 18 in. (457mm) onto the ball repeatedly until some prescribed failure criterion (first crack or fixed extent of deformation) is met. Using this procedure Ramakrishnan et al [22] recorded about 100 to 150 blows to first crack for concrete reinforced with hooked-end fibers.

Constant Strain Rate Test

Although limited in their capacity to achieve high rates of loading, conventional servo-controlled machines have been used to conduct dynamic tests in the intermediate strain rates.

Butler and Keating [23] have studied the effect of rate of load application on the flexural strength of FRC using a hydraulic ram capable of moving at different speeds. They tested 200 x 200 x 1500mm beam specimens under four-point bending (1.25% steel fibers, 50mm long and 0.5mm diameter). They observed a 35% increase in the flexural strength of the composite when the stress rate was increased from 0.017 to 170 MN/m/s. This increase is lower than that observed for plain concrete (75% increase) when the stress rate is increased in the same range.

Kobayashi and Cho [24] using a displacement controlled testing machine obtained load-deflection curves for polyethylene fiber (4%, 40mm long and 0.9mm diameter) reinforced concrete beams (100 x 100 x 400mm) at various strain rates. They observed from four-point bending tests that loading velocity affected both the peak load carrying capacity and the corresponding deflection. When the loading rate was increased from 1mm/min to 200mm/min a 50% increase in the composite strength was observed. Less dramatic increase in first crack stress was also reported. Deflections at peak load were smaller at the higher velocities. This increase in stiffness at higher rates was attributed by them to visco-elastic property of the polyethylene fibers.

Dynamic Tensile Test

Birkimer and Lindemann [25] have reported results of tensile tests conducted on steel and nylon fiber and reinforced concrete specimens. The nylon fibers used were 0.25mm in diameter and the steel fibers 0.43mm. 1% (volume fraction) of these 25.4mm long fibers were used in tests conducted by striking cylindrical specimens at one end with high velocity projectiles. The compression wave generated at the striking end was reflected as a tensile wave when it reached the far end of the cylindrical specimen, causing the specimen to spall. Measuring the fly-off velocity of an impedence matched pellet placed at the far end, enabled the particle velocity to be determined. From this the stresses and strains induced in the specimen were calculated. For strain rates of about 30/s, they recorded a 4 to 5 fold increase in fracture strains/stresses over the corresponding static values (strain rate of about 10^{-6}/s).

Bhargava and Rehnstrom [26] used the "split Hopkinson bar test" to study the dynamic tensile behavior of FRC. The specimens used in the tests were reinforced with 0.2% by volume of polypropylene fibers. Specimens were sandwiched between two very long (5 meters each) aluminum bars. These bars were used to measure the incident and transmitted pulses. The impact strength of the specimen was assumed to have been reached when the transmitted pulse showed no increase with increasing amplitude (impact velocity) of the impact pulse. For observed pulse rise-times of about 50 μs the dynamic strength was found to be about 50% greater than the static tensile strength.

Explosive Test

Explosive tests on FRC slabs have been carried out by Williamson [27]. He observed that shock loading when applied to slabs of plain concrete resulted in the complete disintegration of the specimens. A considerable reduction in spall velocity of the fragments was obtained by him when the matrix was reinforced with 1.75% nylon fibers. The explosive tests conducted on slabs by Robins and Calderwood, [28] also show that inclusion of steel and polypropylene fibers significantly reduces the size and particle velocity of the fragments. Such types of tests are ideally suited for situations where one would expect a structure to be subjected to rapidly rising pressures resulting from blasts or explosives.

Instrumented Impact Test

All the conventional tests described earlier offer only insights into relative merits of different fiber cement composites. They lack in their usefulness as they do not provide information on the basic properties of the material such as stress-strain or load deflection relations at high rates of loading. These relations should be independent of the test setup, that is, they must be reproducible on any setup. It is believed that instrumented impact testing, as described in this section satisfies these requirements.

While retaining the conventional mechanisms to apply impact loads, instrumented impact tests permit monitoring of load, deflection, strain and energy histories during the impact event. This allows one to compute basic material properties such as ultimate strength, strain at peak loads, energy absorbed and fracture toughness at the different strain rates.

Hibbert [29] modified a conventional Charpy type test by instrumenting the pendulum striker. He obtained continuous load-time and energy-time histories for plain and fiber reinforced concrete beams (100 x 100 x 500mm). He observed that for all specimens (unreinforced as well as FRC) the peak load under impact loading (impact velocity of 2.85 m/s) was about 10 times that under static loading (displacement rate of 0.05mm/s). This increase is quite high when compared with a less than two-fold increase generally reported by other investigators for similar strain rates (Fig. 1). This high load recorded by him is not representative of true material response but is a consequence of specimen inertia effects. This parasitic effect and solutions to minimize it will be discussed in greater depth elsewhere in this article. Using specially designed supports, Hibbert was able to compute the kinetic energy imparted to the broken halves of the specimen on impact and thus obtain the fracture energy of plain concrete (3.3 kJ/m^2). This value is an order of magnitude higher than the energy absorbed by plain concrete when fractured under static loading (0.28 kJ/m^2). For FRC beams he then calculated the energy absorption solely due to fiber debondings, pull-out and fracture, by subtracting the kinetic energy of the specimen and fracture energy of plain matrix from the total energy loss of the pendulum. But, since fracture energy of matrix was overestimated, this procedure resulted in lower energy absorption values for fiber debonding, pull-out and fracture. Hibbert as a result concluded that there is no improvement in energy absorbed due to the fibers in FRC beams under impact, compared to the corresponding value under static loading. Contrary to this conclusion, later studies [4-10] have documented enhanced energy absorption at dynamic rates.

Radomski [30] has used a rotating impact machine for performing instrumented tests on FRC. Impact in his tests is simulated by releasing a striker from a rotating fly-wheel when it has attained the desired velocity. On release, the striker hits a simply supported beam specimen (15 x 15 x 105mm). The load-time response is recorded using piezo-electric gages at one of the specimen supports. The author does not report details of such load histories. However, he observes that measurements of energy absorption obtained from his tests do not correlate with those obtained with conventional Charpy tests.

More recently Suaris and Shah [6], Gopalaratnam and Naaman [7] and Gopalaratnam and Shah [10] have conducted series of tests on concrete, mortar and FRC specimens using both an instrumented drop-weight set-up and a modified instrumented Charpy set-up. Results from these studies and a series of more systematic tests on FRC will be presented in later sections.

Results compiled by Suaris and Shah [5] from several earlier investigators, on the behaviour of plain concrete at different strain rates is presented in Fig. 1. It is apparent that the rate sensitive behaviour of plain concrete is different in tensile, flexural and compressive modes of loading. Similar trend is to be expected for FRC based on results from some of the impact studies on FRC, cited earlier. Results from static tensile and flexural tests on FRC too would suggest similar behaviour as a result of the fact that the behaviour of FRC prior to "first-crack" is not too different from that of plain matrix. In contrast, however, the important contribution of the fibers to the composite behaviour is in the post-cracking regime and hence knowledge of the entire load-displacement behaviour is essential to the accurate quantification of energy absorption under impact loading.

Koyanagi et al [31] have studied the deformation and fracture of mortar and FRC beams (75 x 75 x 660mm) subjected to 3 point bending under static and impact-loading. Instead of recording loads and deflections directly, they have used accelerometers on the striker and specimen in a drop weight type of instrumented impact set-up. Comparing deflections computed from strain gage readings with those obtained from the accelerometers, they observed a good correlation between the two until cracking. Impact force was computed from the mass of the striker and its acceleration. Energy required to fracture the plain mortar specimens or deform the FRC specimens to prescribed deflection was computed by subtracting the kinetic energy of the specimen from the applied impact energy. From reported values of strains, an average strain-rate of 0.09 - 0.2/s was computed for their impact tests. They report of a 50% increase in energy absorbed in the fracturing process under impact loading when compared to static loading. Flexural strength increases from 1030 psi (7.1MPa) to up to 1750 psi (12.1MPa) for plain mortar and 1262 psi (8.7MPa) to 3278 psi (22.6MPa) for FRC have been reported at the static loading rates and impact loading rates respectively. Deflections at peak load reported by them increase by around 50% at impact loading rates compared to the respective values at static rates for both mortar and FRC specimens. Results from their study are compared with those from this investigation elsewhere in this article.

INERTIAL EFFECTS IN THE INSTRUMENTED IMPACT TESTS OF BRITTLE MATERIALS

Discussion in this section is restricted to inertial effects in the instrumented impact tests of brittle materials although much of the body of knowledge comes from earlier studies on the impact testing of metals [32-37].

Several investigators have, in the past, recognized that during the initial period of the impact event, the load measured by the striker (tup) and that resisted by the specimen undergoing bending, are not identical

[4-10, 32-37]. This is attributed to specimen inertial effects, which manifest themselves as oscillations on the load-time records. Cotterell [33] observed a linear relationship between inertial load recorded and impact velocity. Inertial oscillations in his tests with mild steel specimens were small compared to bending loads. A mean path correction was used to compute ultimate strength and other material properties. However, for more brittle metals, Venzi, Priest and May [34] have concluded that such a mean-path correction would result in large errors. Server, Wullaert, and Sheckhard [35] have recommended that errors due to inertial oscillations can be neglected after three half-periods of oscillations. While this guide-line has been accepted more or less as a standard practice for metal testing [32], this was shown by Suaris and Shah [4] to be insufficient for the impact testing of asbestos cement composites. They attributed this to the brittleness and the relatively lower strength-to-weight ratios of such composites compared to those of metals. Consequently, they suggested that instances could be realized where inertial loads could overshadow true bending loads sustained by such cementitious composites.

Kalthoff, et al, [36] have shown, using direct optical measurement of the fracture parameter K_{Id} of Araldite specimens, that the tup load records can overpredict peak loads in instrumented impact tests by as much as an order of magnitude. They also stated that this overprediction is more pronounced for larger specimens and higher impact velocities. The unusually high peak loads and fracture energies recorded by Hibbert [29] for plain concrete is similar to that reported by Kalthoff, et al., [36]. The influence of inertial loads is likely to be more pronounced in Hibbert's tests because of the larger specimen size used by him.

Suaris and Shah [4] have analyzed this problem and have illustrated by means of a two-degree-of-freedom spring mass-system that the introduction of a rubber pad between the striker and the specimen is an effective way of reducing specimen inertial oscillations. This model has been used in interpreting and evaluating results from the modified Charpy test. The model has also been developed further to provide suitable guidelines for the selection of the various test parameters while conducting instrumented impact tests of brittle materials [8], so that the inertial oscillations can be minimized.

Winkler, et al, [37] too have experimentally observed that the introduction of an aluminum damping pad effectively reduces the peak load recorded by the striker. While testing "pressure vessel" steel specimens. This reduced peak load, according to them, provides improved correlation with crack tip strains directly recorded in the vicinity of the notch.

The testing scheme used in the present study consequently, has been designed with a view to shed more light on the parasitic effects of inertia and to provide some general guidelines on the selection of the various test parameters. A block diagram showing salient features of the modified instrumented Charpy test developed during the course of this study is illustrated in Fig. 2. Details of the test set-up and associated instrumentation is presented in the next section.

MODIFIED INSTRUMENTED CHARPY SET-UP

Modifications Effected

A conventional Charpy tester, Tinius Olsen Model 64, was modified and instrumented to facilitate tests on concrete, mortar and FRC specimens at different impact velocities. Among the three primary modifications were: (a) instrumentation of the striker and the two supporting anvils, (b) seating arrangement to accommodate large sized specimens and, (c) low-blow fixture to enable tests at different impact velocities.

It was felt that recording of anvil and striker loads simultaneously was essential to a proper interpretation of inertial loads, and to assess the influence of parameters like test system compliance, specimen size and impact velocity on the test results. The anvils and the striker were designed to serve as compression load cells capable of recording dynamic loads transmitted through them during an impact event. They were made from hardened tool steel (oil hardened, SAE 01, Rockwell C55) to ensure elastic behaviour even under high loads. They were sufficiently rounded at the specimen contact points so as to avoid local compressive damage to the specimen on impact, and at the same time facilitate smooth specimen rotation during bending. Semi-conductor P-N type gages (Kulite M (6) CEP-120-500, gage factor 220, 120 Ω) were used in full-bridge configuration within protective recesses provided on either side of all the load cells (2 anvils and the striker). Besides providing a high signal to noise ratio, the configuration also allowed temperature compensation. The dual (P-N) gages were bonded with M-Bond 610, and a post-curing protective acrylic coating M-Coat D (both from Micro-measurements) was later applied. The load cells were calibrated statically using an MTS servo-controlled testing machine, after they were subjected to low-amplitude cyclic pre-loading to eliminate initial gage-seating effects. A 10v D.C. bridge excitation was used for all the load cells. Output from the two anvils were tied in series to monitor total load recorded by the supports.

Commonly recommended size for impact testing of metal specimens is 10 x 10 x 50mm. The heterogeneity of cement based composites necessitates use of larger specimens (76 x 25 x 229mm used here). Consequently, it was necessary to modify the support mechanism of the impact machine. The dimensions of the support and the depth of the specimen did not allow impact to occur when the pendulum reached its lowest position. As a result, the beam and its supports were adequately inclined to ensure a flush contact between the beam and the striker at the moment of impact. While designing the striker, it was ensured that the center of percussion of the pendulum was retained at the center of the striking face so that adverse vibrations on the pendulum were avoided. The larger specimens also did not allow the pendulum to clear the broken halves of the specimen, unlike in a Charpy test. A hydraulic shock absorber mechanism facilitated arresting the motion of the pendulum after the beam had deflected about 50mm, Fig. 3. Peak loads were reached for fiber reinforced concrete specimens, while unreinforced mortar and concrete specimens totally fractured much prior to this deflection. Hence, arresting the pendulum motion did not affect the test results in any adverse manner.

To allow for impacting the specimen at different velocities, a low-blow fixture was designed. This allowed impact velocities in the range

0.5 - 3.0m/s. A safety lock-latching mechanism held the hammer in its raised position and assured a vibration-free release when activated. A photograph showing an overall view of the test set-up including the modified Charpy machine and associated instrumentation is presented in Fig. 4.

Instrumentation

(a) Digital Storage Oscilloscope: A 4-channel digital oscilloscope (Nicolet 4094) with 2 two-channel differential amplifier plug-in units (4562) of high resolution and frequency response (500 nano seconds per point sampling rate) was used for storing the load, strain and deflection histories. The 16K (points) main-frame storage was augmented by a dual-disk recorder (XF-44) that used 133mm diskettes. Results from 20 tests (20 tracks of 16K each) were stored on these diskettes for permanent records. Hard copies of these records were obtained on a conventional x-y recorder by playing back the stored wave-forms at much slower rates. Fig. 2 shows a block-diagram of the instrumentation.

(b) Strain Measurements: Strains were measured by directly bonding foil gages (Precision Measurements, type F400, 120 Ω) at desired locations on the specimen. For the preliminary test strains at quarter-point (and mid-point in some tests) on the tension-face of the unnotched 3 point bend specimen were monitored. Bridge completion network, using dummy gages, was used to provide temperature compensation. Bridge excitation and signal amplification was provided for by a HP-Accudata 218 bridge amplifier. During the later phase of the study, strains at the quarter-point (tension-face) of unnotched plain mortar specimens and notch tip strains (0.5 in., 13mm ahead of the saw cut notch) on the FRC specimens were measured.

(c) Deflection Measurements: Deflections of the beam mid-point were measured using a Schaevitz LVDT (050 MHR). A.C. excitation and output amplification was provided for by a Schaevitz high frequency (20kHz nominal frequency) signal conditioner (CAS-200). A 1:2 hardened steel wedge attached to the beam mid-point (close to mid-point, for notched specimens) drove a plunger that held the LVDT core. The transducer and core assembly was securely enclosed in an aluminum contraption to protect them from possible damage during the impact event. A set-screw arrangement enabled the transducer to be displaced with respect to the core to allow for the initial zeroing operations.

(d) Load Measurements: Detailed description of the dynamic load cell construction was presented earlier. Load outputs from the striker and the support (both anvil outputs tied in series) were fed into one of the plug-in units of the digital scope.

Beam deflection and strain were recorded using the other plug-in unit of the scope. Simultaneous triggering of both the plug-in units was accomplished externally using amplified signal from a fiber-optic block and flag assembly.

GUIDELINES FOR SELECTION OF TEST PARAMETERS

Before embarking on an elaborate test program it was necessary to thoroughly evaluate the performance of the test set-up developed in this study. This was successfully accomplished by testing plain concrete specimens for which sufficient reliable data were available from an earlier study [6] on its impact behavior. Results from this preliminary study are detailed in [9]. The guidelines for developing a reliable instrumented impact test scheme for concrete materials is proposed here.

Period of Inertial Oscillations

The theoretical model used here to evaluate test results obtained from the modified Charpy set-up was originally developed for a drop-weight type impact test by Suaris and Shah [4]. The impact system is represented by a two degree of freedom (d.o.f.) lumped mass system. The test beam is represented as mass m_b of stiffness k_b, the hammer-tup assembly by a mass m_t and the tup-specimen contact zone by an effective stiffness k_e. The governing equation of motion of the system is given by

$$m_t\ddot{x}_1 + k_e(x_1 - x_2) = 0, \qquad m_b\ddot{x}_2 + k_e(x_2 - x_1) + k_b x_2 = 0 \tag{1}$$

where x_1 and x_2 are displacements of the masses m_t and m_b respectively. Equation 1 is solved by applying proper initial conditions for x_1, x_2, $\dot{x}_1$ and $\dot{x}_2$. The two natural frequencies thus obtained, assuming $m_t \gg m_b$, are:

$$\omega_1 = \left(\frac{k_b\, k_e}{m_t(k_e + k_b)}\right)^{\frac{1}{2}}, \qquad \omega_2 = \left(\frac{k_e + k_b}{m_b}\right)^{\frac{1}{2}} \tag{2}$$

Generally, the mass of the hammer is much larger than that of the specimen (m_t/m_b = 60, in this investigation). Consequently, ω_1 is an order of magnitude smaller than ω_2. The frequency of oscillations observed on the load-time traces in instrumented impact tests corresponds closely to ω_2.

It is interesting to note that the formula proposed by Server, et al, [35], to empirically calculate the half-period of inertial oscillation (τ , given in Eq. 3), is analogous to the half-period computed from the two d.o.f. model (π/ω_2). Their formula is given by:

$$\tau = 1.68\ (E\ L\ W\ d\ C)^{\frac{1}{2}}/S \tag{3}$$

where E = Modulus of elasticity of the specimen, L, W, d are the length, width and depth of the specimen, C the compliance of the specimen and S is one dimensional longitudinal wave velocity in the specimen = $\sqrt{E/\rho}$ (ρ = density of the specimen.)

Server, et al, have suggested that if the time to fracture t_f is larger than 3τ , the effects of oscillations beyond this time become negligible. However, unlike metallic specimens, concrete specimens are of lower strength and larger sizes. This results in situations where amplitudes of inertial oscillations may over-shadow true bending loads. To compound the problem, fracture times are also comparatively small for such composites. Consequently the need to know the amplitude of inertial oscillations is the primary motivation for seeking a more elaborate guideline than that proposed by Server, et al.

Amplitude of Inertial Oscillations

Using the 2 d.o.f. model, the load measured by the tup $P_t(t)$ and that measured by the anvil $P_b(t)$ are given by:

$$\left.\begin{aligned} P_t(t) &= k_e\, v\, [(A_1 - A_2)\sin(\omega_1 t + \phi_1) \\ &\quad + (B_1 - B_2)\sin(\omega_2 t + \phi_2)] + m_t g \\ P_b(t) &= k_b\, v[A_2 \sin(\omega_1 t + \phi_1) + B_2 \sin(\omega_2 t + \phi_2)] + m_t g \end{aligned}\right\} \quad (4)$$

where v = impact velocity, ϕ_1 and ϕ_2 are constants characterizing the phase shifts corresponding to ω_1 and ω_2, A_1, A_2, B_1, B_2 are constants chosen to satisfy the initial conditions (all being functions of m_t, m_b, k_e, and k_b), and g = acceleration due to gravity.

To analytically predict P_t and P_b, it is necessary to know v, m_t, m_b, k_e, and k_b. Assuming that the beam vibrates in its first mode, expressions for m_b and k_b can be easily obtained. While v can be experimentally determined, m_t is normally provided by the manufacturer. k_e, the effective stiffness of the contact zone can be experimentally determined as described below. In the limiting case where beam deflections are restrained, i.e., $x_2(t) = 0$, Eq. 1, becomes

$$m_t \ddot{x}_1 + k_e x_1 = 0 \quad (5)$$

Using the initial conditions $x_1(0) = 0$, $\dot{x}_1(0) = v$, and neglecting the static deflection and weight due to m_t, peak load recorded by the striker is

$$P_{tmax} = V\sqrt{k_e m_t} \quad (6)$$

Thus, k_e can be evaluated for both the cases, with and without the rubber pad between the striker and the specimen, once the corresponding peak loads are recorded. This procedure is equivalent to the compliance calibration of the test set-up. Since this model accounts for the stiffness of the test set-up (contact zone) through k_e, predicted trends of peak load and energy absorbed are in line with observations made by Bluhm [16], Abe, et al, [15], and others. That is, increased values of k_e, which is a measure of machine stiffness, would yield larger values of recorded peak-loads and energy absorbed.

The ratio of the amplitude of oscillations of the load about the mean can be approximated for the tup and anvil loads as:

$$\left.\begin{aligned} R_t(t) &= \left[\frac{(B_1 - B_2)}{(A_1 - A_2)\sin(\omega_1 t + \phi_1)}\right] = \frac{m_b}{m_t}\left(\frac{1}{\xi(1+\xi)^2}\right)^{1/2} \frac{1}{\sin(\omega_1 t + \phi_1)} \\ R_b(t) &= \left[\frac{B_2}{A_2 \sin(\omega_1 t + \phi_1)}\right] = \frac{m_b}{m_t}\left(\frac{\xi}{(1+\xi)^2}\right)^{1/2} \frac{1}{\sin(\omega_1 t + \phi_1)} \end{aligned}\right\} \quad (7)$$

where $\xi = k_b/k_e$, R_t and R_b are the ratios of the amplitude of oscillations of the load about the mean tup and anvil loads. If fracture times can be estimated a priori, then the error in using tup and anvil loads can be predicted by $R_t(t_f)$ and $R_b(t_f)$, quite precisely. Otherwise, lower-bounds assuming $\sin(\omega_1 t + \phi_1) = 1$, can be evaluated to give some rough idea of errors due to oscillations for a particular set of m_b, m_t, k_e, and k_b.

It can be observed from Eq. 7 that for small values of ξ, R_t can be very large. For example, from the results reported by Hibbert [29], where loads were measured only using the tup, a value of around 0.5 was estimated. This can explain the erroneously large values of peak loads recorded by him in the earlier cited impact tests. Characteristic values of ξ used in the present study with and without the damping pad are 38.5 and 5.8 respectively.

It can be shown that the difference between the tup and anvil loads is given by

$$\Delta P = \left\{k_e B_1 - (k_e + k_b) B_2\right\} (\sin(\omega_1 t + \phi_1)) \tag{8}$$

If it is assumed that $\xi >> 1$ and $m_b/m_t << 1$ then the maximum value of this difference becomes

$$(\Delta P)_{max} = \left| P_t(t) - P_b(t) \right|_{max} = v\sqrt{k_e m_b/(1 + \xi)} \tag{9}$$

Both Eqs. 7 and 9 suggest that if ξ is large and m_b/m_t is small, the errors in load measurements due to oscillations of the load-time traces can be minimized. A comparison of model predicted load-time traces with experimentally observed results with and without the damping pad showed that predicted trends were accurate [9]. The following sections include details of a systematic experimental program carried out to study the behavior of FRC when subjected to different rates of flexural loading.

EXPERIMENTAL DETAILS

The scope of the test program is presented in Table 1. Flexural beam specimens (3 in. deep, 1 in. wide and 9 in. long, 76 x 25 x 229mm) were tested so as to obtain 5 different strain rates (1×10^{-6}/s to about 0.3/s). Four different mix-proportions (unreinforced and reinforced with 3 different amounts of fibers) were used. For each rate of strain and each mix-proportion, 4 flexural specimens were tested. For each mix-proportion, compression tests were conducted at the slowest loading rate using 3" x 6" (76 x 152mm) cylinders.

The composition of matrix and some characteristics of smooth brass coated steel fibers (length = 1 in. diameter = 0.016 in.; 25.4 and 0.41mm) are presented in Table 2. Three different volume fraction of fibers; 0.5, 1.0 and 1.5% were used.

A vertical mixer with a 1 cu. ft. ($0.03m^3$) capacity was used to mix the constituent materials. For the FRC mixes, cement and sand were first dry mixed and then water and fibers were alternately added in several increments and mixing continued until uniform dispersion and desired amounts of reinforcement were obtained.

Cylinders were cast in cardboard molds while beams were fabricated in plexiglas molds. Beams were cast in two 1½ in. (37.5mm) layers with approximately 15 seconds of vibration after each placement. The casting procedure and the dimensions of the beams were such that distribution of fibers was two-dimensionally random (rather than three-dimensionally random). A reasonably uniform dispersion of fiber was evidenced from the rather small scatter in observed behavior within an identical series of specimens.

FRC beam specimens had a 0.5 in. (12.5mm) deep saw cut notch at mid-span. The notch was provided to avoid any appreciable reduction in pendulum velocity (and thus in strain-rate) during impact. Since the energy consumed during fracture was substantially lower for plain mortar beams, no notch was necessary and none was provided. A comparison of the notched and unnotched mortar specimens at the slowest loading rate showed negligible difference in overall response of the beam. Notch when provided was introduced by a circular diamond saw just prior to testing (after the specimens were cured). Microscopic observation (50X) showed no damage ahead of the notch due to the cutting process.

Specimens were demolded after 24 hours and were then stored in a curing room (80°F, 27°C, 98% R.H.) for around 26 days. Subsequently they were stored in laboratory (70°F, 21°C, 50% R.H.) environment for 2 days before testing to facilitate sawing of notches and gluing of strain gages.

TEST PROCEDURE

Compression Test

Compression tests were performed in a 120 Kip (534 kN) closed-loop universal testing machine at a strain rate of approximately 1×10^{-6}/s. Average axial displacement was recorded using 2 LVDTs (gage length of 5 in., 127mm). This signal was also used for the feedback control.

Flexural Test - Static Rates

Three-point-bend tests at the slowest two strain rates: $\dot{\varepsilon}_1 = 1 \times 10^{-6}$/s and $\dot{\varepsilon}_2 = 1 \times 10^{-4}$/s were conducted in a closed-loop universal testing machine of a capacity of 40 Kips (178 kN). The central deflection of the beam was used as the feed-back signal. The deflection was measured by a specially designed device consisting of a strain-gage extensometer mounted between a fixed arm and a spring loaded arm of the device. The device was mounted between the tension face of the beam and a fixed cross-bar that held the beam supports. For unnotched mortar specimens, strain gages were glued on the extreme tension face at the quarter-points (points half-way between the supports and the load). The rate of deflection for the test was selected so as to obtain the desired strain-rates at the mid-point (twice the recorded strain rate at the quarter point). The notched beam specimens were tested at these identical deflection rates. Deflections for the notched beams were monitored at the point as close to the center of the beam as possible. Strains were recorded using resistance type foil gages (gage length 0.4 in.; 10mm) mounted at a point 0.5 in. (13mm) ahead of the notch.

Flexural Test - Dynamic Rates:

Three-point bend-tests for the highest three strain-rates : $\dot{\varepsilon}_3 = 0.09$,

$\dot{\varepsilon}_4$ = 0.17 and $\dot{\varepsilon}_5$ = 0.3/s were conducted in the instrumented Charpy impact machine (Fig. 3, 4). Tests were conducted at three different impact velocities; 130, 185 and 245 cm/s. The strain-rates at the load-point (center) at these three velocities was computed as twice those measured at the quarter point of unreinforced, unnotched specimens. The deflection at the center (or close to it for notched FRC specimen) was measured using a Schaevitz LVDT (050 MHR). A-C excitation and output amplification was provided by a Schaevitz high-frequency (20 kHz) signal conditioner (CAS-200).

Load outputs from the two supports and the striker, beam deflection and specimen strain were monitored using a 4-channel digital oscilloscope (Nicolet 4094) (Fig. 2), described earlier.

TEST RESULTS

Observations from Flexural Impact Test Results

Typical results obtained from an impact test on a mortar and a FRC specimen are shown in Fig. 5. The values of the load recorded from the instrumented tup, the sum of the two anvil loads, load-point deflection and the measured strain values are plotted with respect to time. The results are for specimens impacted at a velocity of 2.45 m/s. The peak load is reached within about one millisecond. The loads measured by tup and anvil were comparable. The difference at small times apparent in the figure is because of the presence of the rubber pad. Inertial oscillations were present, as expected, but as designed for, their amplitudes around mean values were not significant.

The strain-time plot for the notched specimens was initially linear after which strain increased very rapidly. This occurred just before the peak load. This rapid increase in strain is probably due to crack propagation within the gage length. Note that the linear part of the strain vs. time plot of the unnotched specimens was used for calculations of strain rate (Fig. 5a) and for calculation of modulus of elasticity. For unnotched specimens since the strain was measured away from the critical section, the strain reduces beyond the peak load because of the elastic unloading of the noncritical sections on cracking. For both sets of specimens, deflections continue to increase, at a higher rate, beyond the peak load.

Observations from the Static Compression Tests

Typical results from compression tests performed at low rates (1 μstr/s) of loading are presented in Fig. 6 for the various mixes tested. The softening behavior of plain mortar could be well documented as a result of the type of servo-controlled testing.

The following observations can be made from the static compression tests.

(i) Inclusion of fibers in the matrix enhances the compressive strength and the corresponding strains. Plain matrix had a compressive strength of 4414 psi (30.44 MPa) while 1.5% FRC had a strength of 5942 psi (40.98 MPa). For the same aspect-ratio of fibers used (62.5), strength increases were observed to be linearly related to fiber content. Peak strain of 3750 μ str. was recorded for the 1.5% FRC specimens compared to2700 μ str. for the unreinforced matrix.

(ii) The initial tangent modulus in compression measured experimentally obeys the law of mixture predictions quite well. This value for plain mortar and 1.5% FRC are 3.97 x 10^6 psi (27.38 GPa) and 4.35 x 10^6 psi (30 GPa) respectively.

(iii) The inclusion of fibers has an effect comparable to confining unreinforced concrete. Larger confining pressures yield higher strengths and greater ductility, analogous to larger volume contents. Although the presence of fibers influences the load-deformation behavior in the ascending portion, its major contribution is realized only beyond peak loads. The inset in Fig. 6 shows a normalized plot of stresses and strains (with respect to corresponding value at peak load) which highlight the increased toughness with increased fiber content.

EFFECTS OF STRAIN RATE

Unreinforced Mortar

Load-deflection curves for unreinforced mortar specimens subjected to 5 different strain rates are shown in Fig. 7. Increasing the strain rate increases the modulus of rupture, and the deflection at peak load as evidenced in Fig. 7. The average values of modulus of rupture and the energy absorbed to fracture are shown in Table 3. The energy to fracture of unreinforced mortar increases somewhat with increasing strain rate. It should be pointed out that energy to fracture was calculated from the observed load-deflection curves. Since the dynamic loading rates were obtained through a free fall test whereas the static-loading rates were obtained through a displacement controlled test, the energy to fracture may have been underestimated for dynamic loading rates.

It can be seen from Fig. 7 that the initial modulus of elasticity is not influenced by the rate of straining and that the load-deflection curves up to the peak becomes more linear at higher strain rates. This is further demonstrated in Fig. 8 where the secant modulus at various strain rates (normalized by the corresponding value at the slowest strain rate) calculated at the peak load and at 40% of the peak load are plotted. This plot shows that the load-deflection curves become progressively more linear with increasing strain rate, perhaps indicating that the extent of slow or subcritical crack growth (or nonlinear process zone [38-39]) decreases with increasing strain rate.

Fiber Reinforced Mortar

The relative values of peak loads (average of 4 specimens, each) at various strain rates (normalized by the corresponding values at the slowest rate) are shown in Fig. 9 for specimens of unreinforced mortar as well as those reinforced with different amounts of fibers. The average values of modulus of rupture and the fracture energy are given in Table 3. From Fig. 9, it can be seen that the effect of strain rate is higher for FRC specimens, the more so the higher the volume fraction of fibers. For example, the modulus of rupture for FRC specimens made with 1.5% fibers at highest strain rate was 2237 psi (15.43 MPa) compared to 1056 psi (7.28

MPa) at the slowest strain rate. For mortar specimens, the values at the highest and slowest rates were 1240 and 747 psi (8.55 and 5.15 Mpa respectively). The higher strain rate sensitivity of FRC specimens is probably due to additional cracking (both transverse matrix cracking and interfacial cracking or debonding) generally associated with fiber reinforced concrete specimens and the observation that the strain rate sensitivity in cement based composites is related to crack growth [5-10].

The strain rate sensitivity of FRC also increases with increasing aspect ratio as shown in Fig. 10 where the results from [7] are plotted. The data are for specimens reinforced with 2% of steel fibers (the same type as used here) and made with fibers of 3 different aspect ratios. For this figure the nondimensionalized strength values are plotted versus strain rate.

Note that the effect of strain rate sensitivity of FRC specimens on fracture energy shows a trend similar to that just discussed for flexural strength (MOR), Table 3. For fiber reinforced specimens fracture energy refers to the energy under load-deflection curve calculated up to a deflection of 0.1 in. (2.5mm). This deflection value is about 10 times the deflection at peak load as observed from Fig. 11 where the results for a set of FRC specimens tested at the different strain rates are presented. The load deflection curve for plain mortar is also shown in Fig. 11 for comparison.

COMPARISON WITH OTHER RESULTS

Instrumented Impact Tests

A summary of results obtained from three other investigations [6,31,40] and the present one is shown in Table 4. In this table static properties refer to a strain rate of about 1 x 10^{-6}/s whereas dynamic properties refer to a rate of about .1 to 1/s, both obtained using a 3 point bending configuration Wherever possible, mix properties, dimensions of the specimens, modulus of rupture (MOR) computed from peak load using elastic analysis, and the energy absorbed (G_f) by unreinforced specimen up to fracture and by steel fiber reinforced specimens (SFRC) up to a fixed value of deflection are reported in the table. From these results it can be seen that: (a) ratio of dynamic to static MOR for unreinforced specimens range from 1.43 to 1.90 and for SFRC specimens from 1.79 - 2.63; (b) ratio of dynamic to static G_f for unreinforced and reinforced specimens range from 1.35 - 1.56 and 1.52 and 1.86 respectively; (c) the static values of G_f for unreinforced specimens are between 0.32 and 0.59 lb./in. (56 - 103 N/m) while those for the reinforced specimens are between 7.97 to 15.83 lb./in. (1396 - 2773 N/m); (d) the lower the static MOR, the higher is the influence of strain rate. These observations are similar to those reported in this investigation even though the sizes of specimens, the type of instrumented impact system and the methods of measurements were not identical. This suggests that reliable and reproducible information on material characteristics can be obtained by the type of instrumented impact testing scheme described here.

Conventional Charpy Tests

Some results from standard Charpy tests for mortar specimens and SFRC specimens are presented in Table 5. One important observation from these results is that considerably higher values of fracture energy for mortar are reported from standard Charpy test. For example, Krenchel [12], reports a value of 16.53 lb./in. (2900 N/m) from Charpy tests on plain mortar specimens. This value is much larger than those observed from the instrumented impact tests ($\simeq$150 N/m). In fact, from static flexural test on the same type of mortar, Krenchel reports a value of 44 N/m (up to peak) which is comparable to static values reported in Table 4.

In the Charpy test, the energy value measured includes not only the energy to fracture the specimen, but also the energy absorbed by the testing system and the kinetic energy imparted to the specimen. Abe, et al [15] have shown that for rate insensitive silicon carbide specimens the energy calculated from the Charpy test is much higher than the true fracture energy and that the higher the true fracture energy of the specimen, the smaller is the discrepancy obtained from the Charpy test. This is also seen from Table 5. The results for SFRC specimens reported by Krenchel (20 to 30 kN/m), are comparable to those observed by Suaris and Shah (16.56 kN/m, for deflection up to 0.5 in.) using drop-weight type of instrumented impact testing system [6]. This would mean that the energy measured from the Charpy test will overestimate the true fracture energy, the more so the lower the true fracture energy of the material (for example, for SFRC composites made with low volume fraction and low aspect ratio of fibers).

Johnston [13] reports of about a 3-fold increase in fracture energy measured by the Charpy test when using about 2% steel fibers with an aspect ratio of 100. In light of the results of the present investigation, this unusually low recorded improvement is likely to be due to the parasitic effects of the Charpy testing method - which would overestimate the energy absorption values for the unreinforced matrix.

Comparison of Relative Performance of SFRC

The relative improvements in impact resistance, and in static fracture energy measured by several methods are shown in Table 6. A drop-weight test using 6 in. (152mm) diameter, $2\frac{1}{2}$ in. (64mm) high cylinder has been proposed by Schrader [41]. Number of blows required to produce first crack and to induce a fixed amount of diametrical expansion are used as criteria for quantifying impact resistance. Area under the load-deflection curve up to a fixed value of deflection for FRC specimens obtained by testing beams under static loading rates has been suggested as a measure of fracture toughness by Johnston [42] and Henager [43]. This area for FRC specimens when compared to the area up to the peak load of unreinforced matrix (which is sometimes taken as equal to that up to the first cracking load for FRC specimens) is termed toughness index and is taken as an indication of the relative performance of fiber reinforced concrete specimens. The relative performance obtained using a drop-weight method, the standard Charpy test, the static toughness index, and as observed from the instrumented impact tests - for somewhat comparable amounts of reinforcement, aspect ratios and the type of steel fibers are reported in Table 6.

It can be observed that the values of the toughness index are close to the relative performance as measured accurately by the instrumented impact test. This is not surprising since both methods use the area under the load-deflection curve as a criteria for evaluating the performance of FRC.The other methods generally underestimate the relative performance of fiber reinforced concrete.

Theoretically the area under the complete load-deflection curve should correspond to the energy required to fracture the critical cross-section (under the load-point in a 3-point bend test). However, there are some experimental difficulties in accurately obtaining this value as detailed below.

1. The post-peak load-deflection response is sensitive to the relative stiffness of the testing machine. Testing in a closed-loop mode as performed in this investigation can reduce the parasitic testing-system interaction.

2. If cracks and nonlinear deformations occur in the regions other than the critical section, then the area under the load-deflection curve will overestimate the true fracture energy of the material, (Jenq and Shah, [44]).

3. Unless the deflections are recorded in such a manner that the local deformations occurring under the load-point and supports are eliminated, the measured energy (area under the load-deflection curve) may overestimate the true fracture energy. This is especially critical for unreinforced specimens or for reinforced specimens up to the first-cracking load. Kobayashi, et al, [45] have shown that due to the load-point deformations, the area of the load-deflection curve up to first-cracking load for FRC composites was as much as 200% larger than its true value.

A similar observation was also made in the present study. The mid-point deflection was measured as a relative displacement between a fixed support and the center of the beam. The average modulus of elasticity calculated from the linear part of the load-deflection curve (after making corrections due to shear deformations) for mortar specimens tested at the slowest rate was 2.48×10^6 psi (17.1 GPa). For the same set of specimens, the modulus calculated from the strain measurements (quarter-point, (see Fig. 5) was 3.97×10^6 psi (27.4 GPa). This value was identical to that observed from uniaxial compression and uniaxial tension tests, (Gopalaratnam [9]).

These points should be considered when using the area under the load-deflection curve to evaluate the relative performance of FRC.

CONCLUSIONS

1. The modified instrumented Charpy test described here is useful in studying the dynamic behavior of brittle cement based composites. With the experience gained in this study, it is possible to design a test-system-independent impact test for low-strength brittle materials.

2. Adverse effects due to inertial loads observed in impact testing of tension weak cement composites can be significantly reduced by (a) reducing the impact velocity, (b) increasing the ratio of the tup (hammer) mass to the beam mass, and (c) increasing the ratio of the beam stiffness

to the effective stiffness of the tup-beam contact zone.

3. Mortar, concrete and FRC all exhibit increased flexural strengths at the higher rates of loading. An increase of 65% for mortar and 50% for concrete was observed in this study, when the rate of straining was increased from 1 x 10^{-6}/s to 0.3/s. The weaker mortar mix exhibits greater rate sensitivity than concrete as observed in the earlier studies. FRC is more rate sensitive than plain matrix, showing improvement in flexural strengths of 79, 99 and 111% over respective static flexural strengths for the 0.5%, 1.0%, 1.5% (fiber volume content) composites (aspect ratio of 62.5) at identical loading rates. In addition to improved strengths at the higher rates of loading, the deflection at ultimate load at these rates were consistently higher than the corresponding values at static loading rates. Up to a 50% increase in these deflection values were recorded for the various plain and reinforced composites tested.

4. Energy absorption during the dynamic fracture of the unreinforced composites increased by 40% over comparative static values. Energy absorption of FRC (generally a couple of orders of magnitude larger than that of the unreinforced composites) up to fixed deflection value (of 0.1 in., 3mm) at the dynamic rate of loading increased by 70-80% over the corresponding static value. For the same aspect ratio of fibers used, composites made with higher fiber content showed larger rate sensitivity, perhaps due to the characteristics of cracking in these composites, and the rate sensitivity associated with such a process.

5. Changes in the cracking process at the static and dynamic rates are perhaps primarily responsible for the rate sensitive behavior of cement composites. Several observations reinforce this hypothesis:

(a) Prepeak non-linearities (and micro-cracking which account for these non-linearities) reduce at the dynamic loading rates. While the initial tangent modulus of cement composites shows no significant change at the different rates of loading, the secant modulus (evaluated at ultimate load) becomes stiffer at the higher rates of loading.
(b) Cement composites exhibit a non-isotropic rate sensitivity with specimens subjected to tension, flexure and compression showing descending order of rate sensitivity at comparable rates of loading.
(c) Weaker matrix mixes are more rate sensitive than stronger ones.
(d) FRC is more rate-sensitive than the unreinforced matrix, with fibrous composites made with higher fiber contents or fibers of higher aspect ratio exhibiting a greater rate-sensitivity.

6. Static flexural toughness tests on FRC provide a conservative estimate of the impact strength and toughness of such composites. Until a standard impact test for FRC comes into effect, results from the static flexural toughness test can be used to interpret the dynamic behavior of the composite.

ACKNOWLEDGEMENT

The research reported here is being supported by a grant (DAAG-29-82-K-0171) from the U.S. Army Research Office to Northwestern University. The authors gratefully acknowledge Mr. John Schmidt for his assistance in the development of instrumentation for the project. The authors are also indebted to Denise Cable for her flawless typing of the manuscript.

REFERENCES

1. Shah, S.P., and Rangan, B.V., "Fiber Reinforced Concrete Properties," ACI Journal, Vol. 68, No. 2, Feb. 1971, pp. 126-135.

2. Hannant, D.J., "Fibre Cements and Fibre Concretes," John Wiley and Sons, Ltd., 1978.

3. Swamy, R.N., Mangat, P.S., and Rao, C.V.S.K., "The Mechanics of Fiber Reinforcement of Cement Matrices," Publication SP44, ACI, Detroit, 1974, pp. 1-28.

4. Suaris, W., and Shah, S.P., "Inertial Effects in the Instrumented Impact Testing of Cementitious Composites," ASTM Journal of Cement, Concrete and Aggregates, Vol. 3, No. 2, Winter 1981, pp. 77-83.

5. Suaris, W., and Shah, S.P., "Strain-Rate Effects in Fibre Reinforced Concrete Subjected to Impact and Impulsive Loading," Composites, Vol. 13, No. 2, April 1982, pp. 153-159.

6. Suaris, W., and Shah, S.P., "Properties of Concrete Subjected to Impact," Journal of Structural Engineering, (ASCE), Vol. 109, No. 7, July 1982, pp. 1727-1741.

7. Naaman, A.E., and Gopalaratnam, V.S., "Impact Properties of Steel Fiber Reinforced Concrete in Bending," International Journal of Cement Composites and Lightweight Aggregates, Vol. 5, No. 4, Nov. 1983, pp. 225-233.

8. Gopalaratnam, V.S., Shah, S.P., and John, R., "A Modified Instrumented Charpy Test for Cement Based Composites," Experimental Mechanics, June 1984, pp. 102-111.

9. Gopalaratnam, V.S., "Fracture and Impact Resistance of Steel Fiber Reinforced Concrete," Ph.D. Thesis, Northwestern University, 1985.

10. Gopalaratnam, V.S., and Shah, S.P., "Properties of Steel Fiber Reinforced Concrete Subjected to Impact Loading," Submitted for Publication.

11. Battelle Development Corp., "Two-Phase Concrete and Steel Material," USA Patent No. 3429094, Feb. 1969.

12. Krenchel, H., "Fiber Reinforced Brittle Matrix Materials," Publication SP44, American Concrete Institute, Detroit, 1974, pp. 45-77.

13. Johnston, C.D., "Steel Fiber Reinforced Mortar and Concrete - A Review of Mechanical Properties," Publication SP44, American Concrete Institute, Detroit, 1974, pp. 127-142.

14. Radomski, W., "Application of the Rotating Impact Machine for Testing Fiber Reinforced Concrete," International Journal of Cement Composites and Lightweight Concrete, Vol. 3, No. 1, Feb. 1981, pp.3-12.

15. Abe, H., Chandan, H.C., and Brandt, R.C., "Low Blow Charpy Impact of Silicon Carbides," Bulletin of the American Ceramic Society, Vol. 57, No. 6, 1978, pp. 587-595.

16. Bluhm, J.E., "The Influence of Pendulum Flexibilities on Impact Energy Measurements," ASTM STP 167, American Society for Testing and Materials, 1955, pp. 84-92.

17. Nanda, V.K., and Hannant, D.J., "Fibre Reinforced Concrete," Concrete Bldg. and Concrete Prods., XLIV, No. 10, October 1969, pp. 179-181.

18. Dixon, J., and Mayfield, B., "Concrete Reinforced with Fibrous Wire," Concrete, March 1971, pp. 73-76.

19. Jamrozy, Z., and Swamy, R.N., "Use of Steel Fibre Reinforcement for Impact Resistance and Machinery Foundations," International Journal of Cement Composites, Vol. 1, No. 2, July 1979, pp. 65-76.

20. Bailey, J.H., Bentley, S., Mayfield, B., and Pell, P.S., "Impact Testing of Fiber Reinforced Concrete Stair Treads," Magazine of Concrete Research, Vol. 27, No. 92, September 1975, pp. 167-170.

21. ACI Committee 544, "Measurement of Properties of Fiber Reinforced Concrete," ACI Journal, Vol. 75, No. 7, July 1978, pp. 283-289.

22. Ramarkrishnan, V., Brandshaug, I., Coyle, W.V. and Schrader, E.K., "A Comparative Evaluation of Concrete Reinforced with Straight Steel Fibers with Deformed Ends Glued Together in Bundles," ACI Journal, Vol. 77, No. 3, May-June 1980, pp. 135-143.

23. Butler, J.E., and Keating, J., "Preliminary Data Derived Using a Flexural Cyclic Loading Machine to Test Plain and Fibrous Concrete," Materials and Structures, June 1980, Vol. 14, No. 79, pp. 23-33.

24. Kobayashi, K., and Cho, R., "Flexural Behaviour of Polyethylene Fibre Reinforced Concrete," International Journal of Cement Composites and Lightweight Concrete, Vol. 3, No. 1, Feb. 1981, pp. 19-25.

25. Birkimer, D.L., and Lindemann, R., "Dynamic Tensile Strength of Concrete Materials," ACI Journal, Jan. 1971, pp. 47-49.

26. Bhargava, J., and Rehnstrom, A., "Dynamic Strength of Polymer Modified and Fibre-Reinforced Concretes," Cement and Concrete Research, Vol. 7, 1977, pp. 199-207.

27. Williamson, G.R., "Response of Fibrous Reinforced Concrete to Explosive Loading," Technical Report 2-48, U.S. Army Corps of Engineers, Ohio River Division Laboratories, Jan. 1966.

28. Robins, P.J., and Calderwood, R.W., "Explosive Testing of Fibre Reinforced Concrete," Concrete, Jan. 1978, pp. 76-78.

29. Hibbert, A.P., "Impact Resistance of Fibre Concrete," Ph.D. Thesis, University of Surrey, 1977.

30. Radomski, W., "Application of the Rotating Impact Machine for Testing Fiber Reinforced Concrete," International Journal of Cement Composites and Lightweight Concrete, Vol. 3, No. 1, Feb. 1981, pp. 3-12.

31. Koyanagi, W., Rokugo, K., Uchide, Y., and Iwase, H., "Energy Approach to Deformation and Fracture of Concrete under Impact Load," Transaction of the Japan Concrete Institute, Vol. 5, 1983, pp. 161-168.

32. Electric Power Research Institute, "Instrumented Precracked Charpy Testing," Proceedings of the C.S.N.I. Specialist Meeting, Edited by R.A. Wullaert, California, Nov. 1981.

33. Cotterell, B., "Fracture Toughness and the Charpy V-Notch Test," British Welding Journal, Vol. 9, No. 2, Feb. 1962, pp. 83-90.

34. Venzi, S., Priest, A.H., and May, M.J., "Influence of Inertial Load in Instrumented Impact Tests," Impact Testing of Metals, ASTM STP 466, American Society for Testing and Materials, 1970, pp. 165-180.

35. Server, W.L., Wullaert, R.A., and Sheckhard, J.W., "Evaluation of Current Procedures for Dynamic Fracture Toughness Testing," Flaw Growth and Fracture, ASTM STP 631, 1977, pp. 446-461.

36. Kalthoff, J.F., Winkler, S., Klemm, W., and Bienert, J., "On the Validity of K_{1d}-Measurements in Instrumented Impact Tests," Proceedings, 5th International Conference on Structural Mechanics in Reactor Technology, Berlin, 1969, G4/6, 1-11.

37. Winkler, S., Kalthoff, J.F., and Gerscha, A., "The Response of Pressure Vessel Steel Specimens on Drop Weight Loading," Proceedings 5th International Conference on Structural Mechanics in Reactor Technology, Berlin, 1979, pp. G4/6, 1-9.

38. Wecharatana, M., and Shah, S.P., "A Model for Predicting Fracture Resistance of Cement Composites," Cement and Concrete Research, Vol. 13, 1983, 819-829.

39. Ballarini, R., Shah, S.P., and Keer, L.M., "Crack Growth in Cement Based Composites," Engineering Fracture Mechanics, Accepted for Publication.

40. Zech, B., and Wittmann, F.H., "Variability and Mean Value of Strength as a Function of Load," ACI Journal, Vol. 77, No. 5, Sept.-Oct. 1980, pp. 358-362.

41. Schrader, E.K., "Impact Resistance and Test Procedure for Concrete," ACI Journal, Vol. 78, No. 2, March-April 1981, pp. 141-146.

42. Johnston, C.D., "Definition and Measurement of Flexural Toughness Parameters for Fiber Reinforced Concrete," Cement, Concrete and Aggregates CCAGDP, Vol. 4, No. 2, Winter 1982, pp. 53-60.

43. Henager, C.H., "A Toughness Index for Fibre Concrete," Testing and Test Methods of Fibre Cement Composites," RILEM Symposium 1978, pp. 79-86.

44. Jenq, Y.S., and Shah, S.P., "A Fracture Toughness Criterion for Concrete," Engineering Fracture Mechanics, Accepted for Publication.

45. Kobayashi, K., and Umeyama, K., "Method of Testing Flexural Toughness of Steel Fiber Reinforced Concrete," Japanese Society of Civil Engineers, Proceedings, May 1980, pp. 251-254.

Table 1: Details of the Experimental Program

Mix[1]	Fiber Content v_f, (%)	Flexural Tests[2]							Compression Tests[3] Number of Specimens
		Net Cross Section Depth x Width in. x in. (mm x mm)	Span in., (mm)	Number of Specimens					
				$\dot{\varepsilon}_1$	$\dot{\varepsilon}_2$	$\dot{\varepsilon}_3$	$\dot{\varepsilon}_4$	$\dot{\varepsilon}_5$	$\dot{\varepsilon}_1$
1	0.0	3 x 1, (76 x 25)	8, (203)	4	4	4	4	4	4
		5/2 x 1, (64 x 25)	8, (203)	4	-	-	-	-	
2	0.5	5/2 x 1, (64 x 25)	8, (203)	4	4	4	4	4	4
3	1.0	5/2 x 1, (64 x 25)	8, (203)	4	4	4	4	4	4
4	1.5	5/2 x 1, (64 x 25)	8, (203)	4	4	4	4	4	4

1 Matrix 1:2:0:0.5 (C:S:A:W, by weight), Fibers ℓ = 1 in. (25mm), d = 0.016 in. (0.4 mm)

2 $\dot{\varepsilon}_1 = 1 \times 10^{-6}$/s, $\dot{\varepsilon}_2 = 1 \times 10^{-4}$/s, $\dot{\varepsilon}_3$ = 0.09/s, $\dot{\varepsilon}_4$ = 0.17/s, $\dot{\varepsilon}_5$ = 0.3/s

Flexural tests at $\dot{\varepsilon}_1$ and $\dot{\varepsilon}_2$ were conducted using a closed loop machine under displacement control.

Flexural tests at $\dot{\varepsilon}_3$, $\dot{\varepsilon}_4$ and $\dot{\varepsilon}_5$ were conducted using the modified instrumented Charpy set-up.

3 Compression tests were conducted on 3 x 6 in. (76 x 152 mm) cylinders.

Table 2 : Some Physical and Mechanical Properties of the Constituent Materials.

Mortar Matrix	Composition*		
	Cement, lb/ft^3 (kg/m^3)	43.0	(690.3)
	Sand, lb/ft^3 (kg/m^3)	86.0	(1380.6)
	Water, lb/ft^3 (kg/m^3)	21.5	(345.2)
	Mechanical Properties (Static Loading)		
	Tensile Strength σ_p, psi (MPa)	405	(2.8)
	Compressive strength f'_c, psi (MPa)	4414	(30.4)
	Modulus of elasticity E_c, x 10^6 psi (GPa)	3.97	(27.4)
	Modulus of rupture MOR, psi (MPa)	747	(5.2)
Steel Fibers	Physical Characteristics		
	Smooth brass coated steel fibers		
	Length ℓ, in. (mm)	1.000	(25.4)
	Diameter d, in. (mm)	0.016	(0.41)
	Volume fractions used in the mix, v_f (%)	0.5, 1.0, 1.5	
	Mechanical Properties (Static Loading)		
	Ultimate tensile strength σ_u, ksi (MPa)	90-120	(621-828)
	Modulus of elasticity E_s, x 10^6 psi (GPa)	29.0	(200)

* Type I Ordinary Portland Cement and River Sand Maximum Size 5 mm used

Table 3 : Average[1] Material Properties at Different Strain Rates

Material	Fiber Content v_f(%)	Property[2]	Strain Rate[3] (1/s) $\dot{\epsilon}_1$	$\dot{\epsilon}_2$	$\dot{\epsilon}_3$	$\dot{\epsilon}_4$	$\dot{\epsilon}_5$
Mortar	0.0	MOR, psi (MPa)	747(5.15)	800(5.52)	1113(7.68)	1153(7.95	1240(8.55)
		G_f, lb/in (N/m)	0.43(75)	0.42(73)	0.52(92)	0.56(97)	0.58(101)
FRC	0.5	MOR, psi (MPa)	806(5.56)	864(5.96)	1334(9.20)	1392(9.60)	1450(10.00)
		G_f^*, lb/in (kN/m)	8.01^4(1.40)	8.09^4(1.42)	12.09(2.12)	12.89(2.26)	13.54(2.37)
	1.0	MOR, psi (MPa)	864(5.96)	989(6.82)	1603(11.06)	1670(11.52)	1718(11.85)
		G_f^*, lb/in(kN/m)	9.77^4(1.71)	10.06^4(1.76)	15.24(2.67)	15.93(2.78)	17.20(3.01)
	1.5	MOR, psi (MPa)	1056(7.28)	1181(8.15)	2054(1416)	2150(14.83)	2237(15.43)
		G_f^*, lb/in (kN/m)	14.32^4(2.50)	15.18^4(2.66)	21.20(3.71)	21.83(3.82)	24.50(4.29)

1. Average of four specimens for each reported value of MOR and G_f
2. Modulus of rupture (MOR) computed using elastic theory and net beam depth. Fracture energy (G_f) computed as area under the load deflection curve (for fracture of plain mortar and up to a central deflection of 0.1 in. for FRC) for unit net cross-sectional area. G_f is represented as G_f^* for FRC because area is computed only up to a deflection of 0.1 in.
3. $\dot{\epsilon}_1 = 1.0 \times 10^{-6}$/s $\dot{\epsilon}_3$ = 0.09/s $\dot{\epsilon}_5$ = 0.3/s
   $\dot{\epsilon}_2 = 1.0 \times 10^{-4}$/s $\dot{\epsilon}_4$ = 0.17/s
4. Stopped test at δ = 0.075 in., G_f^* up to δ = 0.1 in. estimated by extrapolating P-δ curve between δ = 0.075 in. and δ = 0.1 in.

Table 5 : Results* from Conventional Charpy Tests on Mortar and FRC

Reference	Specimen Size: Cross-Section (mm x mm)	Specimen Size: Span (mm)	Mix Proportions: Material	Mix Proportions: C:S:A:W (By Weight)	Impact Velocity (m/s)	Pendlum Energy	Impact Resistance (kJ/m^2)
Battele Devp. Corp.	25 x 25	102	Mortar	-	-	-	2.2
			SFRC v_f=2%, d=0.15mm	-	-	-	21.7
Krenchel	40 x 40	140	Mortar	1:2:0:0.45	3.5-4.0	50	2.9
			SFRC v_f=2%,ℓ/d=100	1:2:0:0.45	3.5-4.0	50	20.0
			SFRC v_f=0.9%,ℓ/d=170	1:2:0:0.45	3.5-4.0	50	30.0
Radomski	15 x 15	50	Mortar	1:3:0:0.60	5.2	300	23.2

1mm = 0.0394 in., 1m/s = 3.28 ft/s, 1 J = 0.737 ft-lbs, 1 kJ/m^2 = 5.7 lb/in.

*Johnston, based on conventional Charpy impact tests on 22 x 22 x 100mm specimens reports of relative impact resistance by fiber inclusion of between 2 - 10 times that of the unreinforced matrix (1:2.4:0:0.5) depending upon the reinforcing parameters. Details of the reinforcing parameters or absolute fracture energy values from these tests are, however, not reported.

Table 4 : Results from Instrumented Impact Tests on Mortar, Concrete and FRC

Reference	Set-up	Specimen Size		Mix Proportions		Static Properties		Impact Properties		
		Net Cross-Section Depth x Width (in. x in.)	Span (in.)	Material	C:S:A:W (By Weight)	MOR (psi)	G_f (lb/in)	Strain Rate	MOR_d/MOR_s	G_{fd}/G_{fs}
Zech and Wittmann	Instrumented Drop Weight	0.79 x 0.79	7.9	Mortar	1:4.7:0:0.57	1813	-	1.00	1.50	-
				Mortar	1:8.2:0:0.90	1030	-	1.00	1.90	-
Suaris and Shah	Instrumented Drop Weight	3 x 1.5	15	Mortar	1:2:0:0.5	1060	0.43	0.27	1.67	2.35
				Concrete	1:2:3:0.5	1430	0.59	0.27	1.43	-
				SFRC* v_f = 1%, ℓ/d = 100	1:2:0:0.5	1370	15.83	0.27	2.02	1.86
Koyanagi et al.	Instrumented Drop Weight	3 x 3	24	Mortar	-	1030	0.59	0.20	1.10(1.74+)	1.56
				SFRC* not reported	-	1262	7.97	0.20	2.63	1.52
Present Study	Instrumented Charpy	3 x 1	8	Mortar	1:2:0:0.5	747	0.43	0.30	1.65	1.35
				Concrete	1:2:2:0.5	1400	0.32	0.30	1.50	1.47
		2.5 x 1	8	SFRC* v_f=0.5% ℓ/d = 63	1:2:0:0.5	806	8.01	0.30	1.79	1.69
				SFRC* v_f = 1.0%, ℓ/d = 63	1:2:0:0.5	864	9.77	0.30	1.99	1.76
				SFRC* v_f = 1.5%, ℓ/d = 63	1:2:0:0.5	1056	14.32	0.30	2.11	1.71

1 in. - 2.54 cm, 1 psi = 0.0069 MPa, 1 lb/in. = 175.2 N/m.

* G_f for all FRC specimens is reported up to a 0.1 in. mid-point deflection.

+ When sufficient potential energy available in the impact.

Table 6 : Relative increase in energy absorption capacity due to the incorporation of fibers measured by conventional tests and instrumented impact tests.

Ramakrishnan et al.[1]	First Crack		Ultimate Failure	
	No. of Blows	Relative Increase	No. of Blows	Relative Increase
Mortar 1:3.66:0:0.46	10	1.0	13	1.0
SFRC v_f=1%, ℓ/d=70	48	4.8	155	11.9

Krenchel[2]	Charpy Impact Resistance	
	kJ/m^2	Relative Impact
Mortar 1:2:0:0.45	2.9	1.0
SFRC v_f=2%, ℓ/d=100	20.0	6.9

Static Flexural Toughness	Energy Absorbed	Toughness Index
Henegar[3] SFRC v_f=1.5%, ℓ/d=61	-	30.4
Ramakrishnan, et al.[4] SFRC v_f=1%, ℓ/d=70	-	16.0

Present Study[5]	Static Flexural Toughness		Instrumented Impact Toughness	
	kJ/m^2	Relative Increase	kJ/m^2	Relative Increase
Mortar 1:2:0:0.5	0.075	1.0	0.102	1.0
SFRC* v_f=1%, ℓ/d=63	1.711	22.8	3.013	29.5
Suaris and Shah[6]				
Mortar 1:2:0:0.5	0.075	1.0	0.177	1.0
SFRC* v_f=1%, ℓ/d=100	2.773	37.0	5.158	29.1

*Toughness computed up to 0.1 in. deflection

1 Ramakrishnan, et al. — ACI recommended drop weight test in a compressive configuration on 6" diameter, 1½" height cylinders (152 x 38mm).

2 Krenchel, H. — Conventional Charpy test on flexural specimens, cross-section 40 x 40mm, span 140mm, Impact Velocity 3.5 - 4m/s, Pendulum energy 50J.

3 Henegar, C. — Third point loading, cross-section 102 x 102mm, span 305mm, loading rate 0.5 - 1mm/min, matrix composition 1:2.4:0:0.47, σ_{MOR} = 6.13 MPa, ACI Toughness Index up to δ = 1.9mm.

4 Ramakrishnan, et al. — Third point loading, cross-section 76 x 76mm, Span 305mm, Matrix Composition 1:3.66:0:0.46, ACI Toughness Index up to δ = 2.5mm.

5 Present Study — 3 point flexural tests at static strain rate of 1 x 10^{-6}/s and impact strain rate of 0.3/s.

6 Suaris and Shah — 3 point flexural tests at static strain rate of 0.67 x 10^{-6}/s and impact strain rate of 0.27/s.

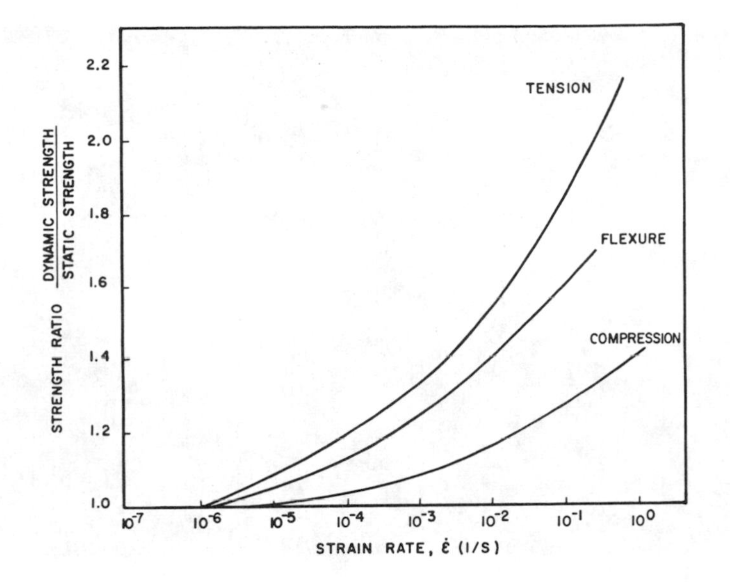

Fig. 1. Strain rate behavior of plain concrete in the different simple response modes. Ref. [5]

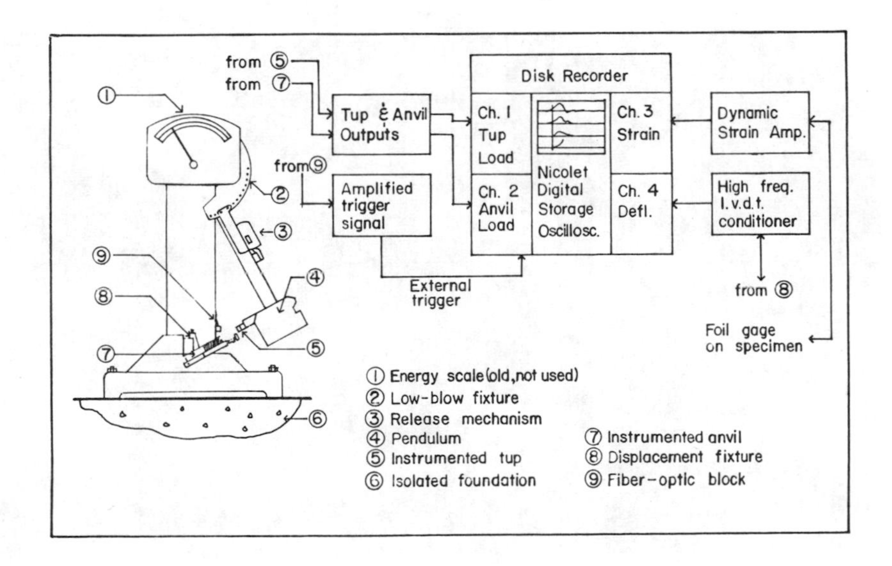

Fig. 2. A schematic showing general features of the modified instrumented Charpy test.

Fig. 3. Close up of the modified supports and the loading configuration, showing the instrumented anvils, striker, dynamic displacement measuring device and the shock absorber mechanism.

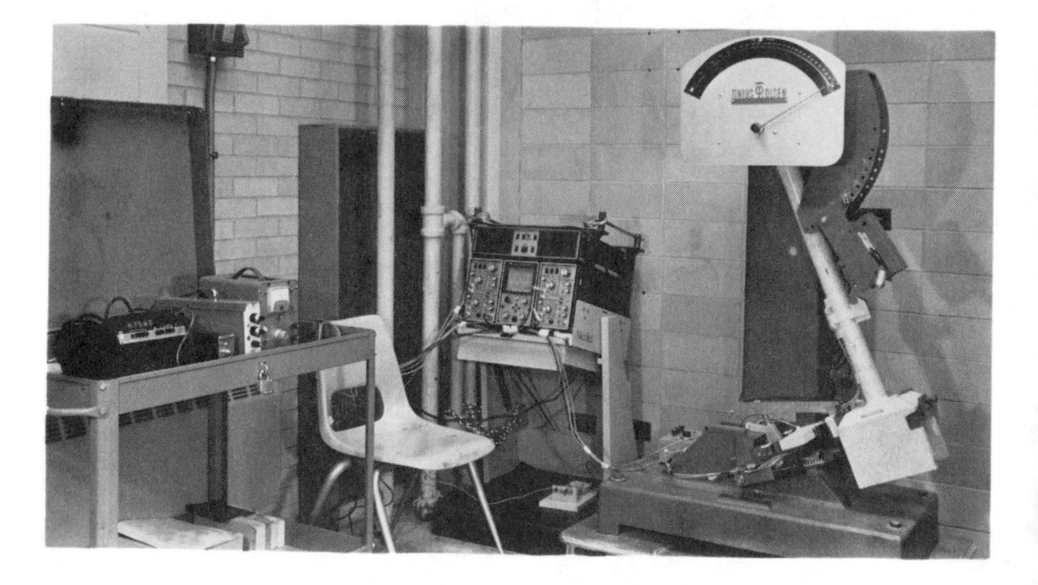

Fig. 4. An over-all view of the modified Charpy impact machine and associated instrumentation.

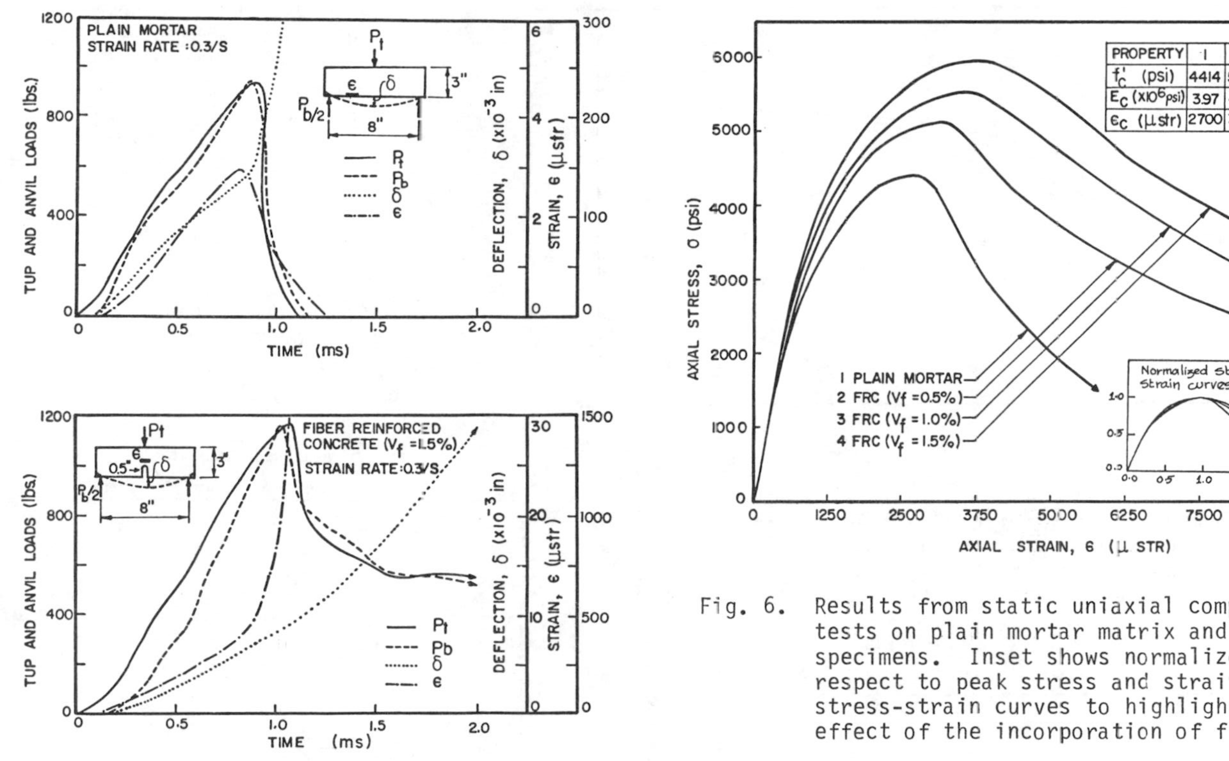

Fig. 5. Typical histories of tup load, anvil load, mid-point deflection and beam strain recorded during an impact event at a strain rate of 0.3/s. Strain rate is computed from 5a as twice the slope of the quarter-point strain-time response (a) Unnotched plain mortar beam, (b) Notched FRC (v_f = 1.5%, ℓ/d = 63) beam.

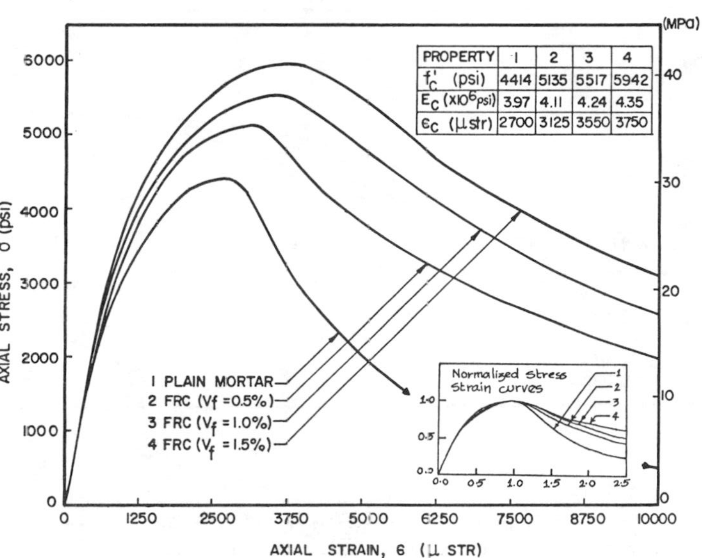

PROPERTY	1	2	3	4
f'_c (psi)	4414	5135	5517	5942
E_c (x10^{6}psi)	3.97	4.11	4.24	4.35
ϵ_c (μstr)	2700	3125	3550	3750

Fig. 6. Results from static uniaxial compression tests on plain mortar matrix and FRC specimens. Inset shows normalized (with respect to peak stress and strain values) stress-strain curves to highlight the effect of the incorporation of fibers.

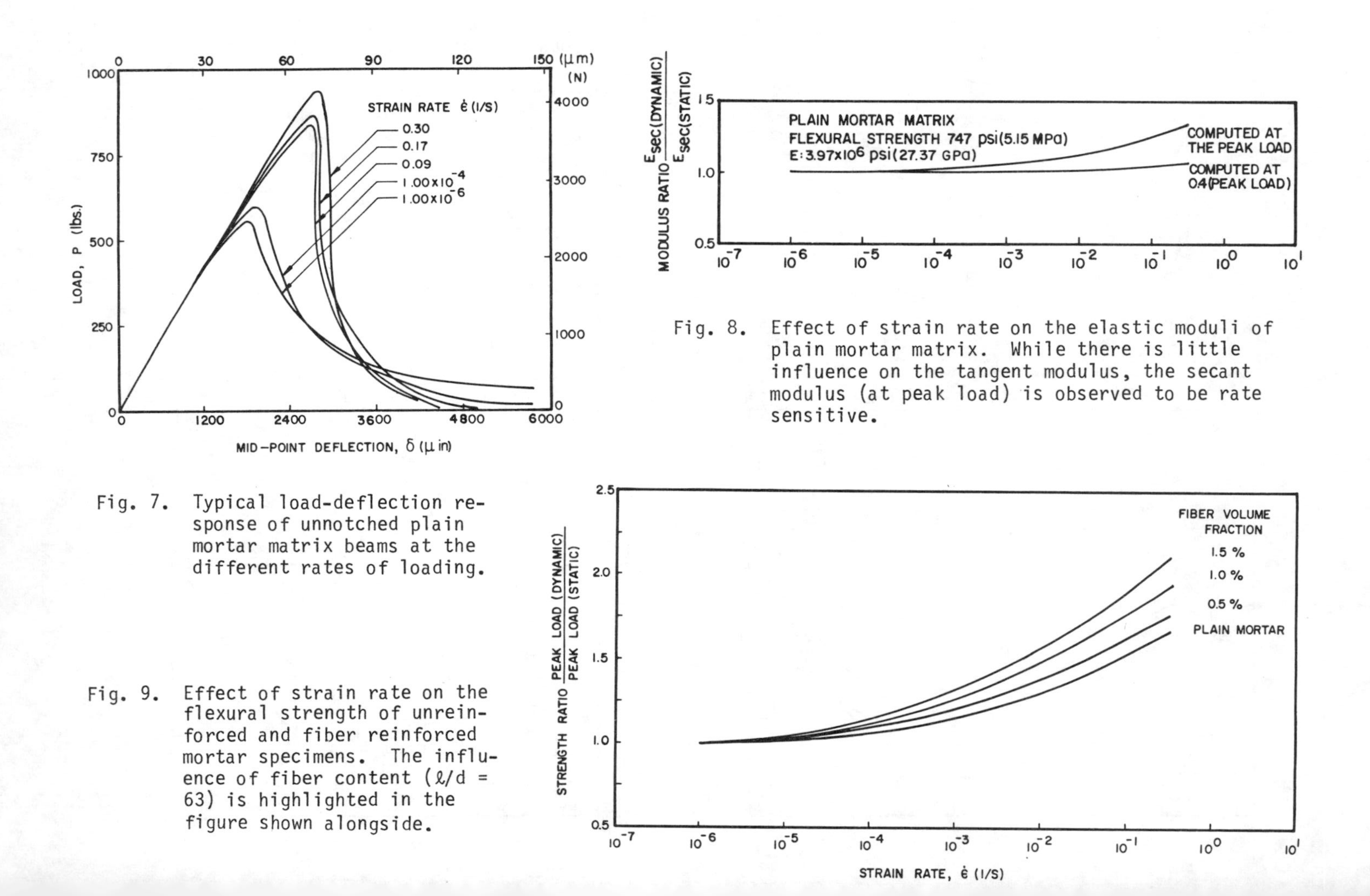

Fig. 7. Typical load-deflection response of unnotched plain mortar matrix beams at the different rates of loading.

Fig. 8. Effect of strain rate on the elastic moduli of plain mortar matrix. While there is little influence on the tangent modulus, the secant modulus (at peak load) is observed to be rate sensitive.

Fig. 9. Effect of strain rate on the flexural strength of unreinforced and fiber reinforced mortar specimens. The influence of fiber content (ℓ/d = 63) is highlighted in the figure shown alongside.

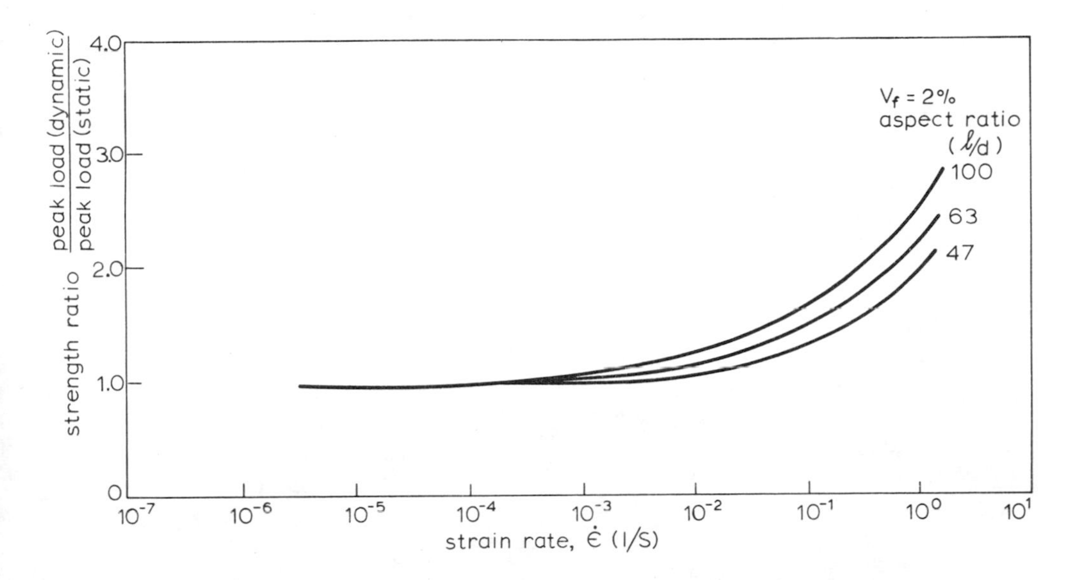

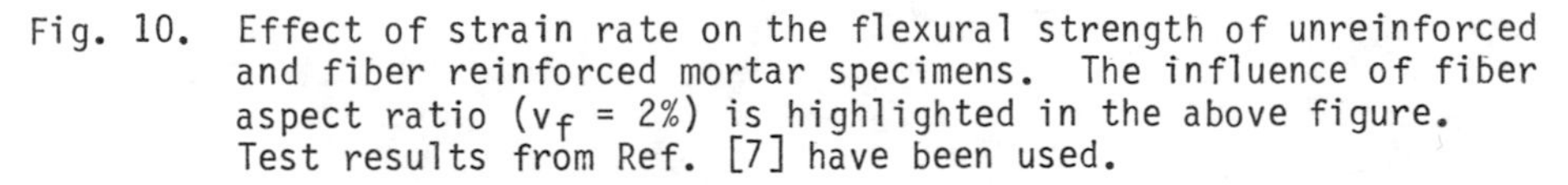
Fig. 10. Effect of strain rate on the flexural strength of unreinforced and fiber reinforced mortar specimens. The influence of fiber aspect ratio (v_f = 2%) is highlighted in the above figure. Test results from Ref. [7] have been used.

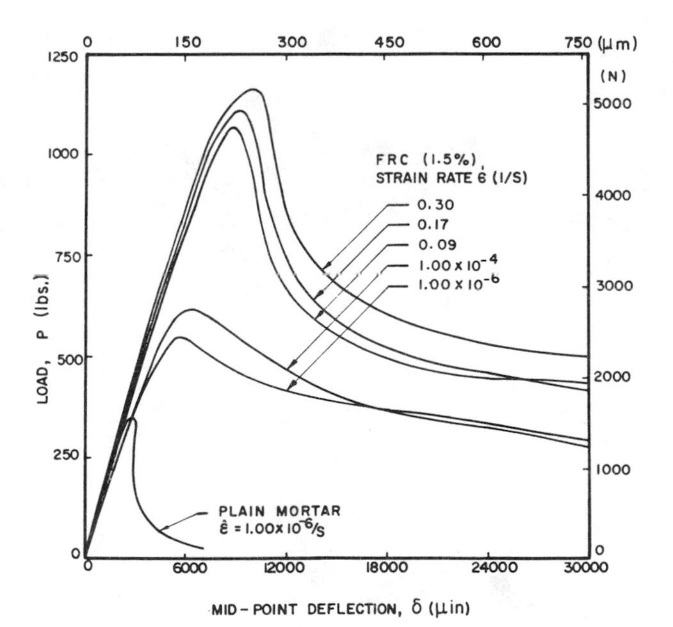

Fig. 11. Typical load-deflection response of notched FRC beams (notch : depth 0.5", width 0.1", 13mm, 2.5mm) at the different rates of loading (v_f = 1.5%, ℓ/d = 63).

STEEL FIBER CONCRETE
US-SWEDEN joint seminar
(NSF-STU)
Stockholm 3-5 June, 1985
S P Shah and Å Skarendahl,
Editors

TOUGHNESS OF STEEL FIBER REINFORCED CONCRETE

by Colin D Johnston, Prof, Civil Engineering Dep,
University of Calgary, Calgary, Alberta, Canada

ABSTRACT

Methods of quantitatively determining the toughness of steel fiber-reinforced concrete are judged and compared in terms of four proposed criteria which reflect not only the primacy of the need to obtain a meaningful and reliable indication of the behavior of the material under conditions appropriate to its main applications, but also the desirability of accomplishing this objective with equipment and techniques which are not precluded from standardization and routine use in the concrete industry on grounds of excessive complexity and expense. The methods discussed and compared include instrumented weighted-pendulum impact, instrumented drop-weight impact, empirical drop-weight impact (ACI 544 procedure), and four alternative forms of slow flexure (including ACI 544, Japan Concrete Institute, and ASTM recommended procedures) in which toughness parameters are derived from the load-deflection curve for a simply supported beam subjected to third-point loading. On the basis of conformance with the established criteria, slow flexure is judged superior to high-rate loading techniques involving impact or fatigue, or combinations of both. The question of how best to analyze and interpret data from the load-deflection curve in slow flexure is addressed in terms of the four previously proposed criteria, and the alternative methods of analysis are compared on this basis. It is concluded that the ASTM C 1018 method conforms most closely to these criteria mainly because the concept of specifying the end-point deflection for assessment of toughness as a multiple of the first-crack deflection, coupled with identification of the stress corresponding to first crack as the first-crack strength, makes it possible to distinguish between the strengthening effect of steel fibers, which is usually small, and their toughening effect, which is much larger and occurs after cracking has started. The method is also shown to be the least likely to yield misleading indications of toughness arising when commonly used but nevertheless inaccurate procedures are employed for measuring deflection. The problem of how to properly measure deflection is discussed in terms of actual deflections reported by various investigators, predictions from the theory of elasticity, and possible effects on computed toughness parameters.

Colin D. Johnston is Professor of Civil Engineering at the University of Calgary, Alberta, Canada. He is currently President of the Alberta Chapter of the American Concrete Institute and a member of ACI Committee 544 on Fiber-Reinforced Concrete. In 1976, he was awarded the Institute's Wason Medal for Materials Research for papers on steel fiber-reinforced concrete. He has been chairman of ASTM Subcommittee C 09.03.04 on Fiber-Reinforced Concrete since its inception in 1980, and is chairman of Canadian Standards Association Subcommittee A266.3 on Superplasticizing Admixtures.

INTRODUCTION

Toughness was recognized very early in the development of steel fiber-reinforced concrete as the characteristic which above all others most clearly distinguishes it from concrete without fibers. In the context of this discussion, toughness means the amount of energy needed to produce a specified damage condition or complete failure of the material (usually by separation into two or more parts).

Under impact conditions, toughness can be qualitatively demonstrated simply by trying to break through a thin section with a manually operated hammer. For example, a thin fiber-reinforced mortar flower pot withstands multiple hammer blows over a period of time before a hole is punched at the point of impact. Even then, the rest of the pot retains its structural integrity. In contrast, a similar pot made of mortar without fibers fractures into several pieces after a single hammer blow, totally losing its structural integrity as a pot.

Under slow flexure conditions, toughness can be qualitatively demonstrated by observing the behavior of simply supported beams loaded in bending. A concrete beam containing fibers suffers damage by gradual development of single or multiple cracks with increase in deflection, but retains some degree of structural integrity and post-crack strength even when deformed to a considerable deflection. In contrast, a similar beam without fibers fails suddenly at a small deflection by separation into two pieces, totally losing its structural integrity as a beam.

These two simple manifestations of toughness serve not only to identify the characteristic of toughness in a qualitative sense, but also exemplify the two categories into which fall most testing techniques for quantifying toughness, namely techniques involving either high-rate single or multiple applications of load, or a single slow-rate application of load.

In the high-rate category are impact tests in which toughness is determined as the energy needed to produce complete fracture of the test specimen into separate pieces by a single application of load using a falling weight or a weighted pendulum. Also in this category are tests in which toughness is empirically determined by the number of repeated impact load applications needed to produce first visible crack or failure, but there is an element of fatigue as well as impact in such techniques. Conventional fatigue tests involving repeated load cycles applied at high frequency also fall in this category.

In the slow-rate category are the single-cycle so-called static loading tests in which toughness is determined as the energy represented by the area under the stress-strain curve in uniaxial tension or compression, or by the area under the load-deflection curve in flexure.

The merits and disadvantages of the available techniques in each category are examined subsequently, but first, criteria for judging and comparing them must be established.

CRITERIA FOR SELECTING A METHOD OF DETERMINING TOUGHNESS

With the proliferation of commercially available sizes and types of steel fiber comes the need not simply for a qualitative demonstration of toughness but for quantitative methods of determination which are precise enough to clearly distinguish between levels of toughness obtainable with different type-amount combinations. Such methods are essential for effectively optimizing job mix proportions, monitoring concrete quality to verify compliance with specifications, or assessing the quality of concrete in service. Considering the needs of the concrete industry as a whole, as opposed simply to the furthering of research, the following criteria are proposed as the four most important:

1. Material Behavior - The method should produce results which quantitatively define material behavior relative to some readily understandable and fundamentally significant reference level of behavior, preferably under loading conditions appropriate to the application considered.

2. Precision - To clearly distinguish levels of performance of different type-amount combinations of fibers, the variability inherent in the method should be low enough to give satisfactory repeatability (single-operator precision) using a single operator-equipment combination and satisfactory reproducibility (multilaboratory precision) between different operators in different laboratories using essentially the same equipment. (ASTM has made the inclusion of quantitative precision statements mandatory in all test methods.)

3. Specimen Adaptability - The method should accommodate various sizes and shapes of test specimens representing the diversity of established applications for steel fiber-reinforced concrete from thick mass concrete to thin overlays or shotcrete linings, and the results should reflect material behavior independent of specimen size and shape. The method should also be capable of demonstrating the effects of preferential fiber alignment on toughness if appropriate in the application considered, or excluding such effects in applications

where the sections are thick enough to ensure essentially random fiber orientation.

4. Standardization Potential - Excessively complex and expensive experimental arrangements, although often very useful in research, are not readily accepted for standardization by the concrete industry. Therefore, any method of determining toughness for routine use by the industry should as far as possible utilize equipment and techniques already employed for testing conventional concrete, keeping new or expensive equipment and techniques which are suitable only for testing fiber-reinforced concrete to a minimum.

REVIEW OF ALTERNATIVE TESTING TECHNIQUES FOR DETERMINING TOUGHNESS

The available alternative methods of determining toughness differ mainly in two respects, rate of loading and consequent possible rate-sensitivity of results, and mode of loading (tension, compression, flexure etc.).

Slow-Rate versus High-Rate Techniques

High-rate loading techniques involve either single-cycle impact(1-5) or a multiple-cycle combination of impact and fatigue(6,7). The main problem even with the simpler single-cycle test is that very sophisticated equipment and instrumentation are required to accurately distinguish between the strain energy actually input to the test specimen, the kinetic energy imparted to the specimen (or parts thereof) after it has failed, and the energy lost to the impacting device and support system. The complexities involved in making this distinction have been very thoroughly documented by Sauris and Shah(1), and impact testing is the subject of a separate paper forming part of these proceedings. Such testing is discussed herein only to the extent that it relates to the four criteria previously identified and provides comparable high-rate and slow-rate determinations of toughness.

As a result primarily of the work of Hibbert and Hannant(2,3) and Sauris and Shah(1,4), two categories of impact test can be identified, those which are sophisticated enough to record the energy actually input to the test specimen, and those which are largely empirical and do not attempt to measure the energy actually input to the test specimen.

To quote Hibbert and Hannant(2), "The impact test is superficially one of the simplest material tests in use today, but, in fact, it is probably the most complicated if accurate results, reproducible on other machines, are to be obtained".

Realizing the difficulties, Hibbert and Hannant(2) built a carefully instrumented weighted-pendulum impact tester which allowed them to determine the effect on energy absorption of varying the loading rate in flexure on 100 mm square specimens from impact to slow flexure. Compilation and analysis of a selection of their results (3)(TABLE 1) reveals that:

(1) The component of energy associated with fiber pullout alone (excluding matrix fracture) is not highly rate-sensitive, and rate-sensitivity from slow flexure to impact is more marked for higher volume fractions

of hooked or crimped fibers with greater mechanical bond than for lower volume fractions of straight less mechanically bonded fibers. Net energy increases from slow flexure to impact range from 0 to 70% with an average of 29% for the fibers tested.

(2) Due to inertial oscillations acknowledged in the source report(3), the energy absorption of the matrix alone is probably not nearly as rate-sensitive as indicated by the more than ten-fold increases shown for mixes without fibers. For the mixes with fibers, the inertial effects are greatly reduced, giving total energy increases (matrix fracture plus fiber pullout) from slow flexure to impact ranging from about 50% to 130% with an average of 84% for the fibers tested.

(3) The level of repeatability of data achievable in an impact test can be comparable to, or in this case better than, the level achievable in a slow flexure test. The average within-test coefficient of variation is 21% for slow flexure and 14% for impact, values not much more than expected in conventional concrete testing.

TABLE 1. Energy Absorption of Fiber Reinforced Concrete (Max. Agg. 10 mm) Loaded in Slow Flexure and Weighted Pendulum Impact

Test Age	Fibers Type	Fibers Size	V_f -%	Slow Flexure Data Total[a] Energy	Slow Flexure Data C of V[b] of Total	Slow Flexure Data Net[c] Energy	Impact Data Total[a] Energy	Impact Data C of V[b] of Total	Impact Data Net[c] Energy	Energy Increase Total[d]	Energy Increase Net[e]
2 mth.	Hooked	50x0.5 mm	1.2	85	(27%)	82	142	(15%)	111	67%	35%
2 mth.	Hooked	50x0.5 mm	0.6	57	(27%)	54	95	(14%)	64	67%	19%
2 mth.	Crimped HC	50x0.5 mm	1.2	66	(13%)	63	138	(18%)	107	109%	70%
2 mth.	Crimped HC	50x0.5 mm	0.6	53	(23%)	50	102	(9%)	71	92%	42%
2 mth.	Duoform	38x0.38 mm	1.2	30	(19%)	27	>57	(11%)	>26	>90%	0%
2 mth.	Melt Ext.	25x0.7 mm	1.2	47	(12%)	44	89	(12%)	58	89%	32%
2 mth.	None		0.0	2.8	(18%)	0	31	(10%)	0	1007%	-
2.5 yr.	Crimped HC	50x0.5 mm	1.2	95	(16%)	93	140	(21%)	111	47%	19%
2.5 yr.	Crimped LC	50x0.5 mm	1.2	80	(30%)	78	130	(19%)	101	63%	29%
2.5 yr.	Duoform	38x0.38 mm	1.2	24	(23%)	22	>55	(13%)	>26	129%	18%
	None		0.0	2.1	(18%)	0	29	(13%)	0	1280%	-
				Average	[21%]		Average	[14%]		84%[f]	29%[f]

a - Total absorbed to 10 mm deflection (matrix + fiber pullout)
b - Based on minimum of five specimens
c - Total minus matrix component (fiber pullout alone)
d - Impact relative to slow flexure based on total energy
e - Impact relative to slow flexure based on fiber pullout alone
f - Average for fiber-reinforced mixes only

Considering the difficulties of building the impact tester, and no doubt recognizing the varying effect of loading rate with type and amount of fibers, Hibbert and Hannant(3) concluded that slow flexure is a more useful and relevant means of measuring energy absorption, more relevant presumably because rate effects are eliminated.

Sauris and Shah (1,4) built a carefully instrumented drop-weight impact tester which produces flexural loading, and used it with slower flexural tests in a strain-rate controlled Instron machine to evaluate the rate-sensitivity of energy absorption data. Compilation and analysis of a selection of their results (4)(TABLE 2) indicates that:

(1) The component of energy associated with matrix (mortar) fracture is rate-sensitive, but much less so than indicated by the results of Hibbert and Hannant (3), the slow-to-impact increase being only 132% (TABLE 2) compared with 1000-1300% (TABLE 1).

(2) The total energy absorption of the composite representing the

combination of matrix fracture and fiber pullout is less rate-sensitive than the matrix component alone. The slow-to-impact increases vary from 0 to 99% depending on the type and amount of fibers, with an average of 40% for the five combinations tested, numbers which are consistently lower than the range 50% to 130% with an average of 84% obtained by Hibbert and Hannant (3).

(3) Although the energy component associated with fiber pullout alone is not specifically identified, reference is made to Gokoz and Naaman (5) who used a similar drop-weight test to show that this energy component is not appreciably rate-sensitive for steel fibers, even the hooked variety.

TABLE 2. Sensitivity of Energy Absorption in Flexure to Strain Rate for Plain and Fiber-Reinforced Mortars (Max. Agg. 5 mm)

Rate	Strain	No Fibers		1.0% Long Fibers[d]		1.0% Short Fibers[e]		Rate	Strain	1.0% Long Fibers[d]	
Category	Rate-s^{-1}	Energy-Nm	Inc.-%[f]	Energy-Nm	Inc.-%[f]	Energy-Nm	Inc.-%[f]	Category	Rate-s^{-1}	Energy-Nm	Inc.-%[f]
Slow[ac]	0.67×10^{-6}	0.22	0	24.0	0	7.2	0	Slow[a]	7.5×10^{-6}	43.4	0
Moderate[a]	0.27×10^{-4}	0.18	-18%	31.2	30%	7.4	3%	Impact[b]	0.41	61.7	42%[g]
Fast[a]	$.7 \times 10^{-3}$	0.26	18%	24.8	3%	6.1	-15%	Impact[b]	0.83	56.0	29%[g]
Impact[b]	0.27	0.51	132%	47.8	99%[g]	7.0	-3%[g]	Impact[b]	0.87	58.5	35%[g]

a - Produced in a conventional Instron machine
b - Produced in an instrumented drop-weight impact tester
c - Rate comparable to that specified in ASTM C78 for normal flexural strength test of plain concrete
d - 25x0.25 mm straight wire, L/D = 100
e - 6x0.25 mm straight wire, L/D = 25
f - Percentage increase relative to slow rate category
g - Used to calculate average slow-to-impact energy increase of 40%

In attempting to rationalize the results of the two separate independent studies, bearing in mind the inertial effects acknowledged by Hibbert and Hannant (3) as causing the extraneously large slow-to-impact energy increases for matrix alone, it is noted that the difference in slow-to-impact increases for fiber-reinforced matrices between the two investigations is much less than for the matrix alone. This is consistent with the observed low rate-sensitivity of the energy associated with fiber pullout (3, 5). However, the difference between the two studies in terms of slow-to-impact increases in total energy, an average 84% (TABLE 1) compared with 40% (TABLE 2) is still appreciable, and, although inertial effects may be partly responsible, the writer wondered if the impact strain rates in each study were comparable. On examining the strain-time plots (3), it seems that the impact strain rate applicable to the TABLE 1 data was 7 to 8 compared with 0.27 to 0.87 in TABLE 2. Accordingly, it is reasonable to conclude that the difference in the average magnitude of the slow-to-impact energy increases in the two sets of data (3, 4) is probably partly due to the more than ten-fold difference in impact strain rates.

The consensus which emerges from these two very thorough studies of impact is that the energy absorption of steel fiber-reinforced mortars or concretes is rate-sensitive from slow flexure to impact, and that the magnitude of the difference between the two depends not only on the inertial effects and strain rates associated with the specific impact test but also on the type-amount combination of fibers employed. Therefore, slow flexure seems preferable for determining toughness not only because the results are insensitive to rate effects and are lower bound values safe for use in design, but because the fully instrumented impact test, even it meets criterion 2 for precision, is so complex that the potential for standardization and routine use (criterion 4) is not good.

In the category of empirical impact tests, one of the more common, which employs a modification of equipment already standardized for soil and asphalt testing, is the multiple-cycle drop-weight technique developed by Schrader (6) and recommended by ACI Committee 544 (7). In terms of the Hibbert and Hannant (2) quotation, it is simple in terms of the procedural and equipment requirements and superficial because the quantity measured bears no relationship to the energy input to the test specimen due to energy losses and inertial effects. Furthermore, unlike the single-cycle high-rate testing techniques which produce true impact conditions, this technique produces some combination of impact and fatigue loading.

In terms of the established criteria, a disadvantage of this drop-weight technique is that the stress condition of axial compression combined with lateral splitting tension bears little resemblance to that encountered in most applications (criterion 1). However, the most serious difficulty with the test as presently performed is the inherently high variability of results which far exceeds normally acceptable levels and is therefore of major concern (criterion 2). Based on available published data, computation of coefficients of variation for sets of three nominally identical specimens reveals averages frequently more than 30%, and worst-case individuals exceeding 50% both for small aggregate shotcrete (8) (TABLE 3) and lightweight concrete (9) (TABLE 4). Coefficients of variation are apparently lower for the failure condition than the first-crack condition with the average approaching 20% for the failure condition (TABLE 4). However, it is clear that the repeatability achievable in this procedure is far inferior to that achieved by Hibbert and Hannant (3), where the comparable average is 14% with a worst individual case of 21% (TABLE 1). This demonstrates that reasonable levels of repeatability can be achieved in some forms of impact test but not in the empirical multiple-cycle drop-weight test as presently performed (criterion 2).

TABLE 3. Mean Number of Blows for Drop-Weight Impact Tests on 38x152 mm Shotcrete Specimens (Maximum Aggregate 5 mm). Within-Set Coefficients of Variation for Sets of Three Specimens in Parenthesis

Test Age - days	No Fibers	Hooked Wire, L/D=60 V_f=0.6%	1.0%	1.3%	Slit sheet, L/D=60 0.6%	1.0%	Wire, L/D=72 1.0%	Test Condition	Average C of V
2	5	11	40	169	14	13	37	First Crack	
	(40.0%)	(20.4%)	(21.1%)	(36.9%)	*	(7.7%)	(21.6%)	First Crack	(24.6%)
14	6	17	63	107	8	80	45	First Crack	
	(72.6%)	(33.1%)	(30.5%)	(38.3%)	(13.9%)	(70.1%)	(67.8%)	First Crack	(46.6%)
28	10	20	82	206	16	42	48	First Crack	
	*	(48.2%)	(7.9%)	(14.1%)	*	(37.4%)	(25.3%)	First Crack	(26.6%)
								First Crack	[33.7%] [a]
2	8	83	215	536	19	39	93	Failure	
	(32.9%)	(17.2%)	(82.6%)	(45.7%)	*	(24.7%)	(14.0%)	Failure	(36.2%)
14	9	89	283	549	24	129	147	Failure	
	(60.1%)	(26.2%)	(33.5%)	(24.3%)	(8.8%)	(45.7%)	(35.8%)	Failure	(33.5%)
28	13	97	366	457	36	66	155	Failure	
	*	(30.2%)	(35.9%)	(14.2%)	*	(17.5%)	(3.9%)	Failure	(20.3%)
								Failure	[30.7%] [a]

* Only two results reported
a Overall average for all ages (18 sets)

TABLE 4. Mean Number of Blows for Drop-Weight Impact Tests at Age 28 Days on 64x152 mm Lightweight Concrete Specimens (Maximum Aggregate 13 mm). Within-Set Coefficients of Variation for Sets of Three in Parenthesis

Concretes Without Fibres by Compressive Strength							Test	Average
17 MPa	35 MPa	46 MPa[a]	58 MPa	41 MPa[a]	46 MPa[a]	44 MPa[a]	Condition	C of V
6	7	12	4	8	6	14	First Crack	
(33.3%)	(67.3%)	(18.7%)	(31.5%)	(59.2%)	(36.8%)	(35.7%)	First Crack	[40.4%]
11	16	17	8	20	10	22	Failure	
(24.1%)	(41.0%)	(22.7%)	(18.3%)	(27.1%)	(26.5%)	(29.8%)	Failure	[27.1%]
Corresponding Concretes with 1.0% Wire Fibers[b]							**Test**	**Average**
Paddle	Paddle	Paddle[a]	Paddle	Crimped[a]	Plain[a]	Hooked[a]	Condition	C of V
12	24	122	70	48	66	61	First Crack	
(4.7%)	(77.2%)	(47.8%)	(76.6%)	(46.0%)	(67.7%)	(80.6%)	First Crack	[57.2%]
227	269	536	958	124	192	793	Failure	
(24.2%)	(13.0%)	(9.9%)	(9.5%)	(35.4%)	(27.7%)	(10.8%)	Failure	[18.6%]

a - Identical concrete matrix proportions
b - Paddle type 53x0.76 mm, crimped type 25x0.42 mm, plain type 25x0.5 mm, hooked type 50x0.5 mm

The proponents of the empirical drop-weight impact test (7, 10) argue that it is useful both for demonstrating the improved impact resistance of fiber reinforced concrete over conventional concrete and for comparing the relative merits of concretes with different type-amount combinations of fibers. Clearly, it can distinguish between plain and fiber reinforced concrete (TABLE 4), but so can the manually operated hammer, possibly with comparable repeatability. Whether it can reliably distinguish between the merits of different type-amount combinations of steel fibers precisely enough to be useful in materials selection, mix proportioning, and quality control is less clear. For example, in column 5 of TABLE 3 we find first-crack values of 169, 107, and 206 at ages of 2, 14, and 28 days and failure values of 536, 549, and 457 at the corresponding ages, and again in column 7 first-crack values of 13, 80, and 42 with failure values of 39, 129, and 66 at 2, 14 and 28 days. These sequences are significantly contrary to the generally visible pattern of increase in values with age. Considering the data in TABLE 4, given the uniformly low impact resistance of the four matrices of compressive strengths 17, 35, 46 and 58 MPa, it seems surprising that for the same type and amount of fibers the toughness could really be increased by over four times, (227 to 958 according to this test) just by strengthening the matrix while keeping the amount and type of fiber constant. The reason in this case may be due to crushing, which takes place under the impacting steel ball, causing the result to be unduly influenced by matrix compressive strength.

In pursuing the question of whether empirical drop-weight impact values truly reflect toughness, very limited comparable data for drop-weight impact and slow flexure are available for examination. TABLE 5a shows the 28-day impact data (failure condition) from TABLE 3 (8) compared with a modified ACI slow flexure toughness index. TABLES 5b and 5c show comparisons (6, 11) of impact values (failure condition) with standard ACI 544 (7) slow flexure toughness indices (350x100x100 mm beams). While the toughness index values are not without inconsistencies, it is clear that the impact values vary widely, 97 to 457 for a comparatively narrow range of index values, 14.6 to 23.2 (TABLE 5a), remain very similar, 516 and 564, for quite different index values, 6.0 and 3.0 (TABLE 5b), and vary widely, 217 to 622 for nearly identical index values (TABLE 5c). Clearly, each test is evaluating different material characteristics with different emphasis. The contrast between the poor correlations obtained for the results in TABLE 5 and the good correlation obtained for fully instrumented impact and slow flexure data in TABLE 1 is illustrated in FIG. 1.

TABLE 5. Comparison of Empirical Drop-Weight Impact Blows to Failure With Slow Flexure Toughness Indices

Table 5a				
Fiber V_f-%	Hooked Wire, L/D=60 0.6% 1.0% 1.3%	Slit Sheet, L/D=60 0.6% 1.0%	Wire, L/D=72 1.0	No Fibers
Impact (blows)[a] Toughness Index[a]	97 366 457 19.5 10.1 23.2	36 66 1.0 5.0	155 14.6	13 1.0
Table 5b				
Fiber V_f-%	Hooked Wire, L/D=60 0.6% 0.8%	Slit Sheet, L/D=60 1.1%		No Fibers
Impact (blows)[b] Toughness Index[b]	516 564 6.0 3.0	82 2.0		38 1.0
Table 5c				
Fiber V_f-%	Hooked Wire, L/D=100 0.3% 0.3% 0.4% 0.5%			No Fibers
Impact (blows)[b] Toughness Index[b]	217 297 414 622 6.14 6.22 6.02 6.38			47 1.0

a - Impact and toughness index at 2.5 mm deflection (modified ACI index) on 38 mm thick shotcrete specimens.
b - Impact on 64 mm thick concrete specimens and toughness index (standard ACI index) on 100 mm square specimens.

Finally, brief reference should be made to multiple-cycle fatigue tests in which the strain rates do not approach impact rates but are nevertheless high compared with normal static loading conditions, typically more than 10 cycles of loading per second. Such tests are again likely to give rate-sensitive results with all the attendant difficulties of interpretation already noted for single-cycle high-rate tests. The results also tend to be highly variable, and therefore offer poor repeatability (criteria 2). Furthermore, the test equipment is inevitably sophisticated and expensive, and consequently is not readily suited for standardization and routine use (criterion 4).

Mode of Loading

Since steel fiber-reinforced concrete is used mainly in pavements, overlays and shotcrete linings where it is loaded primarily in flexure, the preference for having the method of determining toughness reflect the stress conditions in the application (criterion 1) favors a flexural loading technique. In such applications, uniaxial tensile stresses induced by temperature and/or moisture changes, when present, are usually of secondary importance. However, when uniaxial tension is the primary stress condition, and the circumstances justify extra effort and expense in testing, toughness in this mode can be evaluated from a properly established stress-strain curve (12). The problem in practice is that the devices for gripping the concrete specimen to apply tension are relatively complex (12, 13), special attachments to stiffen the testing machine are usually required to obtain the complete stress-strain curve (12), and several strain measurements on the specimen are required with a system of averaging individual readings. Consequently, the testing equipment becomes so complex that it is not well suited for standardization and routine use (criterion 4).

At first glance, uniaxial compression appears the least complex mode of loading for establishing a stress-strain curve and determining measures of toughness from it. However, it has been clearly demonstrated (12) that determining specimen strain simply from movements of the machine cross-head (stroke) during loading of a concrete cylinder or prism is inaccurate, and that testing in stroke control can produce uncontrolled failure of the specimen because of deformations within the testing machine frame. Instead, strain measurements must be made on the specimen using several transducers with a system of averaging individual readings, and testing

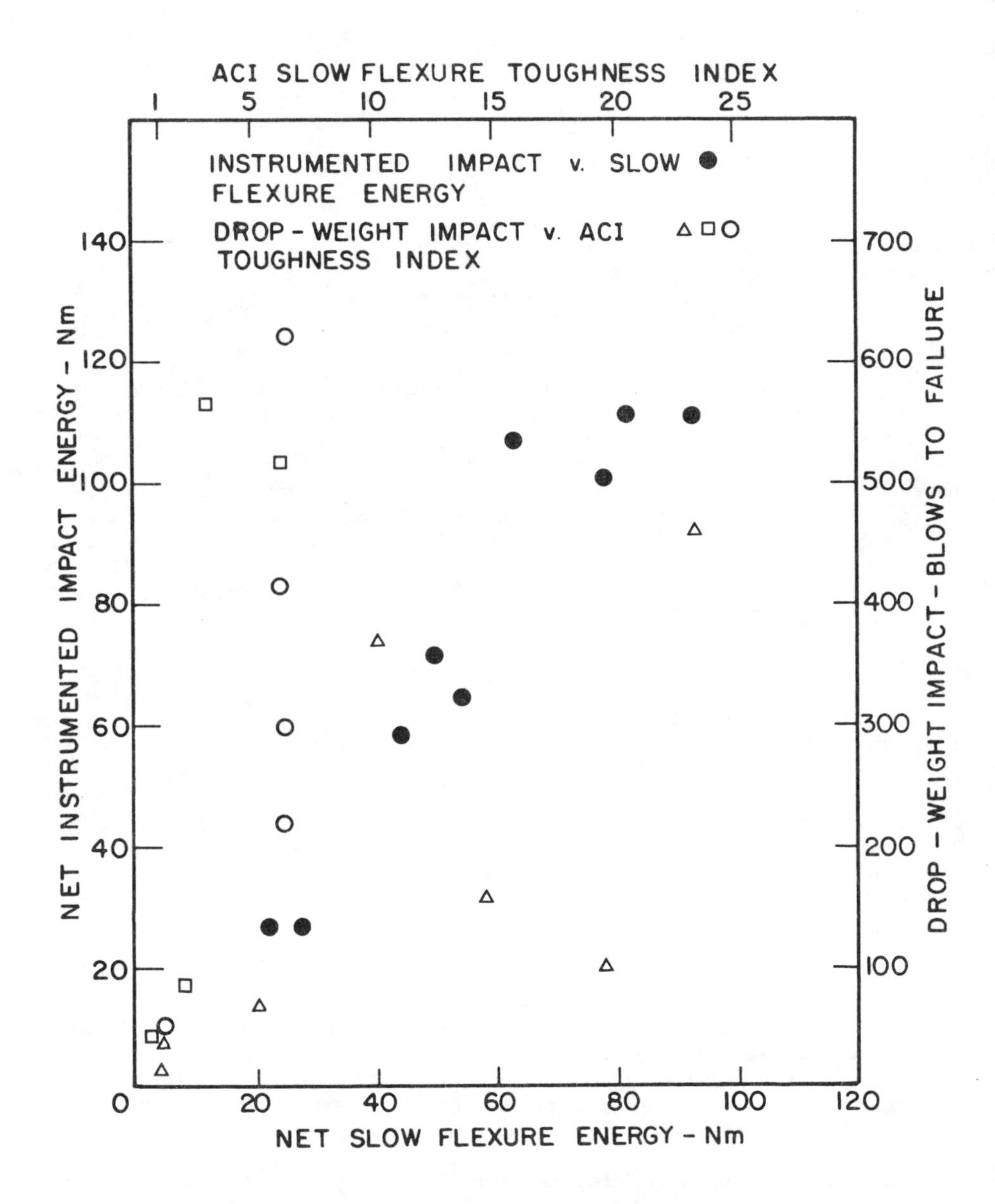

FIG 1. Correlation between Net Instrumented Impact Energy and Corresponding Slow Flexure Energy (3), and between ACI 544 Empirical Drop-Weight Impact Blows to Failure and Corresponding ACI 544 Toughness Index (6, 8, 11).

must be performed in strain control to avoid uncontrolled failure of the specimen (12). However, compared with uniaxial tension, the compression mode is less complex and expensive, and is therefore more amenable to standardization and routine use (criterion 4). Its main disadvantage is that it does not correspond to the primary stress condition of concern in most applications (criterion 1). Obviously, when circumstances warrant, it could be used to evaluate toughness.

On balance, the flexural mode of loading is preferable not only because it reflects the primary stress condition imposed on fiber-reinforced concrete in most applications (criterion 1), but also because the facilities needed are no more complex and expensive than are needed for compression, and are therefore amenable to standardization and routine use (criterion 4).

Summary Comparison of Alternatives

In terms of the four criteria proposed as most important in selecting a method of determining toughness, the arguments favoring some form of slow flexure technique over impact or fatigue loading techniques are as follows:

1. Sensitivity of test results to strain rate effects can be eliminated under slow loading.

2. Each instrumented impact technique has its own inherent strain rate and associated rate-sensitive effect on test results.

3. Properly instrumented impact tests are too complex and expensive for standardization and routine use (criterion 4).

4. The empirical multiple-cycle drop-weight impact test recommended by ACI 544(7) provides little indication of true material behavior (criterion 1) and lacks a level of precision appropriate for standardization (criterion 2)

5. The ability of the impact tests discussed to adapt to different specimen sizes and shapes is quite limited (criterion 3).

6. The flexural mode corresponds to the primary stress condition in many applications of fiber-reinforced concrete (criterion 1).

7. The facilities needed for flexural testing are relatively simple and inexpensive, and are therefore amenable to standardization (criterion 4).

REVIEW OF SLOW FLEXURE METHODS OF DETERMINING TOUGHNESS

In all slow flexure methods which have been proposed, a measure of toughness is derived from analysis of the load-deflection curve. The differences in the methods are in how the curves are analyzed (FIG. 2).

ACI Committee 544 Method (7, 14)(1978)

In this method, the area under the load-deflection curve up to a mid-span deflection of 1.9 mm (originally 0.075 in.) is computed for a 100 mm square beam loaded at the third points over a 300 mm span, and is then divided by

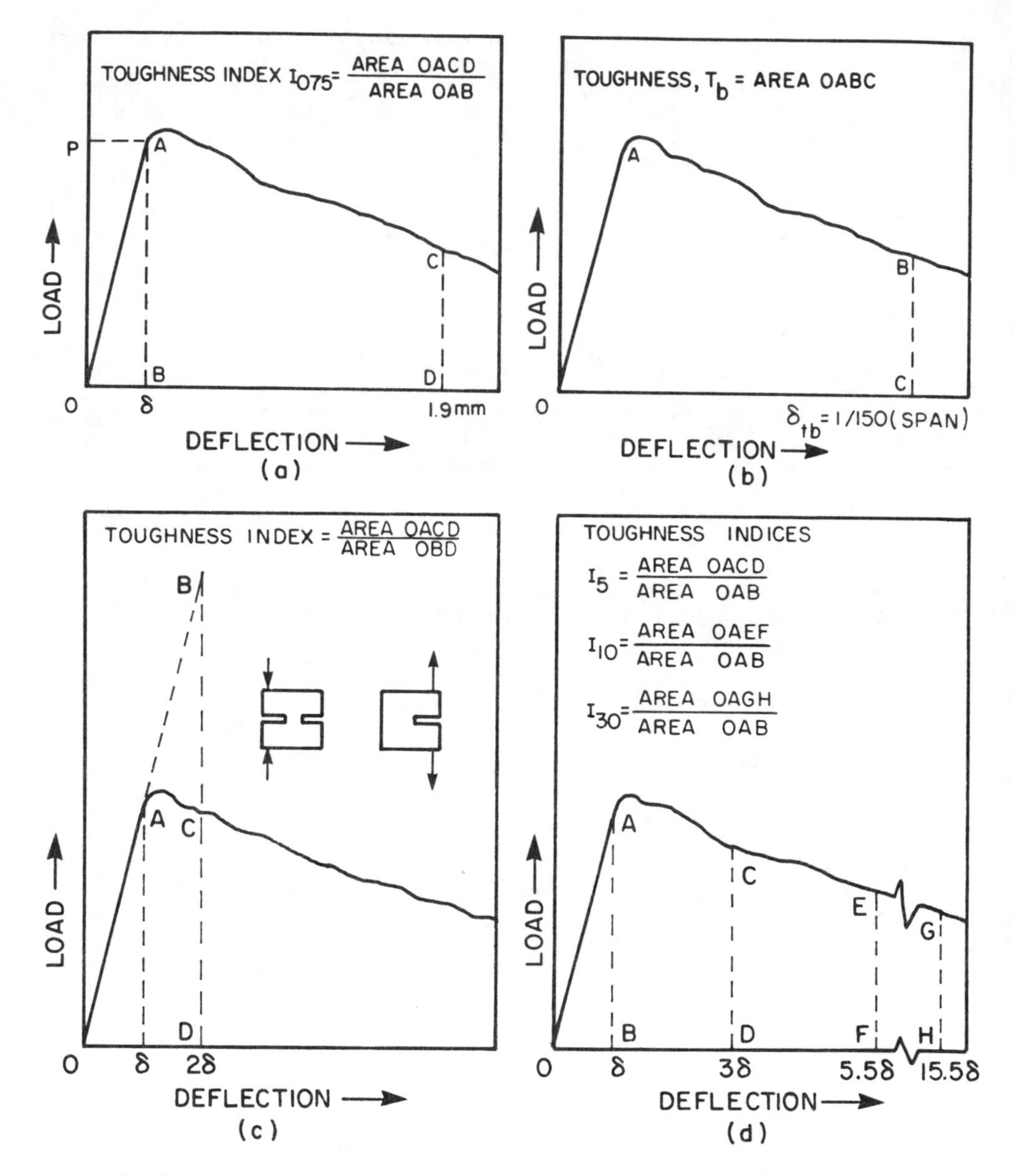

FIG 2. Schematic Load-Deflection Curves and Derived Toughness Parameters (a) ACI Method (7), (b) JCI Method (15), (c) Method Barr et al (16), (d) Method of Johnston (17) and ASTM C 1018 (19).

the area under the same curve up to first crack (identified as the point at which the initially linear portion of the curve becomes nonlinear) to obtain a dimensionless toughness index (FIG. 2a) subsequently identified as I_{075}. Only 100 mm square specimens tested over a 300 mm span are considered regardless of fiber length or maximum aggregate size.

Japan Concrete Institute Standards SF2, SF3, and SF4 (15)(1983)

In this method, flexural toughness, T_b, is the area under the load-deflection curve for 100 mm or 150 mm beams tested under third-point loading up to a deflection, δ_{tb}, of 1/150 of the span (2 mm and 3 mm respectively for the specified 300 mm or 450 mm spans) expressed in absolute units of energy (FIG. 2b). In addition, an equivalent flexural strength, $\bar{\sigma}_b$, given by

$$\bar{\sigma}_b = \frac{T_b}{\delta_{tb}} \frac{\ell}{bd^2}$$

is computed, where ℓ, b, and d are the span, breadth, and depth of the specimen. The reasoning is that the ratio of toughness to corresponding end-point deflection gives the average load up to the end-point deflection, and multiplying it by ℓ/bd^2 gives an equivalent average flexural strength up to the specified end-point. This equivalent flexural strength is considered as an index of flexural toughness (15). Fiber length is taken into account by specifying 150 mm square beams on a 450 mm span for fibers longer than 40 mm and 100 mm square beams on a 300 mm span for fibers of length 40 mm or less. These size restrictions apply to both molded concrete specimens and sawn shotcrete specimens.

Method of Barr et al (16)(1982)

In this method, the specimen is a double edge-notched cube subjected to eccentric compressive loading or a single edge-notched cube subjected to eccentric tensile loading (Fig. 2c). In both cases flexural conditions prevail, and a load-deflection curve is generated from 100 mm cubes with either sawn or molded notches. A toughness index is computed by establishing an end-point deflection of twice the first-crack deflection (as previously defined), and dividing the actual area under the curve up to this deflection by the total area up to this deflection produced by extrapolation of the linear portion of the curve up to first crack. This is of course four times the area up to first crack, so the index is 0.25 for plain concrete. It varies from 0.34 to 0.84 for the steel fiber concrete tested, with, according to the proponents, a maximum value of unity (16). (Just how first crack is identified for the case of an index of unity is unclear, but this is trivial because practical steel fiber-reinforced concretes do not behave in this manner.)

Toughness indices were similar for both the tensile and compressively loaded specimens with sawn notches, even though the notch depths were different, suggesting that the index is independent of the specimen geometry. Index values were about 15% higher (polypropylene fibers) for inbuilt molded notches compared with sawn notches, consistent with the expected preferential alignment and concentration of fibers near the toe of molded notches.

Recently, the same basic concept has been applied by the proponents to 100 mm square beams tested under third-point loading over a 400 mm span. They use both regular molded specimens and geometrically identical specimens with a single notch located at the mid-span to reduce or eliminate the effects of preferential fiber alignment and possible nonuniform fiber distribution at molded surfaces. Publication of results is pending.

Author's Method (17, 18)(1982) incorporated into ASTM C 1018(19)in 1984

The rationale for this method, developed prior to 1982, is based primarily on overcoming perceived deficiencies in the ACI 544 system of determining a toughness index. At the time, it was the only system in print, as the JCI (15) and Barr (16) systems had not yet been published and were unknown to the writer. The deficiencies considered most important in the ACI system were:

(1) The need for a toughness index providing a fundamentally significant indication of material behavior relative to a readily understood reference level (criterion 1).

(2) The inability of the ACI 544 system to adapt to different specimen sizes and shapes appropriate for the current range of applications of steel fiber-reinforced concrete.

(3) The possibly misleading nature of results sometimes obtained with the ACI 544 system as a consequence of the fixed end-point deflection.

In accordance with (1) above, two alternatives were formulated in the original paper (17) on definition of toughness parameters. These are the so-called base-zero and base-unity systems in which the toughness index of plain concrete is either zero or unity. Both systems are conceptually identical and differ only in end-point deflections. After weighing the pro and cons (17), the available load-deflection curves were analyzed using the base-unity system in which a toughness index is calculated by dividing the total area under the curve up to a specified multiple of the first-crack deflection by the area up to first crack (FIG. 2d). The equipment needed is the already standardized third-point flexural strength testing facility, but it must be capable of operating in stroke control rather than load control, and instrumentation for measuring the mid-span deflection must be added.

Clearly, for any specified end-point deflection the greater the slope of the load-deflection curve after first crack, the greater the index. The slope is at a minimum when the portion of the curve after first crack is vertical (plain concrete), and increases gradually from a high negative value through zero to a positive value with increasing toughness. Thus, the shape of the curve after first crack reflects material behavior over the range from brittle elastic (linear) for plain concrete to elastic-plastic for some highly reinforced fibrous concretes. To quantify material behavior relative to a fundamentally significant shape of curve, the reference chosen is elastic-plastic behavior, a reference easily understood because it is typical of mild steel, the behavior of which is familiar to most engineers, and because it is a level of behavior achievable in fibrous concretes only with optimum selection of amount and type of fiber (FIG.3). Consequently, it is a reference to which the designer concerned with maximizing energy absorption in fibrous concrete can readily aspire. Once elastic-plastic behavior is rationalized as the reference, the end-point deflection for area measurement is selected as the multiple of first-crack deflection, δ, which gives a convenient "round number" index value corresponding to elastic-plastic behavior. End-point deflections which give toughness indices of 5, 10, and 30 for elastic-plastic behavior are 3δ, 5.5δ, and 15.5δ respectively (FIG. 2d). Thus, the range of material behavior from elastic (plain concrete) to elastic-plastic (relatively tough FRC) corresponds to ranges of 1 to 5, 1 to 10, and 1 to 30, and each index is subscripted accordingly (FIG. 3). Material behavior tougher than elastic-plastic corresponds to index values greater than the 5, 10, and 30 reference levels.

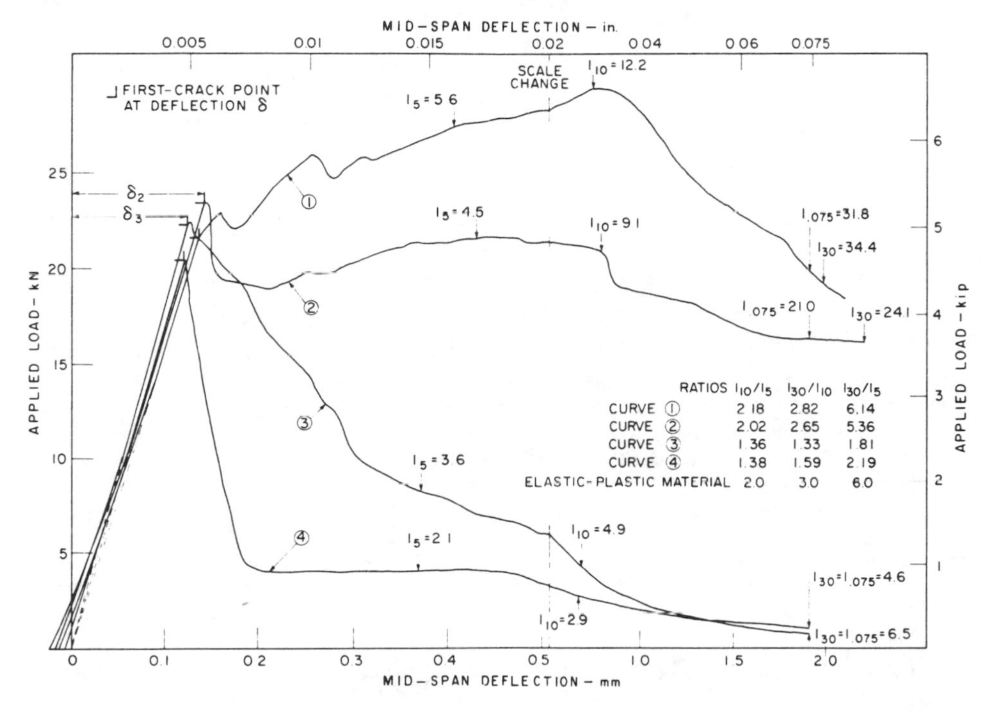

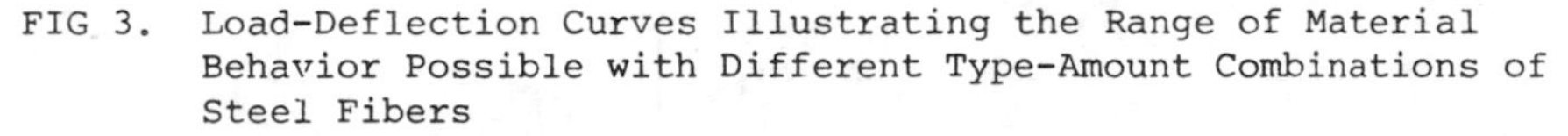
FIG 3. Load-Deflection Curves Illustrating the Range of Material Behavior Possible with Different Type-Amount Combinations of Steel Fibers

Why one or other end-point and which is best is the obvious question. In the answer lies a major advantage of the system, namely that the end-point deflection can be chosen according to the level of serviceability (in terms of deflection and cracking) appropriate for the application considered. For example, the I_5 or I_{10} indices may be appropriate for applications requiring low deflection and crack width, while the I_{30} index may be more appropriate for applications where much greater deflection and higher crack widths are permissible. In cases when preservation of structural integrity is the only consideration, eg. designs to resist earthquake, another index representing almost the complete load-deflection curve, say I_{50}, may be more appropriate.

Another advantage of this system, and any system with an end-point deflection specified as a multiple of the first-crack deflection, is its ability to adapt to specimens of different sizes and shapes (criterion 3). This allows the real benefits of preferential fiber alignment in thin sections to be reflected in appropriately thin test specimens, eg. for shotcrete or overlays. Conversely, the possibility of artificially high index values associated with long fibers in small specimens being considered representative of thicker more massive sections, eg. pavements, can be taken into account by specifying that width and depth of specimen shall be large enough relative to the fiber length to minimize the effect of preferential fiber alignment. ASTM C 1018 requires that specimen width and depth be at least three times the fiber length in such circumstances.

To further support the rationale for this method and address the controversy in ACI Committee 544 regarding whether the denominator in the index should be the area under the load-deflection curve up to first crack for the fibrous concrete, or the area to fracture for a purportedly equivalent plain concrete mix, the possibility that the first-crack stress and deflection may be higher than the corresponding plain concrete values for at least some type-amount combinations of steel fibers is examined schematically (FIG. 4). In FIG 4a is the case of relatively high toughness corresponding to the elastic-plastic reference where $I_{10} = 10.0$, and in FIG 4b is the case of any lower level of toughness intermediate between the elastic behavior of plain concrete ($I_{10}=1.0$) and the elastic-plastic reference. The line OA represents plain concrete. The curve OAE (FIG. 4a) or OAF (FIG. 4b) represents concrete with a type and amount of fibers which does not significantly change the first-crack point. The curve OCG (FIG. 4a) or OCH (FIG. 4b) represents the same concrete with a type and amount of fiber which increases the first-crack strength and corresponding deflection by 20%. Regardless of behavior, calculation of area components needed to determine the toughness index, as illustrated on both figures, shows that, when using the first-crack area as the

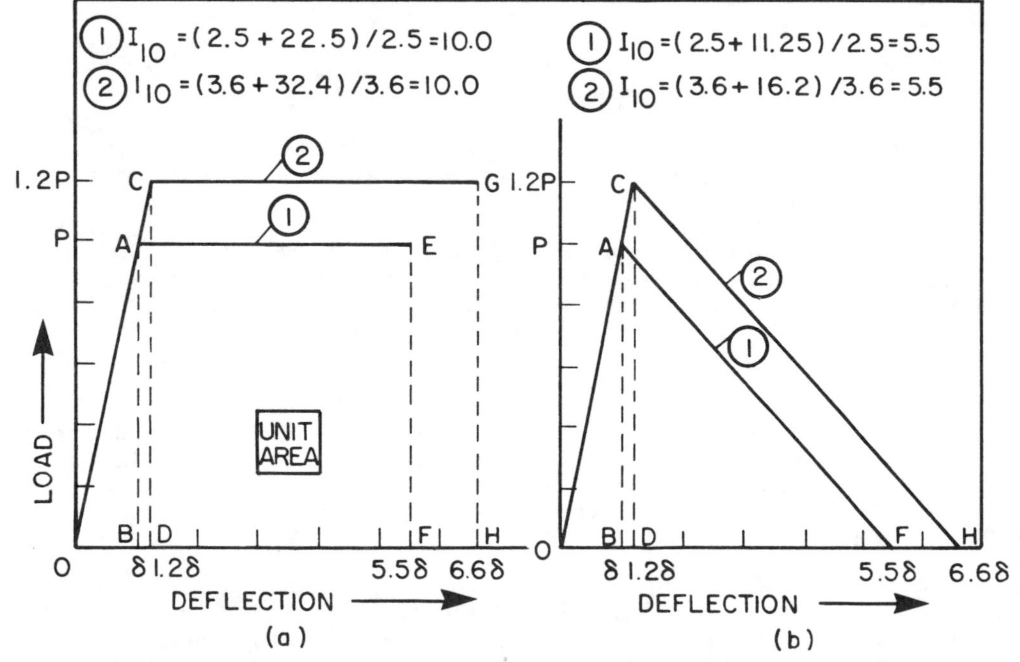

FIG 4. Schematic Load-Deflection Curves Showing that the Toughness Index I_{10} Indicates Curve Shape after First Crack Regardless of Change in First-Crack Strength

denominator, curves of the same shape give the same toughness index. On the other hand, if the denominator is taken as the plain concrete area (2.5 area units) for both curves, the indices for curve 2 would increase to 14.4 (FIG. 4a) and 7.9 (FIG. 4b), quite different from the values for curve 1 even though the curves have the same shape. Also, the correspondence between the index value of 10.0 and the elastic-plastic reference level of material behavior would no longer hold. Therefore, if the index is to indicate toughness by quantifying the shape of the load-deflection curve, the first-crack area must be used as the denominator.

From the foregoing, it is apparent that only the degree of toughening is indicated by the I_5, I_{10}, I_{30} indices. To fully document the behavior of a fiber-reinforced concrete relative to a plain concrete, or indeed any other fiber-reinforced concrete, the first-crack strength must be reported to indicate level of strength (just as flexural strength is reported for plain concrete), and a toughness index or indices must be reported to indicate level of toughness relative to the elastic-plastic reference. Reporting of first-crack parameters was optional in ASTM C 1018-84, but revisions incorporated into ASTM C 1018-85 make the reporting of both first-crack strength and toughness parameters mandatory.

The ability of the I_5, I_{10}, I_{30} indexing system to distinguish different categories of behavior can be seen in terms of load-deflection curves for individual specimens in FIG. 3, where curves 1 and 2 represent respectively 0.75% and 0.5% volume of 50x0.5 mm hooked fibers, and curves 3 and 4 represent 1.0% and 0.5% volume of 18x0.6x0.3 mm enlarged-end slit-sheet fibers. However, the full body of data from the original 121 tests (17) adds weight to this assertion (TABLE 6), and shows that, although differences in first-crack strength are small or insignificant for these particular type-amount combinations of fibers, differences in toughness indices are significant and consistent. As expected, the longer hooked wire fibers are clearly superior to an equal volume of shorter slit-sheet fibers with slightly enlarged ends, or to the same volume of slightly longer slit-sheet fibers without enlarged ends (FIG. 5). Also, toughness increases with increase in volume fraction of fiber up to the maximum amount which the mix can accommodate without excessive loss of workability.

TABLE 6. Summary of Flexural Test Data (121 tests)

Fibers Type	Fibers Volume -%	w/c+f	Flexural Strength-MPa First Crack	Flexural Strength-MPa Maximum	Mean Value of Toughness Index I_5	I_{10}	I_{30}	I_{075}	Ratios I_{10}/I_5 [e]	I_{30}/I_{10} [f]	I_{30}/I_5 [g]	Number of Specimens (batches)
HW[a]	0.5	.474	6.07	6.49	4.57	9.36	25.99	27.12	2.05	2.78	5.69	12(3)
HW	0.5	.471	5.74	6.09	4.32	8.66	25.12	25.11	2.00	2.90	5.81	19(3)
HW	0.75	.509	5.68	7.08	4.98	10.67	32.92	32.87	2.14	3.09	6.61	19(4)
EESS[b]	0.5	.443	5.90	6.00	2.49	3.68	5.58	5.61	1.48	1.52	2.24	14(3)
EESS	1.0	.467	6.48	6.63	3.71	5.33	7.45	7.40	1.44	1.40	2.01	17(3)
EESS	1.5	.515	5.81	6.24	4.66	7.86	13.50	13.66	1.69	1.72	2.90	20(4)
SS[c]	1.1	.494	6.59	6.86	4.17	5.88	-	-	1.41	-	-	4(1)
	0	.445	6.65[d]	6.74[d]	1.0	1.0	-	-	-	-	-	16(2)

a - 50 x 0.5 mm hooked wire
b - 18 x 0.3 x 0.6 mm enlarged-end slit sheet
c - 25 x 0.25 x 0.55 mm uniform slit sheet
d - x-y plotter and digital load readout compared
e - 2.0 corresponds to plastic behavior between applicable deflections
f - 3.0 corresponds to plastic behavior between applicable deflections
g - 6.0 corresponds to plastic behavior between applicable deflections

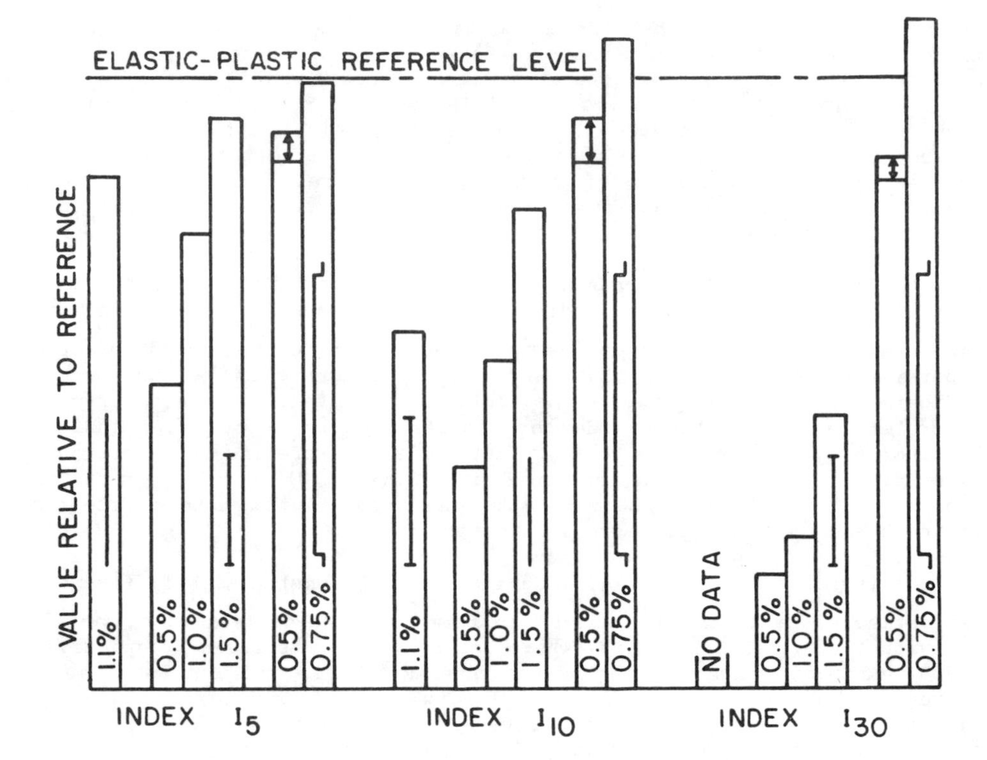

FIG 5. Toughness Indices I_5, I_{10}, and I_{30} Achieved with Hooked Wire, Enlarged-End Slit Sheet, and Uniform Slit Sheet (Mean Values from 121 Tests, TABLE 6).

The comparative levels of the I_5,I_{10}, and I_{30} indices in FIG. 5 also show that material behavior relative to the elastic-plastic reference changes in different ways for different fibers. Toughness relative to the reference drops noticeably as the end-point deflection is increased for both types of slit-sheet fibers, while the level for the longer hooked fibers remains essentially constant regardless of end-point deflection, an important difference in terms of total energy absorption capability. Illustrative of this difference are the ratios I_{10}/I_5, I_{30}/I_{10}, and I_{30}/I_5 given in FIG 3 and TABLE 6. Values of 2.0, 3.0, and 6.0, repectively for these ratios correspond to plastic behavior between the deflection limits for the corresponding indices in each ratio. The further the ratios are below 2.0, 3.0 and 6.0 respectively, the more inferior the material behavior is in terms of toughness. Reporting of these ratios therefore helps to indicate material behavior at higher deflections, particularly when load-deflection curves are not included in the report. In ASTM C 1018, testing beyond the I_{10} end-point deflection is at the option of the user, so only I_5, I_{10}, and the ratio I_{10}/I_5 must be reported.

Comparative Evaluation of Slow Flexure Methods

Following detailed examination of each method separately, they are summarily compared in terms of criteria 1, 2, and 3 previously identified and two other relevant factors, level of serviceability and the possible effect of preferential fiber alignment (TABLE 7). The following points

TABLE 7. **Summary Comparison of Slow Flexure Methods for Determining Toughness**

Index Characteristic	ACI 544	JCI-SF4	Johnston (C 1018)	Barr et al
Magnitude indicates shape of load-deflection curve and therefore material behavior (criterion 1)	No	No for T_b Yes for $\bar{\sigma}_b$	Yes	Yes
Magnitude relates to a readily understood and fundamentally significant reference (criterion 1)	No	No	Yes	Not clearly (possible in principle)
Choice of alternative indices according to serviceability required	No	No	Yes	Yes
Repeatability or single-operator precision[a] (criterion 2)	17.6%[18] 17.7%[b] 11.0%[11]	No data	11.5% for I_5 [18] 13.8% for I_{10} [18] 16.4% for I_{30}	10.2%[16]
Obtainable using specimens of different sizes and shapes (criterion 3)	No	Yes (two sizes only)	Yes	Yes
Value independent of specimen size and shape alone, exclusive of the effect of preferential fiber alignment (criterion 3)	Not Applicable	No for T_b No data for $\bar{\sigma}_b$	Limited supportive data[16]	Limited supportive data[16]
Possible effect of preferential fiber alignment recognized in provisions relating specimen size to fiber length	No	Partially	Yes	No

a - Expressed as average within-batch values for sets of at least three specimens.
b - Unpublished data made available through Battelle Memorial Institute detailed in reference 18.

clarify the table responses with regard to the ability of the stipulated toughness parameters to uniquely quantify material behavior in terms of the shape of the load-deflection curve and to do so relative to a fundamentally significant reference level of material behavior (criterion 1).

(1) Using the ACI 544 method (7), it is possible to construct load-deflection curves with the same elastic modulus to first crack which, when analyzed at the same specified end-point deflection, give the same toughness index (FIG. 6). However, depending on the first-crack strength, the post-crack behavior can range infinitely from strain-softening to plastic to strain-hardening. Therefore, the ACI 544 index cannot uniquely define the shape of the load-deflection curve and the level of material behavior associated with it. It follows that no reference level of material behavior can correspond to a single unique value of the index.

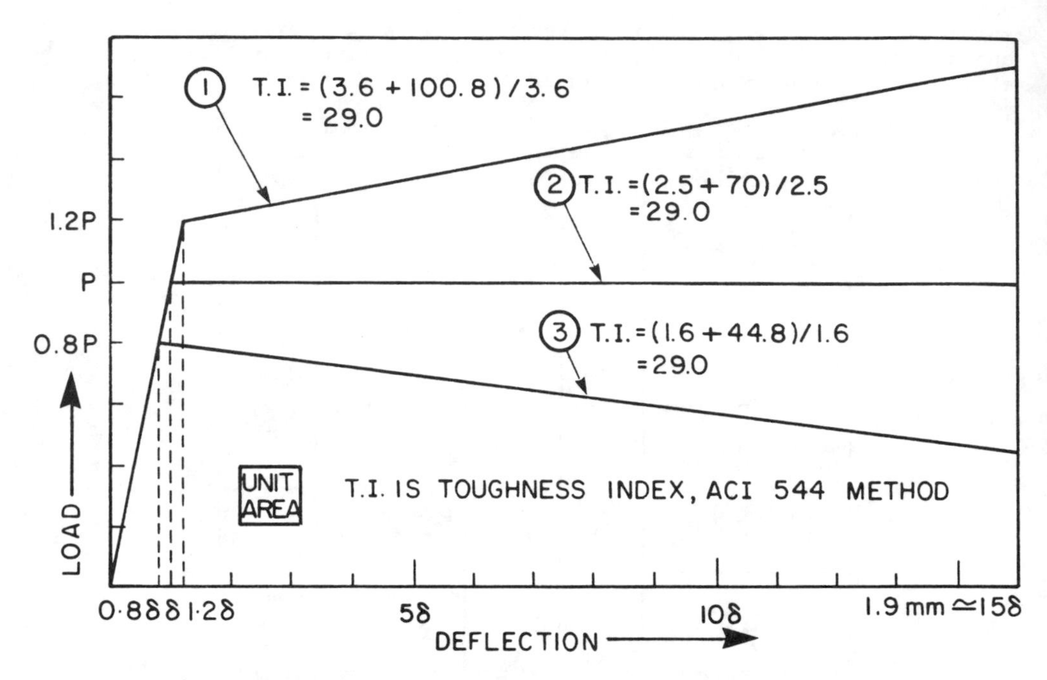

FIG. 6 Schematic Load-Deflection Curves with the Same Toughness Index Derived by the ACI 544 Method for Different First-Crack Strengths.

(2) Using the JCI method (15), it is possible to construct load-deflection curves (FIG. 7) with the same total area, T_b, at the specified end-point deflection, δ_{tb} (1/150 of the span), and therefore the same $\bar{\sigma}_b$. Once again, depending on the first-crack strength, the post-crack behavior can vary infinitely from strain-softening to strain-hardening. Therefore, neither T_b nor $\bar{\sigma}_b$ can uniquely define the shape of the load-deflection curve and the level of material behavior associated with it. Both this method and the ACI method fail in this regard for the same reason, namely that the end-point deflection is fixed in magnitude (for any given span of beam in the JCI method) instead of being defined relative to the magnitude of the first-crack deflection.

(3) Any method in which the end-point deflection is a multiple of the first-crack deflection automatically quantifies the shape of the load-deflection curve in average or steady state terms. However, curves with oscillations (FIG. 8) which produce the same post-crack area as the steady state curve can give the same values of toughness index in such methods. Fortunately, even though actual curves for fiber-reinforced concretes do not exactly follow the straight line schematics shown (FIG. 4, 6, 7), they are usually close to a steady state shape without wide and frequent oscillations (FIG. 3). Therefore, indices such as I_5, I_{10}, and I_{30} are good quantitative indicators of the steady state shape of the load-deflection curve, and

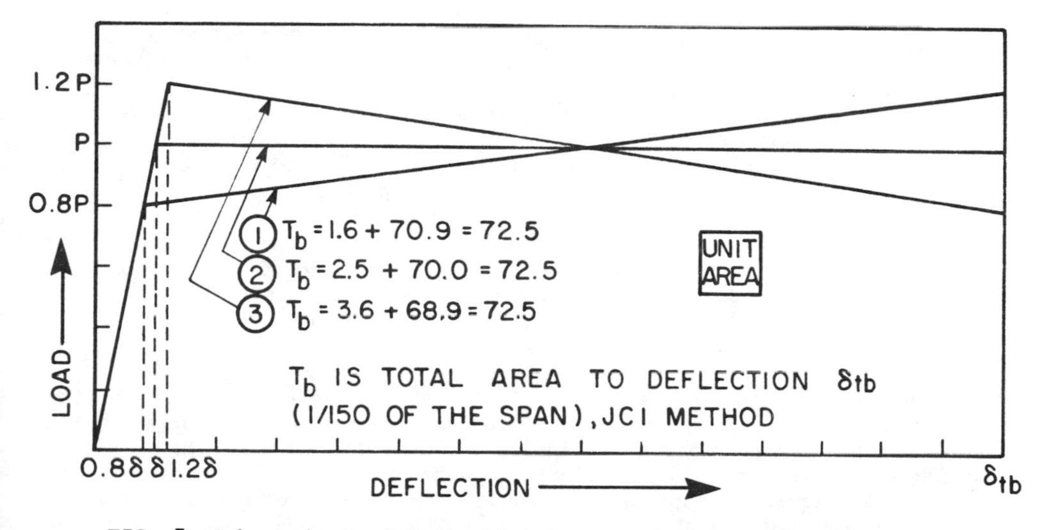

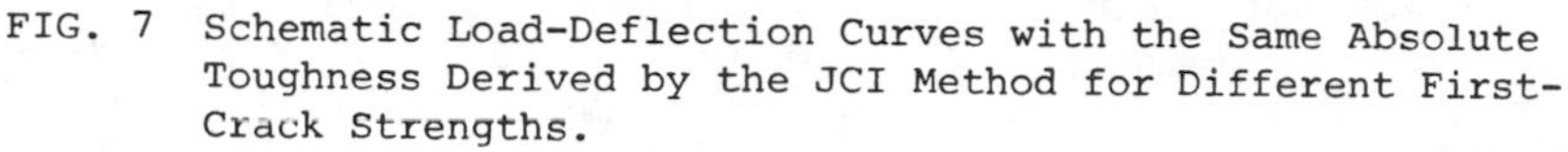
FIG. 7 Schematic Load-Deflection Curves with the Same Absolute Toughness Derived by the JCI Method for Different First-Crack Strengths.

tend to nullify the effect of minor oscillations in the curve which sometimes occur with individual specimens but have little significance in relation to the general curve shape for a set of nominally identical specimens(FIG. 8). The same advantages and limitations apply to the index obtained in the method of Barr et al (16) in which the end-point deflection is twice the first-crack deflection.

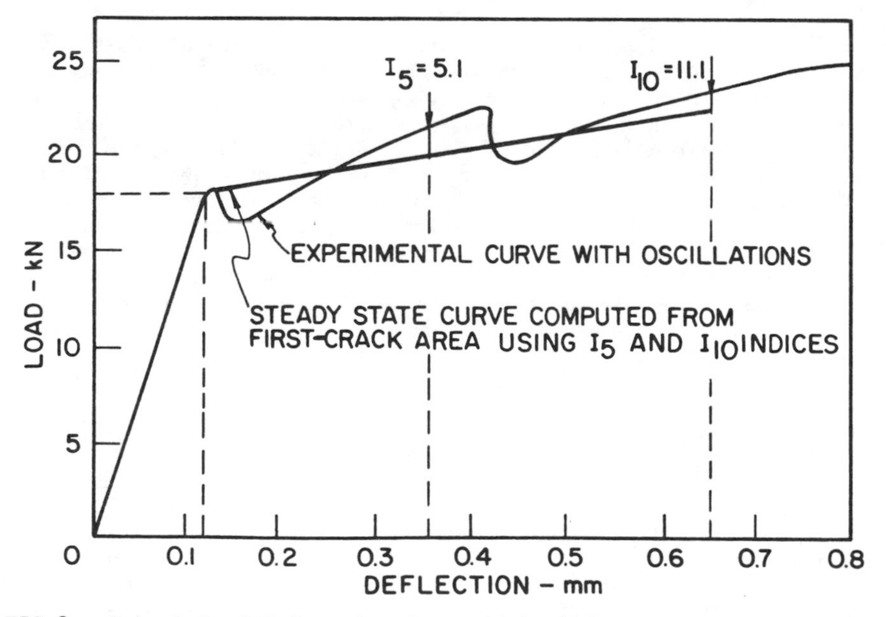

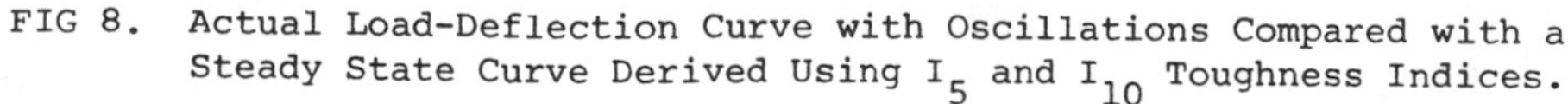
FIG 8. Actual Load-Deflection Curve with Oscillations Compared with a Steady State Curve Derived Using I_5 and I_{10} Toughness Indices.

(4) The relationship of the actual value of the toughness index to a fundamentally significant reference level of material behavior is not immediately obvious in the method of Barr et al (16), although a simple calculation shows that on the index scale of 0.25 (plain concrete) to 1.00, 0.75 represents elastic-plastic behavior. However, 0.75 is a somewhat odd number for such a significant and relevant level of material behavior for steel fiber-reinforced concretes, and 1.0, which might be regarded as an appropriate round number for a reference, is unattainable for such concretes. (Glass fiber-reinforced cements may approach an index of 1.0.) In contrast, in the writer's method, the subscript to the index obviously identifies the index value corresponding to the elastic-plastic reference level of material behavior, a level readily attainable with certain type-amount combinations of steel fibers.

In attempting to place the various sources of data on repeatability (criterion 2) in perspective, it should be noted that the greater the portion of total area under the load-deflection curve involved in computing the index the greater the coefficient of variation of results for nominally identical specimens from the same batch. Thus, the coefficient increases from an average of about 10% for an end-point of twice the first-crack deflection to an average of 16 to 17% for an end-point of 15.5 times the first-crack deflection. For the 1.9 mm end-point deflection used in the ACI 544 method, it approaches 18%, and, for an end-point deflection of 10 mm, it reaches 21% (5)(TABLE 1). Generally, these levels of repeatability are not much in excess of what is normal in tests on conventional concrete, and in the writer's experience can be improved upon with better equipment and more precise control of test procedure.

LIMITATIONS AND REFINEMENTS OF SLOW FLEXURE TECHNIQUES

To precisely determine energy absorption, the distance actually moved (and work done) by the applied loads must be determined. Experimentally, this is difficult because the loading devices interfere with the placement on the concrete specimen of dial guages, tranducers, or similar devices. Moreover, the third-point loading mode with its two loading points requires a minimum of two devices to measure deflection and a system of manual or electrical averaging of the two to produce a single load-deflection curve. Also, there is the problem of localized concrete deformation at the beam supports and the loading points together with deformations in the support system. To obtain the true deflection of the test specimen requires at least two additional deflection-measuring devices, one at each support, a system of averaging the measurements to obtain the mean support deflection, and a system of manually or electrically subtracting this average from the average deflection at the loading points.

Problems Associated with the Deflection-Measuring System

For most investigators, a system requiring a minimum of four deflection-measuring devices has proven too complex, and a single device located at the mid-span is a common but imperfect compromise which is simpler and more convenient for routine use (criterion 4). However, it is clear from examining various sources of data that this simplified system is responsible for wide variations in experimentally established mid-span deflections at first crack, so it is necessary to derive an expected mid-span deflection, δ_{ms}, assuming elastic behavior up to first crack. The required equation follows:

$$\delta_{ms} = \frac{23\ P\ L^3}{1296\ EI}\ [1 + \frac{216\ d^2(1 + \mu)}{115\ L^2}] \qquad (20)$$

The expression in parenthesis allows for shear and has the effect of increasing the deflection due to bending by 25% for a span/depth (L/d) ratio of 3 and a Poisson's ratio, μ, of 0.2. For example, with a good quality concrete matrix typically having a modulus of 35 GPa and a first-crack load of 20 kN using a 300 x 100 x 100 mm specimen, δ_{ms} is 0.041 mm. Yet, in various references the typical experimental values for such specimens are 0.12 mm (FIG. 3), 0.13 mm(14), 0.11 mm(21), 0.13(22), 0.06(23), ranging up to 0.32 mm (unpublished committee document) and as high as 0.64 mm(11). Clearly, there is good agreement between values obtained by the writer(17) (FIG. 3), Henager(14), and Kobayashi(21,22), but they exceed the theoretically expected value by a factor of three. Kobayashi(21) in attempting to measure what he calls true deflection, ie. with provision to measure deformation at the supports, obtained a value of about 0.063 mm at the 20 kN load level (21,23), still appreciably higher than the theoretical. In recent tests to specifically address the problem, the writer has obtained net mid-span deflections in the range 0.048 - 0.060 mm using a mid-span transducer and a transducer at each support. The outputs from the support transducers are electrically averaged and substracted from the output of the mid-span transducer to give an electrical output of net deflection which is plotted directly on the X-Y plotter. While this is termed the net deflection, rather than the true deflection, it does appear to closely approach the theoretically expected value of 0.046 mm (based on a measured instantaneous tangent modulus in compression of 31 GPa for this concrete). Load-deflection curves based on net deflection are compared with curves obtained using only a single deflection-measuring device at the mid-span in FIG 9. Further work is in progress to refine and simplify what is a rather complex system for deflection measurement.

The implications of the foregoing are as follows:

(1) Estimates of toughness or energy absorption based upon absolute areas under the load-deflection curve using a single mid-span deflection-measuring device are likely to be grossly inaccurate. Therefore, accurate determination of the T_b value in the JCI method is only possible using devices to account for deflection at the supports. Also, deflection measurement at the loading points is preferable to measurement at the mid-span. Up to first crack, loading point deflections are about 87% of mid-span deflection, but the difference after first crack may be more or less. Kobayashi(21) has also shown that the location of the hinge which develops after first crack relative to the position of the deflection-measuring devices may have an effect on deflection readings, but the magnitude is uncertain.

(2) Any toughness index based on a fixed end-point deflection, such as the ACI index, can be grossly underestimated when deflection of the supports is ignored. Kobayashi(21) gives an example where the ACI index was determined as 46 using net (true) deflection and 18 when deflection at the supports was not considered. Likewise in FIG 9, the

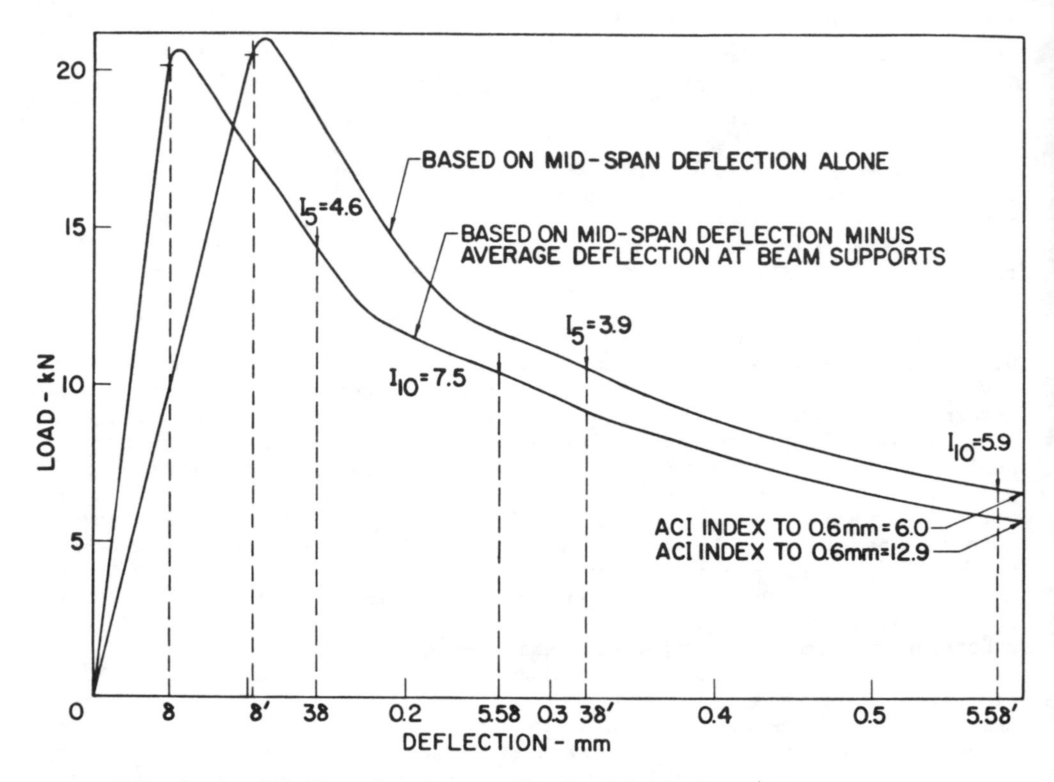

FIG. 9 Load-Deflection Curves Obtained by Deflection Measurement at the Mid-span Only Compared with Curves for the Same Mixture Obtained by Deflection Measurement at the Mid-span and Each Support.

first-crack area for the curve based on net deflection is seen to be much smaller than the corresponding area based upon mid-span deflection alone, so no matter what value is chosen as the end-point deflection, the index is much lower in the latter case.

(3) Toughness indices, such as the ASTM C 1018 indices, where the end-point adjusts to remain a constant multiple of the first-crack deflection, are not affected nearly so severely as indices based on a fixed end-point deflection. While more data will soon be available and further refinements may be made to the deflection-measuring system, the preliminary results indicate that for concretes of relatively low toughness the I_5 and I_{10} values based on mid-span deflection alone are up to 10% and 20% respectively lower than the values based on net mid-span deflection with deflections of the supports taken into account. Clearly, as toughness increases this difference will decrease to negligible for materials that approximate plastic behavior after first crack. Obviously, if in future we all aim at developing deflection-measuring systems which give net deflections corresponding closely to the predictions of elastic theory, the problem will cease to exist.

Rate-Sensitivity of Results

Another aspect of the proposed slow flexure technique which may require refinement to expedite routine testing for job mix evaluation, quality control etc., is the rate of increase of mid-span (or loading point) deflection. Limited data obtained by Kobayashi and Cho (22,23) at rates of 0.5 to 200 mm/minute suggest that the greater the rate the greater the first-crack strength and the more pronounced the drop in load immediately after first crack, although at deflections above about 1 mm all curves correspond fairly closely. Accordingly, an increase in first-crack strength and a drop in the I_5 and I_{10} toughness indices can be expected at rates of increase of mid-span deflection above 1 mm/minute. Such differences appear to be quite small between the 1 mm and 20 mm/minute rates, and lend credibility to the proposition that rate-sensitivity of results is neglible below a rate of increase of 1 mm/minute in mid-span deflection. However, more supportive data is needed before the rate of 0.05 to 0.10 mm/minute presently required in ASTM C 1018 is substantially raised. In any case, this appears unnecessary because, if the equipment can be operated to control the rate of increase of net mid-span deflection at the permitted maximum of 0.10 mm/minute, testing to the I_{10} end-point deflection requires only about 3 minutes, little more than required for a compressive strength test on conventional concrete at the maximum permitted rate of 20 MPa/minute. However, if the test is performed using "soft" supports with deflection monitored only at the mid-span, considerably more time may be required because the 0.10 mm/minute applies to a larger range of deflection.

CONCLUSIONS

Recognizing that no single method of determining toughness is necessarily appropriate for all circumstances or applications involving steel fiber-reinforced concrete, it is proposed that the various methods should be evaluated primarily in terms of (1) ability to define material behavior relative to a fundamentally significant reference, (2) repeatability or precision of results attainable, (3) capability of the method to adapt to specimens of various sizes and shapes representing the diversity of established applications (mass concrete, thin overlays, shotcrete linings etc.), and (4) potential for standardization for routine use in the concrete industry as opposed to just research.

Based on the foregoing four criteria, which may be weighted differently, or indeed supplemented with others perhaps more important according to the reader's perspective, the writer from the perspective primarily of the chairman of a committee responsible for standards development has reached the following conclusions:

(1) Impact tests which necessarily involve high strain rates give results which are rate-sensitive depending on the specific characteristics of the equipment used, require equipment which is too complex and expensive for standardization if meaningful results with satisfactory precision are to be obtained, and may yield largely meaningless and highly variable results when sophisticated equipment is not employed.

(2) Slow flexure tests employing low strain rates give results which are

essentially free of rate-sensitive effects, require the relatively simple equipment already standardized for flexural strength testing supplemented with instrumentation for measuring mid-span deflection, and are consequently more suitable for standardization and routine use than impact tests. However, considerable care is required in measuring deflection to ensure that the measured mid-span deflection used in developing the load-deflection curve represents the actual beam deflection exclusive of deformations at the supports or within the support system.

(3) Whether or not the load-deflection curve obtained in slow flexure can yield a meaningful indication of toughness and a correspondingly meaningful indication of material behavior depends on how it is analyzed. Of the four alternative methods of analyzing the load-deflection curve to determine toughness parameters, the American Concrete Institute and Japan Concrete Institute methods have serious deficiences which can lead to misleading results, largely because the end-point deflection in each case is fixed in magnitude. The methods of Barr et al (16) and the writer (17), in which the end-point deflection is a multiple of the first-crack deflection, more accurately represent material behavior relative to a fundamentally significant reference level of behavior. The choice of elastic-plastic material behavior as the reference in a system of toughness indices which clearly quantify material behavior relative to the reference, coupled with other refinements such as allowing specimen size and shape to adapt to reflect the type of construction (thick mass concrete or thin sections including shotcrete), permitting the end-point deflection to be selected according to the level of serviceability appropriate to the application, and requiring the reporting of the magnitude of the first-crack strength to distinguish between the strengthening and toughening effects of fibers, favor the method developed by the writer with the refinements to it recently incorporated into ASTM Standard C 1018-85.

(4) The principle of defining the test end-point as a multiple of the first-crack deflection taking into account serviceability requirements, and computing toughness indices on this basis as specified in ASTM C 1018 for flexure, is equally applicable to the stress-strain curve for uniaxial tension where the first-crack point is clearly identifiable (12). For uniaxial compression, the first-crack point is not easily identifiable due to the marked curvilinearity of the stress-strain curve compared with tension or flexure. Employing the strain corresponding to the maximum stress to define the denominator area of the toughness index and an appropriate multiple of it to define the numerator area would permit distinction between the strengthening and toughening effects of fibers, and would result in an index indicative of the shape of the post-peak portion of the stress-strain curve. However, the possibility that the degree of curvilinearity varies from one mixture to another may make it difficult to establish end-points which can be uniquely linked to the elastic-plastic reference level of material behavior. Comparison of toughness parameters obtained in flexure, uniaxial tension, and compression is obviously warranted in future investigations.

ACKNOWLEDGEMENT

Preparation of this paper and investigations attributed to the writer from which data are reported were funded by an operating grant from the Natural Sciences and Engineering Research Council of Canada.

REFERENCES

(1) Suaris, W. and Shah, S.P., "Inertial effects in the instrumented impact testing of cementitious composites", ASTM, Cement, Concrete, and Aggregates, CCAGDP, Vol. 3, No. 2, Winter 1981, pp. 77-83.

(2) Hibbert, A.P. and Hannant, D.J., "The design of an instrumented impact test machine for fibre concretes", RILEM Symposium, Testing and Test Methods of Fibre Cement Composites, Apr. 1978, The Construction Press Ltd. (U.K.), pp. 107-120.

(3) Hibbert, A.P. and Hannant, D.J., "Impact resistance of fibre concrete", U.K. Transport and Road Research Laboratory, Report SR 654, 1981, 25 pp.

(4) Suaris, W. and Shah, S.P., "Properties of concrete subjected to impact", ASCE, Journal of Structural Engineering, Vol. 109, No. 7, July 1983, pp. 1727-1741.

(5) Gokoz, U.N. and Naaman, A.E., "Effect of strain rate on pull-out behavior of fibers in mortar", International Journal of Cement Composites, Vol. 3, No. 3, Aug. 1981, pp. 187-202.

(6) Schrader, E.K., "Impact resistance and test procedure for concrete", ACI Journal, Vol. 78, No. 2, Mar.-Apr. 1981, pp. 141-146.

(7) ACI Committee 544, "Measurement of properties of fiber reinforced concrete", ACI Journal, Vol. 75, No. 7, July 1978, pp. 283-289.

(8) Ramakrishnan, V. Coyle, W.V., Dahl, L.F., and Schrader, E.K., "A comparative evaluation of fiber shotcretes", ACI Concrete International, Vol. 3, No. 1, Jan. 1981, pp. 59-69.

(9) Swamy, R.N. and Jajagha, A.H., "Impact resistance of steel fibre reinforced lightweight concrete", International Journal of Cement Composites and Lightweight Concrete", Vol. 4, No. 4, Nov. 1982, pp. 209-220.

(10) Schrader, E.K., "Formulating guidance for testing of fibre concrete in ACI 544", RILEM Symposium (See reference 2), pp. 9-21.

(11) Ramakrishan, V., Coyle, W.V., Kulandaisamy, V., and Schrader, E.K., "Performance characteristics of fiber reinforced concretes with low fiber contents", ACI Journal, Vol. 78, No. 5, Sept.-Oct. 1981, pp. 388-394.

(12) Shah, S.P., Stroeven, P., Dalhuiser, D., and van Stekelenburg, P., "Complete stress-strain curves for steel fibre reinforced concrete in uniaxial tension and compression", RILEM Symposium (See reference 2), pp. 399-408.

(13) Johnston, C.D. and Gray, R.J., "Uniaxial tensile testing of steel fibre reinforced cementitious composites", RILEM Symposium (See reference 2), pp. 451-461.

(14) Henager, C.H., "A toughness index for fibre concrete", RILEM Symposium (See reference 2), pp. 79-86.

(15) Japan Concrete Institute, "Method of test for flexural strength and flexural toughness of fiber reinforced concrete", Standard SF4, JCI Standards for Test Methods of Fiber Reinforced Concrete, 1983, pp. 45-51.

(16) Barr, B.I.G., Liu, K., and Dowers, R.C., "A toughness index to measure the energy absorption of fiber reinforced concrete", International Journal of Cement Composites and Lightweight Concrete, Vol. 4, No. 4, Nov. 1982, pp. 221-227.

(17) Johnston, C.D., "Definition and measurement of flexural toughness parameters for fiber reinforced concrete", ASTM, Cement, Concrete, and Aggregates, CCAGDP, Vol. 4, No. 2, Winter 1982, pp. 53-60.

(18) Johnston, C.D., "Precision of flexural strength and toughness parameters for fiber reinforced concrete", ASTM, Cement, Concrete, and Aggregates, CCAGDP, Vol. 4, No. 2, Winter 1982, pp. 61-67.

(19) American Society for Testing and Materials, "Standard method of test for flexural toughness of fiber-reinforced concrete", ASTM Standards for Concrete and Mineral Aggregates, Vol. 04.02, Standard Number C 1018, August 1984, pp. 637-644.

(20) Irwin, L.H. and Galloway, B.M., "Influence of laboratory test method on fatigue test results for asphaltic concrete", Fatigue and Dynamic Testing of Bituminous Mixtures, ASTM STP 561, 1974, Appendix 1, pp. 40-45.

(21) Kobayashi, K. and Umeyama, K., "Method of testing flexural toughness of steel fiber reinforced concrete", Reprint UDC 691.328.2:620.174, Dept. of Building and Civil Engineering, Institute of Industrial Science, University of Tokyo, 1980.

(22) Kobayashi, K. and Cho, R., "Flexural behavior of polyethylene fiber reinforced concrete", International Journal of Cement Composites and Lightweight Concrete, Vol. 3, No. 1, Feb. 1981, pp. 19-25.

(23) Kobayashi, K. and Cho, R., "Flexural characteristics of steel fibre and polyethylene fibre hybrid-reinforced concrete", International Journal of Cement Composites and Lightweight Concrete, Vol. 4, No. 2, Apr. 1982, pp. 164-168. 1018

STEEL FIBER CONCRETE
US-SWEDEN joint seminar
(NSF-STU)
Stockholm 3-5 June, 1985
S P Shah and Å Skarendahl,
Editors

ADVANCED TUNNEL SUPPORT USING STEEL FIBRE REINFORCED SHOTCRETE

by Jonas Holmgren, DSc(Eng), Bergsäker Konsult, Bandhagen, Sweden

---

ABSTRACT

Advanced tunnel support is discussed in principle. The expression is defined and there are given examples of cases where advanced solutions for the support are applicable. The key words are found to be strong and flexible, bolt supported fibre reinforced shotcrete linings.

The flexibility in bending of steel fibre reinforced shotcrete is studied in tests and comparisons between the test results and solutions of the yield line theory. It is found that the load-deflection curve for a statically indeterminate structure is "more favourable" than the curve for the corresponding simply supported beam.

The flexibility in shear is studied in small scale shear tests. They showed that the shear strength of steel fibre reinforced shotcrete after cracking is considerable and that the flexibility is also noticeable.

It is assumed that the same slip mechanism of the fibre governs the behaviour in both bending and shear. Therefore it is suggested that the requirements on the fibre concrete are defined using the load-deflection curve for a well-defined test beam.

The main results from static and dynamic tests on bolt supported steel fibre reinforced shotcrete linings are presented.

Finally some practical applications of bolt supported steel fibre reinforced shotcrete linings are presented.

Jonas Holmgren is Associate Professor of Structural Mechanics and Engineering at the Royal Institute of Technology in Stockholm.

He has been working within the field of reinforced concrete since 1968 and during the last decade on shotcrete and its interaction with hard rock.

INTRODUCTION

During the twentieth century shotcrete has become widely used for rock support. In the beginning it was used without reinforcement at fairly good rock conditions. Later on it was also used together with mesh reinforcement at worse rock conditions. Today it is possible to replace the mesh with steel fibres so the shotcreting and the reinforcing work take place at the same time and in the same operation.

This paper deals with advanced tunnel support using steel fibre reinforced shotcrete. What is advanced tunnel support then? According to the opinion of the author advanced tunnel support is at hand in all cases when thin layers of unreinforced shotcrete are not used. The reason for this definition is purely practical and economical: One thin layer of unreinforced shotcrete is a comparatively cheap and quick way of rock reinforcement. The decision to use it is normally made at the construction site by the head of the tunnelling team. Provided the rock surface is mapped before shotcreting and the work is properly done on surfaces which are judged to give reasonable adhesion between the shotcrete and the rock there is no reason to treat the shotcrete lining as an advanced design problem.

Advanced tunnel support is at hand for instance when

i) high demands are put on the lining regarding tightness and crack distribution e.g. pressure water tunnels, waste water tunnels and gas storages.

ii) it is impossible to obtain a sufficient adhesion between the rock and the shotcrete e.g. when the mineral composition of the rock is unsuitable, when a lot of drainage channels are covering the rock surface or when the rock is severely crushed and/or consists of altering minerals. An other important example is when shotcrete has to be combined with thermal insulation of the rock surface.

iii) even small downfalls of shotcrete pieces are unacceptable e.g. road tunnels, railway tunnels, underground public areas of all kinds.

iv) heavy rock loads and/or large deformations are expected e.g. when the crack pattern indicates that considerable loosening may occur, especially at large spans, or when high rock pressure is at hand as in deeply situated caverns or mines. Large weak zones is an other example.

vi) the lining can be loaded dynamically e.g civil defence shelters, fortifications, tunnels in the vicinity of production areas in mines where large volumes are blasted or tunnels in seismic regions.

STEEL FIBRE REINFORCED SHOTCRETE

General viewpoints.

In all the above mentioned examples a reinforced shotcrete lining has to be used. In cases i) and iii) the interaction between the rock mass and the shotcrete may be secured by the adhesion only. In all the other cases an other statical principle has to be used. When the geometry is regular sometimes arches or closed rings may be suitable. Otherwise rock bolt support of the reinforced shotcrete lining has to be used. The static principle for this system is the same as for a concrete plate on columns i.e. the concrete plate is subjected to negative bending moments in the vicinity of the supports and positive moments in the fields. Moreover the plate is subjected to concentrated shear stresses around the supports.

In most loading cases a shotcrete lining is probably subjected to a combination of forced deformation and static loading. The initial deformation of a rock mass cannot be prevented by a comparatively weak lining and when the main part of this deformation has taken place some loosening may occur. The latter causes a certain load on the lining. It is of the utmost importance that the lining can follow the deformations of the rock and that there is still remaining a certain carrying capacity. These requirements make it necessary to investigate to what extent the yield line theory is applicable to steel fibre reinforced shotcrete used in plates and statically indeterminate beams.
Forced deformations can be both of the bending type and the shear type so the shear performance of steel fibre reinforced shotcrete is also of great importance. In this paper, firstly, the plasticity of steel fibre reinforced shotcrete in flexure and shear will be treated. Secondly the requirements on the

fibre and the concrete will be illustrated. Thirdly results from static and dynamic tests on bolt supported steel fibre reinforced shotcrete linings will be presented. At last examples of solutions for advanced tunnel support using steel fibre reinforced shotcrete will be given.

APPLICATION OF THE YIELD LINE THEORY ON STEEL FIBRE REINFORCED SHOTCRETE IN FLEXURE

One fundamental assumption of the yield line theory is the rigid-plastic behaviour of the yield line. This assumption is sufficiently true for conventionally reinforced shotcrete, see Fig 1. Steel fibre reinforced concrete on the other hand has a more or less pronounced decrease of the moment-deflection curve after the maximum load. As can be seen from Fig 1 simply supported fibre reinforced beams are inferior to mesh reinforced ones considering the plastic behaviour. If the comparison is made between clamped beams another result is obtained, however. Since mesh reinforced shotcrete for practical reasons very seldom is double reinforced, the carrying capacity of a clamped shotcrete beam with mesh reinforcement is nearly the same as for the simply supported one. Fibre reinforced concrete on the other hand has the same reinforcement over the whole section height and these beams take advantage of the fact that the carrying capacity of a clamped beam is dependent on the sum of the field moment capacity and the support moment capacity. For an idealized case the crack width at midspan is double the crack width at the supports. Thus the support moment does not reach its peak value at the same time as the field moment and so the load-deflection curve becomes more horisontal than the curve for the corresponding simply supported beam.

This effect becomes still more obvious for point loaded circular plates. In this case the ratio between a radial crack angle and a tangential one is dependent on the angle between two adjacent radial cracks, see Fig 2. Assuming load-deflection curves of three different types for steel fibre reinforced simply supported beams load-deflection curves for point-loaded circular plates have been calculated. The centre angle between two adjacent radial crack then becomes a parameter in such calculations. For a steel fibre which gives a fairly brittle behaviour a small value of the centre angle is required for a plastic behaviour of the plate to be obtained, see Fig 4. This can be achieved for a large amount of fibres which allows for a good crack distribution. In practice this will probably mean that for a fibre like the EE-fibre amounts around 1.5 % are required. For long high strength fibres with end hooks like for instance the Bekaert zl 45x.35 the initial load-deflection curve itself has a more favourable course. Even 0.6 % of these fibres gives a sufficiently plastic behaviour of the plate when compared to a plate containing a mesh with 6 mm diameter and 100 mm bar spacing. Figs 5 and 6.

Some limited laboratory tests on this subject were made. The results were disturbed by dome action effects and a gradual punching failure at large deflection of the plate. Both the plates containing EE-fibres and the ones containg Bekaert fibres got very favourable load-deflection curves. Before any definite conclusions are drawn further tests and theoretical considerations have to be made. The theoretical treatment mentioned above is, however, believed to be correct enough to be used as an indication that fibres giving an unfavourable beam curve might be used for rock reinforcement if the amount is large enough.

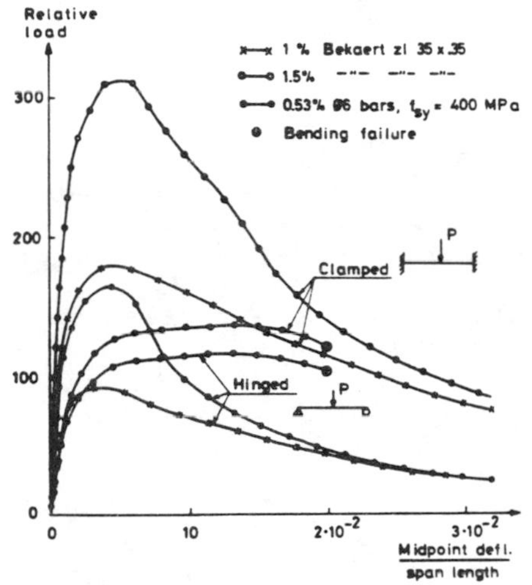

Fig 1. Load-deflection curves for tested simply supported steel fibre reinforced and mesh reinforced shotcrete beams. Curves for clamped beams are calculated from the test curves for the simply supported beams.

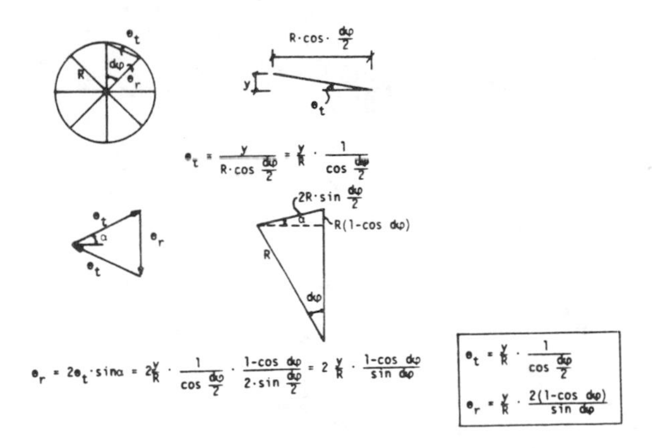

Fig 2. Relationship between angular displacements of radial and tangential cracks and the deflection of a cracked concrete plate.

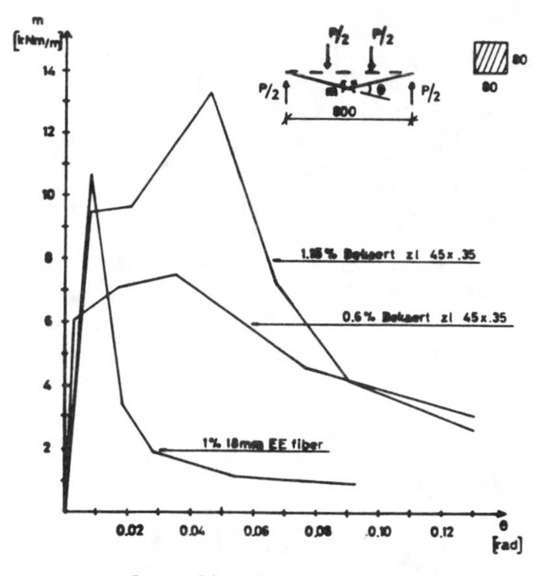

Fig 3. Moment-angular displacement relationship for steel fibre reinforced shotcrete beams.

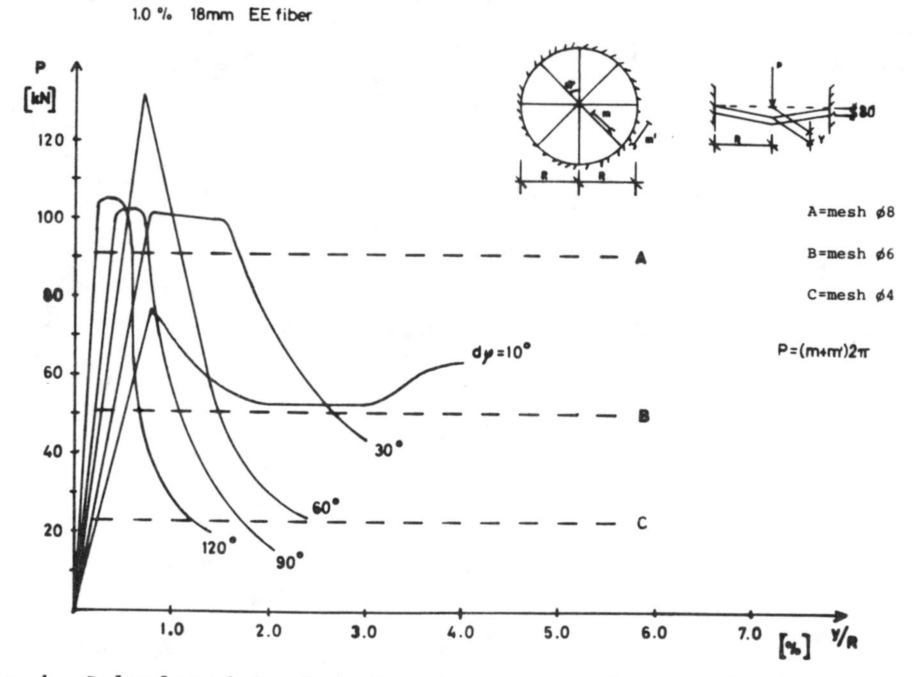

Fig 4. Calculated load-deflection curves for steel fibre and mesh reinforced clamped shotcrete plate subjected to centric point load. "Brittle" fibre concrete assumed.

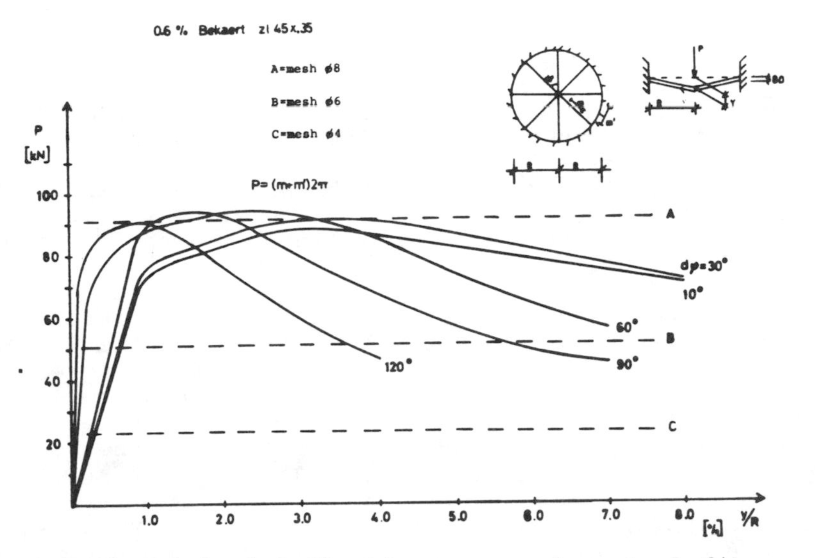

Fig 5. Calculated load-deflection curves for steel fibre and mesh reinforced clamped shotcrete plate subjected to centric point load. Small fibre amount giving fairly "plastic" fibre concrete assumed.

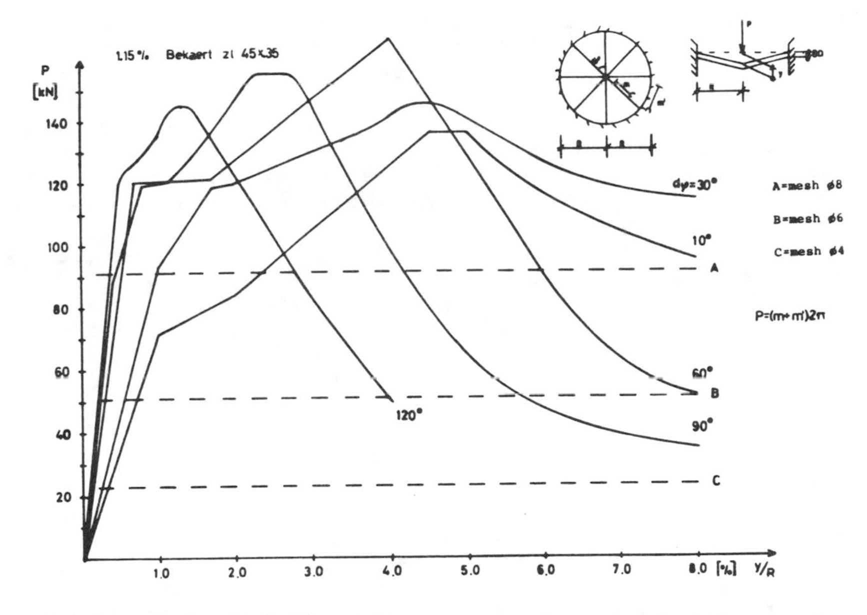

Fig 6. Calculated load-deflection curves for steel fibre and mesh reinforced clamped shotcrete plate subjected to centric point load. Large fibre amount giving fairly plastic fibre concrete assumed.

SHEAR PERFORMANCE OF STEEL FIBRE REINFORCED SHOTCRETE

The most interesting property considering shear of a shotcrete lining is the shear performance after cracking. At small crack widths the shear capacity consists of contributions from the aggregate interlock and from the fibres and eventually also from an uncracked compression zone. In the small scale investigation reported here the aim was limited to the contribution from the fibres alone.

From shotcreted boxes there were sawn out steel fibre reinforced prisms 150x80x50 (mm). The fibres used were the Bekaert zl 35x.35. The 28 days compressive strength of the concrete was 51 MPa measured on 15 cm cast cubes and 72 MPa measured on 8 cm sawn out cubes. The shear tests on the prisms were performed when they were 49 days old as an average. Until the testing day the prisms were stored under water.

Before testing of the prism a one mm deep track was sawn around the middle of it. After that it was subjected to positive and negative flexure so that a crack all through it developed. The two parts were then pulled apart. 1,2,3,4 and 5 mm crack widths were tested. During testing the crack width was kept constant. It was judged that six of the tests were succesful i.e. no aggregate interlock took place.

In the table below only the test results of the successfully tested specimens are listed. The specimens are listed as follows: the first two figures give the fibre content in promille by volume, the third figure gives the crack width in mm and the fourth one is a serial number. The maximum shear stress is expressed as the maximum shear force divided by the area of the cracked sektion ($\tau = V/A$). The shear displacement is expressed as the dimensionless quantity u/l where u is the displacement and l the fibre length.

Test specimen No	τ max MPa	u/l at τ max -	τ at u/l=0.5 MPa
1023	0.63	0.085	0.14
1041	0.66	0.10	0.04
1523	0.81	0.16	0.19
1531	1.54	0.17	0.26
1551	1.35	0.11	0.18
1553	1.34	0.12	0.20

In order to get an idea of the strength level the peak values of τ above are compared with bar reinforced shotcrete. A 80 mm thick shotcrete layer is considered. Such a layer is normally reinforced with 6 mm bars spaced 100 mm. According to the Bauman-Rusch formula for the dowel force of a steel bar in concrete each bar contributes with 3.6 kN to the shear resistance. This means a formal shear stress of 0.45 MPa when the force is "spread out" over the whole section.

The level 0.45 MPa is shown in the stress-displacement diagram of the shear tests. It is easily seen that the maximum strength of the fibre reinforced shotcrete is significantly higher than what would be expected from conventionally reinforced shotcrete. It is also interesting to observe that there is some carrying capacity remaining even when the "average fibre" is completely pulled out i.e. when u/l=0.5.

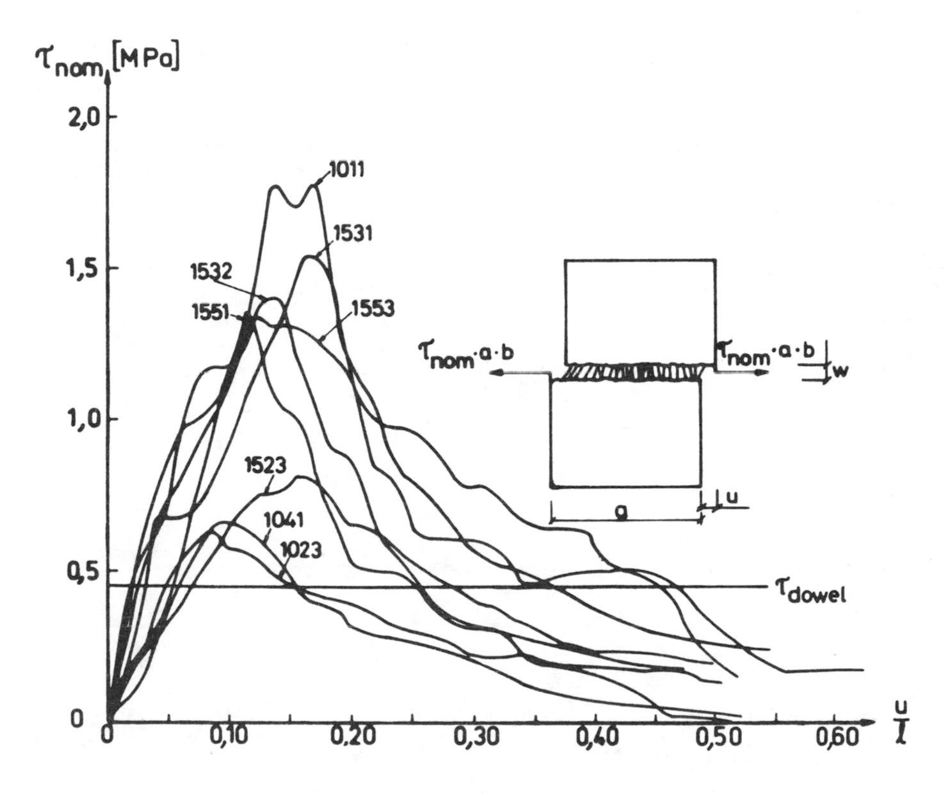

Fig 7. Shear stress versus displacement for small scale shear tests. τ is the shear force divided by the sheared section. u/l is the shear displacement divided by the fibre length. N.b. only tests 1023, 1041, 1523, 1531, 1551 and 1553 were free from aggregate interlocking effects.

The shear tests are of a small scale but they indicate that the shear performance of steel fibre reinforced shotcrete is probably better than of bar reinforced shotcrete. Provided an effective fibre is used it is to be expected that steel fibre reinforced shotcrete linings will withstand forced shear displacements without great damage or loss of the carrying capacity.

A PRACTICAL WAY TO FORMULATE THE REQUIREMENTS ON STEEL FIBRE REINFORCED SHOTCRETE

One conclusion from the discussion above is that a reinforced shotcrete lining should have a certain carrying capacity left after the occurence of a forced deformation. It is of course also important that the maximum carrying capacity reaches a certain level. Which one of these two properties that is the most important one will be varying from one case to another. For the time being it has to be assumed that high strength and yielding capacity in flexure also will give acceptable properties in shear. Therefore the requirements on the steel fibre reinforced could be expressed using the force-deflection curve for a well-defined test beam, see Fig 8. These requirements are based on actual tests with steel fibre reinforced beams and they ensure that the properties of the fibre reinforced shotcrete are at least as good as for mesh reinforced shotcrete in flexure and better in shear. By this approach the need for specifying the fibre type is avoided.

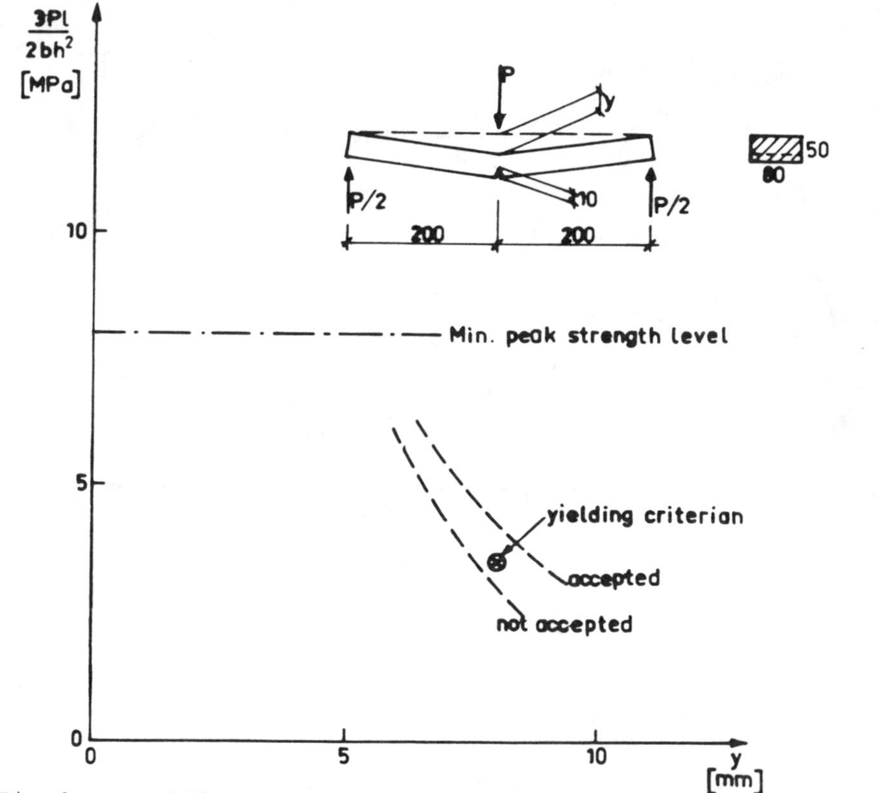

Fig 8. Specification of the requirements on steel fibre reinforced shotcrete. The load-deflection curve of the specified beam must reach the minimum strength level and also pass to the right of the yield criterion point.

LARGE SCALE TESTS ON BOLT ANCHORED STEEL FIBRE REINFORCED SHOTCRETE LININGS

Tests of this kind were performed by Holmgren /1979/ using bar reinforced shotcrete. The actual tests were performed in order to find the best design of the anchorage of the rock bolt in the steel fibre reinforced shotcrete, and to make a comparison between mesh reinforcement and fibre reinforcement for shotcrete. Details are given in Holmgren /1985:1/ so here will be given only the most important results.

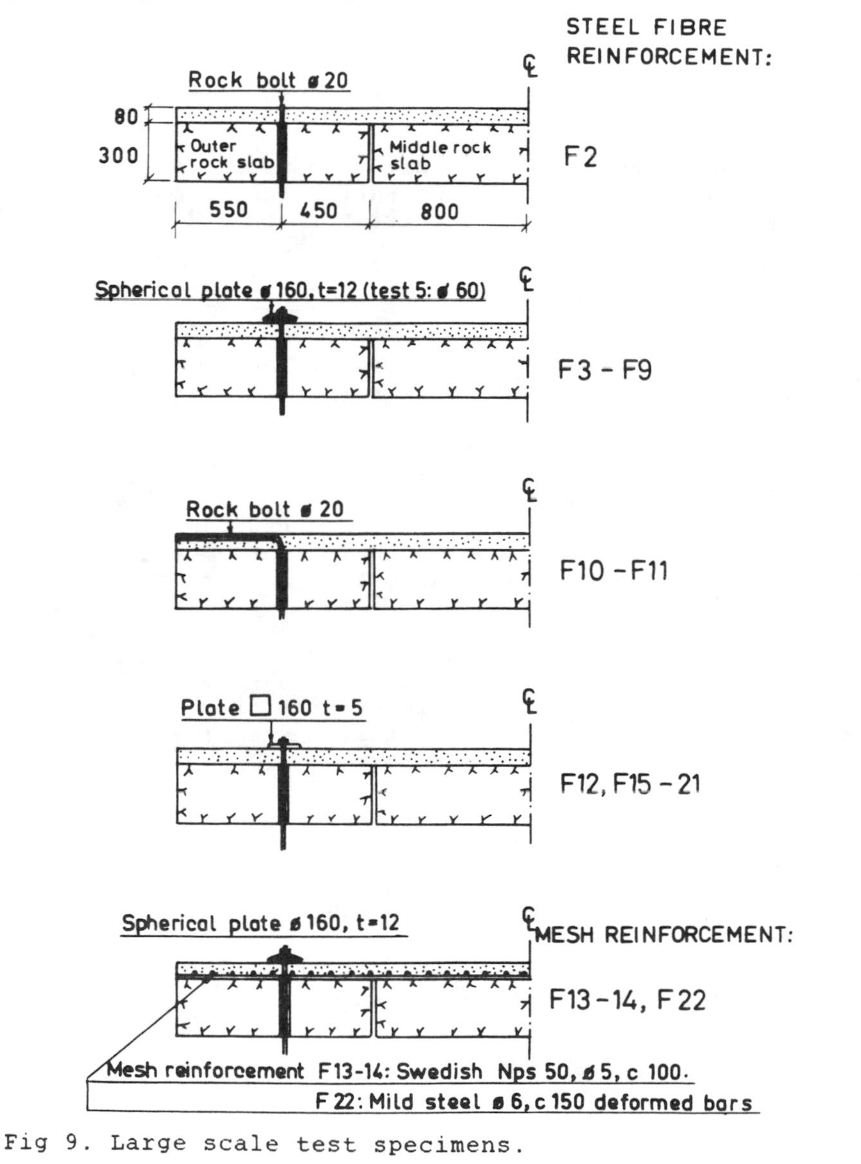

Fig 9. Large scale test specimens.

The main results of these tests were:

(i) It is possible to produce steel fibre reinforced shotcrete linings which are at least equally strong and ductile in flexure as conventionally reinforced ones.

(ii) Cheap washers manufactured of mild steel can be used for the anchorage of the rock bolt in the shotcrete.

(iii) Bent rock bolts provide poor anchorage in steel fibre reinforced shotcrete.

(iv) Wire mesh with cold tensioned bars is not suitable for rock reinforcement where ductility is desired. Mild steel should be used.

It is to be observed that these results are valid for a steel fibre with end anchors and a suffficiently high strength which allows the fibre to slip in the matrix at failure.

BOLT SUPPORTED REINFORCED SHOTCRETE LININGS SUBJECTED TO DYNAMIC LOADING

In Holmgren /1985:2/ there are presented static and dynamic tests on circular shotcrete plates centrically supported by a rock bolt. These test specimens simulated the portion of the plate which is situated between the line of zero bending moment and the bolt support. There were tested plates containing 1.0 % EE-fibres and Bekaert zl 35x.35, 1.5% of these two fibres and also mesh reinforced ones. The tests showed that long high-strength fibres with end hooks are to be preferred for this kind of linings. The tests showed also that the energy which can be absorbed by the shotcrete at collapse deflections is of about the same magnitude as the energy absorbed by a 25 mm rock bolt of 2 m free length at 5% elongation. Since such a bolt has an elongation at failure of about 15% this means that a lining which is designed to resist the failure force from the rock bolt can endure several loadings and that the main energy absorber will be the bolt.

The test specimen is shown in Fig 10. In Fig 11 there are shown registrations from one test with an over-strong bolt which remained elastic during testing and one test with a yielding bolt. In the first case the load on the plate reached a level which was determined by the carrying capacity of the plate. At that level severe cracking occurred. In the second case the maximum load was governed by the yield force of the bolt and the plate remained nearly unaffected by the test.

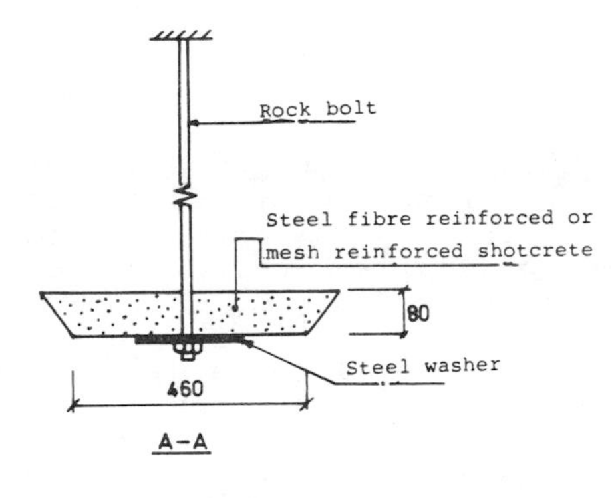

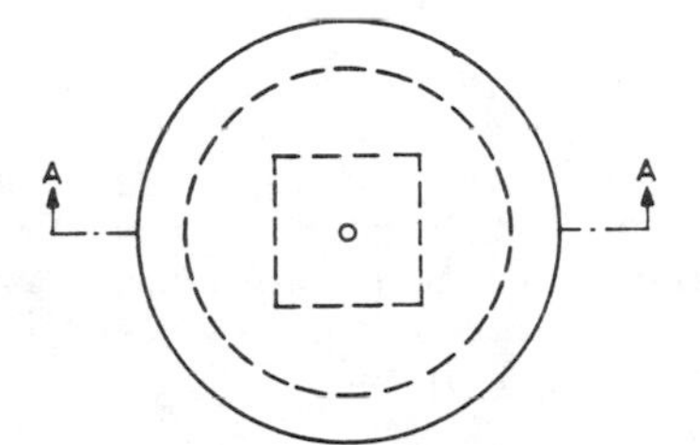

Fig 10. Test specimen for dynamic testing of bolt supported reinforced shotcrete.

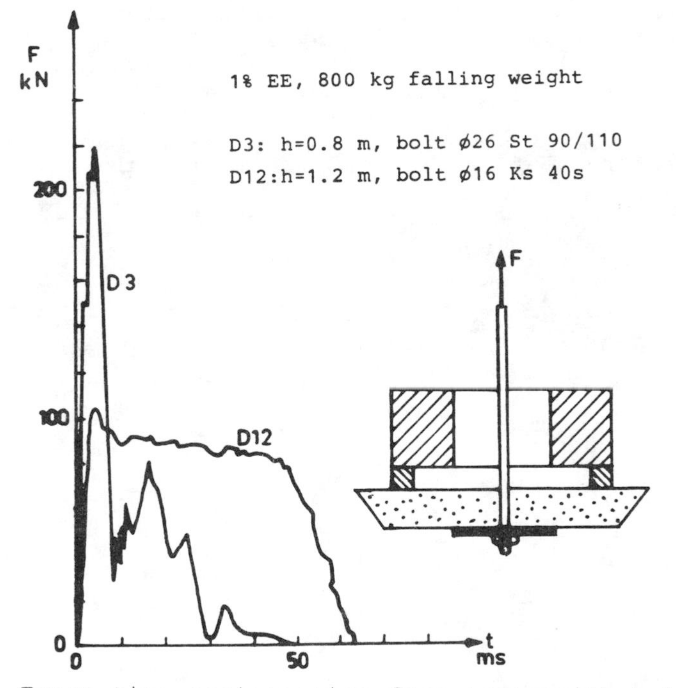

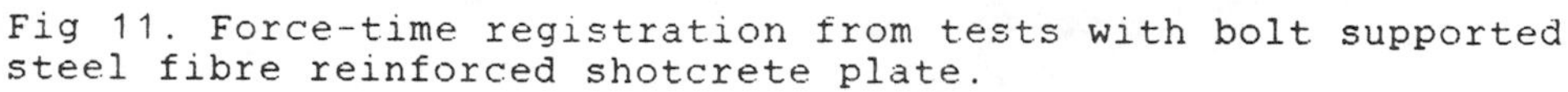

Fig 11. Force-time registration from tests with bolt supported steel fibre reinforced shotcrete plate.

PRACTICAL APPLICATIONS

In Figs 12-15 there are shown some examples where the strength and flexibility of steel fibre reinforced shotcrete are taken advantage of. The shotcrete is combined with fully grouted or grouted end anchored rock bolts. In one special case a yielding device which is put on the end of the bolt is suggested.

In Fig 12 a weak zone and the surrounding rock is reinforced with fibre reinforced shotcrete and grouted end anchored bolts. At some distance from the zone fully grouted bolts are used. This design allows the weakest parts of the rock to move a little without damages on the lining.

In Fig 13 the design for a civil defence cavern is shown. The bolt system is designed for the absorption of the energy from a bomb detonation on the surface. The steel fibre reinforced shotcrete lining is designed to withstand the yield forces from the rock bolts without severe cracking. Where the vertical crack system is parallell to a wall the wall lining has the same capacity as the the ceiling lining.

In Fig 14 some suggestions for swelling rock conditions are shown. When the rock gives fair anchorage for the bolts end anchored grouted bolt of small dimensions are suggested. When this is not the case a more or less fully grouted bolt in combination with a yielding device is suggested. This design allows the swelling to take place at a minimum risk for collapse of the lining. Some damages cannot probably be avoided.

Fig 15 shows a suggestion for road and railway tunnels where both rock reinforcement and thermal insulation are needed.

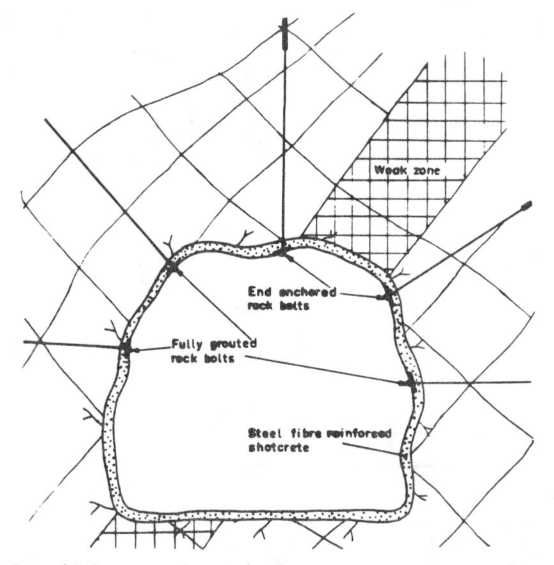

Fig 12. Flexible rock reinforcement around weak zone.

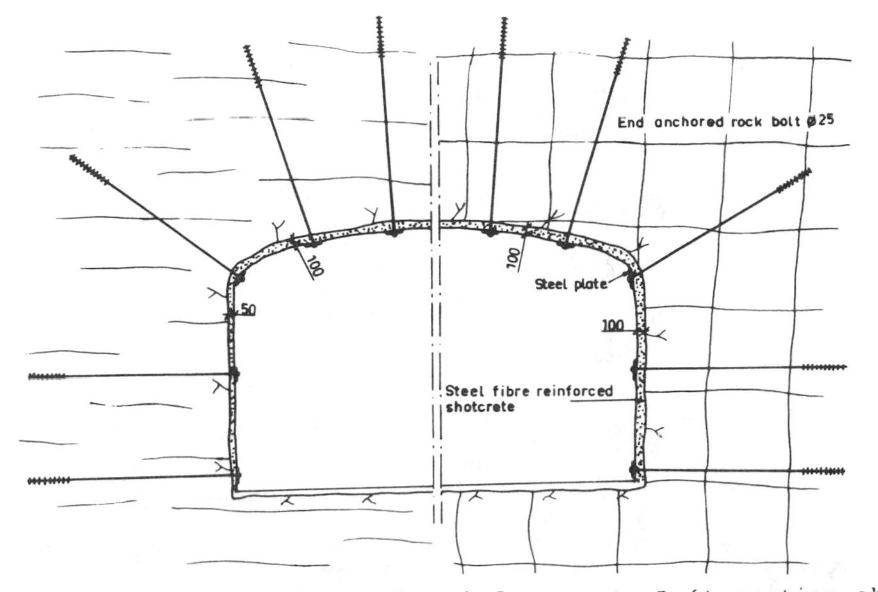

Fig 13. Energy absorbing rock reinforcement. Left section shows wall reinforcement where vertical crack system deviates from wall direction. Right section where it is fairly parallel to the wall.

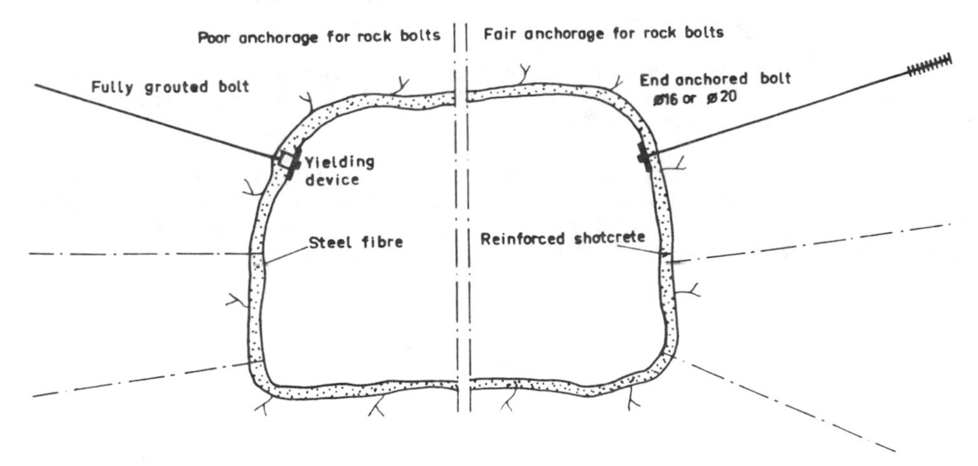

Fig 14. Suggested design of rock reinforcement for swelling rock conditions. Left section shows design for poor anchoring conditions when the bolt itself cannot be expected to yield.

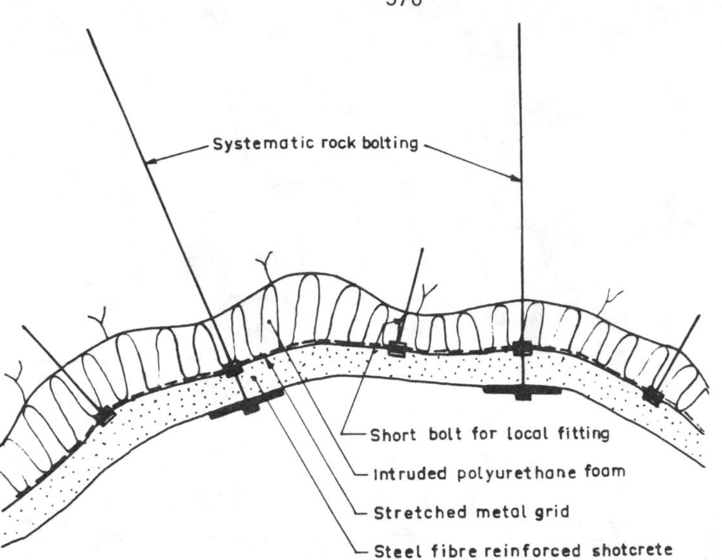

Fig 15. Suggested design for road and railway tunnels where rock reinforcement and thermal insulation are required.

REFERENCES

Holmgren J, "Punch-loaded shotcrete linings on hard rock", Swedish Fortifications Administration, Research Dept, Report 121:6, Stockholm 1979. Also Report no 7:2/79, Swedish Rock Engineering Research Foundation, Stockholm 1979. Dissertation Royal Inst of Technology, Stockholm 1979.

Holmgren J, "Bolt anchored steel fibre reinforced shotcrete linings", to be printed by the Swedish Rock Engineering Research Foundation, Stockholm 1985.

Holmgren J, "Dynamiskt belastad bergförstärkning av sprutbetong", (Shotcrete linings subjected to dynamic loading), to be printed by the Swedish Fortifications Administration, Stockholm 1985. In Swedish with English summary.

STEEL FIBER CONCRETE
US-SWEDEN joint seminar
(NSF-STU)
Stockholm 3-5 June, 1985
S P Shah and Å Skarendahl,
Editors

USE OF STEEL FIBERS FOR SHEAR REINFORCEMENT AND DUCTILITY

by Gordon B Batson, Dr, Clarkson University, Potsdam, New York, USA

ABSTRACT

Steel fibers are effective as shear reinforcement beacuse the random dispersion of the ductile high strength fibers throughout the brittle low strength concrete improves the tensile strength,ductility and fracture toughness of the plain concrete.A review of the published laboratory test data demonstrates the effectiveness of steel fibers to resist bending,torsional and shear stresses induced by static and dynamic loads acting on beams and beam-column joints.This paper is a review of the use of steel fibers for shear reinfocement and for improving the ductility of concrete.Formulas are summarized for predicting the load capacity and stiffness of steel fiber reinforced concrete. Most of the suggested equations for design are identical to existing design codes for reinforced concrete modified for the particular strength properties due to the inclusion of steel fibers in the concrete. A wider range of parameters associated with the use of steel fibers in concrete must be investigated before design procedures and equations that reliable predict the shear capacity and the ductility can or will be adopted by code writing organizations.

GORDON B. BATSON is Professor of Civil and Environmental Engineering at Clarkson University,Potsdam,New York, USA,13676. He is a member and past chairman of ACI Committee 544,Fiber Reinforced Concrete and Secretary of ACI Committee 549, Ferrocement.

INTRODUCTION

There is little reason to doubt that the random distribution of small volume percentage,less than 2.5 percent,of high strength ductile fibers in low strength brittle concrete significiently improves many of the mechanical properties of plain and reinforced concrete. The fibers create an effective crack arresting mechanism which improves the pre and post cracking properties. The enhanced mechanical properties of steel fiber reinforced concrete (SFRC) can be used to improve the structural response of beams and beam-column joints subjected to static and dynamic loads.A very promising application is the use of steel fibers for shear reinforcement and for improving the ductility o concrete.The current state-of-the-art of steel fibers as shear reinforcement is reviewed for beams subjected to flexure or combinations of flexure,torsion and shear loads, flexural fatigu and for beams and beam-column joints subjected to low cycle high amplitude loads. A review of published data and on-going researc has demonstrated the effectiveness of steel fibers for shear reinforcement and for improving the ductility of concrete,but less information on equations for predicting reliably the shear capacity and ductility of SFRC. The review shows that much of the test data evaluated and the methods developed to predict the strength of SFRC are comparable with the design procedures for normal reinforced concrete found in codes,such as ACI 318-83 (1) or CIP 110 (2).This is a logical approach given the importance of design codes for public safety and the usual reluctance of code authorities to make significient changes unless the public safety is endangered.The inclusion of steel fibers as an accepte shear reinforcing material for concrete will be slow in adoption by code authorities if new concepts or new more efficient materials conflict with long established methods as recently pointed out by Dr. N.Hawkins(3).This will surely be the fate of SFRC if the mechanics and limitations on steel fibers as shear reinforcing are not thoroughly understood and well developed.

SHEAR REINFORCEMENT IN BEAMS

Extensive laboratory testing has substantiated the increased shear capacity of concrete or mortar beams.The increase can be attributed to the random distribution of the fibers throughout the volune of the beam.The average spacing of the fibers is less than pratical for the smallest reinforcing bars and thereby increases the first crack and ultimate strength.Bridging of fibers across cracks provides post cracking strength and increases the shear friction of the crack surfaces.Test data show that steel fibers by themselves or in combination with conventional shear reinforcement can provide sufficient shear capacity for beams to fail in a ductile flexural manner. Batson,Jenkins,and Spatney(4) tested 96 simple span beams 4x6x78 inches(100x150x2000 mm) with four point loading.Fiber size,shape and volume percentage were varied as well as the shear-depth span ratio,a/d,where a = the distance from the applied load to the beam support and d = the depth to the flexural steel.All beams had conventional flexural steel except the control specimens without stirrups designed to fail shear and those with vertical stirrups designed to fail in flexure.Beams tested with an a/d ratio of 4.8 failed in shear with and without stirrups,but the addition of a small volume percentage of steel fibers between 0.44 and 0.88 was sufficient to produce moment failures. Additional tests were conducted by decreasing the a/d ratio until a shear failure occurred for each percentage of each different fiber.Figure 1 shows the shear stress,f_v,versus the a/d ratio of the beams that failed in shear.The values of $f_v = 0.063V_u$ plotted were computed for the uncracked transformed section and would be reduced in magnitude if computed by

$$f_v = V_u/bd = 0.05V_u \qquad \text{Eq(1)}$$

where V_u is the ultimate shear,b=width of beam and d=depth to centroid of flexural steel.In Figure 1 there is a significient deviation in the trend of the data for a/d values less than 3.0. Zutty (5) noted a similar occurrence for conventionally reinforced concrete beams which he attributed to the direct transfer of the applied load by compession to the support.

Paul and Sinnamon (6) conducted tests to predict the shear capacity of segmented tunnel liners made of SFRC.The results wer similar to those of Batson,Jenkins,Spatney for the same a/d ratios.Paul and Sinnamon developed an equation to predict the ultimate shear cacapity following ACI 318-71 recommendations.Their equations for the shear capacity were:

$$V_c = 0.255\ f_r\ b_w\ d \qquad \text{(English units,psi) Eq(2}$$

or

$$V_c = (0.253\ f_r - 2500\ p_w V_u d/M_u) b_w d \quad \text{(English units,psi) Eq(3}$$

where:

V_c = nominal shear strength provided by the concrete
d = effective depth of the beam
b_w = web or width of a rectangular
$p_w = A_s/b_w\ d$
A_s = area of flexual reinforcing steel
V_u = factored shear force at section
M_u = factored moment at section
f_r = modulus of rupture ($7.5\sqrt{f'_c}$ for plain concrete).

The modulus of rupture,f_r,was used instead of the square root of the twenty eight day commpressive strength,f_c,because it is more sensitive to the fiber content than the ultimate compressive strength.The value of $7.5\sqrt{f'_c}$ was suggested by Kobayashi (7). Moustafa (8) suggested that the split cylinder tensile strength ,f_{st} ,be used inplace of the modulus of rupture or ultimate compressive strength,but there is some uncertainty about its use in an equation to predict shear strength because of the unknown stress distribution after the first crack in the split cylinder test.

Jindal (9) applied four point loading to 44 simple span beams 4x6x30 and 4x6x60 inches (100 x150 x762 and 100 x150 x 1525 mm) with two #4 (12 mm) bars for flexural reinforcement ,no stirrps and a constant 1.0 fiber volume percentage.The fibers were of tw different tensile strengths;a high tensile strength of about 300 ksi (2100 MPa) and a mild steel with an yield stress of 54 ksi (377 MPa).The high strength fibers had three different aspect ratios (length divided by equivalent diameter)and the mild stee fibers had seven different aspect ratios ranging between 10 and 100.The test results are shown in Table 1 indicating that the fibers developed sufficient shear capacity for the beams to fail in flexure. Jindal's data indicates there is an optimal aspect ratio for the fibers between 75 and 80.

TABLE 1 Jindal Test Results

Beam No.	Shear Span Ratio = a/d	Ult. Shear Vu kips	Ult. Moment Mu Inch-kips	Strength Ratio Vuf/Vuo	Strength Ratio Muf/Muo	Maxim Shear Stress = psi	Mode of Failure
(1)	(2)	(3)	(4)	(5)	(6)	(7)	(8)
A	2.0	4.4	44.0	1.0	0.46	433	S
	2.4	4.4	52.0	1.0	0.55	433	S
	3.6	4.4	79.0	1.0	0.83	433	S
	4.8	4.0	96.0	0.91	1.0	394	M
B	4.8	3.6	86.4	0.82	0.9	354	M
C	3.6	4.7	84.6	1.07	0.88	463	M-S
	4.8	5.55	133.0	1.26	1.35	546	M
D	3.6	5.5	99.0	1.25	1.03	541	M
	4.8	6.35	153.0	1.44	1.59	625	M
E	3.6	6.05	109.0	1.38	1.13	595	M
	4.8	6.35	153.0	1.44	1.59	625	M
F	3.6	5.25	94.5	1.19	0.98	517	M
	4.8	5.0	120.0	1.14	1.25	492	M
G	2.0	6.63	66.3	1.51	0.96	652	S
	2.4	6.63	80.0	1.51	0.83	652	M-S
	3.6	6.50	117.0	1.48	1.22	640	M
	4.8	4.4	106.0	1.0	1.1	432	M
H	2.0	9.0	90.0	2.05	0.94	880	M-S
	2.4	8.55	103.0	1.94	1.07	837	M
	3.6	6.45	116.0	1.47	1.21	632	M-S
	4.8	5.5	132.0	1.25	1.38	540	M-S
J	2.0	7.4	74.0	1.68	0.77	725	S-M
	2.4	7.75	93.5	1.76	0.97	760	M
	3.6	5.5	99.0	1.25	1.03	540	M
	4.8	5.0	120.0	1.14	1.25	490	M

Notes: 1. S - Shear failure; M - Moment failure; M-S - Initial failure by moment and ultimate failure in shear; S-M - Initial failure in shear and ultimate failure in moment.

2. All the above results are average of 2 test specimens.

Williamson (10) tested a few large reinforced concrete beams 12 x 21.5 inches by 23 ft long (305 x 546 x 7010 mm) with and without vertical stirrups and with two different types of fibers,a straight fiber and a fiber with deformed ends,replacing the stirrups.The beams with a 1.6 volume percentage of straight fibers had a 45 percent increase in shear capacity compared to the beams without stirrups,but the fiber reinfroced beams failed in shear. The beams with a 1.1 perecent volume percentage of fibers with deformed ends had an increase in shear capacity of 45 to 67 percent compared to the beams without stirrups and the beams failed in flexure.

Bollana (11) tested two span continuous reinforced concrete beams 4 x 6 x 120 inches (100 x150 x 3048 mm) with equal spans of 58 inches (1473 mm) and two concentrated loads 24 inches (610 mm) either side of the center support.Two types of fibers, a straight fiber at 1.5 and 2.0 volume percentage and the a bent or deformed end fiber at 0.75 and 1.5 volume percentage ,were used to replace the vertical stirrups.The value of V/bd for thes beams was between 5 and 8,well outside the normal range of 3 to 4 for reinforced concrete beams.The main purpose of the tests was investigating the effectiveness of steel fibers as shear reinforcement in a region of high shear and a steep moment gradient.The beams with vertical stirrups failed in flexure and the beams without stirrups failed in shear.The beams with the steel fibers as the only shear reinforcement experienced initial flexure cracking but ended in a shear failure either side of the center support at nearly the same load as the beams with vertical stirrups.For the various failure loads the bent or deformed fibers at half the volume percentage of the straight fibers provided equal shear capacity as shown in Table 2.

Table 2 Bollana Test Data

Fiber Content(%)	Load at Failure,lbs(kN)
1.5 straight	13970 (62.1)
2.0 straight	15000 (66.7)
0.75 bent	14750 (65.6)
1.00 bent	16350 (72.7)
stirrups	16200 (72.1)
no stirrups	8000 (35.6)

Craig (12) tested simple span reinforced concrete beams 6 x 12 x 72 inches (150 x 300 x 1829 mm) using crimped (bent end) fibers at 1.0 and 1.5 volume percentage.The a/d ratios were 1.0,1.5,2.0,2.5 and 3.0.The shear capacity increased 130 percent and 140 percent for 1.0 and 1.5 volume percentage of fibers for an a/d ratio of 1.5 and incresed 108 percent for 1.0 percent volume percentage for an a/d ratio of 3.0 compared to the control

beams without stirrups. Flexural failures occurred for fiber contents of 1.0 percent and greater. Figure 2 shows the shear stress plotted against the a/d ratio.
Craig recommends that 0.147 f_{st} or 0.132 f_r should be substituted for $\sqrt{f'_c}$ and thereby places an upper limit on Eq.2 and Eq.3 of 0.51f_{st} b_w d or 0.46 f_r b_w d.
Swamy and Bahia (13) conducted shear tests on T-beams and rectangular beams containing 0.4 , 0.8 and 1.2 volume percentage of deformed (bent end) fibers.The T-beams were 3.4 m(11.1 ft.) long on a simple span of 2.8 m (9.1 ft.).Flange was of 500 x 50 mm (20 x 2 in.),total depth of 250 mm(10 in) and stem width of 175 mm (7 in.).Three parameters were evaluated by ; varying the fiber content for constant percentage of longitudinal steel,varying the percentage of longitudinal steel (1.95 t0 4.00 percent) for constant fiber percentage,and varying the amount of stirrups for constant longitudinal steel and fiber content.All tests were conducted with an a/d ratio of 4.5.Significient results were:ultimate shear strength increased almost linearly with fiber content;T-beams with 1.95 percent flexure steel and 0.8 percent fibers by volume failed in flexure ;fiber volume of 0.8 to 1.2 percent with a nominal amount of stirrups caused the flexural steel to yield and reach the full flexural capacity of the beam; the contribution to the shear strength by dowel action of the flexural steel was enhanced by the steel fibers.
Lafraugh and Moustafa (14) reported data on rectangular beams and T-beams where the steel fibers were not effective in preventing sudden shear failures.Muhidin and Regan(15) tested I-beams with three different volume percentages of fibers and a/d ratios between 3.24 and 4.68 which behaved the same as beams with stirrups.
The shear test data reviewed clearly shows that steel fibers are effective as shear reinforcement at volume percentages of one percent and greater and the flexural strength is equal to or greater than the flexural strength of beams with stirrups. The shear-span depth ratio,a/d,is an important parameter for both conventional and steel fiber reinforced beams.

SHEAR REINFORCEMEMENT FOR COMBINED BENDING,TORSION AND SHEAR

It is reasonable to assume that the mechanism which creates the shear capacity of steel fiber reinforced concrete subjected to flexural loadings should also be effective for beams subjected to combined bending,torsion and shear.During the last five years several investigations have been conducted on the effectiveness of fibers as shear reinforcement for torsion,bending and torsion or torsion ,bending and shear.Design requirements for torsion are relatively new in building codes.The 1971 edition of ACI 318 for the first time contained specific design requirements for torsion which applies when a minimum amount of torsion is exceeded.

Longitudinal top steel with closed stirrups must be added to the reinforcement required for flexure and shear.It was assumed that the steel fibers could partially or completely replace the reinforcement required for torsion in the codes.
There are two types of torsion.Equilibrium torsion is statically determinate and the torsion can not be redistributed by internal forces without collpase. Compatibility torsion is statically indeterminate and the compatibility of connected members allows a redistribution of internal forces after cracking occurrs without collapse.Nearly all the steel fiber reinforced concrete beams tested in torsion or tosion combined with other loads have been of the equilibrium type of torsion.
Narayanan and co-workers(16),(17),(18) have published test data on steel fiber reinforced concrete beams subjected to torsion and combined torsion and bending.Narayanan and Toorani-Goloosalar(16) reported data on plain concrete and steel fiber reinforced concrete beams using cold drawn round steel fibers with an aspect ratio of 150, 0.7 fiber volume percentage and three concrete compressive strengths ranging from 22 to 48 N/mm^2 (3.2 to 7.0 ksi).The torsion test specimens were 100x100x1000 mm (4x4x40 inches) and specimens for combined bending and torsion were 50x100x1300 (2x4x51 inches).The ultimate strength was much greater than predicted by ACI 318 or the CP 110 codes and was closest to the values predicted by Nadai's sand-heap analogy (plastic theory) using the split cylinder test for the tensile strength of the steel fiber concrete.

$$T_u = (0.5 - X/6Y)x^2Y\ f_{st} \qquad \text{Eq(4)}$$

where

X= smaller cross-section dimension
Y= larger cross-section dimension
f_{st} =split cylinder tensile strength

Mansur and Paramasivam (19) conducted two seriers of pure torsion tests.The first series varied the fiber volume from 1 to 3 percent for a constant aspect ratio and the second series varied the fiber aspect ratio between 26 and 77 for a constant 1.0 fiber volume percentage.The 21 torsion specimens were 102x102x760 mm (4x4x30 inches),loaded by applying torques at each end and the angle of twist measured over the central 400 mm (16 inches).The ultimate strength was best predicted by the sand-heap analogy (plastic theory) using the direct tensile strength of the steel fiber concrete,f_t,inplace of f_{st}.The value of f_t was computed by a relationship developed by Mangat (20) based on the law of mixtures of a cement composite.

$$f_t=f_{mt}V_m + 0.34(1/d)V_f \qquad \text{Eq(5)}$$

where

f_{mt}=tensile strength of cement matrix
V_m=volume percentage of cement matrix
V_f =volume fraction of fibers
l/d = aspect ratio of fibers

Mansur (21) extended the range of variables of Narayanan and Toorani-Goloosalar(16).The ratio of the torsion to bending moment,T/M, varied from zero(pure bending) to infinity(pure torsion) for a constant 0.75 volume percentage of a sinlge size and shape of fiber with bent(deformed)ends.Torsion specimens were 100mm(4 inches)wide by 100,150 and 200 mm (4,6 and 8 inches)depth and 1000 mm(40 inches)long.The combined torsion and bending specimens were 100x150x2000 mm(4x6x80 inches). For a T/M ratio greater then 2.57,failure was by crushing on the vertical(wider)beam surface and tension cracking on the opposite surface was inclined about 45 degrees to the beam axis.For T/M less then 2.57,bending dominated with crushing of the concrete on the top (compressive)surface.The two modes of failure are the classical Mode 1(compresion of the top) and Mode 2(compression on the side) associated with skew-bending theory.The ultimate torsional strength for T/M greater than 2.57 was best predicted by an equation similar to Hsu's (22) for plain concrete using a reduced modulus of rupture of the steel fiber reinforced concrete.

$$T_U=X^2Y(0.71\ f_r)/3 \qquad \text{Eq(6)}$$

where

f_r =modulus of rupture

For T/M less than 2.57,bending dominates, the ultimate torsion was given by a more complex equation.

$$T_u = \frac{x^2y}{3} f_r[1+0.29(\frac{1-R}{1+R})][\frac{1}{\Psi^2} + 1 - \frac{1}{\Psi}] \qquad \text{Eq(7)}$$

where

Ψ = T/M

$R = 1+(\frac{\Psi}{3k})^2$, $K = f(\frac{Y}{x})$

Figure 3 shows the good agreement for the interaction diagram between the test data and the predicated values form Eq (6) and Eq (7).The interaction diagram indicates that the torsional strength is unaffeccted if the bending moment is less than about 60 percent of the flexural strength.
Narayanan and Kareem-Palanjian (18) carried out torsion tests on

28 steel fiber reinforced beams in an effort to reconcile the various equations to predict the ultimate torsional strength by considering two different concrete mix designs,three types of fibers(deformed ends,crimped,and round straight)of several aspect ratios.Test specimens had a constant width of 85 mm(3.3 inches) with the depth ranging fom 85 to 178 mm (3.3 to 7 inches) and a length of 1500mm(60inches).Torques were applied at each end of the beam for a test span of 1350 mm (53.1 inches).Two equations were developed,one to predict the cracking torque and the second to predict the ultimate torsional strength.The ultimate torsional strength is given by :

$$T_U=(0.326 -0.035X/Y)X^2Yf_{st} \quad Eq(8)$$

where

$$f_{st} = \text{split cylinder tensile strength}$$

Equation (8) was developed by starting from a modification of the tensile strenth of steel fiber reinforced cement composite suggested by Swamy and Mangat(23).The modification was the inclusion of a bond factor for the various types of steel fibers which varied from 0.5 to 1.0.They reasoned that since the torque is directly proportional to the tensile stress,then the ultimate torque should be directly related to the ultimate torsional strength of the plain concrete matrix multiplied by the same factor that represents the contribution of steel fibers to the increased tensile strength of a fiber composite.Figure 4 shows the predicted values of the ultimate torque plotted against their test data and the data of others (17) ,(19)and (21) appears to be very good.

Mansur and Paramasivan (26) have extended the tests by Mansur (21) to modify Equation (6) for the effect of shear on combined bending and torsion.Test beams were 100 x150 mm in cross-section (4 x 6 inches) and either 1400 mm or 2000 mm (55 0r 79 inches) long. A single fiber volume percentage of 0.75 of deformed (bent ends) fibers was used.Test data showed that the shear had no effect on the mode of failure and both Mode 1 and Mode 2 failures occurred as noted previously in Mansur's tests. Although the shear did not effect the mode of failure ,it had to be considered in the equations for predicting the ultimate torsional strength.For Mode 2 failure the ultimate torsional strength is predicted by:

$$T_u = \frac{x^2 y}{3(1 + \frac{x}{6\alpha})} (0.71\ f_r) \qquad \text{Eq(9)}$$

where

$\alpha = T/V$

The predicted ultimate torsional strength for Mode 1 failures remained the same as given by Eq (7).Figure 5 shows schematically the interaction surfaces for combined bending,torsion and shear.For a given intensity of shear,Mode 2 governs when T/M is large and the torsional strength remains constant.When T/M is low bending dominates, Mode 1 failure occurs and a small increase in bending substantially reduces the torsional strength.
Batson,Terry and Chang (25) and Lim (26) investigated combined bending and torsion of beams in which the torsional reinforcement of top longitudinal steel and closed stirrups required by ACI 318 were partially or completely replaced by steel fibers.The beams were 4 x 6 x102 inches (100 x 150 x 2600 mm).Straight steel fibers 1.0 x 0.022 x 0.01 inches (25.4 x 0.56x 0.25 mm) with an aspect ratio of 100 were used in volume percentages of 0.5,1.0,and 1.5 with either two #5 (15 mm) rebars for flexure or two #5 rebars for flexure and two #4(12 mm) longitudinal top rebars for torsion. Beams without steel fibers had stirrups, flexural steel and top longitudinal steel as required by ACI 318. The testing procedure allowed the torsional loading and flexural loading to be applied independently of each other to specimens that were restrained against twisting but free to rotate with the flexural loads at the end supports.The testing procedure simulated compatibility type of torsion that allows a redistribution of internal forces before final collapse. Beams with 1.0 or more volume percentage of fibers and top longitudinal steel for torsion and beams with 1.5 volume percentage of fibers and no top longitudinal steel for torsion equalled or exceeded the strength the beams with stirrups and top longitudinal steel for torsion to moment ratios ,T/M, between 0.02 and 5.88.Mode 2 failures,torsion dominates, were observed for high T/M ratios and Mode 1 failures,bending dominates, for low T/M ratios.Figure 6 shows the test data plotted on interaction equations beveloped by Martin (27) for combiuned bending and torsion with only flexural steel.Although the interaction equation appears satisfactory of low T/M ratios it is not satisfactory for high T/M ratios. Also, the computed ultimate torsional strength of the steel fiber reinforced beams varied greatly depending on whether the tensile

strength of the concrete was evaluated by the modulus of rupture,split cylinder or by $6.7 \sqrt{f'_c}$ as recommended by ACI.Basing the modulus of rupture on the compressive strength as suggested by Hsu was not appropriate for steel fiber concrete. For a fiber content of 1.5 percent the cracking was confined to the middle third of the beam between the points of load application,but cracking extended into either end thirds of the beams with fiber contents of 1.0 percent and less. It appears that the fiber percentage influences the redistribution of the internal forces for compatibility type torsion.

STEEL FIBERS IN CONCRETE JOINTS

Joints in reinforced concrete structures have to resist axial,bending and shear loads, which may also include seismic foces.To resist these loads large quantities of reinforcing steel have to be placed in very a confined region.Test have been conducted on concrete joints with steel fibers replacing some the reinforcing steel,particularly the reinforcing required for shear. Test results have shown steel that fibers increase the shear capacity ,ductility and reduce the severity of cracking. Henager (28) tested two full size beam-column joints;one designed according to the seismic requirements of ACI 318 -71 and the second was the same except that 1.67 volume percentage of round steel fibers, 1.5 x 0.020 inches (38.1 x 0.051 mm), replaced the column hoops as shown in Figures 7 and 8.The cyclic loading shown in Figure 9 simulated an earthquake.A constant axial load of 175 Kips (765 KN) was applied to the column.A 200 Kip (890 KN) ram mounted as shown in Figure 7 created a bending moment reversal as it followed the simulated earthquake loading.Figure 10 is the moment-rotation curve for the tenth loading cycle for both test specimens. The ductility factor of the steel fiber joint is 13.5 while that for the conventional joint is 8.5.The ductility factor was defined as the ratio of the rotation for a particular load to the rotation for the yield load.Major cracking occurred in the beam and at the intersection of the beam and column of the conventionally designed joint while most of the cracking occurred outside the steel fiber reinforced region execpt for a cracked that developed in the beam.
Jindal and Hassan (29) tested six two span continuous beam column specimens shown if Figures 11,12 and 13.Two were conventionally reinforced,two with steel fibers and a reduced number of stirrups (increased spacing) and two with fibers confined to the joint region.Round steel fibers 1.0x 0.01 inches (25 x 0.25 mm) in 2.0 percent concentration by weight of concrete which is equivalent to about 1.5 volume percentage.The static loading scheme is shown in Figure 14 which included an constant axial load of 40Kips (178 KN).The test results are summarized in Table 3.

Table 3. Jindal and Hassan Test Data

Specimen	Cycles of Loading	Load at Failure	Mode of Failure
A1	1	24 Kips (107 KN)	Shear
A2	4	17 kips (78.3 KN)	Shear
B1	1	24.2 Kips (107.6 KN)	Moment-shear
B2	5	28.6 Kips (127.2KN)	Shear
C1	5	26.4 Kips (117.4 KN)	Moment
C2	5	22.2 Kips (98.7 KN)	Shear

The steel fibers increased the load capacity ,stiffness and ductility of the beam-column joints although only one joint had a moment failure.

Craig,Mahadev,Patel,Viteri,and Kertesz (30) tested ten beam-column joints shown in Figure 15.Five the beam-column joints contained 1.5 volume percentage of bent end (hooked end)steel fibers 30 x 0.50 mm (1.18 x 0.02 in.) or 50 x 0.50 mm (1.97 x 0.02 in).The a/d ratio as well as the amount tension and compression flexural steel were varied as shown in Table 4. The cyclic loading shown in Figure 16 simulated an earthquake.The first three cycles are elastic, followed by groups of three cycles at each increment of the ductility factor until failure occurred.The ductility factor was defined as the ratio of the deflection at the load point for a particular load to the deflection at the load point for the yield load.Ductility factors up to 6 were obtained.Figures 17 and 18 show the load-rotaion at the face of the beam for specimens SP9 and SP10,joints with and without steel fibers respectively.In general the steel fiber joints showed increased shear capacity,greater stiffness,larger ductility factors,better confinement of the concrete and improved structural integrity compared to the conventionally reinforced joints.

Table 4 Craig et al Test Data

Specimen Number	Concrete Strength, PSI Lower Column	Upper Column	Beam	Joint	Column Loading (Kips)	a/d*	A_s in.^{2}	A'_s in.^{2}
SP1	4090	4308	4255	4255	80.16	4.337	0.88	0.40
SP2+	4090	4308	4255	5034	80.16	4.337	0.88	0.40
SP3	3545	4149	3642	3642	80.16	3.614	0.88	0.40
SP4+	3545	4149	3642	3855	80.16	3.614	0.88	0.40
SP5	4881	4303	5080	5080	80.16	2.891	0.88	0.40
SP6+	4881	4303	5080	5464	80.16	2.891	0.88	0.40
SP7	4759	4908	5057	5057	80.16	4.819	1.58	0.88
SP8+	4759	4908	5057	5266	80.16	4.819	1.58	0.88
SP9	4514	5322	5217	5217	80.16	4.337	0.88	0.88
SP10+	4514	5322	5217	5564	80.16	4.337	0.88	0.88

*a = distance from load to face of column

d = structural depth

+ = joint had fibers - 1.5% by volume of concrete

FATIGUE STRENGTH OF STEEL FIBER REINFORCED CONCRETE

The pre and post-cracking properties of steel fiber reinforced concrete for static flexural loadings suggested that steel fibers might improve the flexural fatigue strength of plain or reinforced concrete.
Batson,Ball,Bailey,Landers and Hooks (31) tested a series of beams 4 x 6 x 102 inches (100 x 150 x 2590mm) with 2.0 and 2.98

volume percentage of round steel fibers having an aspect ratio ranging from 75 to 90.No conventional flexural or shear reinforcement was used.The beams were loaded at the center of a 96 inch (2438 mm) simple span by a ram under load control in a closed-loop electro-hydraulic system at a cyclic rate of 3 Hz.Figure 19 is an S-N plot for complete reversal of loading of beams with 2.98 volume percentage of fibers with an aspect ratio of 75.The fatigue strength for 2.98 percent of fibers was 74 and 83 percent of the first crack flexural strength(end of the linear load deflection curve)for two million cycles of nonreversal and complete reversal of the loading respectively.The beams that failed before two millon cycles did so by bond failure and pull out of the fibers,not by breaking of the fibers.An unanticipated result was that the static flexural strength of the beams that did not fail after two million cycles of loading had a greater flexural strength than the virgin beams.
Kromeling,Reinharat and Shah (32) conducted a series of flexural static and fatigue tests on beams with conventional flexural reinforcement and fibers.Forty six test beams were fabricated using three types of fibers in volume percentages of 1.27 straight fibers,0.89 hooked end fibers,and 1.54 paddled end fibers and three flexural steel percentages of 0.17,0.75 and 2.09. Nineteen static and 27 constant amplitude fatigue tests at 3Hz were conducted on the beams with the loading configuration shown in Figure 20.The influence of the fibers was evaluated in terms of the parameter,pl/d,where p=volume percentage of fibers,l=length of fiber and d=equivalent diameter of the fiber. Tables 5 and 6 summarize the static and dynamic test data.

Table 5 Kromeling et al Static Test Data

Fiber Type	$\frac{pl}{d}$	2 φ 4 (ρ=0.17)	4 φ 6 (ρ=0.75)	4 φ 10 (ρ=2.09)
None	0	0.99	-	0.99
Hooked	69	1.17	1.12	1.01
Straight	76	1.26	1.30	1.01
Paddled	96	1.56	1.26	1.08
Calcuated ultimate moment, kN-mm (lb-ft)		2472 (1823)	7346 (5418)	17161 (12657)

ρ = reinforcement ratio percent

Table 6 Kromeling et al Dynamic Test Data

$\frac{pl}{d}$	Beam No.	P_{min},* kN	P_{max},* kN	Calculated σ_s,max, N/mm^2	Cycles to failure 10^3	Measured σ_s,max, N/mm^2
0	9	3.42	19.10	402	265	380
69	63	3.99	19.52	411	453	340
76	51	3.56	18.75	395	600	342
96	42	3.85	19.05	401	>1400	300

*Loads do not include dead weight
1 kN = 224.809 lb.
1 N/mm^2 = 145.038 psi

Increasing values of the parameter,pl/d,increased the fatigue life and static ultimate moment and decreased the deflection and crack width. However,the influence of the fibers decreased with increasing percentage of flexural steel.The separate effects of p,l and d could not be determined from the test data. Measured strain data also indicated that the fibers reduced the average tensile stress in the flexural steel which probably explains the increase in fatigue life of the beams with flexural steel and fibers shown in Table 6.
Currently low cyclic rate high amplitude test data is being evaluated for steel reinforced concrete beams at Clarkson University.Beams were 4 x 6 inches in cross-section (100 x 150 mm) and either 4 or 8 feet (1219 or 2438 mm) long with the reinforcement shown in Figure 21.Eleven beams of each reinforcement arrangement were fabricated;three were tested by complete reversal of the loading at 1 Hz and 0.2 Hz and the remaining by partial reversal of the loading at 1 Hz and 0.2 Hz. A ram controlled by a closed loop electrohydaulic system applied the load at mid-span.Special end supports were constructed that allowed for flexural rotation and axial shortening.A microcomputer controlled the loading system and served as a data acquisition system (33).Figures 22,23,24 and 25 show deflection versus time and load versus deflection for the short length beams with #4 (12 mm) top and bottom longitudinal flexural steel and combinations of steel fibers and stirrups for the four beams listed in Table 7.

Table 7 Shear Reinforcement and Loading for Beams in Figures 22,23,24,and 25

Beam	Reinforcement Stirrups	Fiber	LoadingCycles 9000 lbs @ 1.0 Hz (40.0 KN @ 1.0 Hz)
#1	no	0.75%	15
#10	no	1.50%	50
#33	yes	0.0%	19
#42	yes	0.75%	15

Figures 26,27,29 and 30 show the corresponding computed deflection versus time and load versus defelection using system identification techniques to optimize the model parameters that best define the physical characteristics of the various beam types.The computed curves are based on a single degree of freedom system with varying stiffness using a three parameter model.The varying stiffness is assumed to be a function of the initial stiffness and the number of cycles of loading. During the loading phase

$$K = K_o(1.0- C(i/N)^E) \qquad Eq(10)$$

and during the unloading phase

$$K = K_o \qquad Eq(11)$$

where

K = current stiffness
K_o = initial stiffness
C = coefficient that controls total amount of degradation
E = exponent that controls the rate of degradation
i = current cycle number
N = total number of cycles

System identification techniques are used to determine the values of K_o ,C and E that provide the best agreement between the actual response and the modeled response.The modeled system response is computed using step-by-step numerical integration technique with K from Eq (10) and (11).Table 8 lists the results for the system identification technique applied to the four beams in Table7.

Table 8 Model Parameters from System Indentification

Beam	K_O lbs/in	C	E
#1	$7.11x10^4$	0.088	0.139
#10	$6.36x10^4$	0.395	0.475
#33	$6.72x10^4$	0.885	0.101
#42	$4.89x10^4$	0.778	0.226

The response of the beams is predicted fairly accurately,but the end result is to determine if the constants C and E have particular values for each shear reinfocement combination and loading condition.The model used for the stiffness degradation may not be appropriate and a different model considered.

SUMMARY

The inclusion of steel fibers in concrete provides a pratical method for shear reinforcement of beams subjected to flexural loads, combined bending ,torsion and/or shear loads, and for beam-column joints subjected to fatigue or dynamic loads.Equations that predict the response of steel fiber reinforced concrete beams subjected to various loadings have been developed for a limited range of steel fiber parameters.However,several questions must be resolved before steel fibers for shear reinforcement can be adopted by building code authorities.

The geometric shape of the fiber appears to be an important factor, because data indicate that some fibers,such as the bent and crimped fibers,may be twice as effective as straight fibers for the same volume percentage.The size of the beam also appears to be a factor.The size effect for conventional reinforced beams without shear reinforcement has been demonstrated by Kani(34)and shown in Figure 30.Hawkins (3) has suggested that the size effect may be quantified based on nonlinear elastic fracture mechanics as suggested by the research of Bazant (35).Can this approach be applied to SFRC?

Equations 2 and 3 proposed to predict the shear capacity of steel fiber reinforced concrete beams are modifications of the those in ACI 318-83 ,but more data are needed to establish an upper limit on the capacity.Also, is the modulus of rupture the appropriate

strength test to characterize the effect of fiber volume in predicting the shear capacity of a SFRC beam?
Another general question to be answered is how do steel fibers improve the shear capacity of reinforced concrete beams without stirrups.The truss analogy does not appear to be appropriate since there are no stirrups or less the minmum required.One possibility is the tied arch concept proposed by Kani (38) with the flexural steel acting as the tie and the fibers providing for transfers of tension and compression forces between arches as shown in Figure 30. Another approach would be to consider the upper and lower bound shear capacity of beams without shear reinforcement based on limit analysis and plasticity theory of concrete (36).
Most of the data for combined bending and torsion showed increased shear capacity,stiffness and ductility compared to concrete beams with conventional reinforcement as required by current codes.Equations for predicting the torsional capacity were developed and some indication of the importance of the size,shape,volume percentage and bonding of the fibers established.
Limited data for beam-column joints show improved shear capacity and much greater ductility when steel fibers are incorporated in the joint compared with conventional reinforcement required for seismic design. However, analytical methods are needed to predict reliably the response SFRC to loads and to design steel fiber reinforced joints.For instance more research is needed to determine the limits on substituting steel fibers for hoops that confine the longitudinal steel in the beam-column joints.
Limited test data for reversal and nonreversal of fatigue loading and for low cyclic high amplitude loading show that steel fibers by themsevles or in conjunction with conventional reinforcement in concrete increases the fatigue life,fatigue strength and ductility. More research is needed to establish the contribution of fiber size,shape,and bonding characteristics for predicting the flexural fatigue strength.Specimen size may be very important with respect to the fatigue strength and needs to be investigated.

CONCLUSIONS

Laboratory test data show that small volume percentages,less than 2.5 percent,of steel fibers increases the shear capacity,ductility, stiffness,fatigue strength and life of concrete beams and beam-column joints.Before code writing authorities will allow the use of steel fibers as shear reinforcement, design procedures and equations that reliably predict the structural response to static and dynamic loads must be developed beyond the present state-of-the-art.

ACKNOWLEDGEMENTS

The author would like to acknowledge the National Science Foundation grant to Clarkson University for the low cycle high a mplitude research on the SFRC beams,my colleague Dr.Levon Minnetyan,and the graduate students,particularly Mr Thomas Glebas for his analytical and experimental contributions.Dr.R.John Craig,Department of Civil and Environmental Engineering ,New Jersey Institute of Technology contributed unpublished data and a draft manuscript for a textbook on SFRC.

REFERENCES

(1) American Concrete Institute,"Building Code Requirements for Reinforced Concrete ACI 318-83",American Concrete Instititute,Detriot,Micigan,1983.

(2) British Standards Institute," Code of Practice for the Structural Use of Concrete",CP110:Part 1:1972,British Standards Institute,London,1972.

(3) Hawkins,N.M.,"The Role for Fracture Mechanics in Conventional Reinforce Concrete Design",Applications of Fracture Mechanics to Cementitious Composites, S.P.Shah,Editor,NATO Advanced Research Workshop, Northwestern University,September 1984.

(4) Batson G.,Jenkins E.,Spatney R.,"Steel Fibers as Shear Reinforcement in Beams".ACI Journal.,Proceedings V.69,No.10,Oct.1972,pg. 640.

(5) Zutty,T.C.,"Beam Shear Strength Prediciton by Analysis of Existing Data",J.American Concrete Institute,Proceedings,V.65,No.11, Nov.1968,pg.943.

(6) Paul,S. and Sinnamon,G."Concrete Tunnel Liners Structural Testing of Segmented Liners",Department of Transportation RFA Final Report,FRA-O&RD-75-93,Aug.1975,Washington,D.C.

(7) Kobayashi K.,"Current Research and Development on Steel Fiber Reinforced Concrete in Japan",Transactions ofJSCM,V.3 ,No.1/2,Dec.1977.

(8) Mosstafa S.E.,"Use of Fibrous Concrete Shear Reinforcemenmt in T-Beam Webs",Steel Fibrous Concrete Short Course,Joint

Center for Graduate Study,Richmond,Washington,July,1974.

(9) Jindal R.L.,"Shear and Moment Capacities of Steel Fiber Reinforced Concrete Beams",Fiber Reinforced Concrete International Symposium,SP-81,American Concrete Institute,Detriot,Michigan,1984,pg.1.

(10) Williamson,G.R.,"Steel Fibers as Web Reinfocement in Reinforced Concrete",Proceedings,U.S.Army Science Conference,V.3,June 1978,West Point,New York.

(11) Bollana R.D.,"Steel Fibers as Shear Reinforcement in Two Span Continuous Reinforced Concrete Beams",M.S. Thesis,Civil and EnvironmentalEngineering,Clarkson University,Potsdam,N.Y.May 1980.

(12) Craig R.J.,unpublished manuscript for textbook of fiber reinforced reinforced concrete,New Jersey Institute of Technology,Newark,New Jersey,1984.

(13) Swamy R.N. and Bahia H.M.,"The Effectiveness of Steel Fibers As Shear Reinforcement",Concrete International,V.7,No.3,March,1985,pg.35.

(14) Lafraugh R. W. and Moustafa S.E.,"Experimental Investigation of the Use of Steel Fibers as Shear Reinforcement",Techical Report, Concrete Technology Associates,Tacoma,Washington,Jan.1975,pg.52.

(15) Muhidin N.A. and Regan P.E.,"Chopped Steel Fibers as Shear Reinforcementin Concrete Beams",Proceedings,Conference on Fiber Reinforced Materials:Design and Engineering Applications,Institute of Civil Engineers,London,1977,pg.149.

(16) Narayanan R. and Toorani-Goloosalar Z.,Fibre Reinforced Concrete in Pure Torsion and in Combined Bending and Torsion",Proccedings of the Institute of Civil Engineers,part 2,1979,67,Dec.,pg.987.

(17) Narayanan R.and Green K.R.,"Fiber Reinforced Concrete Beams in Combined Bending and Torsion"Indian Concrete Journal,V.5,No.8,Aug.1981,pg.222.

(18) Narayanan,R.and Kareem-Palanjian,A.S.,"Steel Fibre Reinforced Concrete Beams in Torsion',The Int.Journal of Cement Composite and Lightweight Concrete,V.5,No.4,Nov.1983,pg.235.

(19) Mansur,M.A. and Paramasivam,P.,"Steel Fibre Reinforced

Concrete Beams in Pure Torsion",The Int.Journal of Cement Composites and Lightweirht Concrete,V.4,No.1,Feb.1982,pg.39.

(20) Mangat,P.S.,"Tensile Strength of Steel Fibre Reinforced Concrete",Cement and Concrete Research,v.6,No.2,March 1976,pg.245.

(21) Mansur,M.A.,"Bending-Torsion Interaction for Concrete Beams Reinforced with Steel Fibers',Magazine of Concrete Research,V.34,No.121,Dec.1982,pg.182.

(22) Hsu,T.T.C.,"Torsion of Structural Plain Concrete Rectangular Sections",American Concrete Instutute Special Publication SP18,1968,pg.203.

(23) Swamy,R.N. and Mangat,P.T.,"Flexural Strength of Steel Fibre Reinforcrd Concrte",Proceedings of the Institute of Civil Engineers,London,v.57,part 2,Dec.1974,pg.701.

(24) Mansur,M.A. and Paramasivam,P.,"Fiber Reinforced Concrete Beams in Torsion,Bending and Shear"Journal of the American Concrete Institute,Proceedings V.82,No.1,Jan.-Feb.1985,pg.33.

(25) Batson,G.,Terry,T.and Chang,M."Fiber Reinforced Concrete Beams Subjected to Combined Bending and Torsion',Fiber Reinforced Concrete Symposium,SP81,American Concrete Institute,Detriot,Michigan,1984,pg.51.

(26) Lim,E.L.,"Steel Fibner Reinforced Concrete Breams Subjected to Combined Bending and Torsion",M.S.Thesis,Clarkson University,Potsdam,New York,Aug.1983.

(27) Martin,L.H.,"Torsion and Bending in Longitudinally Reinforced Concrete Beams",Building Science,V.8,1973,pg.339.

(28) Henager,C.H.,"Steel Fibrous,Ductile Concrete Joint For Seismic Resistant Structures",Reinforced Concrete Structures in Seismic Zones,SP 53,American Concrete Institute,Detroit,Michigan,1977,pg.371.

(29) Jindal,R.L. and Hassan,K.A.,"Behavior of Steel Fiber Reinforced Concrete Beam-Column Connections",Fiber Reinforced Concrete Interenational Symposium,SP-81,American Concrete Institute,Detriot,Michigan,1984,pg.107.

(30) Craig,R.J.,Mahadev,S.,Patel,C.C.,Viteri,M.and Kertesz,C.,"Behavior of Joints Using Reinforced Fibrous Concrete", International Symposium,SP-81,American Concrete Institute,Detroit,Michigan,1984,pg.69.

(31) Batson,G.,Ball,C.,Bailey,L.,Landers,E.,and Hooks,J.,"Flexural Fatigue Strength of Steel Fiber Reinforced Concrete Beams",J.of the American Concrete Institute,Proceedings V.69,No.11,Nov.1972,pg.673.

(32) Kormeling,H.A.,Reinhardt,H.W. and Shah,P.S.,"Static and Fatigue Properties of Concrete Beams Reinforced with Continuous Bars and with Fibers",J.of the American Concrete Institute,Proceedings,V.77,No.1,Jan-Feb.1980.pg.36.

(33) Glebas,T.,Minnetyan,L. and Batson,G.,"Computer Based Testing of Dynamic Response",Proceedings of the Fourth Engineering Mechanics Division Specialty Conference,American Society of Civil Engineers,1983,Vol.1,pg.589.

(34) Kani,G.N.J.,"How Safe Are Our Large Reinforced Concrete Beams?",J.of the American Concrete Institute,Proceedings,V.64,No.3,Mar.1967,pg.128.

(35) Bazant,Z.P.,"Size Effect in Blunt Fracture:Concrete,Rock,Metal",Engineering Mechanics Journal,American Society of Civil Engineers,V.110.No.4,Apr.1984,pg.518.

(36) Kani,G.N.J.,"A Rational Theory for the Function of Web Reinforcement",J.of the American Concrete Institute,Proccedings,V.66,No.3,Mar.1969,pg.185.

(37) Chen,W.F.,PlasticityinReinforcedConcrete,McGraw Hill,1982.

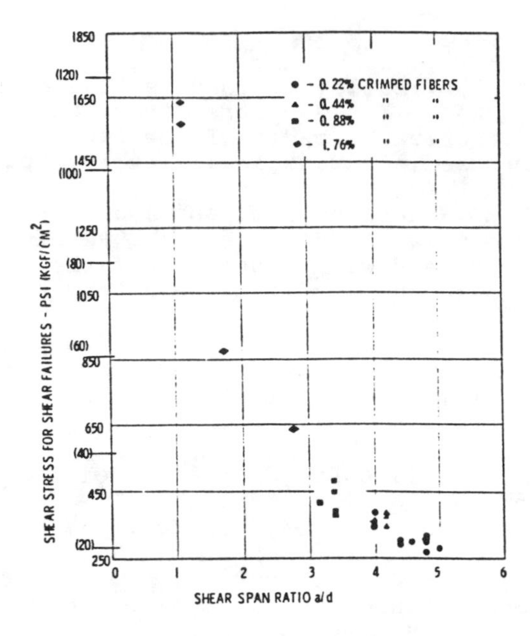

FIG 1. Shear stress versus a/d ratio for steel fiber reinforced beams.Ref.(4)

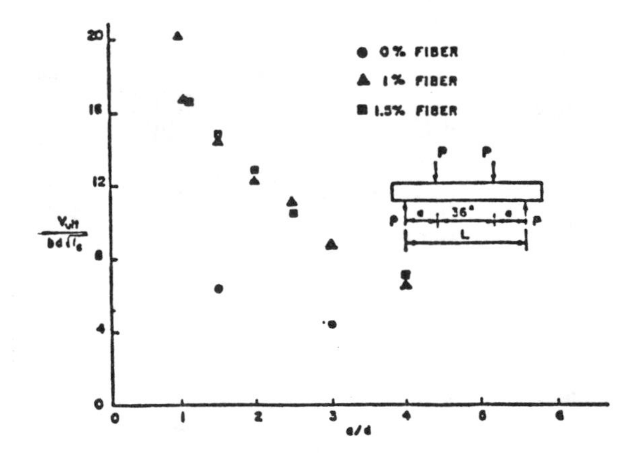

FIG 2. Shear stress versus a/d ratio for reinforced beams with steel fibers. Ref.(12)

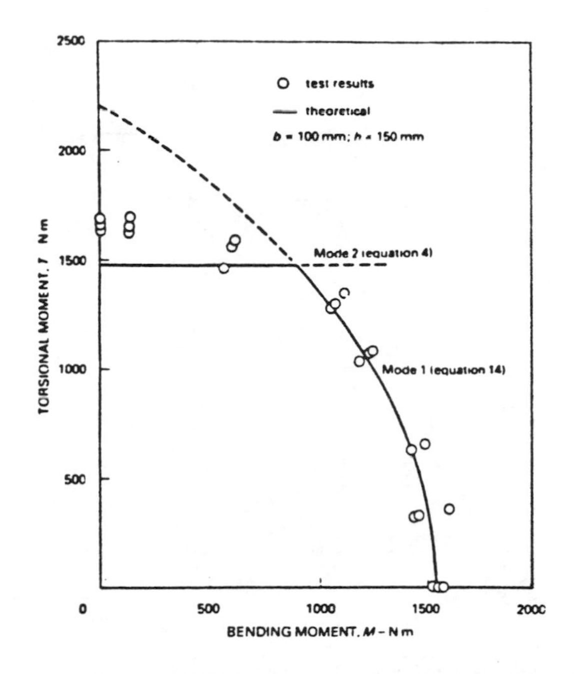

FIG 3. Combined bending and torsion interaction Ref. (21)

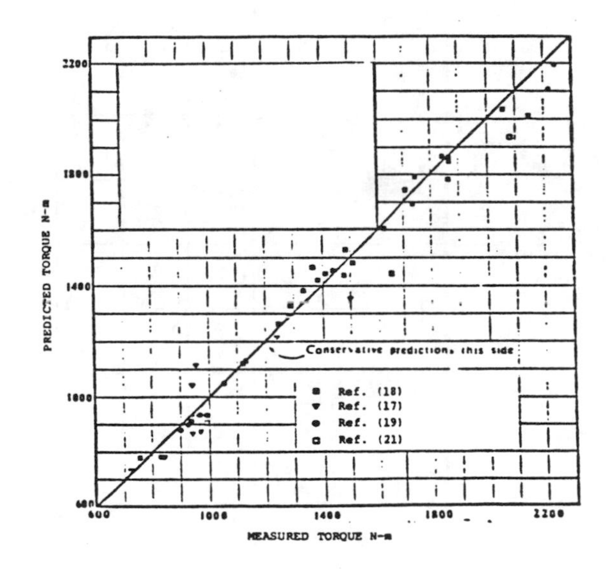

FIG 4. Predicted versus measured torsion. Ref. (18)

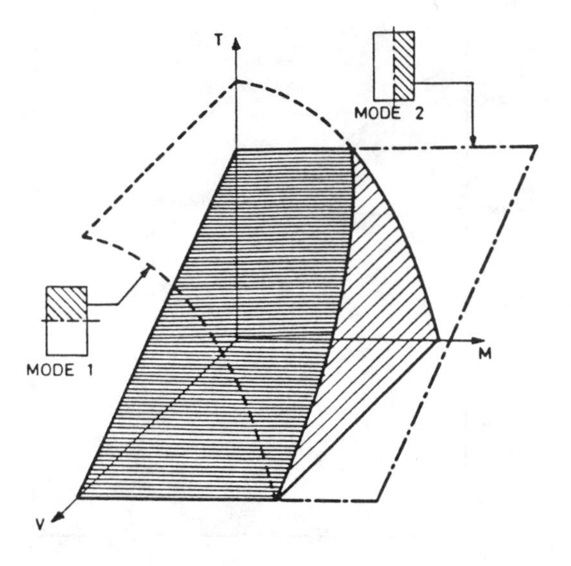

FIG 5. Interaction of bending,torsion and shear.
Ref. (19)

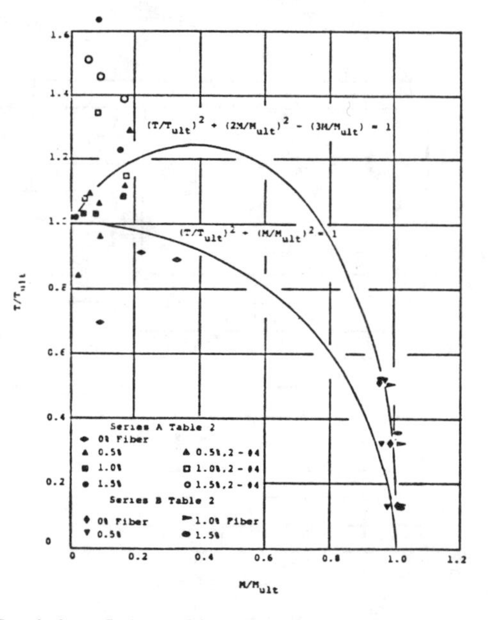

FIG 6. Combined bending and torsion interaction.
Ref. (25)

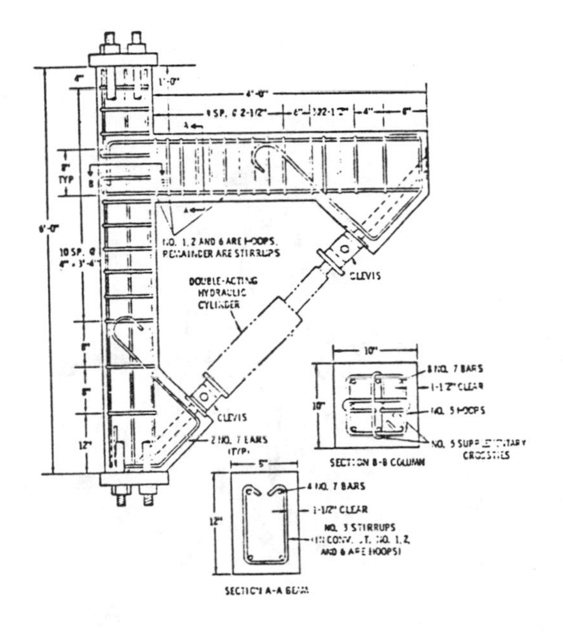

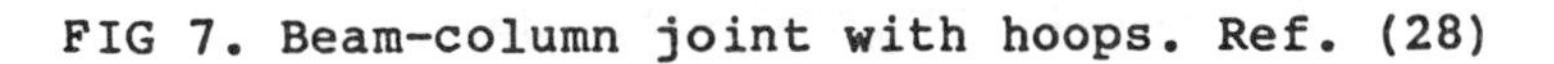

FIG 7. Beam-column joint with hoops. Ref. (28)

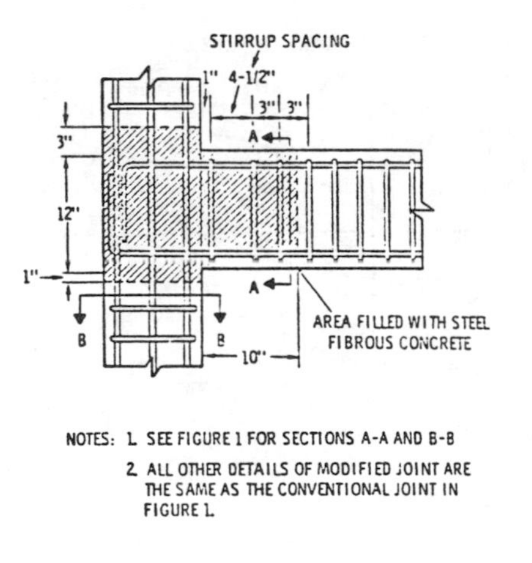

FIG 8. Beam-column joint with steel fibers. Ref. (28)

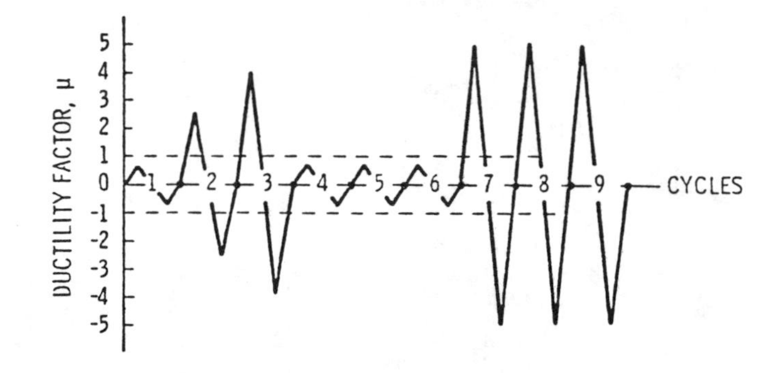

FIG 9. Cyclic loading. Ref. (28)

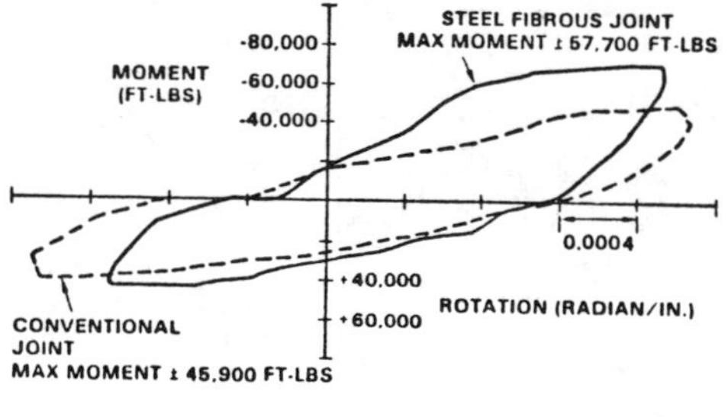

FIG 10. Moment-rotation for beam-column joint with steel fibers. Ref. (28)

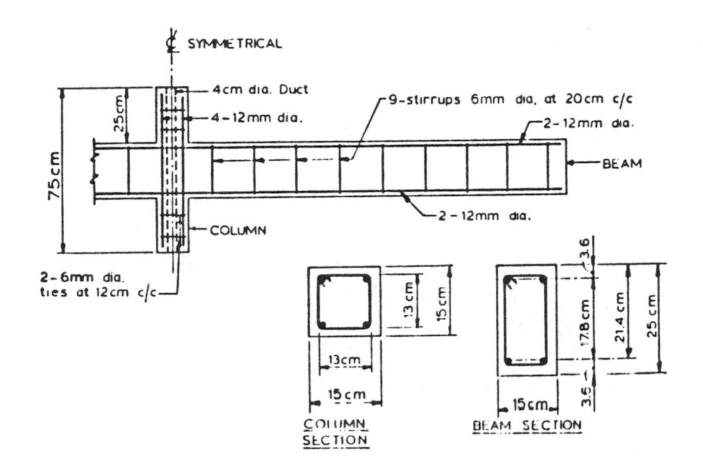

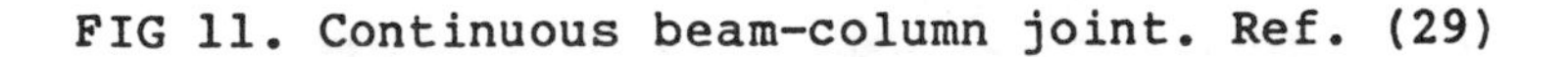

FIG 11. Continuous beam-column joint. Ref. (29)

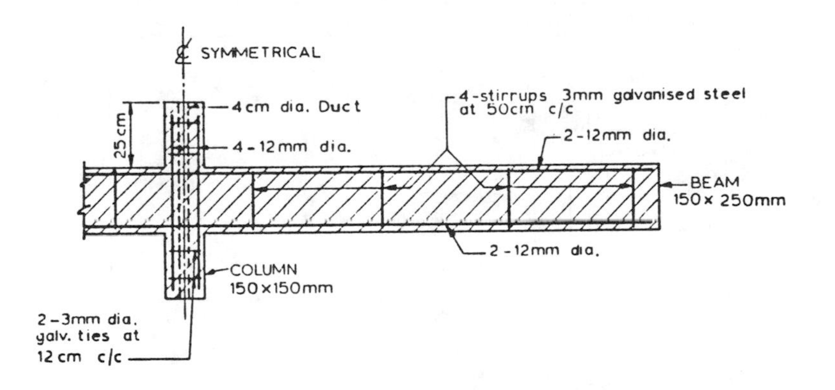

FIG 12. Continuous beam-column joint with steel fibers. Ref. (29)

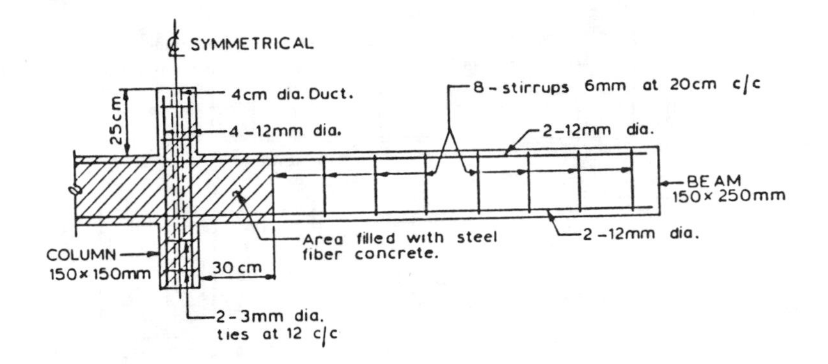

FIG 13. Continuous beam-column with steel fibers in joint region. Ref (29)

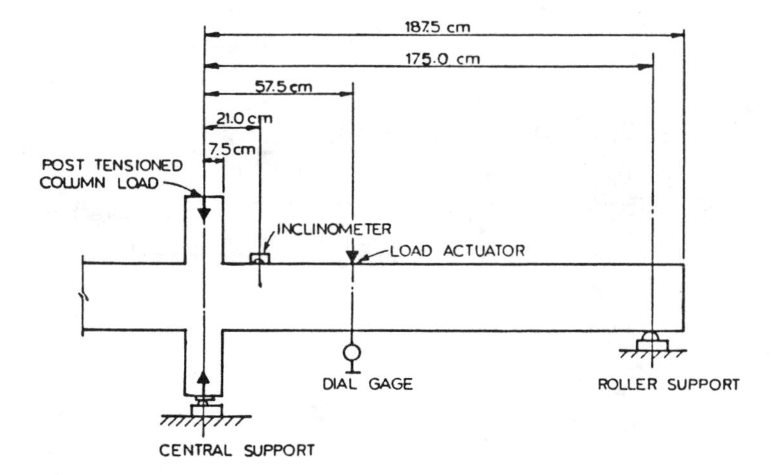

FIG 14. Loading method for continuous beam-column Ref. (29)

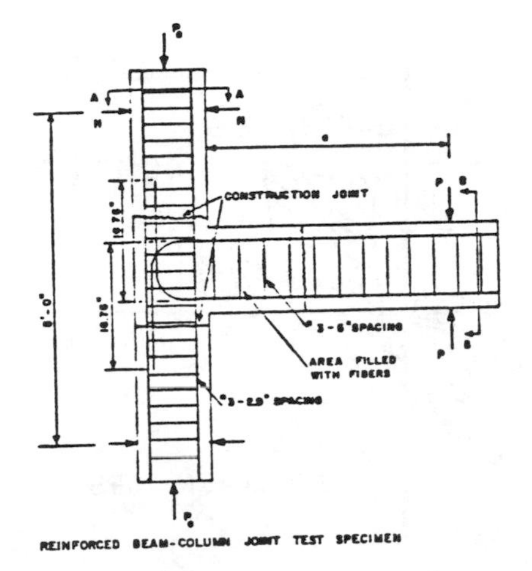

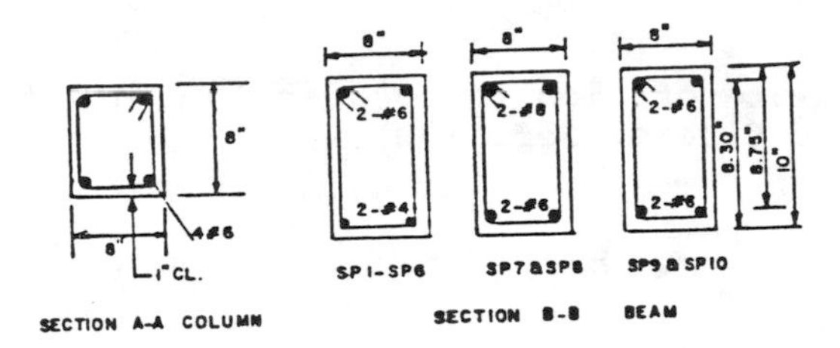

FIG 15. Beam-column joint with steel fibers. Ref. (30)

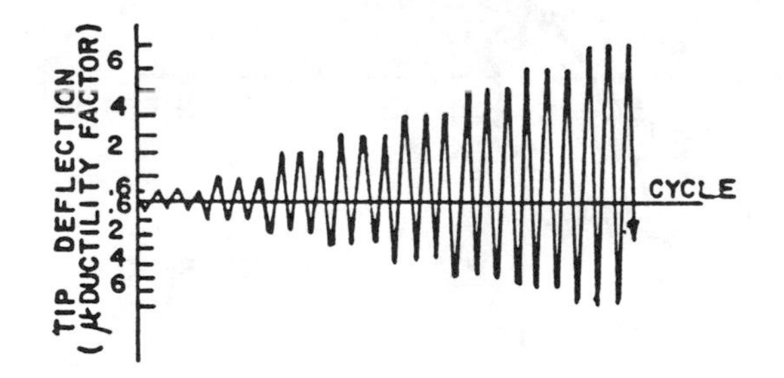

FIG 16. Beam-column loading. Ref. (30)

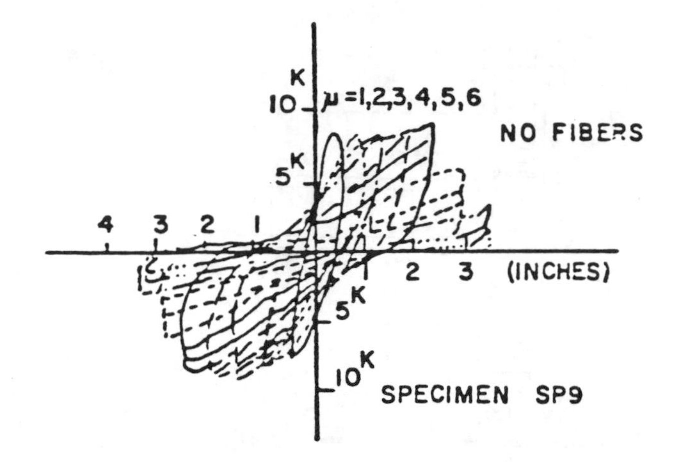

FIG 17. Moment-rotation for beam-column joint without fibers. Ref. (30)

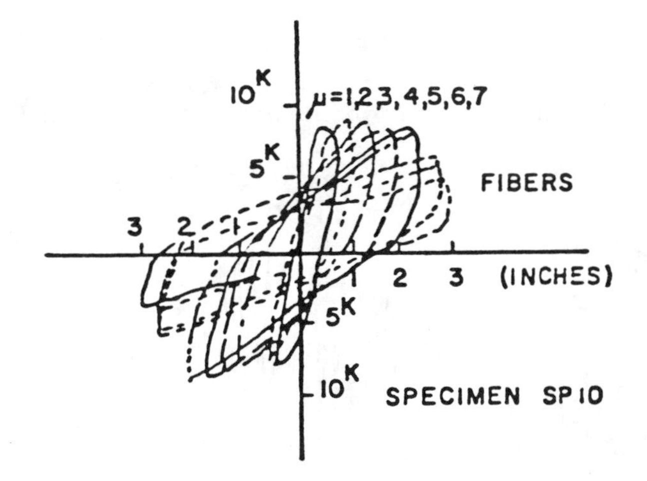

FIG 18. Moment-rotation for beam-column joint with fibers. Ref. (30)

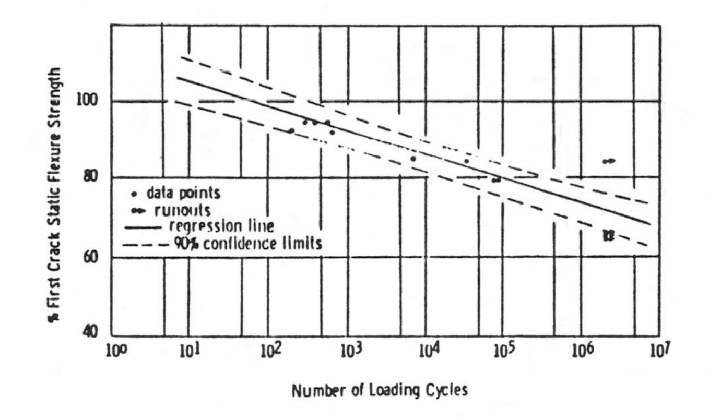

FIG 19. Fatigue life for complete reversal of loading. Ref. (31)

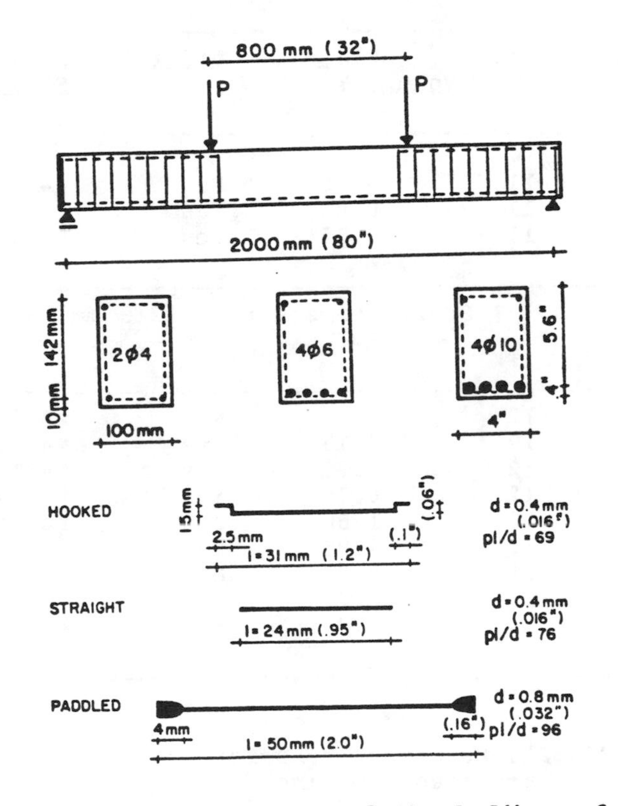

FIG 20. Beams and types of steel fibers for fatigue tests. Ref. (32)

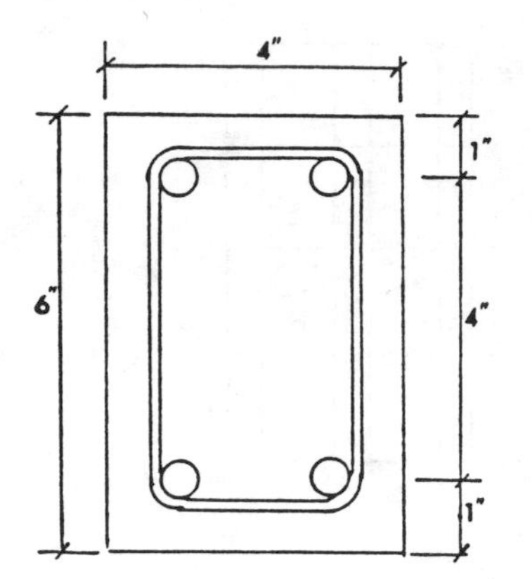

ASTM A615 grade 40 steel
Stirrups #2 @ 2.5 in
(3 mm @ 63.5 mm)

Long Beams 8 ft. (2438 mm)

Type	No.	Fiber Volume %	Stirrups	Flexural Steel
A	11	0.75	No	#4 T&B
B	11	0.0	Yes	#4 T&B
C	11	0.75	Yes	#4 T&B

Short Beams 4 ft. (1219 mm)

Type	No.	Fiber Volume %	Stirrups	Flexural Steel
A	12	0.0	No	#4 T&B
B	12	0.75	Yes	#4 T&B
C	12	1.25	Yes	#4 T&B
D	12	0.75	Yes	#4 T&B
E	10	1.50	Yes	#4 T&B
F	12	1.25	Yes	#4 T&B
G	11	0.75	No	#3 T&B
H	11	0.75	Yes	#3 T&B
I	11	0.0	Yes	#3 T&B

FIG 21. Beam specimens for low cycle high amplitude loading.

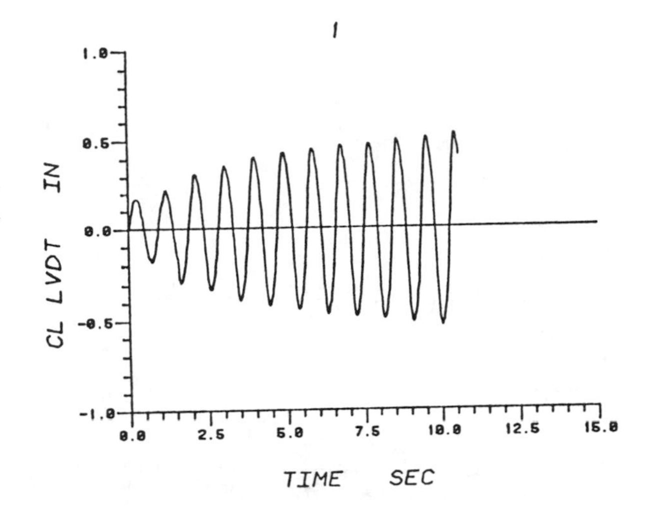

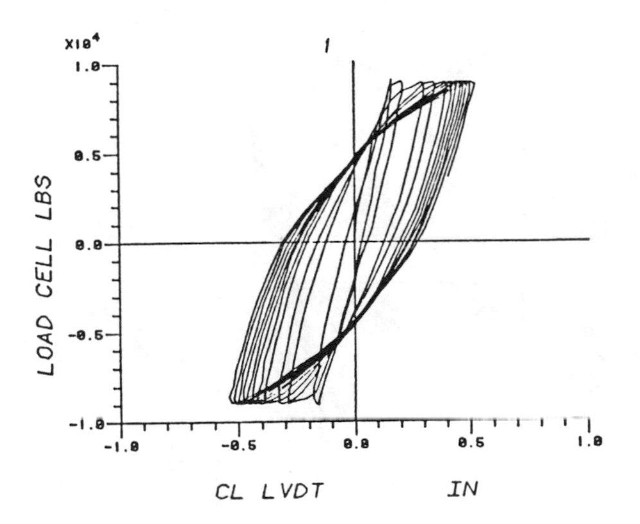

FIG 22. Experimental displacement-time and load-displacement for beam 1 in Table 7.

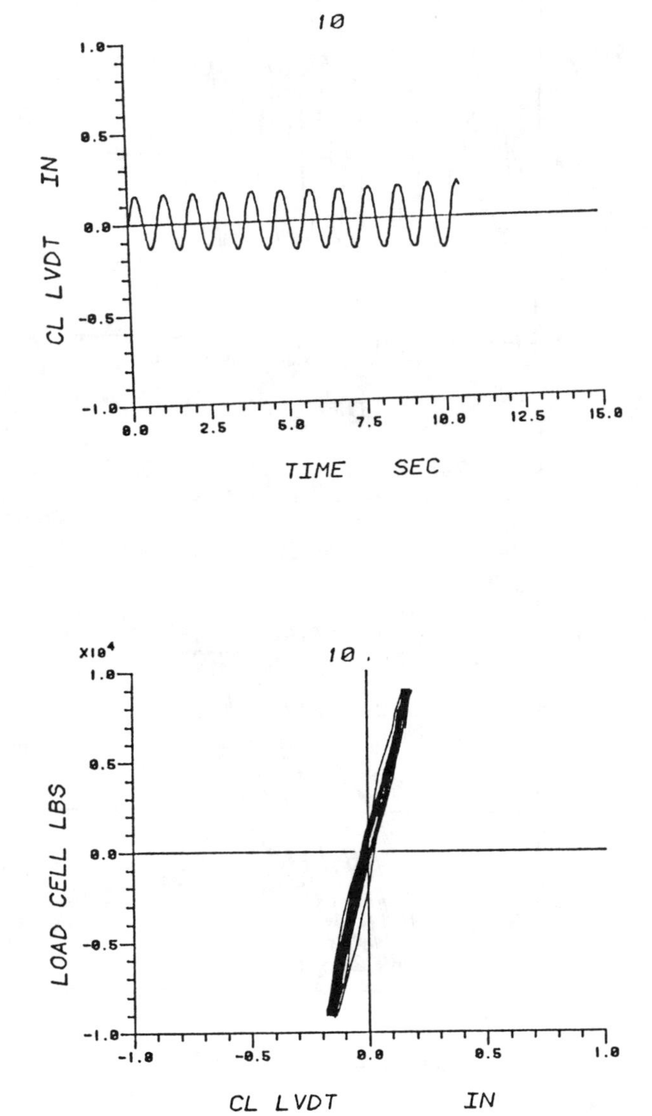

FIG 23. Experimental displacement-time and load-dispalcement for beam 10 in Table 7.

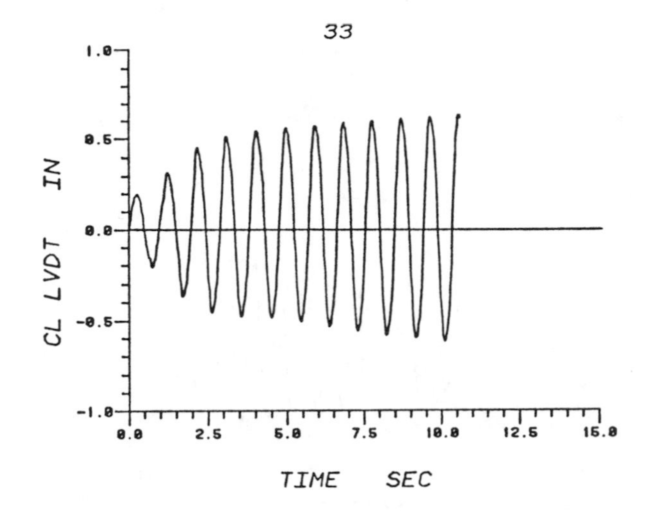

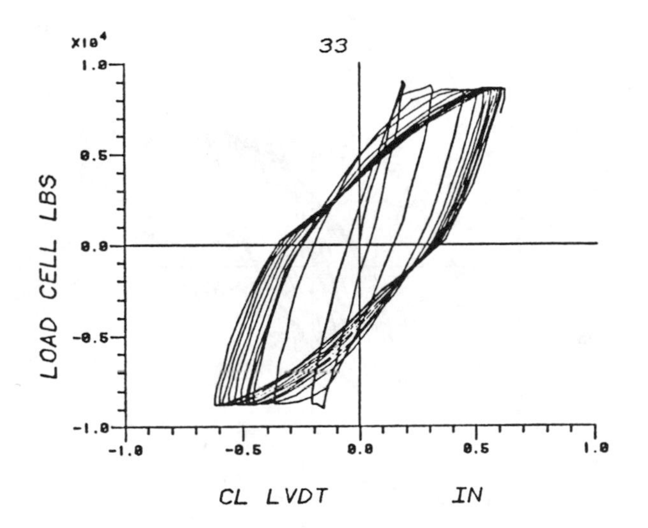

FIG 24. Experimental displacement-time and load-displacement for beam 33 in Table 7.

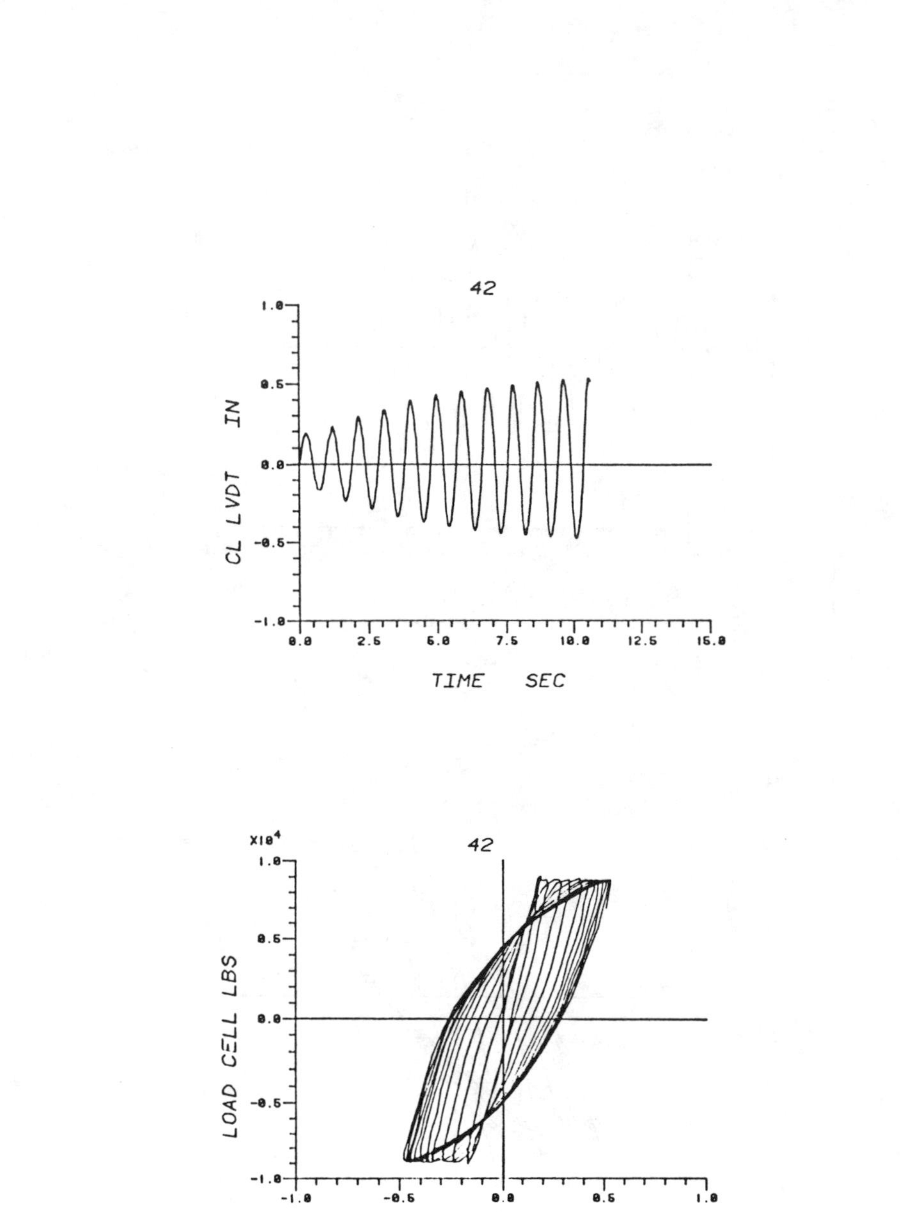

FIG 25. Experimental displacement-time and load-displacement for beam 42 in Table 7.

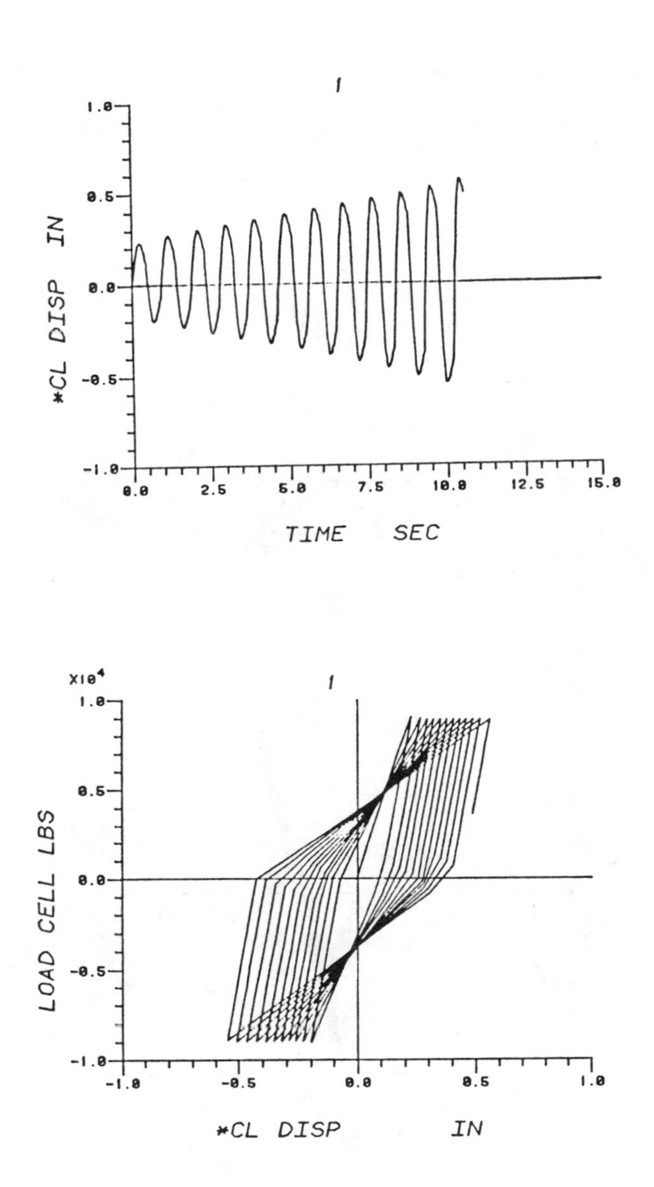

FIG 26. Computed displacement-time and load-displacement for beam 1 in Table 7.

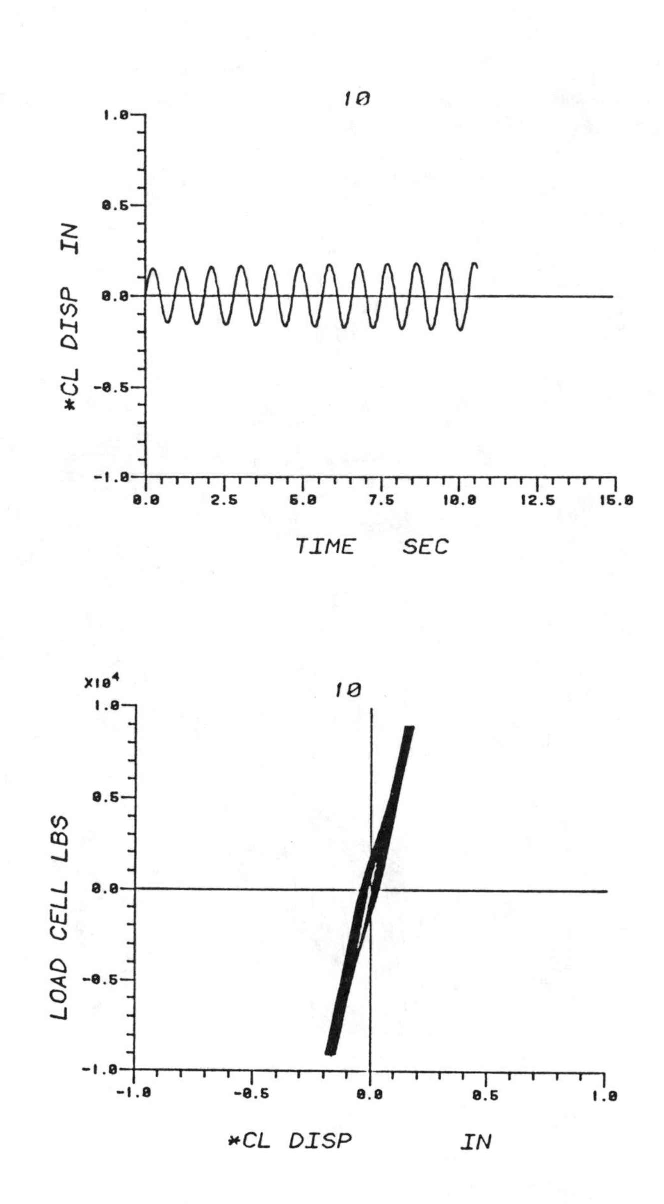

FIG 27. Computed displacement-time and load-displacement for beam 10 in Table 7.

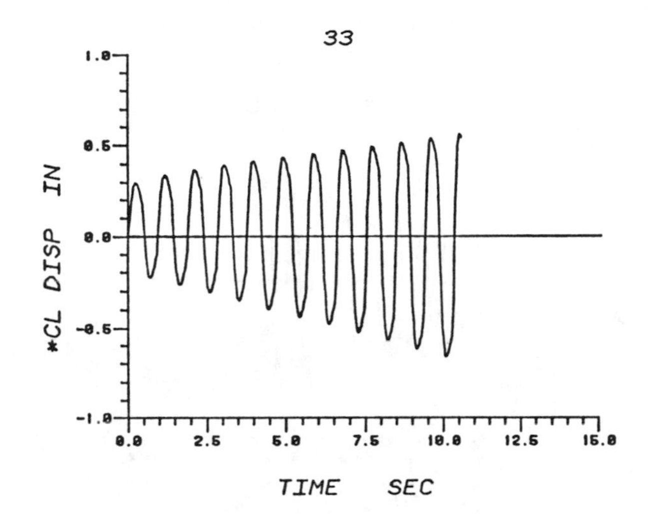

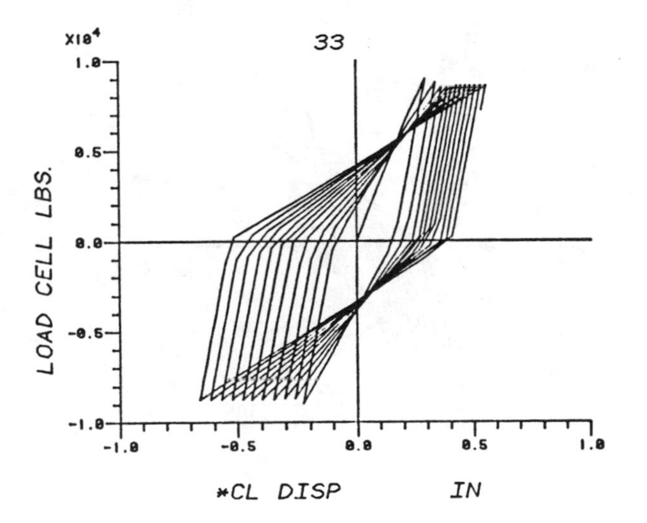

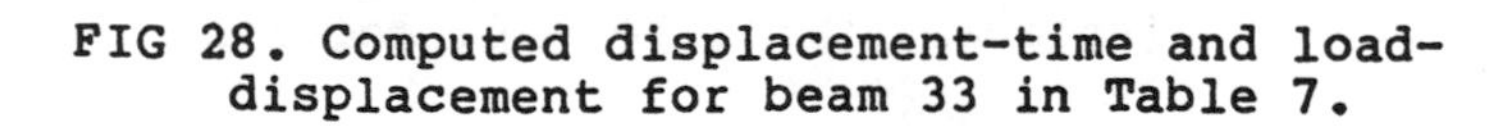
FIG 28. Computed displacement-time and load-displacement for beam 33 in Table 7.

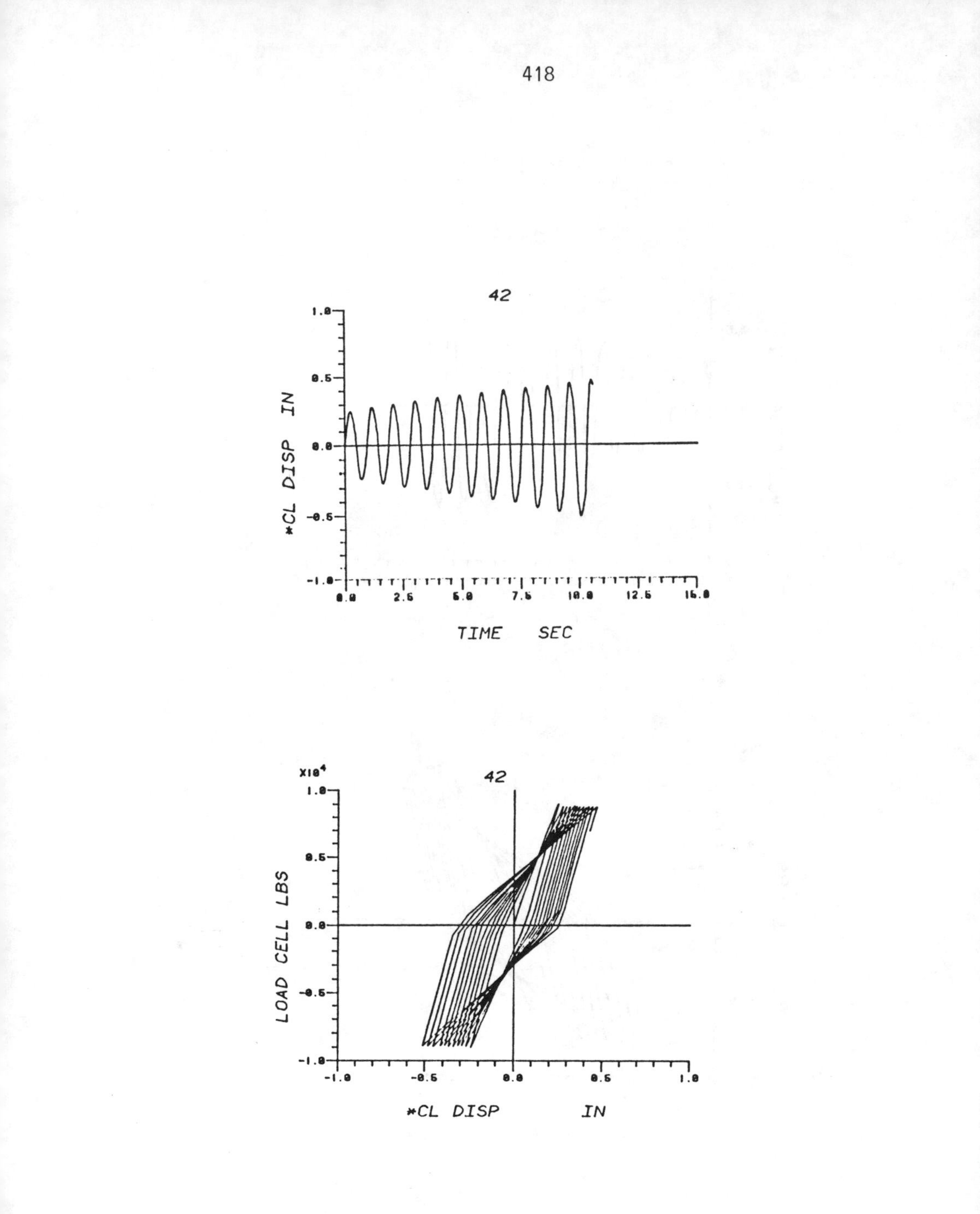

FIG 29. Computed displacement-time and load-displacement for beam 42 in Table 7.

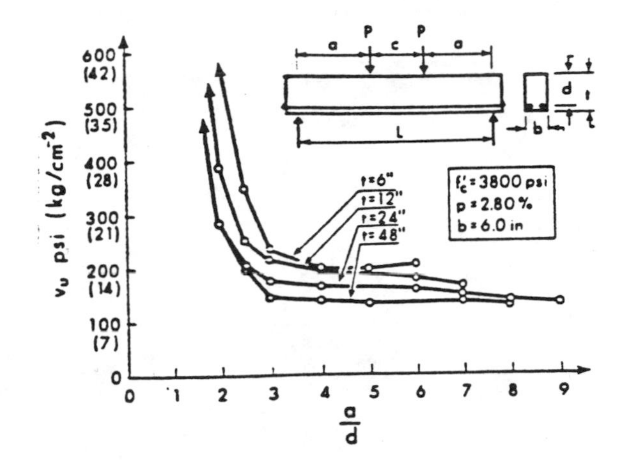

FIG 30. Size effect of large beams without shear reinforcement. Ref. (34)

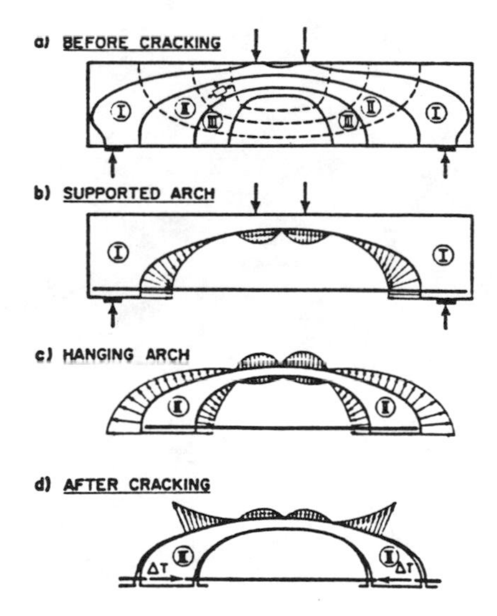

FIG 31. Internal arch concept of Kani. Ref. (36)

STEEL FIBER CONCRETE
US-SWEDEN joint seminar
(NSF-STU)
Stockholm 3-5 June, 1985
S P Shah and Å Skarendahl,
Editors

STEEL FIBRE CONCRETE FOR STRUCTURAL ELEMENTS

by Mogens Lorentsen, Prof, the Royal Institute of Technology, Stockholm, Sweden

---

ABSTRACT

The research programme described in this paper has been run at the Royal Institute of Technology, Dep. f. Bridge Building and Structural Engineering. The programme concentrates on the question on the use of steel fibre reinforced concrete (SFRC) for load-bearing structural elements.

In particular the aim has been to study the prerequisites for an introduction of SFRC on the Swedish market as to necessary standard tests and design methods as related to normal reinforced and prestressed concrete.

It is concluded that SFRC is most advantageous as shear reinforcement and should be used in combination with normal reinforcement - prestressed or nonprestressed - as bending reinforcement.

Mogens Lorentsen is professor of Structural Engineering at the Royal Institute of Technology in Stockholm since 1972. His research programme comprises concrete structures and reliability. He has taken part in national and international code work on concrete structures and is past chairman of the Swedish Concrete Society. He is also active as a consulting engineer.

---

1 INTRODUCTION

The research described in this paper has concentrated on the question how to use steel fibre reinforced concrete (SFRC) for load-bearing structural elements. In particular the aim has been to study how SFRC compares with normal reinforced concrete and the prerequisites of an introduction on the Swedish market, which has shown reluctance towards the use of SFRC for load-bearing structures e.g. beams.

On the basis of existing knowledge at the start of the research programme it seemed clear, that the most interesting property of SFRC was its capacity to increase the shear resistance of beams as compared with ordinary reinforced concrete. On the other hand it was also clear, that the strength properties are susceptible to variations in production method.

For SFRC to be attractive to the designing engineer and to authorities it was also considered necessary to devise methods for material testing and component design that conform as much as possible with existing methods for normal reinforced or prestressed concrete.

In addition to this it was easy to conclude, that the cost of fibres is so high, that it cannot compete with normal bar reinforcement used in bending.

Thus, due to the above considerations this study concentrates on beams with its main reinforcement in the form of bars and with the fibres acting as secondary reinforcement - mostly to improve the shear capacity.

The research programme has comprised mixing tests, moulding tests, compaction tests, strength tests on standard test specimens, strength tests on rectangular beams, I-shaped beams and square panels and bond tests, Jahlenius Å (1). All in all some 500 specimens of different shape have been tested. Cf fig.s 1.1, 4.4, 4.5.

The cube strength has varied between 30 and 60 MPa. Five fiber-types were used some shown in fig. 1.2. End ancored fibres were used in most tests (Bekaert 0,4x40 mm).

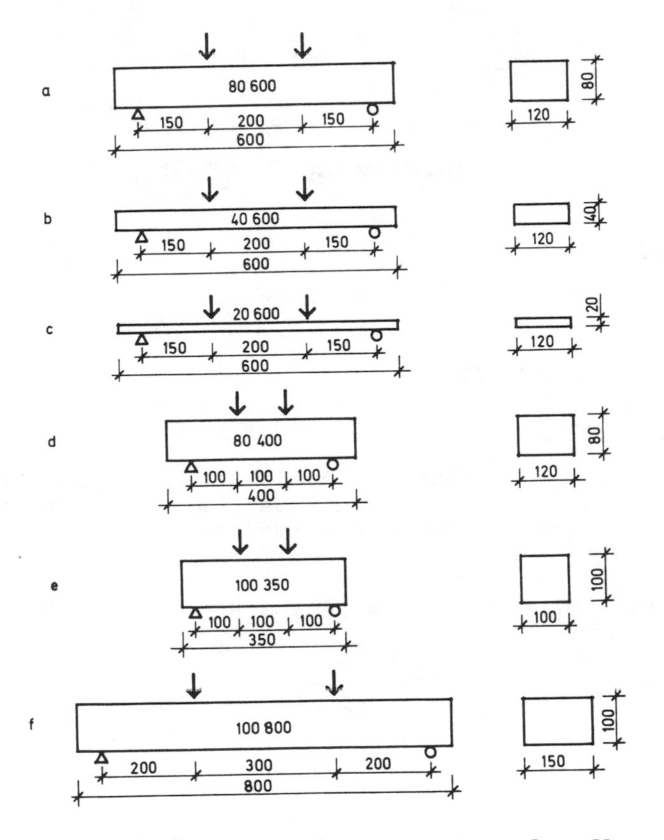

FIG 1.1 Dimensions and test setups for flexure beams.

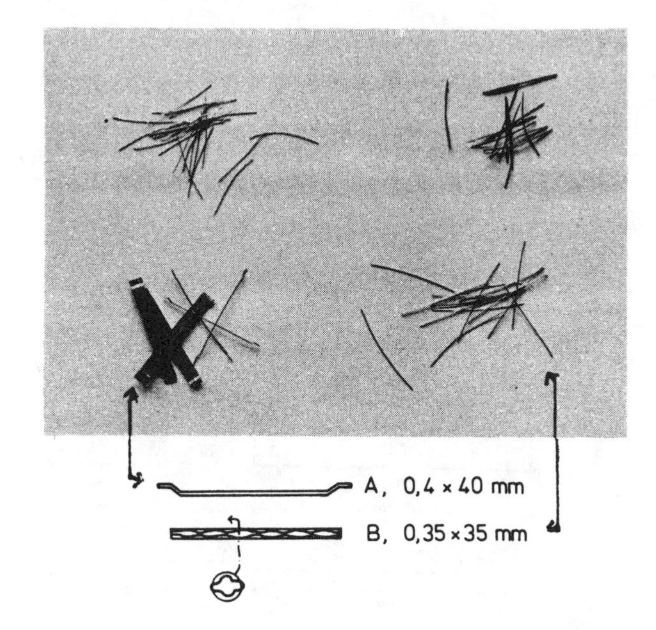

FIG 1.2 Some fibretypes.

2 PRODUCTION METHOD

2.1 Concrete mix

The concrete mix contained 350-600 kg/m^3 cement and maximum gravel size was 8 mm to avoid fiber balls. Mixing tests with 1,5 vol.% fibers showed that 16 mm max. gravel size gave rise to fiber balls.

The majority of the test specimens had the following mix

Portland cement	600 kg/m^3
Gravel 0-8 mm	1500 kg/m^3
Water	240 kg/m^3
Plasticizer	10 kg/m^3

2.2 Fibres

The fibres used appear in fig. 1.2. In most test specimens type a was used (Bekaert 0,4x40 mm).

2.3 Mixing procedure

The ingredients were normally loaded into the mixer in the following order:

1 Gravel
2 Cement
3 Water
4 Plasticizer

- Mixing to a flowing consistency
5 Fibres (strewed in)
Mixing

Total mixing time about 3 minutes.

When "glued-together" fibres were used the mixing time was increased to 5 minutes.

The use of plasticizer is essential.

2.4 Moulding and compaction

The concrete was placed in the mould using a pitchfork, cf fig. 2.1. Compaction was performed using vibration table, vibrating the form. As will be seen from fig. 2.2 the compaction time needed for SFRC is substantially longer than for ordinary concrete. The diagram is valid for 1,5 % Bekaert fiber 0,4x40 mm.

In a series of tests using different procedures for moulding it was found that the strength was highly dependent on the production procedure. Fig. 2.3 summarizes the findings. They are important since the show that the method of production is very important for the properties of SFRC-components.

In practice it is necessary to use very consistent production methods, for instance extrusion under good control, to achieve a reliable strength gain. Tests using extrusion under industrial conditions are under way at Strängbetong, Sweden in cooperation with my institute.

FIG 2.1 Fibre concrete with 0,75 % fibre lying on a pitchfork.

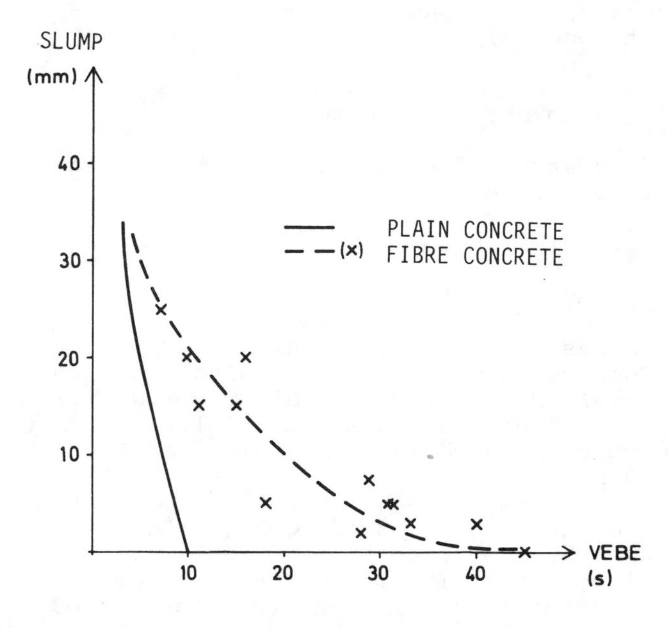

FIG 2.2 Measured relation between slump and Vebe time.

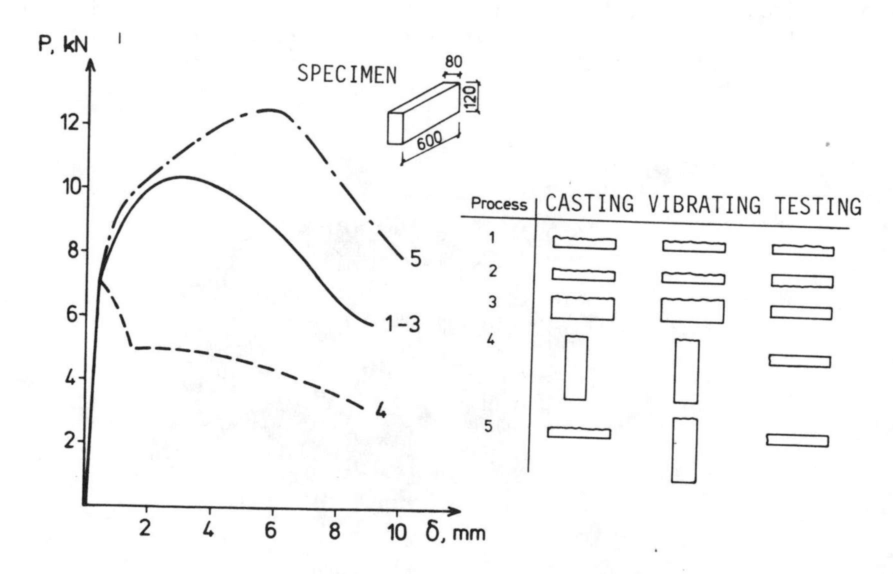

FIG 2.3 Flexure beams cast and vibrated in different ways.

3 MATERIAL TESTING

The specimens for material testing must consist of both concrete and fibres. It does not seem practical to test the concrete matrix separately and correct for the fibres afterwards.

There will also be difference in strength between moulded specimens and specimens sawed out from the component in question due to the fact that fibres near the surface will be cut off in the latter case. The difference will change with fibertype and length.

3.1 Standard compression tests

Compression strength was controlled using 150 mm cubes, the Swedish standard specimen. The over all effect of fibres is small on compressive strength, say 10 % with 1,5 % fibre admixture.

3.2 Standard splitting tests

Standard splitting tests are susceptible to the direction of splitting. Cf fig. 3.1. This effect is pronounced in SFRC since the fibres tend to orientate laterally to the direction of compaction.

Comparing cubes in plain concrete and SFRC with same direction of splitting relative to compaction, steel fibres increased the splitting strength approximately 70 % at a fibre admixture of 1,5 %. (50 mm cubes.)

3.3 Standard bending tests

In Sweden the most popular method to test tensile strength is the use of bending beams 800x150x100 mm. We have studied different sizes of specimens. Fig. 3.2 shows a comparison between 50x50x300 mm specimens and 120x80x600 mm specimens, and it will be seen that there is a size effect, which tends to decrease with increasing fibre admixture.

From a practical point of view the standard bending test is suitable to give a good estimate of the ductility, since this property is at least as important as the tensile strength. A simple procedure is suggested in which the load on a test beam is measured at a specified crack width.

Fig. 3.3 shows some typical load-deformation curves from tests on flexural beams with different fibretypes.

3.4 Choice of ductility measure

There are several possible ways to measure ductility, the most common is to calculate the area under the load-deformation curve (measure of energi). Another way is to measure the load at one (or more) point at the curve.

The method that is suggested here is to measure the load at a specified crack widht. A convenient way is to use theoretical crack width which easy could be transformed to deformation, cf fig. 3.4.a.

$$y_m = \frac{L}{4H} w$$

were y_m = mid point deflection

L = beam length

H = beam hight

w = crack width

The deformation should be measured from the point where one single crack propagates, which for practical purposes could be taken as the point for maximum load. The crack width is choosen to 1 mm because it gives a resonable result and is above the maximum crack width that could be accepted.

The method is illustrated in fig. 3.4.b. A ratio FF gives the ductility

$$FF = \frac{P_w}{P_u}$$

The advantage of this method to measure ductility is:

- Simple to use
- Possibility to compare different beam sizes
- Significant difference between different fibertypes
- The ratio FF is independent of amount of fibre for fibres that gives good ductility

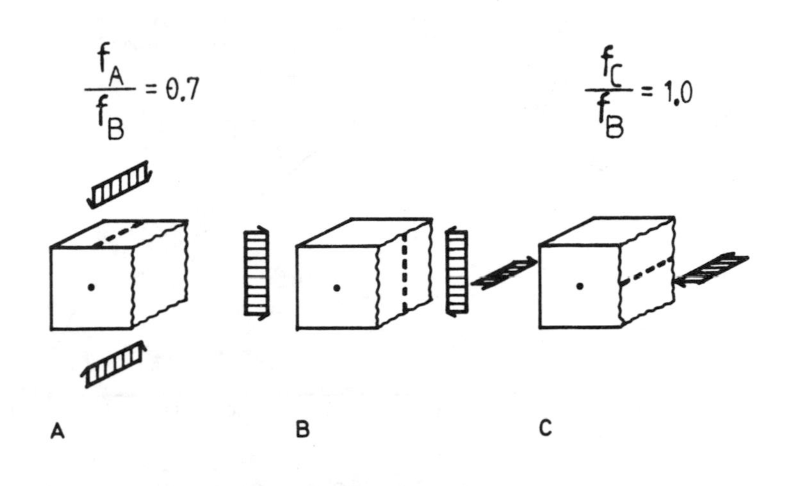

FIG 3.1 Different ways to split a cube.

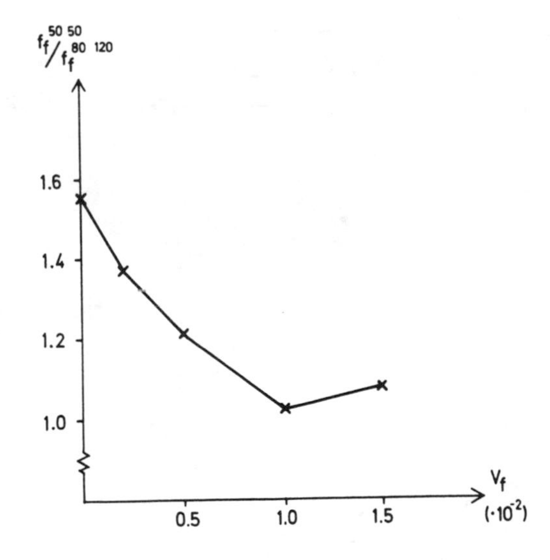

FIG 3.2 Comparison between 50x50x300 and 120x80x600 mm specimens in flexure.

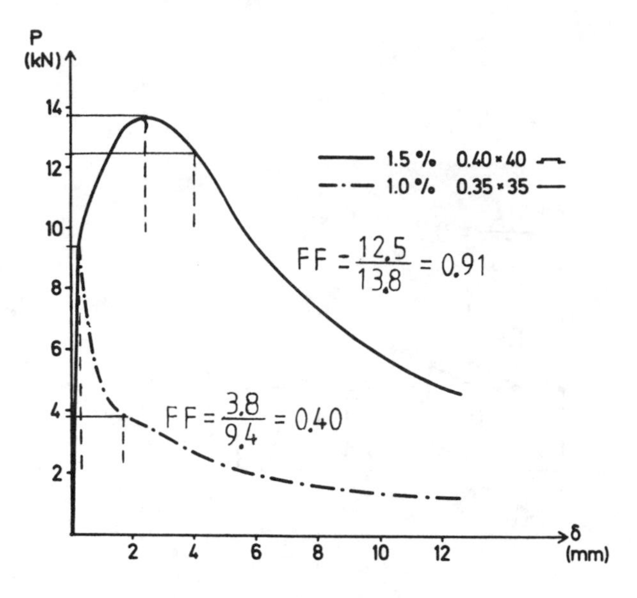

FIG 3.3 Load-deformation curves from tests in flexure.

a)

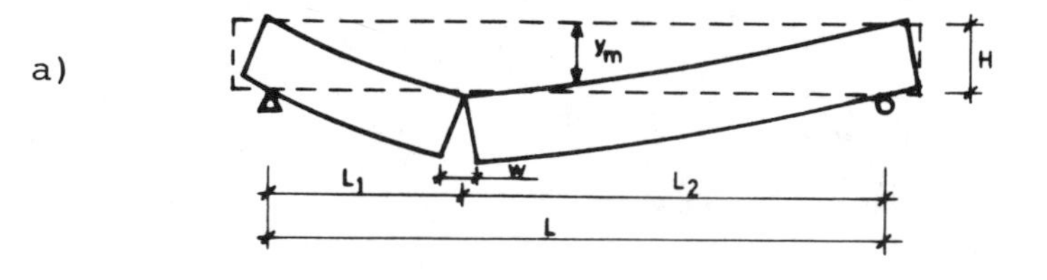

b)

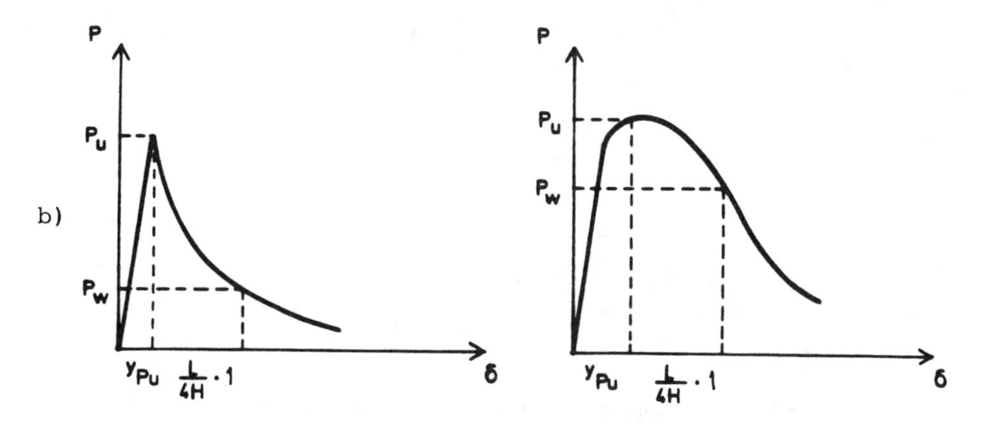

FIG 3.4 Measurement of ultimate load, P_u, and load at 1 mm theoretical crack width, P_w.

4 BEHAVIOUR OF BEAMS

4.1 Bending

As mentioned earlier steel fibres are less cost effective than regular bar reinforcement as main bending reinforcement. On the other hand steel fibre admixture has a favourable influence on crack distribution and deformation. Cf fig. 4.1.

These effects alone hardly justify the use of SFRC, since the same effects can be achieved cheaper by increasing normal bar reinforcement and using suitable bar types and diameters.

On the other hand the use of SFRC can have favourable effects in concrete pavements where limitation of cracking is important.

A rough method to calculate fibres increasing effect in flexure strength is to assume a stress-strain relation as in fig. 4.2.a. This gives a strength distribution over the cross-section as shown in fig. 4.2.b. Only value of α differ from calculations on plain concrete.

The value of α could be estimated from fig. 4.3 that gives the relation between α and the ductility ratio FF. The diagram shows results from tests, with different fibertypes and fibre-admixtures, were two parts of the same spacemen, one in tension and one in flexure have been tested.

4.2 Shear

Our tests confirm earlier findings that steel fibres have a pronounced, favourable effect on the shear capacity of beams.

An interesting question is to what extent the fiber orientation affects the shear resistance. In practice the fibre orientation will vary according to production method.

To study this question a series of tests were performed using square shear panels as shown in fig. 4.4.

Fiber types and orientation were varied according to table 4.1 below.

Nr	Shear reinforcement			Orientation
A	None			-
B	0,75%	0,4x40	endanch.	random
C	1,5%	0,4x40	endanch.	random
D	1,0%	0,35x35	intented	random
E	1,0%	0,35x35	intented	2 directions
F	1,0%	0,35x35	intented	1 direction
G	1,0%	0,35	plain	continuous, 2 dir.
H	1,0%	0,35	plain	continuous, 1 dir.

TABLE 4.1 Data for shear panels.

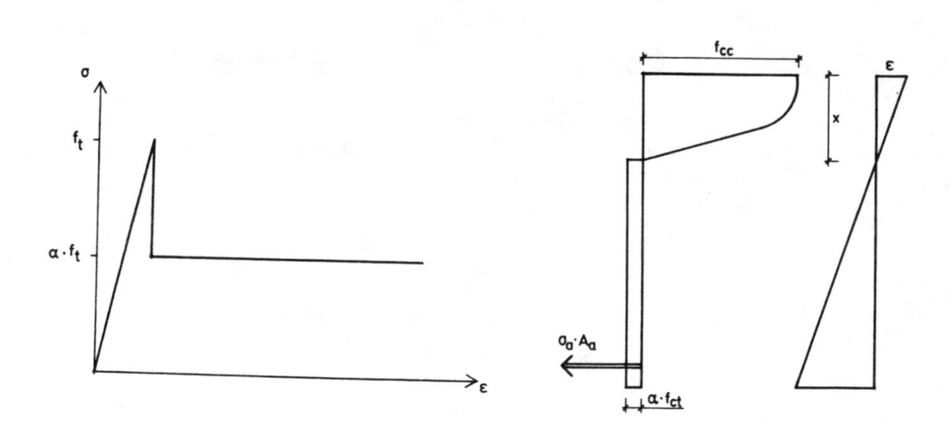

FIG 4.2 Simplified stress-strain relation and corresponding strength distribution over the cross-section of a beam.

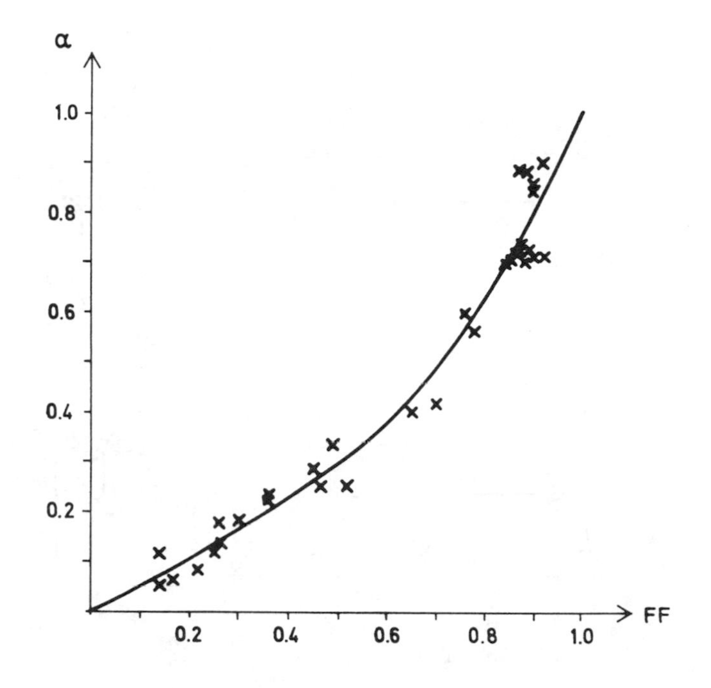

FIG 4.3 Relation between ductility in tension and flexure from tests.

Test results are shown in fig. 4.5.

As might be expected, continuous wires gave the best results, especially if oriented in two directions, but end-anchored fibres oriented at random compare quite well with continuous wire.

It should be noted, however, that the absolute values of the shear strength - 5 to 10 MPa - are not applicable to beams, the used test specimen gives much higher strengths than can be expected in beams.

10 I-shaped beams and 10 rectangular beams, cf fig 4.6 and 4.7 were tested. The bending reinforcement in the I-beams was pre-tensioned and consisted of two 12.7 mm strands with f_y = =1400 MPa. The strands were anchored at the ends with ancors of the type shown in fig 4.8. The bending reinforcement in the rectangular beams consisted of 4 ø 16 deformed bars with f_y=800 MPa.

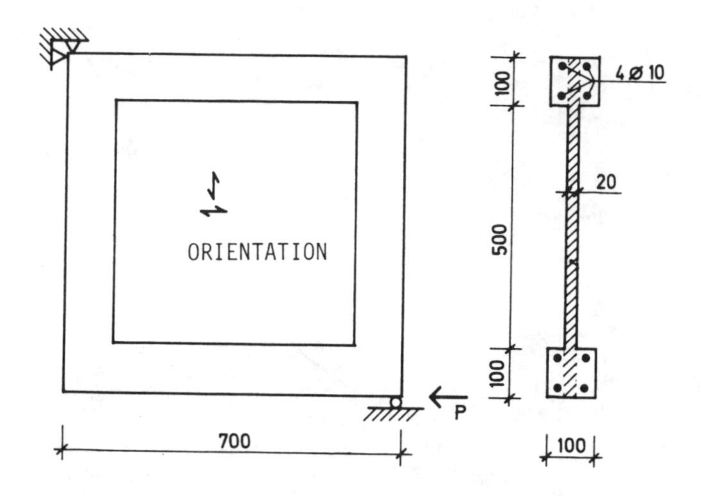

FIG 4.4 Shear panel.

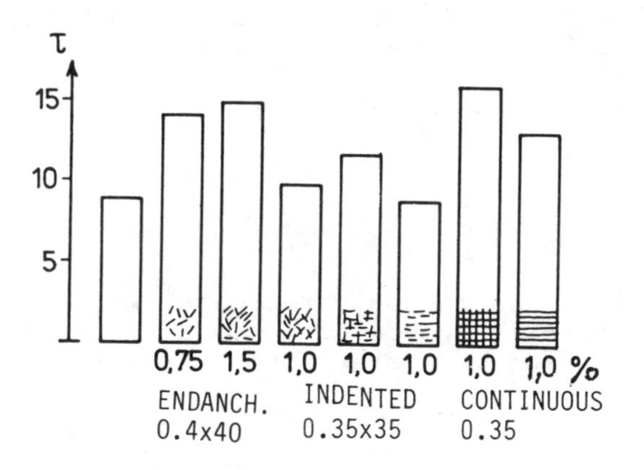

CONCRETE STRENGTH ~ 55 MPa

FIBRE STRENGTH ~1200 MPa

FIG 4.5 Results from test with shear panels.

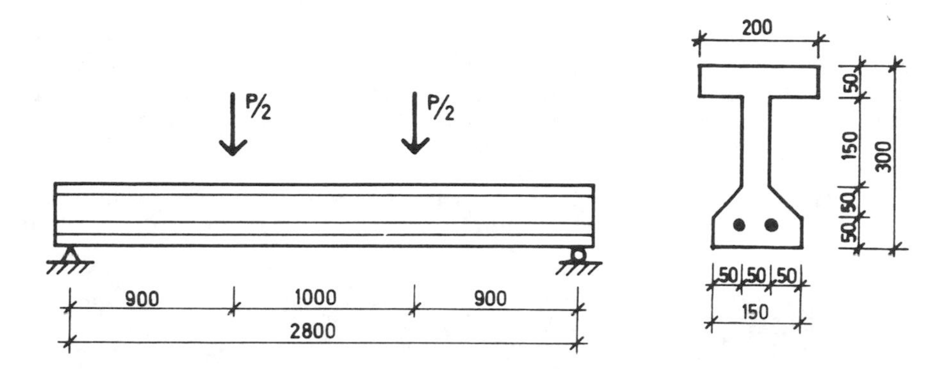

FIG 4.6 I-beams.

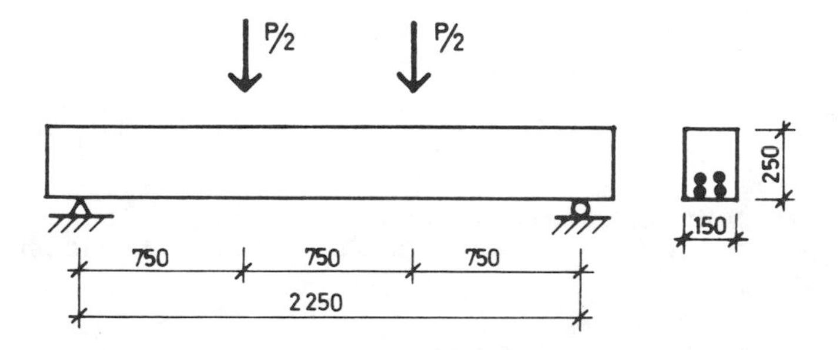

FIG 4.7 Rectangular beams.

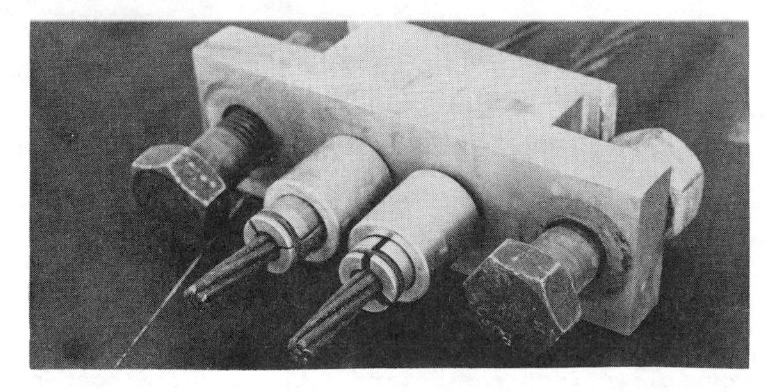

FIG 4.8 End anchore of strands on I-beams.

Tables 4.2 and 4.3 give the data for the different test series, and it will be seen that ordinary stirrups were used in some cases for comparison.

Fig. 4.9 shows a load-deformation curve illustrating the favourable effect of fibres as compared to normal vertical stirrups, and it is seen, that 1,5 % end anchored fibres 0,4x40 equals stirrups ø 6 per 100 mm.

According to the tests, it seems reasonable, in practical design, to use the simple addition theory given in CEB Model Code.

The fibre term in the addition formula may be expressed as

$$V_{fib} = b_w \cdot H \cdot f_{cfib}$$

where b_w = web thickness
H = beam depth

f_{cfib} varies with the fiber percentage (by volume) V_f and fibretype.

Cf fig 4.10. It is important to note, that different fibre types give different result in this respect.

The efficiency of fibres in shear is not only a matter of strength, it is very much a matter of ductility.

An approximate way to account for the ductility is to use the standard bending test mentioned in 3.3, where it was suggested, that the ratio

$$FF = \frac{P_w}{P_u} = \frac{f_{fw}}{f_f}$$

could be taken as a measure of ductility - for practical purposes. f_{fw} is the stress on the declining part of the load-deformation curve at 1 mm theoretical crack width.

Fig. 4.9 shows, that there is a correspondence between the fibre term $V_{fib} = b \cdot H \cdot f_{cfib}$ and f_{fw}.

It is felt, that the development of a prefabricated product, e.g. a beam type, will have to be based on ready product tests, but a curve like fig. 4.11 may serve as a starting point of design.

4.3 Bond

Several bond tests have been performed using test specimens according to fig. 4.12 and 4.13 The tests showed that fibres have a favourable effect for deformed bars, but almost no effect for strands, cf fig. 4.14. The concrete strength and cover was varied. In the tests represented in the figure the concrete cube strength was 50 MPa and the cover 24 mm.

TABLE 4.2 Results from I-Beams
B = Endanchored 0.4x40 mm

LIT	CONCRETE STRENGTH (MPa)	SHEAR REINF.	PRESTR. FORCE	ULTIMATE LOAD (kN)	FAILURE TYPE	NOTE
I1A	40	1.5%B	0	82.6	B	BAD COMPACTION
I1B	40	1.5%B	0	109.6	B	
I1C	40	1.5%B	150	124.6	L	
I1D	40	1.5%B	150	129.6	L	
I1E	40	-	150	84.4	L	
I2A	55	-	80	92.3	B	
I2B	55	Stirrups ø6c100	80	144.8	B	
I2C	55	1.5%B	80	152.7	B	
I2D	55	1.5%B	150	154.6	L	
I2E	55	0.75%B	150	141.3	L	

FAILURE TYPES

L

B

TABLE 4.3 Results from rectangular beams
B = Endanchored 0.4x40 mm
S = Indented 0.35x35 mm

LIT	CONCRETE STRENGTH (MPa)	SHEAR REINF.	ULTIMATE LOAD	NOTE
R2A	55	-	120	
R2B	55	1.5%B/50mm	120	50 mm high with fibre concrete
R2C	55	1.5%B	270	
R3A	55	-	114	
R3B	55	Stirrups ø8c100	252	Bending failure
R3C	55	0.75%B	236	
R3D	55	0.75%S	144	
R3E	55	1.5%B	274	Bending failure
R3F	55	1.5%S	186	
R3G	55	1.5%B	274	Internal vibrator used

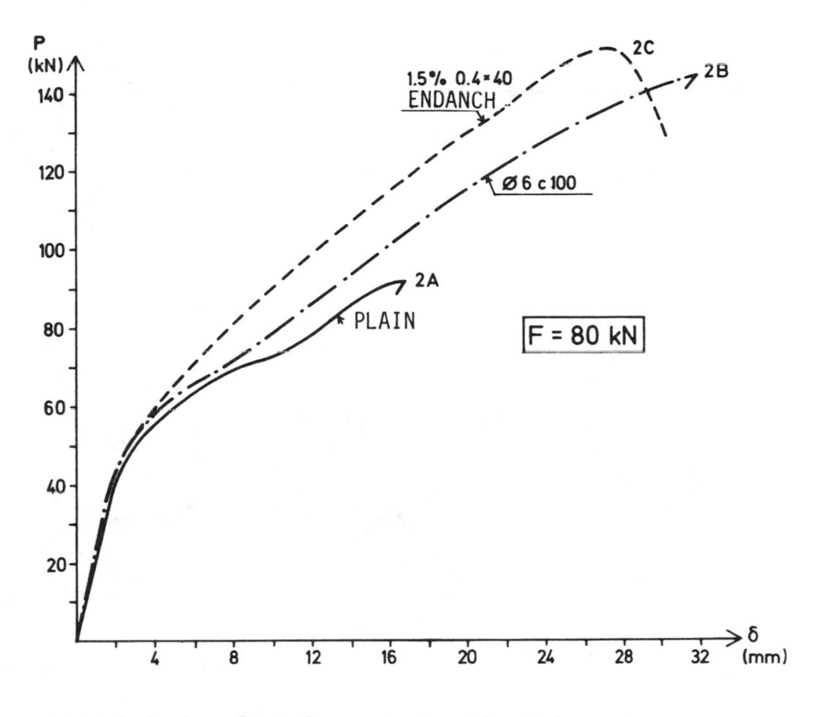

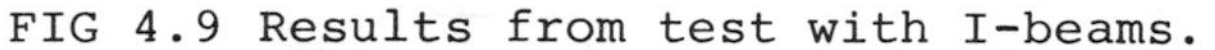
FIG 4.9 Results from test with I-beams.

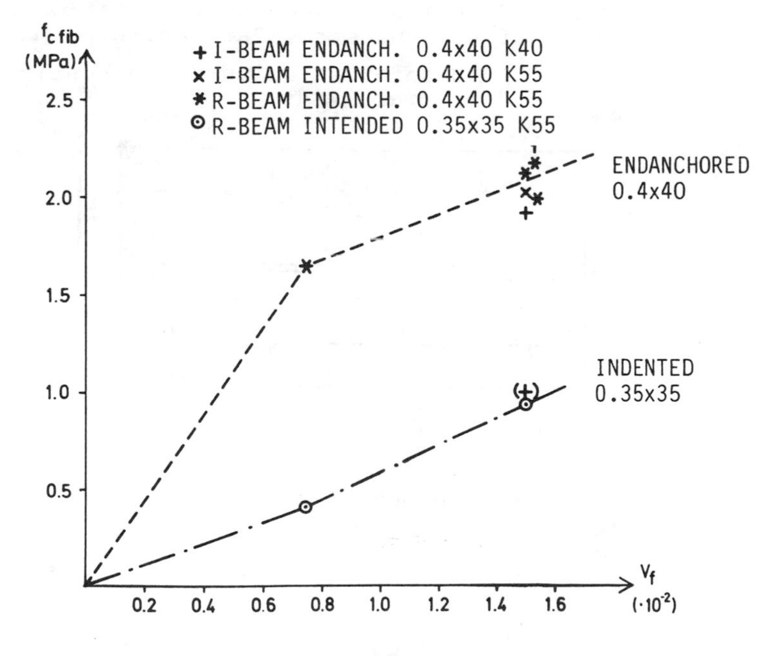

FIG 4.10 Relation between fibre admixture and increase in shear strength.

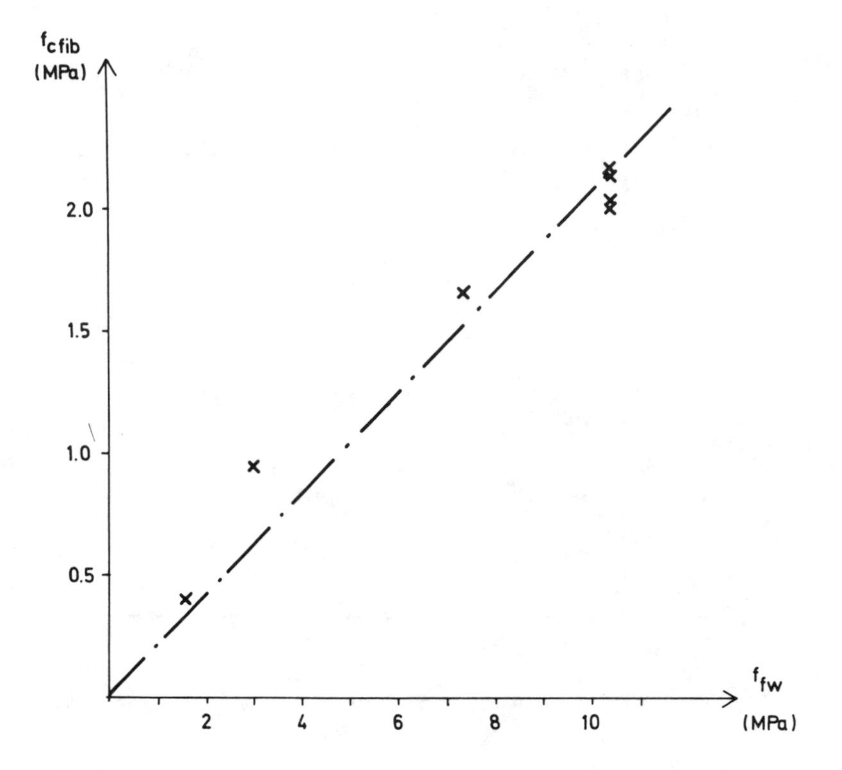

FIG 4.11 Relation between fibre addition term f_{cfib} and stress at declining part of load-deformation curve in bending for 1 mm crack width.

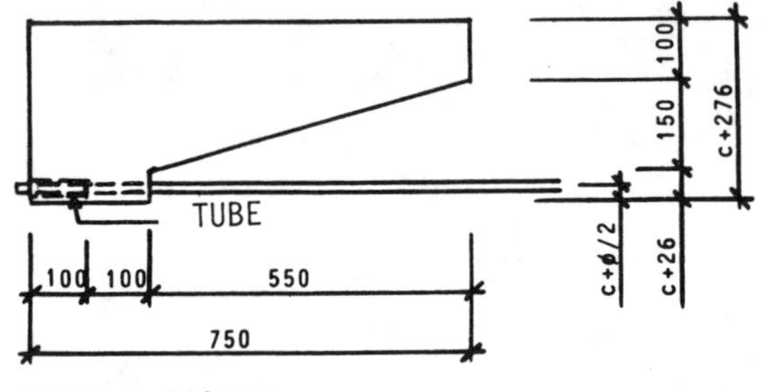

FIG 4.12 Speciment for bond tests.

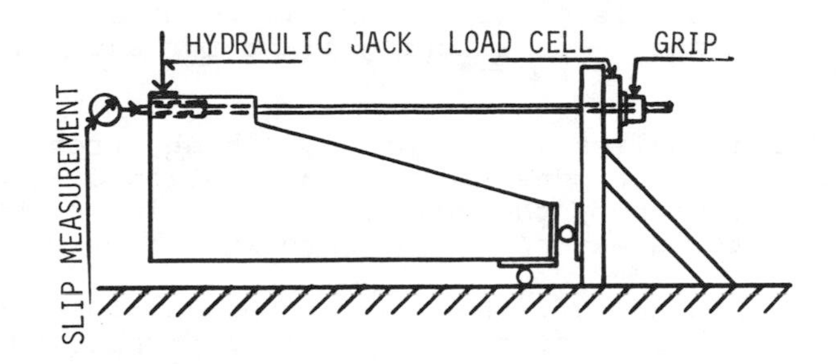

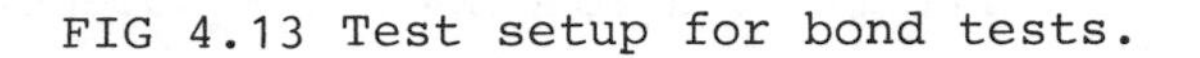
FIG 4.13 Test setup for bond tests.

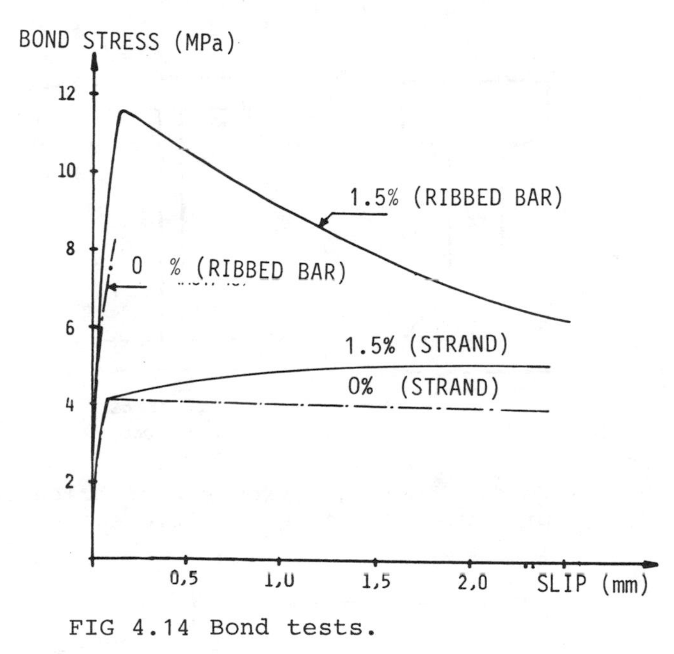

FIG 4.14 Bond tests.

5 CHOICE OF FIBRE TYPE

From our investigation it is clear, that the choice of fibre type must be a compromise. For load-bearing structures the most interesting effect of fibres is the increase in shear strength. Using the end-anchored fibre it is possible to reach a level of shear strength that make very thin webs possible, so that prefabricated, prestressed beams of today could be made much lighter. A weight gain of 40-50 % is within reach at unchanged carrying capacity. Cf fig. 5.1.

To achieve this, a fibre of the type illustrated in fig. 5.2 b must be used. On the other hand, this fiber type gives a concrete that is very difficult to mould and compact. The production method must therefore be adapted to this, if at all possible. It is highly desirable to make production tests, for instance in an existing plant for extrusion of concrete products.

Another compromise lies in the price. A high strength fibre with end anchors is expensive and - as mentioned - not so easy to handle, whereby the achievable gain in weight and the absence of stirrups may not compensate for the fibre cost.

It should be pointed out, that this investigation has not covered the durability properties or fire properties of steel fibre reinforced concrete, questions that must undoubtedly be better illuminated to convince a potential market.

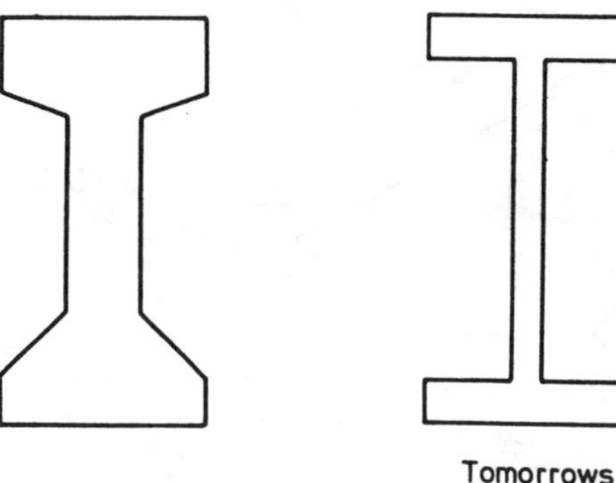

FIG 5.1 To-day´s and tomorrow´s prefabricated prestressed concrete sections.

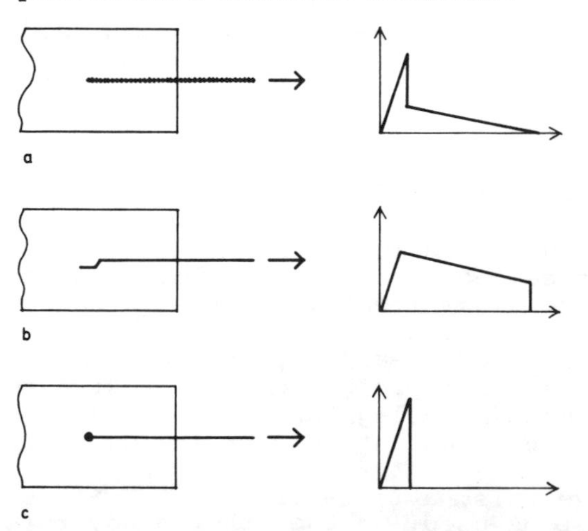

FIG 5.2 Typical load-deformation curves pulling out single fibre. a) High initial bond b) High pull out force c) Fibre rupture

REFERENCES

(1) Jahlenius, Å., "Fibre concrete as structural material", Publication 1/83, Department of Structural Engineering and Bridge Building. Royal Institute of Technology, Stockholm 1983. Published in Swedish.

STEEL FIBER CONCRETE
US-SWEDEN joint seminar
(NSF-STU)
Stockholm 3-5 June, 1985
S P Shah and Å Skarendahl,
Editors

STEEL FIBRE CONCRETE FOR BRIDGE DECK AND BUILDING FLOOR APPLICATIONS

by R N Swamy, MSc(Eng), University of Sheffield, England

ABSTRACT

This paper presents a comprehensive review on the use of steel fibre concrete for bridge decks and building floor slabs. The data shown cover both normal weight and lightweight concrete incorporating steel fibres. Emphasis is given to the mix proportioning of these two concretes to produce workable and compactible mixes without causing fibre bundling, segregation or bleeding. Both material and engineering properties of steel fibre concrete are then discussed in relation to their use in bridge decks and building floors. The structural behaviour of conventional reinforced concrete slabs containing steel fibres is then presented with particular reference to deformation, cracking, service loads and failure behaviour. The significant advantages of fibre concrete in structural members are highlighted. It is shown that durability of steel fibres need no longer pose a threat to their use in concrete structures.

R. Narayan Swamy is on the Staff in the Department of Civil and Structural Engineering, University of Sheffield, England.

His main research interests are in materials and structures. He is the Chairman of RILEM Technical Committee 49-TFR on the testing of Fibre Cement Composites. He is the recipient of the George Stephenson Gold Medal from the Institution of Civil Engineers, and Henry Adams Diploma from the Institution of Structural Engineers. He is the Editor of the International Journal of Cement Composites and Lightweight Concrete and also Series Editor of Concrete Technology and Design. He is past Chairman of the Local Chapters of the Institution of Civil Engineers and Institution of Structural Engineers.

INTRODUCTION

Extensive research on steel fibre concrete has shown that the most significant influence of the incorporation of steel fibres in concrete is to delay and control the tensile cracking of the composite material. Steel fibre reinforcement thus transforms an inherent unstable tensile crack propagation in concrete into a slow controlled crack growth. This crack controlling property of the fibre reinforcement, in turn, delays the onset of flexural and shear cracking, imparts extensive post-cracking behaviour, and significantly enhances the ductility and energy absorption properties of the composite. These three unique properties of the fibre reinforcement can be exploited to advantage in concrete structural members containing both conventional bar reinforcement and steel fibres.

Two factors, however, have tended to inhibit the widespread use of steel fibres in reinforced and prestressed concrete structures. One is the inherent difficulties of mixing, placing and compacting satisfactorily concrete containing short discrete fibres in forms which may be congested with conventional steel reinforcement. The second is the inevitable fear of steel corrosion which in time could deprive the structural member of its enhanced structural properties assumed in design. Fortunately both these problems have been satisfactorily solved - it is now possible to produce steel fibre concrete with adequate flow characteristics, and there is evidence that fibre corrosion does not in fact occur in practice. Further, stainless steel fibres are also now available which would have negligible economic impact to the cost of the structure as a whole.

These factors point out to the possibility of designing and

constructing concrete structures incorporating steel fibres without fear of loss of serviceability life or their major structural characteristics. A particular application where the properties of fibre reinforcement can be used to advantage is in bridge decks and building floors. This paper presents a comprehensive review of this application with special emphasis on the design and fabrication of fibre concrete and the structural behaviour of fibre concrete composite slabs.

FIBRE CONCRETE MIX PROPORTIONING

The properties of fibre concrete in the fresh and hardened states are known to be influenced by numerous factors related to the volume and geometry of the constituents of the mix. 1,2,3,4). The major problem with fibre concrete mixes is a practical one - that of ensuring adequate flowability and compactability that will enable the concrete to be placed and compacted with ease, and uniform fibre distribution, particularly in structural members with bar reinforcement. The basic objectives in mix proportioning should therefore be to produce a cohesive and compactible mix of structural quality which will have enough cementitious matrix to embed the aggregates and the fibres, and at the same time, avoid bundling of fibres, segregation and bleeding during placing and compaction.

The best way to produce such a mix is to use a slight excess of fines, in conjunction with a water-reducing plasticizer or a superplasticizer. An excess of fines can be obtained by using increased cement contents or an inert filler but the best filler is one that has pozzolanic qualitites such as pulverized fuel ash or fly ash (PFA). Fly ash has very favourable particle geometry and particle size distribution compared to portland cement, which leads to reduced water demand and improved cohesion. The pozzolanic property also ensures continued chemical interaction between the ash and the mineral components of the cement clinker. The reduced water demand can result in reduced bleeding, reduction in shrinkage and creep, lower permeability and smaller porosity - qualities all of which can significantly contribute to the long term durability of concrete structures. The continued hydration of the cement-ash mixture can give reduced heat of hydration, increased long-term strength, better resistance to sulphates and other chemical attack.

Because of the nature of the hydration mechanism of the ash - cement mixture, the presence of fly ash inhibits early strength development. It is here that the addition of a water - reducing plasticizer or superplasticizer can help by reducing the water content further and enabling strength development in fly ash cement mixes that is compatible with those of all portland cement mixes. The plasticizing admixtures should be non-toxic and contain no chlorides and nitrates. With superplasiticizers, the amount added can be controlled to produce a cohesive and workable mix: in conditions of limited accessibility, congested reinforcement or thin sections, mixes of high fluidity, with slump in excess of 200mm, can also be produced to suit a given situation.

Incorporation of fly ash

In literature there are various methods available for designing mixes with fly ash (4). Of all the available mix design methods for incorporating fly ash in concrete, the simplest procedure is a direct partial replacement of the portland cement on an equal weight basis. This method has two distinct advantages: because of the lower density of the ash, the volume of the cement paste matrix available to lubricate fibres and aggregates is increased, and consequently, the rheological properties of the mix are enhanced. The effects of enhanced flow characteristics are obtained for both low (250 kg/m^3) and high (450 kg/m^3) cement contents (5, 6). Direct replacement of cement has also the added advantage that it gives a feel for the function and effectiveness of the ash particularly when normal concrete mix designs are generally always available.

Two questions still remain: How much fly ash is to be added, and what type of fly ash should one use? Many tests have shown that an optimum cement replacement to give comparable workability, strength and elastic modulus as concrete without fly ash, is about 30%. With this amount of fly ash and an adequate amount of water-reducing/plasticizing admixture or superplasticizer, early strength development is comparable to that of concrete without fly ash, and the resulting concrete is of structural quality.

A further advantage of this particular approach to fly ash-fibre concrete mix design is that mixes are identical for both types of concrete - i.e. concrete without and with fibres so that no changes in mix design are required if both types of concrete are to be used in the same structure. This gives practical advantages to the designer and the contractor which are not to be underestimated, but above all, this direct partial replacement of cement by weight enables to mobilise the expertise and knowledge of the concrete technologist to adjust the mix constituents to suit the workability and strength requirements of a given application.

Quality of fly ash

One of the factors that has hindered widespread use of fly ash in concrete mixes is their variability. No two ashes are completely alike and from an engineering point of view ashes need to be characterized in terms of their chemical and mineralogical composition, carbon content, particle geometry and size distribution. It is important that quality controlled fly ash is available and it is only such ash that is incorporated in concrete. Poor specifications, variable quality control and substandard concreting practices can lead to poor quality concrete and continuous maintenance problems - whether the cementitious material is portland cement or a mixture of portland cement and fly ash! The ash used in the tests reported here was monitored regularly and contained typically, 50-56% silica, of

26-27% alumina and 7-10% of iron oxide, with a specific surface of 396-399 m^2/kg and a density of 2.17 gm/cm^3.

Workability properties of fibre concrete

To illustrate the workability characteristics of fibre concretes incorporating fly ash two types of concretes are reported - normal weight and lightweight concrete. Ordinary portland cement, fly ash (both from one source) and washed, dried river sand were used in all the tests. For normal weight concrete, a graded river gravel with 10mm maximum size and rounded to angular in shape was used. The lightweight aggregate was also graded with 14mm maximum size, manufactured from sintered pulverized fuel ash. These aggregates had a 24 hr. absorption of 12-14%, about 70% of which generally occurred in the first 30 secs. (7). The lightweight aggregates were used in a partially soaked condition: the water content of the aggregates was determined before each casting, and the water added at the mix adjusted accordingly to suit the workability required. Several types of steel fibres were used, varying in their size, surface configuration and shape of cross-section.

Tables 1 and 2 show typical workability properties of normal weight and lightweight concretes respectively of structural quality. These mixes have been used extensively in T beams, I beams and slabs with bar reinforcement, and were designed for a 28 day compressive strength of 40-45 N/mm^2. Over a period of four years, over a hundred structural members were cast with these concrete mixes, and the slumps varied from 45 to 100mm.

It is of course possible to further enhance the fluidity of these concrete mixes to produce superplasicized fibre concretes, with excellent flow characteristics suitable for spraying or in-situ applications (8). The inverted - slump - cone - time test developed by ACI Committee 544 can also be applied to identify the effects of fibre shape and fibre geometry on the flow characteristics of these concrete mixes (9). The main aim in all these mix designs is to reduce the interparticle friction between and within the solid constituents of the mix, and to ensure that adequate cohesiveness and flowability are mobilised to obtain complete compactability in a structural form.

Because of their nature, the best form of mechanical consolidation suited for fibre concrete mixes is external mould vibration, particularly if the mobility of the mix is restricted. Even with fully superplasticized concrete mixes some compaction is recommended to ensure that the presence of bar reinforcement does not create obstructions to the flow of the fresh concrete mix and cause non-uniform distribution of fibres.

STRENGTH AND ELASTICITY PROPERTIES

The strength and elasticity properties of normal and lightweight concrete mixes incorporating fly ash and fibres are shown in Tables 3 to 6. The data in Table 3 relate to the gravel concrete mix in Table 1 whereas those in Tables 4 to 6 are of lightweight concrete mixes given in Table 2. The results in these tables were obtained from test specimens which were exposed to uncontrolled internal environment after demoulding at about 24 hours.

The data given in Tables 3 to 6 show that with PFA of controlled quality it is possible to design mixes of normal and lightweight aggregate concrete of structural quality with adequate early strength and stiffness properties. It is also significant to note the presence of fly ash does not affect the tensile strength of plain and fibre concrete mixes.

One significant tensile property of fibre concrete mixes, not shown in Tables 3 and 5, is their ability to control tensile cracking: they have a distinct first crack strength as opposed to the ultimate tensile strengths shown in Tables 3 and 5. In small prismatic specimens, this first crack strength may not be vastly different from the ultmate strength, but in structural members with conventional bar reinforcement, this property can impart important structural benefits to its behaviour, as shown later.

Effect of curing regime

Fly ash concrete mixes are more sensitive than normal concretes to temperature and moisture conditions, particularly at early ages. Early and adequate water curing, even if only for a few days, is therefore essential for adequate strength development. The data in Tables 3 to 6 relate to strength development where the specimens were demoulded at 24 hours and then exposed to internal drying environment - i.e. practically no water curing after 24 hours, - the sort of extreme environmental conditions that are likely to occur in practice, where buildings and bridge members are seldom subject to wet curing once the formwork is removed. In fact often the only period when structural members are protected from moisture evaporation is when the formwork is in position. The results in Tables 3 to 6 indicate that even with the extremely unfavourable curing conditions mixes can be designed to develop structural quality strength and elastic modulus.

The effect of lack of curing on the strength properties of fibre concrete containing fly ash is shown in Table 7. There is clearly some slowing down in the development of early strength: and even though the results in Table 7 relate only to the first few weeks, there is some evidence that there may be some longterm retrogression of strength if concrete with fly ash is not given adequate curing at early ages.

Other engineering properties

Strength properties form only one aspect of structural design, and other engineering benefits of incorporating fibres in concrete need to be considered if the use of fibre reinforcement is to be encouraged. Drying shrinkage is an inherent property of all concrete materials; associated with this phenomenon is the development of internal stresses due to differential moisture gradients, and the resulting incidence of cracking. Cracking behaviour is also often further aggravated by the restraints, internal or external, imposed on free shrinkage.

The addition of steel fibres to concrete has been shown to impart several beneficial effects to counteract the effects of movements arising from volume changes taking place in concrete, particularly in the early stages of its life (10-12). Fibres exercise an additional restraint to drying shrinkage to that produced by the aggregate particles (Fig. 1). In fibre concretes, the fibre volume is very low compared to the aggregate volume, and the matrix tends to be slightly richer than that in comparable normal concrete. Nevertheless, fibres have a distinct beneficial influence in reducing shrinkage strains, and also tend to stabilise these movements earlier. If fibres are used laregely to restrain movements alone, then these benefits can be further enhanced by using shorter fibres, because of the larger number of fibres available in a given volume.

Reduction of drying shrinkage strains is only one aspect of the role of fibres in concrete. Fibres also control shrinkage cracking: they delay the formation of the first crack, enable the concrete to accommodate more than one crack, reduce the crack widths substantially, and continue to sustain the shrinkage stresses for up to 12 months. Fig 2 and Table 8 illustrate these properties obtained from restrained shrinkage tests. The data in Table 8 show that fibre concretes also resist higher shrinkage - induced tensile stresses. With a 1% fibre volume, about 50 to 100% higher tensile stresses are resisted; and this resistance continues for as long as one year.

When one considers the overall durability of bridge decks, control of shrinkage cracking is a significant contribution to this preservation of durability. Shrinkage cracking is only one aspect of the spectrum of cracking behaviour; but it does occur in the early life of a structure, and together with load and temperature induced cracking, can lead to steel corrosion. The presence of fibres can thus be a significant beneficial factor to add to the serviceability life of bridge structures.

R.C. SLABS WITH FIBRES: STRENGTH CHARACTERISTICS

A reinforced concrete slab - in the form of a building floor or a bridge deck - may fail in flexure or shear depending primarily on the amount of tensile reinforcement. In design and in practical behaviour, punching shear failures are more critical and undesirable because they do not allow an overall yield mechanism to develop, and lead to failures that are sudden and catastrophic.

The effect of fibre reinforcement on flexural and punching shear failures is illustrated here through tests on slabs loaded through a column stub. A test configuration such as this (Fig. 3) has the advantage of simulating a slab - column connection of a flat - plate structure as well as a bridge deck in the vicinity of a heavily concentrated load. The test slabs were 1800 x 1800 x 125mm thick with an average effective depth of 100mm. The slab reinforcement was designed for a 28 day cube strength of about 40-45 N/mm². In order to enable the test specimens to simulate both building floors and bridge decks, the tests were carried out with all four sides simply supported and the corners free to lift. This is considered to be the best representation in a practical model of the actual boundary conditions in the prototype. The test specimens thus in effect represent one - to - one full scale models of the prototype structures.

Flexural behaviour

From purely theoretical considerations, the presence of fibres is unlikely to increase substantially the ultimate flexural strength of a conventional structural member. Tests generally seem to confirm this. In normal reinforced concrete beams, the increase in ultimate flexural strength is only about 10-15%, and this can be satisfactorily predicted by the conventional reinforced concrete theory taking into consideration the contribution of the steel fibres to steel bar stress and concrete in the tension zone (13, 14). The role of steel fibres in structural members is primarily to arrest advancing cracks and increase the post-cracking stiffness of the member, which result in narrower cracks widths and substantially less deformations. The net result is a substantial increase in the ductility and energy absorption properties of the member.

The influence of steel fibres on the flexural behaviour of reinforced concrete slabs is not so clearly established. To illustrate their effect Tables 9 and 10 are shown which summarise the results of slab tests referred to earlier, which were loaded through a column stub. Table 9 shows gravel concrete slab test data (Table 1) whilst Table 10 referes to lightweight concrete slabs (Table 2). These tables show some interesting aspects of the flexural behaviour of slabs containing both steel fibres and conventional bar reinforcement.

In Table 9, slab S1 failed in punching shear. The addition of 0.9% volume of steel fibres and reducing the steel by 30-40% enabled the slabs not only to fail in flexure but also achieve higher load capacity. Slab S9 with steel fibres took about 40% more load compared to slab S19 without steel fibres, both failing in flexure. Similar behaviour also occurred in lightweight concrete (Table 10) - slabs FS6 and FS7 containing less tension and compression steel and 1% fibre volume failed in flexure compared to slab FS1. These data show that fibres have a very positive role in controlling cracking and transforming a sudden punching shear failure into a ductile flexural failure without loss of load capacity - and be able to achieve this with 30 to 40% reduction in tension steel area and some 50% or more reduction in compression steel.

PREDICTION OF ULTIMATE FLEXURAL STRENGTH

It has been shown that the ultimate flexural strength of slabs (loaded through a column stub) and reinforced with steel bars in two directions and steel fibres can be predicted satisfactorily by yield line theory (15). The actual and assumed strain and stress distributions are shown in Fig. 4. The tensile contribution of the steel fibres is assumed to be given by a rectangular stress block with a value of σ_{cu} given by

$$\sigma_{cu} = \eta_o \; \eta_L \; \sigma_{fu} \; V_f$$

where the ultimate composite strength is related to the fibre orientation factor, η_o, the length efficiency factor η_L, the fibre fracture stress σ_{fu}, and the fibre volume fraction V_f.

To take account of the presence of fibres in the compression zone, the compression stress blocks are modified for both normal weight and lightweight concrete, and these are shown in Figs. 5 and 6. Based on these stress blocks, the ultimate flexural strength of slabs containing steel bars and steel fibres has been computed, and the results are summarised in Table 11. Unfortunately the published test data are limited, and the results in Table 11 include only four slabs tested by other investigators (15).

The results show excellent support for the proposed theory, the ratio of calculated to test strength varying from 0.90 to 0.99 with a standard deviation of 1.3 to 4.3%. The computations showed that there was practically no difference between BS and ACI stress blocks (15). The contribution of the compression reinforcement to ultimate flexural strength is generally negligible and this explains why compression bars can be effectively replaced by steel fibres. On the other hand, fibres have a distinct and beneficial role in resisting shear stress which is not shared by compression steel. The presence of steel fibres has therefore an overall benefit to the ultimate flexural strength of the slabs.

The increase in flexural strength due to the inclusion of fibres in the slabs can be predicted theoretically. In practice, the increase in ultimate strength appears to vary from about 25% to 40%. The results are obviously limited, but there is no doubt about the beneficial nature of the role of fibres in slabs failing in flexure

PUNCHING SHEAR BEHAVIOUR

In slabs, punching shear failures are decidedly more critical and catastrophic, and are, from a design point of view, more important. Apart from punching shear, a second major concern in design is the avoidance of unacceptable deflections at service loads. Both can lead to structural distress and structural failures. There is evidence to show that steel fibres have a major role to play in both cases - in reducing deformations at service loads (as already shown earlier), and in preventing punching shear failures. These and other characteristics are illustrated below.

The results reported in this section are again based on tests on slabs 1800 x 1800 x 125mm thick loaded through a column stub. The slab reinforcement was again designed for a concrete strength of 40-45 N/mm² at 28 days. The data reported relate to slabs that were designed to fail in punching shear, and the effect of fibre addition on the structural behaviour of these slabs is examined.

Influence of fibre reinforcement on deformation behaviour

Probably the first aspect that one ought to look at is the influence of fibres on the deformation behaviour of slabs failing in shear. The two important deformation characteristics are deflection and tensile steel strain - since the former has a major serviceability effect, and the latter has a direct relationship to cracking behaviour. Figs. 7 and 8 illustrate the role of fibre reinforcement in controlling deflection and tensile steel strain. It is clear that fibres have a very positive role in deformation behaviour, and that this influence progressively increases as the load on the structural member increases, particularly in the post-cracking stage (16).

The primary role of the fibre reinforcement is thus to act as a crack arresting mechanism, and as a consequence, its presence reduces all deformations at all stages of loading up to failure. This important contribution of the fibre reinforcement to structural performance under the three important load stages, namely, first crack load, service load and prior to failure is quantified in Tables 12 and 13 for normal weight and lightweight concrete slabs respectively. The first crack load deformations are based on the visually observed cracks, and are therefore not completely reliable; but they nevertheless confirm the trend shown in Figs. 7 and 8. The service load is based on the BS CP110 Code and is for the control slab without fibre reinforcement. Without any doubt, the fibres are seen to exercise significant reductions in steel strain, concrete strain and rotations of the slab under service load conditions.

Prior to failure, the slabs with fibres show substantially higher deformations, obviously due to their increased capacity arising from the presence of fibres. The values quoted in the Tables near failure are necessarily approximate and represent those at about 90-95% of the ultimate load. The actual values at fracture are likely to be much higher since large deformations occur very rapidly during the failure process.

Effect of deformation control on service loads

Figs. 7 and 8 and Tables 12 and 13 show that from a design point of view the most significant effect of fibre reinforcement through crack control is the substantial reductions in the deformation behaviour of the slab particularly at service loads. If, therefore, the deformations of the control slab without fibres at its service load were taken as the criterion of serviceability then it should be possible to increase the service loads in the slabs with conventional bar and fibre reinforcement. Tables 14 and 15 show possible increases in service loads for normal weight and lightweight concrete slabs respectively based on various deformation criteria.

For normal weight concrete containing 1.2% fibre volume, the increases in service load range from 35 to 65%, depending on the deformation criteria used. In addition, it is worth noting that slabs S3 and S4 with 0.9% and 1.2% fibre volume respectively, can have service loads higher than the ultimate design load of 137.5 kN (based on the BS Code CP110) of the plain concrete control slab. The technical advantages of including fibre reinforcement are thus very obvious.

For lightweight concrete slabs containing 1.2% fibre volume, similar increases in service loads were observed, and these varied from about 20 to 30%, depending on the deformation criteria used (Table 15). In practical terms, the cost of fibres can be offset by increases in the service load carried by the slabs.

Diagonal tension cracking load of slabs

For slabs failing in punching shear the load at which the diagonal tension crack occurs is an important stage in their load performance. But this load cannot be easily identified visually in slabs.

From the compressive steel strain distribution, however, it is possible to deduce the load at which internal shear cracking develops within the critical section of the slab. This stage can be related quantitatively to the diagonal tension cracking load, and these loads are shown in Tables 16 and 17 for normal weight and lightweight concrete slabs respectively.

With normal weight concrete, the fibres have increased the diagonal cracking load from about 40% for a control slab without fibres to about 60% for a slab with 1.2% fibre colume. If, in addition, the bar reinforcement is concentrated near the columns, this load is further increased to about 80% of the corresponding maximum load of slabs failing in both punching and flexure. Fibres have thus a second positive role - apart from controlling deformation - in controlling diagonal tension cracking, and in delaying the formation of such cracks.

Similar behaviour is also shown by lightweight concrete slabs (Table 17). With 1% fibre volume, the diagonal cracking load was raised from about 40% to about 60% of the failure load. The geometry of the fibre and its aspect ratio both have an influence on the shear cracking load, but they also affect the ultimate load, so that the overall effect remains largely the same. For the different types of fibres shown in Tables 16 and 17, the diagonal tension cracking load varied from 60 to 75% of the failure load.

BEHAVIOUR AT FAILURE

Apart from controlling deformation at service loads and enhancing the diagonal cracking loads, fibres also have a significant role to play at the failure stage of the slabs. Fibres increase the ultimate punching shear loads of both normal weight and lightweight concrete slabs (Tables 16 and 17). With normal weight concrete slabs, the maximum load increased by 42% for a fibre volume of 1.2%. In lightweight concrete slabs, the same increase in ultimate load was obtained with a fibre volume of 1.0%. As noted earlier the geometry of the fibre and its aspect ratio have an influence on the failure load, although it is not significant.

The notable feature of the failure mode is that whilst the control concrete slabs without fibres failed suddenly with extensive spalling of concrete in the vicinity of tension steel, the fibre concrete slabs showed a gradual and ductile punching failure. In some slabs, the cracking process enabled a second punching shear zone to form (in part or full).

The presence of fibres also imparted further benefits to the failure condition. In plain concrete slabs, the brittle-type of failure resulted in rapid loss of load capacity. With normal weight concrete, the residual slab resistance supported by the tensile membrane action was about 25% of the ultimate load for the plain concrete slabs and about 70-75% for slabs with a fibre volume of 1% (Fig.9). Lightweight concretes also showed similar behaviour, with residual slab resistance of the same order of magnitude.

Ductility and energy absorption characteristics

The post-maximum load behaviour of structural elements has important implications in relation to their stability against dynamic loads and at collapse. The fact that fibres enable higher loads to be carried at failure and also contribute to much higher residual loads to be sustained after collapse has significant influence on their ability to preserve the continuity and integrity of the structural element.

This post-maximum load behaviour can be quantified by considering the load - deflection curves of the slabs as shown in Fig. 10. Here the loading of the slabs is continued until the residual capacity is reduced to 25% of the maximum load.

Ductility is then quantified in terms of the ratio of the deflection at 25% of the maximum load (after peak load) to the deflection at first crack (as visually observed). Energy absorption capacity is defined in terms of the area under the load - deflection curve up to this 25% post-maximum load.

The ductility and energy absorption properties at failure are shown in Tables 18 and 19 for normal weight and lightweight concrete slabs respectively. These data show the dramatic increases in ductility and energy absorption properties that could be obtained from slabs containing fibre reinforcement. These properties have important applications in situations where structures are subjected to dyanmic and impact loads. Fibre reinforcement can be seen to have an important role to play in such situations.

Can punching shear failure be averted?

The data presented earlier show that the presence of fibres along with conventional bar reinforcement can significantly improve the shear resistance of a reinforced concrete slab. This raises an important question as to whether fibres can completely avert a punching shear failure? To illustrate this possible capability of steel fibres, Table 20 is presented which shows the load capacity and mode of failure of normal weight and lightweight concrete slabs with different percentages of tension and compression steel.

These results show some interesting aspects of structural behaviour. For slabs failing in punching shear (slabs S 7 and S 8 and FS 2 and FS 4), a reduction in compression steel has only a minor effect on load capacity when fibres are present. This reduction in compression steel can be as high as 100%. A reduction in tension and compression steel can also transform the mode of failure from punching to flexure without reducing the load capacity below the original punching shear failure capacity (slabs S 1, S 16 and S 10 and slabs FS 1, FS 7 and FS 6). Obviously large reductions in tension steel would induce flexural failure (slabs S 1 and S 9 and FS 1 and FS 6) and this would also result in reductions in the slab load capacity. In practical designs, there may be valid technical considerations which would not allow substantial reductions in tension and/or compression steel. But what Table 20 shows is that for a given load capacity in shear, redistribution of tension and/or compression steel together with steel fibres can avert the undesirable brittle type failures and produce ductile flexural failures without reduction in overall load capacity. This is an important structural benefit, that engineers can utilise in practical designs.

PREDICTION OF ULTIMATE PUNCHING SHEAR STRENGTH

The ultimate punching shear strength of slabs loaded through a small area such as a column can be predicted theoretically (17). Failure is assumed to occur in the compression zone above the inclined cracking when the shear stress is equal to the tensile splitting strength of concrete. The total shear resistance of a slab without shear reinforcement is given by

$$V_u = V_c + V_a + V_d$$

where the components V_c, V_a and V_d are due respectively to the concrete compression zone, aggregate interlock and dowel action, (Fig.11). In a slab with fibres, they increase the punching resistance of the slab by acting as shear reinforcement V_s because of their ability to bridge the shear cracks.

Neglecting the aggregate interlock contribution, the ultimate punching shear strength of a plain concrete slab is given by (17)

$$V^P_{u.p} = v_{cc} \cdot b_p\, x/\sin\theta$$

where v_{cc} = average shear stress at failure (related to the tensile splitting strength)

b_p = critical punching shear perimeter

x = neutral axis depth, and

θ = inclination of the failure surface

For a slab with fibres,

$$V^F_{u.p} = V^F_{u1.p} + V^F_{u2.p}$$

where the second term quantifies the shear resistance offered by the fibres when the compression zone fails by shearing along AB (Fig.11). The details of the theory are given elsewhere (17).

A comparison of the theoretical and experimental ultimate punching shear strengths of slabs without and with fibres is shown in Table 21. The results show excellent support for the theory with the ratio of theoretical to experimental strength varying from 0.940 to 0.972 and the standard deviation from 1.0 to 6.4%. To illustrate the effect of the four major parameters on punching shear strength, namely, fibre percentage, fibre typ concrete compressive strength and tension reinforcement ratio, Table 22 is presented in which the theoretical and experimental ultimate punching strengths are compared. The results again show close agreement between test results and theoretical prediction.

DURABILITY OF STEEL FIBRES

One of the main problems associated with the use of steel fibres is their durability in concrete structures. Corrosion of steel fibres can lead to the loss of their ability not only to arrest and control crack propagation but also to contribute to the load capacity of the member at service and ultimate load conditions. These design and performance characteristics cannot be ignored if one is to ensure the long-term serviceability behaviour of structures.

There is, however, growing evidence that in practice, in good quality concrete, fibre corrosion does not penetrate into the concrete, and that it is confined to fibres that are exposed at the surface (which breakaway in time in any case) (18). Cracked reinforced concrete beams exposed to open atmosphere for over ten years have exhibited negligible carbonation and virtually corrosion free existence of steel bars and steel fibres (18). The lack of durability of steel bars cannot therefore be taken as an automatic corollary of their presence in concrete structures.

In very aggresive environments, it is possible to use melt extracted stainless steel fibres which are totally immune to corrosion. The higher cost of these fibres has practically no effect on the overall cost of a structure bearing in mind the relative cost of materials, fabrication and constructions. The problem of steel fibre corrosion is therefore no longer a design or cost inhibiting parameter.

CONCLUSIONS

The data presented in this paper show that steel fibre concrete of structural quality can be satisfactorily proportioned for use in concrete bridge decks and building floors with adequate flow and compactibility characteristics without the danger of fibre bundling or non-uniform distribution of the fibres. It is shown that such concrete can be designed to have adequate early strength, and can enhance the cracking, deformation and strength behaviour of reinforced concrete slabs made with fibre concrete. Extensive data are presented to show that in both normal weight and lightweight concrete slabs the presence of fibres controls cracking and deformation, enhances serviceability and diagonal cracking loads, and imparts dramatic improvements in ductility and energy absorption properties. All there properties can be exploited to advantage in bridge decks and building floor slabs. Steel fibre corrosion need no longer pose a threat to the long-term stability and service life of concrete structures

REFERENCES

(1) ACI Committee 544, "State-of-the-Art report on fibre reinforced concrete", ACI Publication SP-44, Am.Conc. Institute, Detroit, Michigan, 1974, pp.535-550.

(2) Swamy, R.N., "Fibre reinforcement of cement and concrete", RILEM Materials and Structures, Vol.8, No.45, May-June 1975, pp.235-254.

(3) Swamy, R.N., "The technology of steel fibre reinforced concrete for practical applications", Proc. Instn. of Civil Engrs., Part 1, May 1974, pp.143-159.

(4) Swamy, R.N., Ali, Sami. A.R. and Theodorakapoulos, D.D. "Early strength fly ash concrete for structural applications", Journal, American Concrete Institute, Proc. Vo.80, No.5, Sept-Oct. 1983, pp.414-423.

(5) Ellis, C., "The application of the two-point workability test and British Standard tests to OPC/PFA concretes", Proc. Internat. Symp. on The Use of PFA in Concrete, Leeds, April 1982, Vol.1, pp.121-131.

(6) Banfill, P.F.G., "An experimental study of the effects of PFA on the rheology of fresh concrete and cement paste", Proc. Internat. Symp. on the Use of PFA in Concrete, Leeds, April 1982, Vol.1, pp.103-109.

(7) Swamy, R.N. and Lambert, G.H., "The microstructure of Lytag aggregate", International Journal of Cement Composites and Lightweight Concrete, Vol.3, No.4, Nov. 1981, pp.273-282.

(8) Ramakrishnan, V., Coyle, W.V., Kopac, Peter A, and Pasko, Thomas J.J., "Performance characteristics of steel fibre reinforced superplasticized concrete", Developments in the use of superplasticizers, ACI Publication SP-68, Am. Conc. Inst., Detroit, 1981, pp.515-534.

(9) Swamy, R.N., and Jojagha, A.H., "Workability of steel fibre reinforced lightweight aggregate concrete", The International Journal of Cement Composites and Lightweight Concrete, Vol.4, No.2, May 1982, pp.103-109.

(10) Malmberg B. and Skarendahl, Å, "Method of studying the cracking of fibre concrete under restrained shrinkage", RILEM Symposium on Testing and Test Methods of Fibre Cement Composites, Sheffield, April 1978, Construction Press, 1978, pp.173-179.

(11) Swamy, R.N., and Stavrides, H., "Influence of fibre reinforcement on restrained shrinkage and cracking", ACI Journal, Proc. Vol.76, No.3, March 1979, pp.443-460.

(12) Swamy, R.N., Ali, S.A.R. and Theodorakopoulos, "Engineering properties of concrete composite materials incorporating fly ash and steel fibres", ACI Special Publication SP-79, American Concrete Institute, Detroit, Michigan, 1983, pp.559-588.

(13) Henager, Charles H., and Doherty, Terence, J., "Analysis of reinforced fibrous concrete beams", Proc. ASCE, Vol.102, ST1, Jan. 1976, pp.177-188.

(14) Swamy, R.N. and Al-Ta'an, Sa'ad A., "Deformation and ultimate strength in flexure of reinforced concrete beams made with steel fibre concrete", ACI Journal, Proc. Vol.78, No.5, Sept.-Oct. 1981, pp.395-405.

(15) Swamy, R.N. and Theodorakopoulos, D.D., Unpublished data.

(16) Swamy, R.N., Al-Ta'an, A., and Ali, Sami, A.R. "Steel fibres for controlling cracking and deflection", Concrete International Design and Construction, Vol.1, No.8, Aug. 1979, pp.41-49.

(17) Swamy, R.N. and Theodorakopoulous, D.D., Unpublished data.

(18) Unpublished data, Seminar on Durability of FRC materials, Sheffield, April 1985.

(19) Swamy, R.N., Jiang Ende and Yue Changnian, Unpublished data.

Table 1 Workability Properties of Steel Fibre Concrete Containing PFA. (Normal weight aggregates)

Mix	% of Fibre by volume	Slump mm	Vebe Time sec
Plain	-	103	2.4
FRC-A	0.6	84	2.8
FRC-B	0.9	66	3.8
FRC-C	1.2	45	4.9

Mix : Cement : PFA : Sand : Gravel 10mm : Water/(Cement + PFA) : 0.7 : 0.3 : 1.8 : 2.2 : 0.47

Admixture/(Cement + PFA) : 0.6%

Fibres : steel, crimped 0.5 x 50mm

Table 2 Workability Properties of Steel Fibre Lightweight Concrete Mixes

Mix*	Fibre Type	Fibre Size mm	Fibre Volume %	Admixture cc/1 kg (Cement + P.F.A.)	Slump mm	Vebe time Sec.
A	-	-	0.0	2.5	160	1
B	Crimped	0.5 x 50	0.5	4.0	90	3
C	Crimped	0.5 x 50	1.0	5.5	40	7
D	Crimped	0.418 x 25	1.0	3.5	65	4.5
E	Hooked	0.5 x 50	1.0	3.5	50	5.5
F	Paddle	0.76 x 53	1.0	3.5	55	5.5
G	Crimped	0.425 x 38	1.0	4.0	70	3.5

* Cement : PFA : Sand : Lytag : 287 : 123 : 560 : 696 (kg/m^3)

Effective Water/(Cement + PFA) ratio = 0.40

Table 3 Strength Characteristics of FRC Concrete with PFA and Gravel Aggregates

Mix	Fibre Vol.	Method of Vibration	Comp.St.N/mm^2			E, kN/mm^2			Flex.St. N/mm^2			Split St. N/mm^2			Poisson's Ratio
			1d	7d	28d	1d	7d	28d	1d	7d	28d	1d	7d	28d	
Plain	-	Table	8.7	29.7	43.2	21.0	23.9	33.3	1.6	3.8	3.8	0.7	2.4	3.3	0.19
FRC-A	0.6	Table	11.1	28.7	43.4	24.6	28.1	35.5	2.5	4.7	4.9	1.8	3.4	4.4	0.15
FRC-A	0.6	Internal	7.1	28.5	38.5	-	-	-	2.0	4.4	4.5	1.4	3.1	3.7	-
FRC-B	0.9	Table	11.0	34.7	45.4	26.1	37.7	39.4	2.5	6.0	6.1	1.7	4.3	5.4	0.13
FRC-C	1.2	Table	10.8	31.3	41.5	28.8	36.6	37.2	2.6	4.4	6.7	2.0	4.6	6.0	0.15

Table 4 Compressive Strength of Fibre Concrete with Lightweight Aggregates

Mix \ Age	Compressive Strength N/mm^2						
	1d	7d	28d	90d	180d	350d	540d
A	7.4	37.0	45.3	46.0	51.3	50.5	52.6
B	7.6	37.6	43.8	47.8	49.1	48.8	50.9
C	9.1	40.5	47.1	52.6	-	-	-
D	8.5	34.5	44.2	49.9	47.3	50.8	50.5
E	8.8	38.0	46.1	50.5	49.6	52.4	52.8
F	8.2	35.9	42.8	48.5	47.3	51.4	49.3

Table 5 Tensile Strength of Fibre Concrete With Lightweight Aggregates

Mix	Tensile Strength, N/mm^2							
	1d	3d	7d	28d	90d	180d	350d	540d
A(F)	1.1	-	2.9	3.2	4.8	4.9	4.7	4.6
A(TS)	0.6	-	2.6	3.0	3.1	3.1	3.4	3.5
B(F)	1.8	4.0	5.1	5.7	5.7	5.8	5.5	5.7
B(TS)	1.2	2.2	3.5	4.0	4.0	4.1	4.1	4.3
C(F)	2.4	4.9	6.1	6.7	7.0	-	-	-
C(TS)	1.2	2.6	3.8	4.7	4.8	-	-	-
D(F)	1.7	3.8	4.8	5.4	5.8	5.6	5.8	5.7
D(TS)	1.1	2.3	3.5	3.8	4.1	4.2	4.3	4.2
E(F)	2.3	5.3	6.2	7.1	7.2	7.1	7.2	7.0
E(TS)	1.0	2.1	3.3	4.1	4.6	4.4	4.5	4.7
F(F)	2.0	5.1	6.0	6.8	6.9	7.0	6.9	6.7
F(TS)	1.2	2.3	3.6	4.4	4.5	4.6	4.7	4.7

(F) - flexural strength

(TS) - tensile splitting strength

Table 6 Elastic Modulus of Fibre Concrete Containing Lightweight Aggregates

Age / Mix	Modulus of Elasticity, kN/mm^2						
	3d	7d	28d	90d	180d	360d	540d
A	10.8	16.4	17.4	16.6	17.3	18.9	19.2
B	12.5	16.5	17.2	17.5	17.8	19.9	20.5
C	12.9	17.6	19.4	20.3	-	-	-
D	12.4	17.6	18.9	19.9	20.2	21.3	21.9
E	12.7	17.2	19.5	19.9	20.7	22.1	22.9
F	12.9	17.9	19.5	20.8	21.0	21.8	23.0

Table 7 Effect of Curing Regime on Strength Properties

Series No.	Mix Details+	Age Days	Flexural Strength N/mm^2	Compressive Strength* N/mm^2
M3	Fibre concrete Water-reducing agent Air curing	1	3.2	12.8
		3	4.8	26.8
		8	5.8	37.0
		28	6.7	47.1
		50	6.8	49.7
M6	Fibre concrete Water-reducing agent 3 days water + air curing	1	2.3	9.6
		3	5.5	27.2
		8	6.4	41.9
		28	8.5	52.1
		38	8.6	56.1

+ Mix 0.7 : 0.3 : 1.85 : 2.15 : 0.465 : C : PFA : S : CA : W
Fibres : Crimped 0.50 x 50mm

* Based on 150mm cubes

Table 8 Influence of Fibres on Resistance to Restrained Shrinkage Stresses

Specimen	Type of Fibre	Fibre Vol. %	Composite Strength, N/mm^2 (28 days)			Tensile Resistance to Restrained Shrinkage N/mm^2
			Compressive	Flexural	Direct Tensile	
Series A - Normal Aggregate Concrete						
R1	Unreinforced	-	47.00	4.80	2.60	1.60
R2	Steel-Plain	0.5	48.50	7.10	2.90	2.20
R3	Steel-Plain	1	50.90	8.20	3.00	2.50
R4	Steel-Crimped	1	51.00	8.80	3.00	3.30
Series B - Lightweight Aggregate Concrete						
L1	Unreinforced	-	31.00	3.20	-	1.25
L2	Steel-Plain	1	31.50	4.90	-	1.25

Series A : Fibres 0.51 x 38.1 mm

Series B : Fibres 0.51 x 50.8 mm

Table 9 Flexural Strength of Normal Weight Concrete Slabs with Steel Fibres

Slab No.	Tension Steel	Compression Steel	Concrete Strength N/mm^2	Fibre* Volume %	Failure Load kN	Mode of failure
S1	12-10mm	7-8mm	50.7	0.0	198.0	Punching
S16	8-10mm	3-8mm	39.0	0.9	213.0	Flexure
S10	7-10mm	3-8mm	46.4	0.9	203.0	Flexure
S9	6-10mm	3-8mm	51.6	0.9	179.0	Flexure
S19	6-10mm	3-8mm	47.0	0.0	130.7	Flexure

* Steel fibres confined to a distance of 3.5h from column faces all around

Table 10 Flexural Strength of Lightweight Concrete Slabs with Steel Fibres

Slab No.	Tension Steel	Compression Steel	Concrete Strength N/mm^2	Fibre* Volume %	Failure Load kN	Mode of failure
FS1	12-10mm	7-8mm	44.2	0.0	174.0	Punching
FS19	8-10mm	7-8mm	43.1	0.0	137.0	Punching
FS7	8-10mm	3-8mm	45.8	1.0	192.0	Flexure
FS6	8-10mm	-	44.6	1.0	175.0	Flexure

* Fibres distributed to a distance of 3.8h from column faces all around

Table 11 Comparison of Predicted and Experimental Ultimate Flexural Strength of Slabs

Investigation	No. of Slabs	Type of Concrete	Fibre Type	Fibre Vol. %	Theory/Expt.	
					Av. value	S.D.
Ref. 15	4	LW	Crimped/paddle	1.0	0.988	0.043
Ref. 15	3	NW	Crimped	0.9	0.901	0.013
Ref. 15	3	NW/LW	Plain	0.91 - 1.74	0.952	0.027
Ref. 15	1	NW	Plain	1.0	0.904	-

LW - lightweight; NW - normal weight; SD - standard deviation

Table 12 Deformation Characteristics : Norman Weight Concrete Slabs

Slab Number	Fibre Vol. %	Load kN	Deformations		
			Maximum Tensile Steel Strain x 10^6	Maximum Concrete Strain x 10^5	Maximum Rotation radians x 10^3
First Crack Load					
S1	0.0	35.0	426	24.3	0.95
S2	0.6	43.1	470	22.7	1.07
S3	0.9	56.9	364	25.0	1.41
S4	1.2	53.1	290	33.5	1.45
Service Load					
S1	0.0	118.1	2369	159.3	9.30
S2	0.6	118.1	1877	98.9	7.75
S3	0.9	118.1	1265	82.4	5.38
S4	1.2	118.1	847	79.8	5.38
Prior to Failure					
S1	0.0	197.7	13000	291.0	41.2
S2	0.6	243.6	9890	341.0	29.7
S3	0.9	262.9	10980	367.0	24.7
S4	1.2	281.0	10490	368.0	27.9

All the slabs are identical except for the fibre volume

Table 13 Deformation Behaviour : Lightweight Concrete Slabs

Slab Number	Fibre Vol. %	Load kN	Maximum Tensile Steel Strain x 10^6	Maximum Concrete Strain x 10^6	Maximum Rotation Degrees
First Crack Load					
FS1	0.0	32.0	400	260	0.08
FS2	0.5	42.5	400	340	0.10
FS3	1.0	46.8	600	280	0.10
Service Load					
FS1	0.0	125.0	3950	1620	0.64
FS2	0.5	125.0	2600	1360	0.52
FS3	1.0	125.0	2500	1200	0.49
Prior to Failure					
FS1	0.0	173.5	7600	2700	1.15
FS2	0.5	225.0	10500	4240	1.85
FS3	1.0	247.4	11750	5140	2.60

Slabs FS1, FS2 and FS3 are identical except for the fibre volume

Table 14 Service Loads Based on Deformation Criteria

Slab No.	Fibre percent by volume V_f	Service Load, kN							
		Deflection mm	Service Load kN	Steel Strain x 10^6	Service Load kN	Concrete Strain x 10^5	Service Load kN	Rotation in radians x 10^3	Service Load kN
1	2	3	4	5	6	7	8	9	10
S1	0	6.72	118.1*	2369	118.1*	159.3	118.1*	9.30	118.1*
S2	0.6	6.72	133.4	2369	132.5	159.3	153.7	9.30	132.9
S3	0.9	6.72	147.8	2369	161.7	159.3	173.8	9.30	157.1
S4	1.2	6.72	159.2	2369	195.7	159.3	183.2	9.30	161.9

* Slabs S1 : Service Load 118.1 kN according to CP110

Table 15 Lightweight Concrete Slabs : Service Loads Based on Deformation Criteria

Slab No	Fibre Vol. %	Deflection mm	Service Load kN	Steel Strain x 10^6	Service Load kN	Concrete Strain x 10^6	Service Load kN	Rotation Degrees	Service Load kN
FS1	0.0	9.15	125.0	3950	125.0	1620	125.0	0.64	125.0
FS2	0.5	9.15	147.0	3950	156.0	1620	142.5	0.64	147.0
FS3	1.0	9.15	150.0	3950	160.0	1620	147.0	0.64	150.0

Table 16 Normal Weight Concrete Slabs - Diagonal Tension Cracking Load

Slab No	Fibre percent by volume V_f	Load at which diagonal tension cracking developed, kN	Maximum Load kN	percent of $\frac{(3)}{(4)}$	Mode of failure	Remarks
1	2	3	4	5	6	7
S1	0	78.0	197.7	39.45	Punching	Equal reinforcement distribution
S2	0.6	121.9	243.6	50.04	Punching	
S3	0.9	140.1	262.9	53.29	Punching	
S4	1.2	162.6	281.0	57.86	Punching	
S7	0	180.3	221.7	81.33	Punching	Reinforcement concentration near the column
S11	0.9	214.5	262.0	81.87	Punching	
S18	1.37	195.3	265.7	73.50	Punching	
Effect of Type of Fibre						
S11*	0.9	170.1	262.0	64.9	Punching	Fibres continued to 3h from face of column all round
S12+	0.9	152.1	249.0	61.0	Punching	
S13x	0.9	150.0	236.7	63.4	Punching	

* crimped fibre

\+ hooked fibre

x straight fibre

Table 17 Lightweight Concrete Slabs - Diagonal Tension Cracking Load

Slab No	Fibre Vol. %	D.T. cracking load kN	Max. load kN	D.T. load / Max. load	Mode of failure
FS1	0.0	69.4	173.5	40.0	Punching
FS2	0.5	120.0	225.0	53.3	Punching
FS3[a]	1.0	150.0	247.4	60.6	Punching
Effect of Type of Fibre					
FS12*	1.0	160.0	217.5	73.6	Punching
FS13+	1.0	177.5	235.5	75.4	Punching
FS14x	1.0	170.0	239.5	71.0	Punching

a crimped fibre, 0.50 x 50mm, aspect ratio 100
* Fibre rect. section, 0.418 x 25mm, aspect ratio 60
\+ Round, hooked fibre, 0.50 x 50mm, aspect ratio 100
x Paddle fibre, elliptical section, 0.76 x 53mm, aspect ratio 70

Table 18 Post-yield Ductility and Energy Absorption Capability - Normal Weight Concrete

Slab No	Fibre Vol. %	Ductility Δ_2/Δ_1	Increase in ductility %	Energy absorption capacity kNm	Increase in energy absorption capacity
S1	0.0	33.0	1.00	4.10	0
S2	0.6	53.7	1.63	11.00	2.68
S3	0.9	62.3	1.89	17.07	4.17
S4	1.2	70.9	2.15	16.83	4.11
S11*	0.9	58.8	1.78	17.38	4.24
S12+	0.9	51.7	1.57	13.91	3.39
S13x	0.9	48.6	1.47	11.93	2.91

* crimped fibre confined to 3h around column
\+ hooked fibre confined to 3h around column
x straight fibre confined to 3h around column

Table 19 Post-yield Ductility and Energy Absorption - Lightweight Concrete

Slab No.	Fibre Vol. %	Ductility Δ_2/Δ_1	Increase in ductility %	Energy absorption capacity kNm	Increase in energy absorption capacity
FS1*	0.0	23.5	1.00	3.30	1.00
FS4*	1.0	53.0	2.25	11.13	3.37
FS19+	0.0	21.2	1.00	2.40	1.00
FS20+	1.0	76.7	3.62	8.86	3.69

* steel ratio 0.5574, paddle fibres confined to 550mm all round column

+ steel ratio 0.3716, crimped fibres over entire slab

Ductility ratio based on deflection at 30% of maximum load

Table 20 Shear Capacity of Steel Fibre Reinforced Concrete Slabs

Slab No	Steel reinf. Tension	Steel reinf. Compression	Fibre Vol. %	Failure Load Punching kN	Failure Load Flexure kN
Normal Weight Concrete Slabs					
S1	12-10mm	7-8mm	0.0	197.7	-
S2	12-10mm	7-8mm	0.9	262.9	-
S8	12-10mm	3-8mm	0.9	255.7	-
S16	8-10mm	3-8mm	0.9	-	213.0
S10	7-10mm	3-8mm	0.9	-	203.0
S9	6-10mm	3-8mm	0.9	-	179.3
Lightweight Concrete Slabs					
FS1	12-10mm	7-8mm	0.0	173.5	-
FS2	12-10mm	7-8mm	1.0	225.0	-
FS4	12-10mm	-	1.0	224.4	-
FS5	8-10mm	7-8mm	1.0	198.1	-
FS7	8-10mm	3-8mm	1.0	-	192.4
FS6	8-10mm	-	1.0	-	174.5
FS20	8-10mm	-	1.0	211.0	-

Table 21 Comparison of Predicted and Experimental Ultimate Punching Shear Strength of Slabs

Investigation	No. of Slabs	Type of concrete	Fibre type	Fibre Vol. %	Theory/Expt.	
					Av. value	S.D.
Ref. 17	4	LW	-	-	0.970	0.010
Ref. 17	12	LW	Crimped, hooked, and paddle	0.5 - 1.0	0.940	0.064
Ref. 17	9	NW	Plain, crimped and hooked	0.6 - 1.2	0.972	0.040

Table 22 Influence of Test Variables on Predicted Shear Strength of Slabs

Variable	Range	No. of slabs	Theory/Expt.	
			Av. value	S.D.
Fibre volume	0.5 - 1.2%	5	0.945	0.059
Fibre type	Plain, crimped, hooked, paddle	9	0.942	0.077
Compressive Strength	17.8 - 51.4 N/mm^2	8	0.985	0.042
Tension Steel Ratio	0.3716 - 1.83	5	0.905	0.104

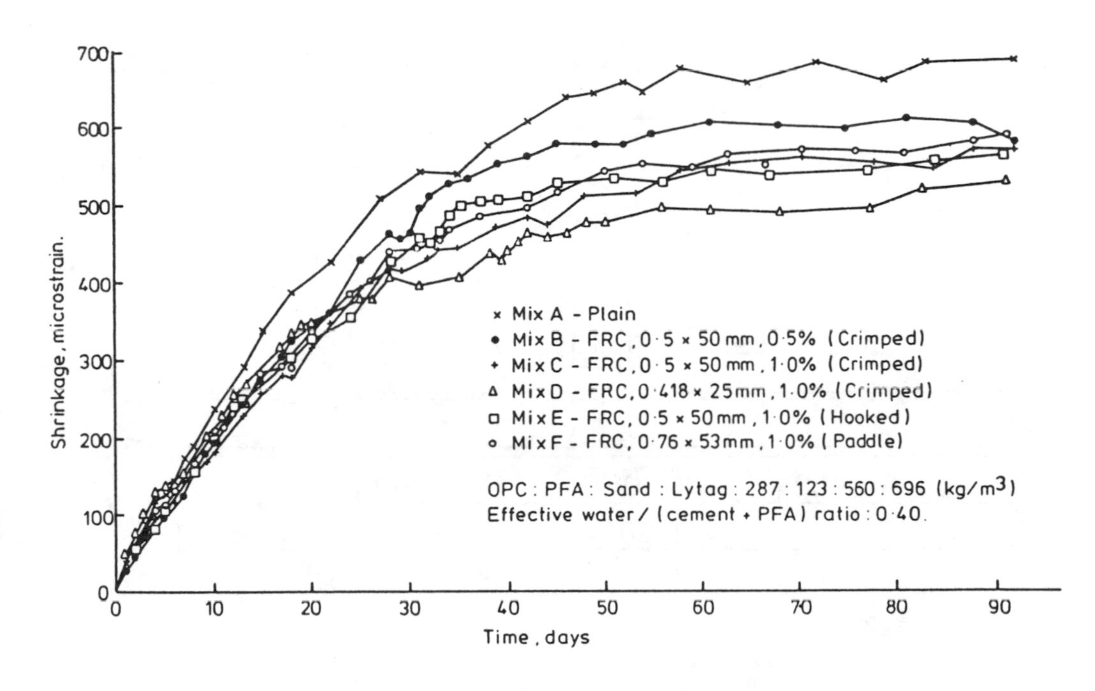

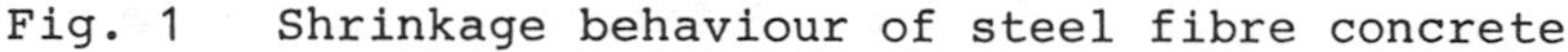

Fig. 1 Shrinkage behaviour of steel fibre concrete

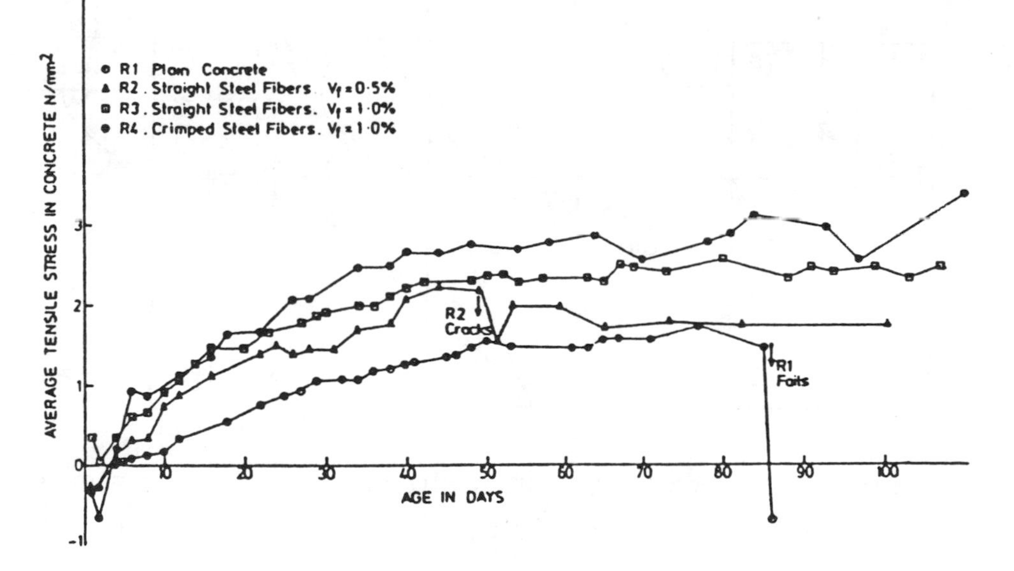

Fig. 2 Influence of steel fibres on restrained shrinkage stresses

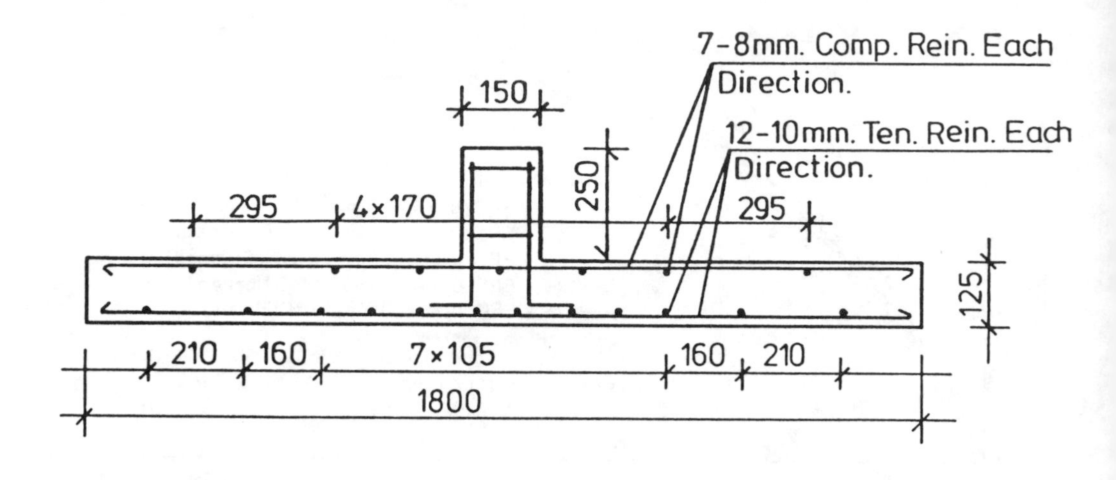

Fig. 3 Typical slab reinforcement details with load configuration

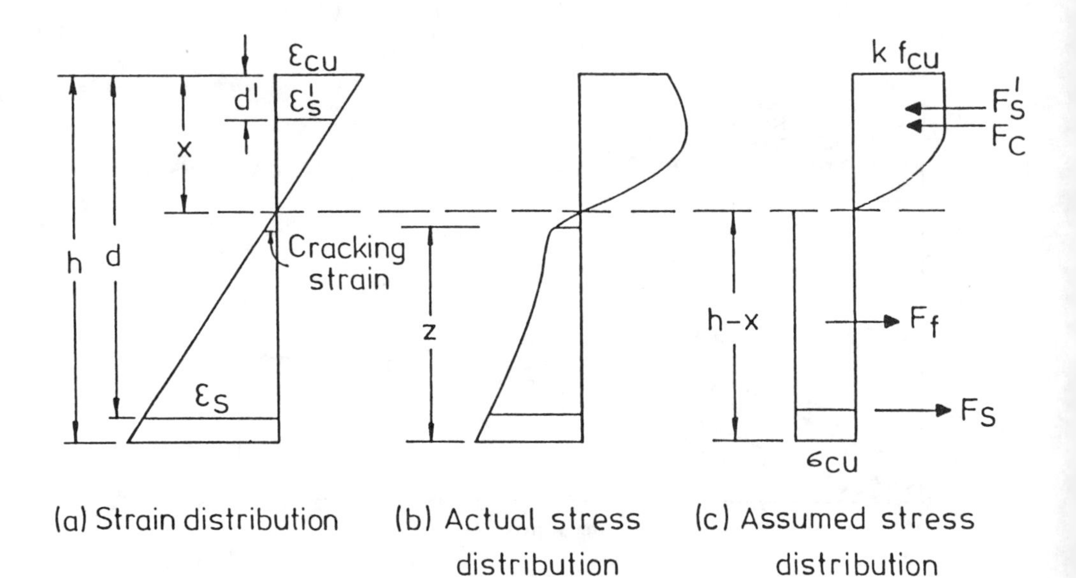

Fig. 4 Strain and stress distribution at failure

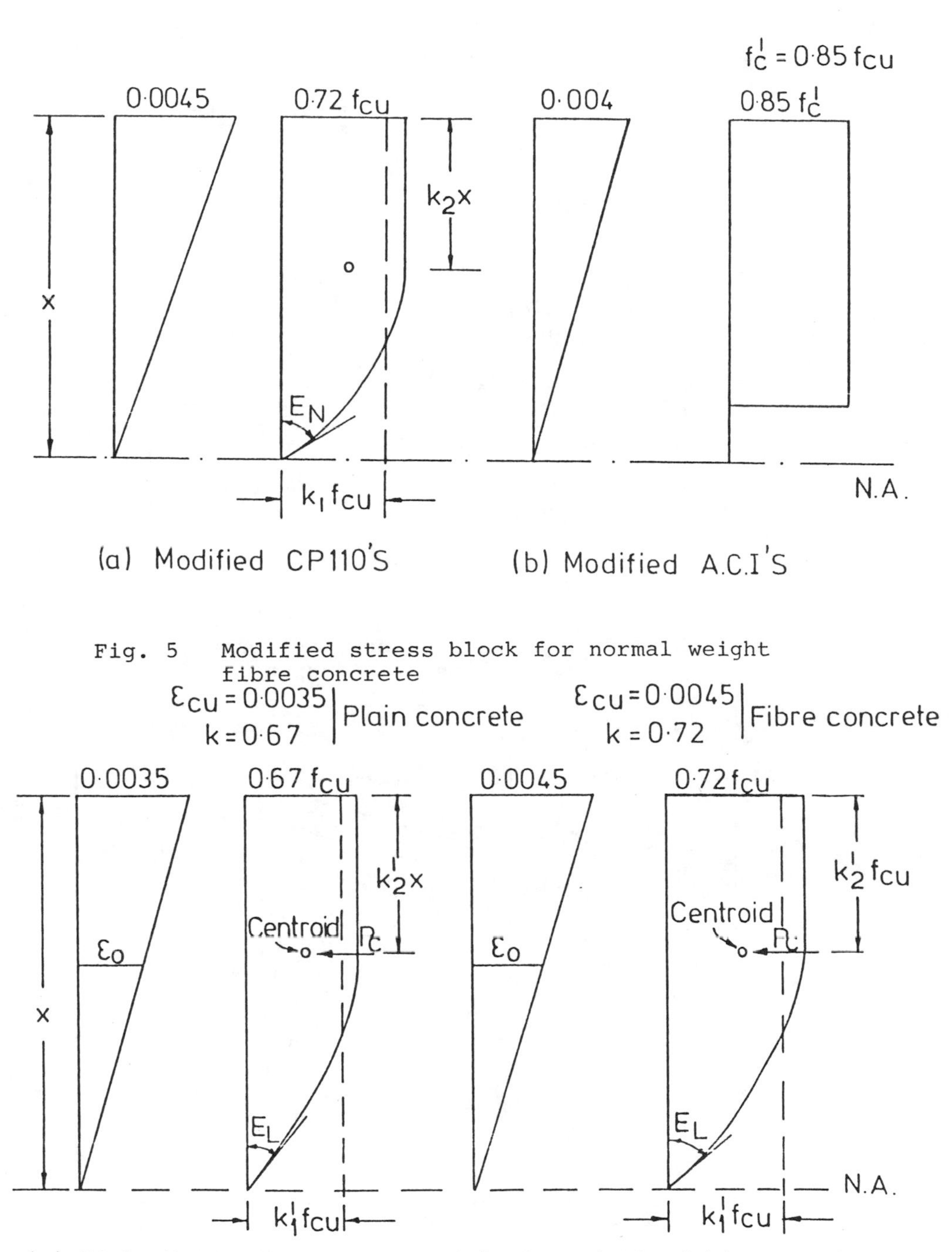

Fig. 5 Modified stress block for normal weight fibre concrete

Fig. 6 Modified stress block for lightweight fibre concrete

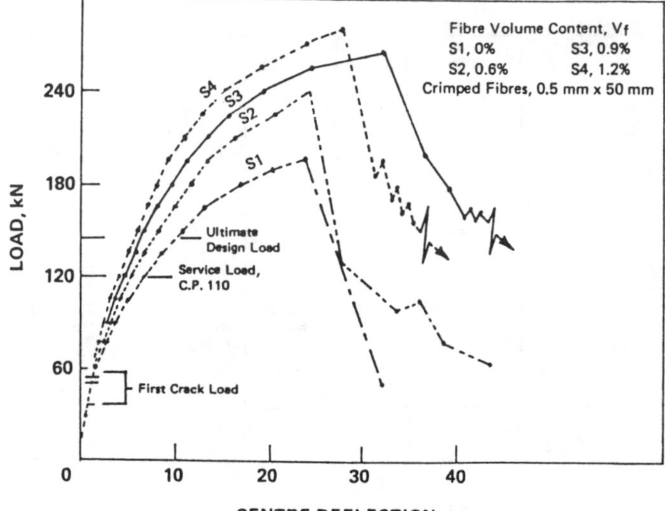

Fig. 7 Influence of steel fibres on deflection behaviour of slabs at mid-span

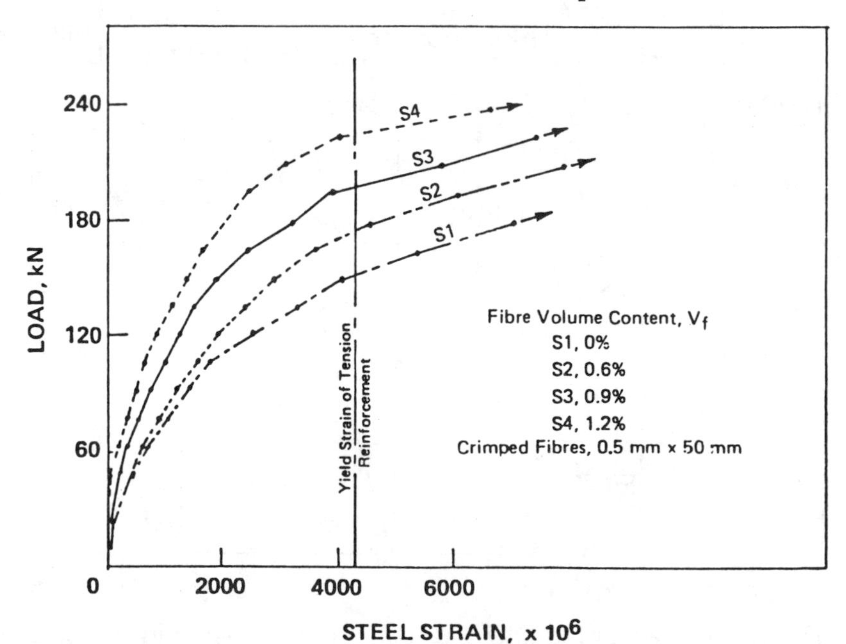

Fig. 8 Influence of steel fibres on tension steel strain near mid-span

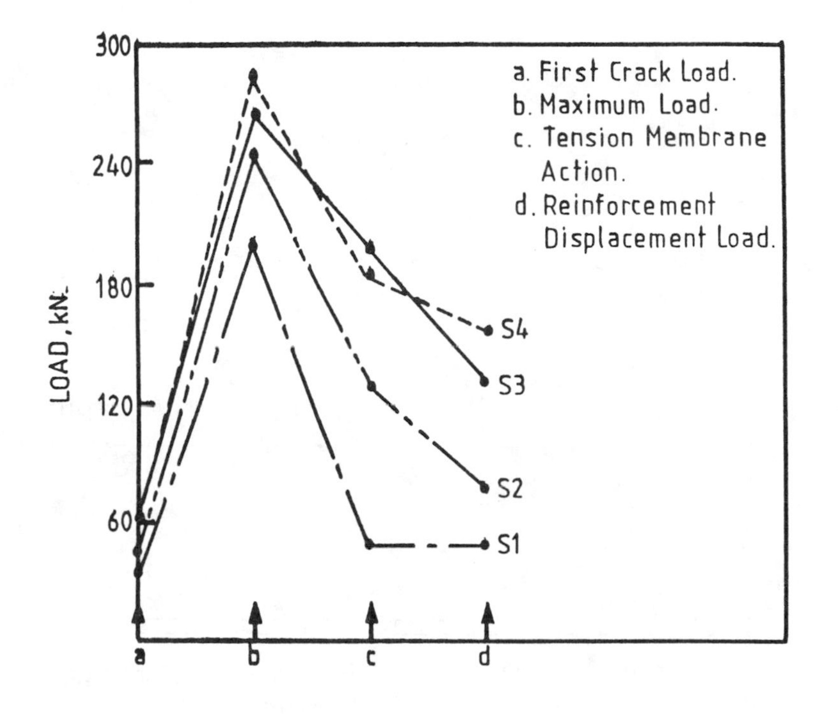

Fig. 9 Residual load capacity of slabs after failure

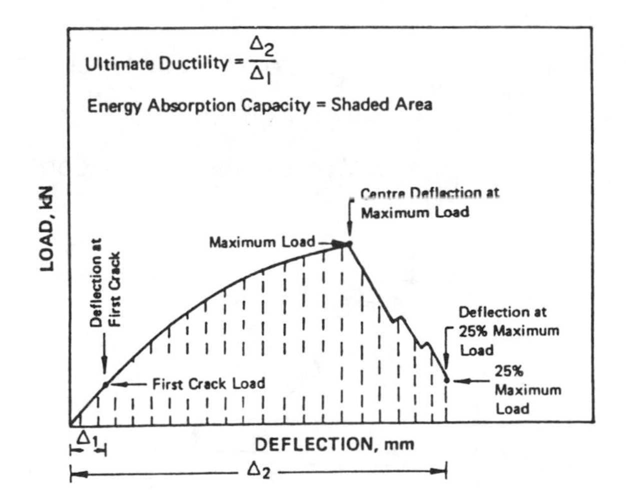

Fig. 10 Determination of ductility and energy absorption capability

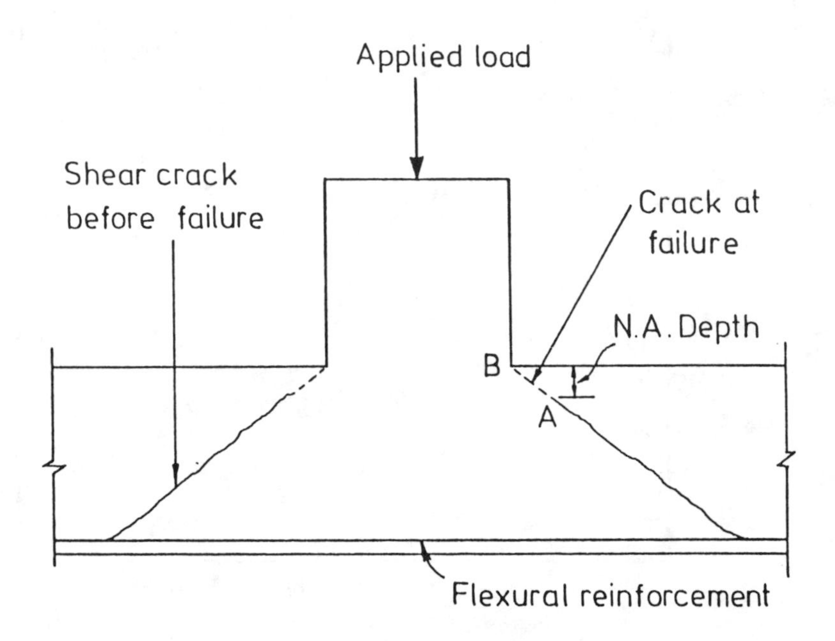

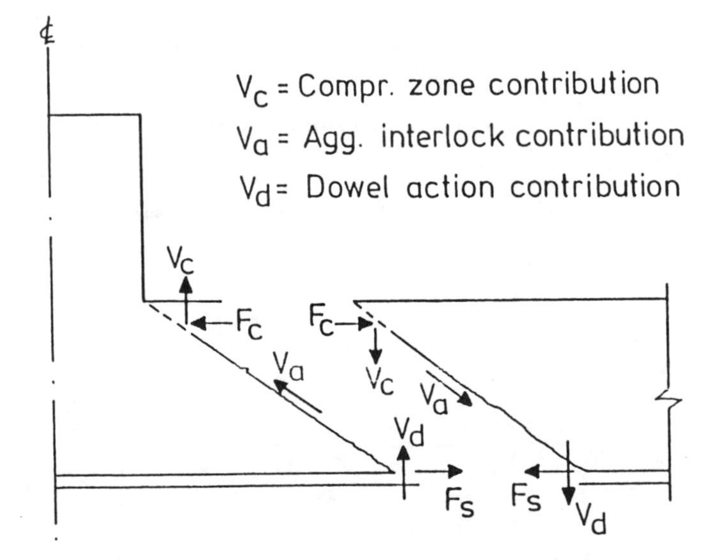

Fig. 11 Failure model for punching shear failure in slabs

STEEL FIBER CONCRETE
US-SWEDEN joint seminar
(NSF-STU)
Stockholm 3-5 June, 1985
S P Shah and Å Skarendahl,
Editors

DURABILITY OF SFRC EXPOSED TO SEVERE ENVIRONMENTS

by Morris Schupack, President, Schupack Suarez Engineers, Inc Structural Consultants, South Norwalk, Connecticut, USA

ABSTRACT

The corrosion performance experience of SFRC in three different aggressive environments, and the results of laboratory tests performed, are described. In general, the results indicate that, in about a ten year period, in a freezing and thawing moderate marine environment, when uncracked, good quality concrete (water cement ratio under 0.5, entrained air, good consolidation) is used, the corrosion of steel fibers is limited to the surface of the uncracked concrete. After 10 years exposure in a severe freezing and thawing environment, SFRC has about the same durability as a non-reinforced reference specimen. In a marine tidal exposure, cracked SFRC, with time, tends to show corrosion on the wires passing through the cracks.

Morris Schupack, president of Schupack Suarez Engineers, and a U. S. pioneer in prestressed concrete, has long term involvement studying the performance of concrete reinforcing materials. As a practicing structural engineer, he has worked on developing fibrous concrete products, has developed various proprietary concrete composites, and has 12 patents related to concrete structures and material. He serves on numerous committees dealing with the promulgation of codes and practices including the ACI Fibrous Concrete Committee. He is a director of the Post Tensioned Institute, a past director of the American Concrete Institute and author of 35 concrete related papers.

INTRODUCTION.

The corrosion behavior of steel fibers in SFRC has not been as methodically studied as those of reinforcing steel and prestressing steel. The very serious corrosion performance problems occurring in reinforced concrete structures in the United States and Canada, particularly those subjected to deicing salts, point out the need to understand the long term corrosion problems of SFRC. This paper will attempt to bring together some observations of the writer and others. There are other tests and information available which should be compiled and made the subject of a more extensive and scientific study.

In general, the findings regarding the corrosion behavior of SFRC have been good. Corrosion of the fibers at the surface seems to stop there. However, when cracks occur in SFRC, the corrosion of the wires passing through the cracks can be expected with time.

As a structural consultant, my practice frequently involves the evaluation of the performance of prestressed concrete structures where corrosion of the prestressing and reinforcing steel has occurred. This experience has made me take an interest also in the durability of SFRC. My evaluation of the behavior of SFRC has been made subjectively rather than following an analytical scientific approach. Hopefully, the researchers in this area will take a more direct scientific approach regarding the long term behavior of SFRC.

If one has the time, obviously the most dependable test of the durability of SFRC is to expose it to the actual environment in which it is to be used. Since the "time" factor makes this impractical for us mortals, either accelerated laboratory tests or severe natural tests can be used to obtain behavior information which then has to be "interpreted". Laboratory tests can reveal valuable information of probable durability. However, the numerous disruptive mechanisms in nature are so diverse and complex, that laboratory tests alone may be misleading. To obtain realistic natural weathering information, the use of extreme natural weathering environments can give accelerated data on performance.

Assembled in this paper are results from some very severe weathering sites and from less severe sites. Laboratory information, as reported by Battelle, is the only laboratory data included. From the writer's unsuccessful attempt to obtain more long term laboratory controlled data, it is obvious that more work in this area is desirable. It is hoped that work that may be ongoing will be reported in the literature in the near future.

WATERWAY EXPERIMENT STATION (WES) SFRC SPECIMENS EXPOSED AT THE U. S. CORPS OF ENGINEERS SEVERE WEATHERING STATION, TREAT ISLAND, MAINE.

Description of exposure station.

In July 1975, 39 SFRC beams and 11 reference plain concrete beams were installed at half-tide elevation on the exposure rack at Treat Island, Maine (Ref. 1), to determine the effects of sea water and freezing and thawing action on the flexural strength and other properties of various fiber concretes. This test site is located in the Cobscook Bay area near Eastport, Maine, the most easterly point in the United States. This almost ideal test site has been in use since 1936. It provides twice daily tidal submergence, the mean tidal range being about 6m (18 ft.), with a maximum of 8.5m (28 ft.) and a minimum of 4m (13 ft.)

In winter, the combination of air and water temperatures creates a condition in which specimens at the mean tide elevation are thawed to a temperature of about 2.8^{o} C (37 F) when covered with water, and are frozen to temperatures as low as -23°C (-10 F) when exposed in air. A recording thermometer, the bulb of which is embedded in the center of a reference concrete specimen, records these temperatures. A cycle of freezing and thawing consists of the reduction of the temperature at the center of a concrete specimen to below -2° C (28 F), and its subsequent rise to above that figure. The number of annual cycles has ranged from 71 to 188, with the average being 130. The effect of the relatively cool summers is to lessen, in general, autogenous healing and chemical reactions in the concrete.

The composition of sea water at Treat Island, Maine, in parts per million are: total solids - 35,275; calcium - 370; magnesium - 1,175; sodium - 9,500; potassium - 370; chloride - 17,100; and sulfate - 2,385.

This station has tested many types of concrete for the Corps of Engineers, the Reinforced Concrete Research Council, various U. S. government agencies and the Government of Canada. It is interesting to point out that similar more severe exposure stations exist in Siberia (Ref. 2). At these two Siberian stations, the temperature is colder and the freezing-thawing cycles are reported to be over 350 per year.

The author was involved in the design of a test program, and has been observing the behavior of pretensioned and post-tensioned beams placed at half tide at Treat Island, for the last 24 years (Ref. 3). He has had no direct involvement, however, with the fibrous concrete beams being reported upon, except as a casual observer on occasion of his visits to the exposure station every other year. The data on these tests have been extracted from the Corps of Engineers' report (Ref. 4) and other information was supplied by Henry Thornton of the Corps of Engineers, whose contribution is acknowledged with thanks.

FIELD EVALUATION METHODS.

Frequency readings.

The concrete specimens at all installations are subjected to a test for fundamental transverse frequency (test method CRD-C 18-59, Ref. 5) at regular intervals during exposure, unless prevented by their shape or size. The specimen is supported in a horizontal position at the nodes and caused to vibrate in its fundamental flexural mode. The resonant frequency is obtained by observing the maximum reading on a suitable meter as the applied frequency is varied. From this value, together with the size, shape and weight of the specimens, the dynamic modulus of elasticity (E) is determined. The moduli so determined are expressed as percentages of the initial dynamic modulus obtained at installation (%E). A specimen is considered as having failed if this percentage (%E) drops below 50 during the exposure.

Pulse velocity readings

The concrete specimens at all installations are subjected to pulse velocity tests (test method CRD-C 51-72, Ref. 5) at regular intervals during exposure, unless prevented by their size or shape. The test instrument measures the time of travel of a sound pulse through a concrete specimen. From the travel time and the path length, values for velocity of sound in concrete (V) are calculated. The square of the velocity thus determined is expressed as a percentage of the square of initial velocity obtained at installation ($\%V^2$). Example:

V_o = velocity of sound in a certain specimen at installation.
V_t = velocity of sound in this same specimen at at later date.

Therefore;

$$\%V^2 \text{ (at time t)} = \frac{V_t^2}{V_o^2} \times 100$$

Visual inspection.

All specimens are visually inspected periodically to ascertain their condition. Those specimens not suitable to quantitative testing are inspected more thoroughly to determine their condition and to permit comparisons of the durability of these specimens.

Criteria of failure.

Specimens are regarded as having failed when they separate into pieces, when %E is 50 or less, or when deterioration has progressed to such a point that reliable determinations of fundamental frequency and pulse velocity cannot be obtained. Specimens which have been broken in handling are so listed and noted as "failed".

Description of beams.

The beams were made from nine different mixtures. The fine and coarse aggregates were manufactured limestone sand and 19mm (3/4 in.) maximum limestone, respectively. All mixtures contained a water-reducing admixture (admixture B), and five mixtures contained an air-entraining admixture (admixture A). Type II portland cement was used in the amount of 468 Kg per M^3 (7.89 cwt per cu. yd.) except for mixtures N and O, which contained 653 Kg per M^3 (11.0 cwt per cu. yd.). The water cement ratio was 0.45 for all mixtures.

The number and types of beams exposed were: twelve 152 x 152 x 762mm (6 by 6 by 30 in.), twenty-one 152 x 152 x 914mm (6 by 6 by 36 in.) and seventeen 229 x 229 x 1143mm (9 by 9 by 45 in.). The 229 x 229 x 1143mm (9 by 9 by 45 in.) beams were yoked and stressed by third-point loadings to working loads of 35 percent of ultimate, see Fig. 1. Table 1, (WES-FC) gives the exposure record of the specimens. Additional mixture data is shown in Table 2.

FIG 1

229 x 229 x 1143mm SFRC beams subjected to bending, exposed at Treat Island, Maine.

TABLE 1. (WES-FC)

Record of Testing of Concrete Beams for WES Fibrous Concrete Program

Installed July 1975

Beam	Jul 1975 0 Cycles		1976 146 Cycles		1977 223 Cycles		1978 370 Cycles		1979 463 Cycles		1980 612 Cycles		1981 677 Cycles		1982 710 Cycles		1983 736 Cycles	
No.	Load, lb	fps	%E	$\%V^2$	%E	$\%V^2$	%E	$\%V^2$	%E	$\%V^2$	%E	$\%V^2$	%E	$\%V^2$	%E	$\%V^2$	%E	$\%V^2$
		9- by 9- by 45-in. Beams																
H-3	2720	16,095	•	103	•	102	•	101	•	106	•	107	•	105	•	95	Broke	
I-1	4340	15,560		97		97		97		94		102		100		79	NR	70
I-3		16,375		101		98		97		101		105		95	Cracked		Broke	
J-1		14,315		103		102		100		97		105		102		99		84
J-3		14,590		106		105		101		102		108		106		103		86
K-1		14,590		104		102		100		101		108		104		103		85
K-3		14,260		103		100		100		99		107		103		108		84
L-1		16,520		94		93		94		82		95	Failed		Failed		Broke	
L-3		16,305		99		101		98		95		103		92		86		75
M-1		14,590		98		99		98		100		114		103		99		82
M-3		15,060		101		100		99		103		109		105		102		85
N-1		14,765		108		103		104		106		108		108		106		88
N-3		14,705		103		103		102		109		107		107		105		88
O-1		14,150		108		107		105		110		110		108		105		89
O-3		14,370		104		105		102		105		108		106		104		88
P-1		14,940		106		103		102		107		106		106		98		85
P-2		15,245		109		105		105		111		109		110		99		88
		6- by 6- by 30-in. Beams																
H-7	None	16,235	98	105	106	103	102	103	101	103	100	107	100	107	102	95	102	75
H-8		15,725	102	105	109	100	100	101	98	100	98	100	100	103	100	96	100	74
H-15		15,825	106	107	110	101	102	103	102	105	102	103	96	104	106	96	98	74
H-16		15,825	102	108	108	99	102	101	103	104	107	100	107	101	106	99	108	76
I-8		15,925	120	105	116	97	102	103	102	101	110	100	112	101	106	97	106	76
K-8		14,970	109	109	125	99	116	101	107	104	107	99	111	105	100	100	96	78
L-7		16,130	94	108	109	100	105	101	103	99	99	94	99	94	98	76	102	74
L-8		16,130	103	108	103	101	101	101	102	97	100	89	98	103	91	93	96	70
M-7		14,125	100	102	103	94	103	97	101	93	103	106	103	71	98	94	96	76
O-8		13,890	100	103	103	101	105	97	100	99	107	105	89	100	102	75	100	103
O-16		15,245	88	110	97	100	95	101	97	102	108	111	115	110	100	77	92	112
P-8		14,705	106	105	106	100	104	100	109	100	112	113	112	106	95	104	104	79
		6- by 6- by 36-in. Beams																
I-7	None	15,875	119	103	96	98	94	99	94	86	94	100	96	100	109	93	127	76
I-15		16,130	105	109	114	102	111	103	111	103	103	102	111	104	110	99	128	80
J-7		14,495	103	106	112	99	100	101	97	103	97	106	99	106	100	96	103	82
J-8		14,495	109	104	118	96	97	98	190	98	190	101	145	100	108	94	326	102
J-15		14,495	100	104	100	97	100	99	97	99	94	100	97	104	86	96	.91	80
J-16		14,565	115	109	106	103	112	101	100	102	94	109	105	107	100	101	102	83
K-7		14,780	103	106	109	98	115	100	95	100	101	120	103	111	95	100	83	80
K-15		14,565	106	106	103	101	97	103	154	105	124	103	124	106	100	101	87	82
K-16		14,285	102	109	100	102	100	104	101	102	97	115	100	111	97	105	135	89
L-15		14,635	104	104	108	101	102	101	107	98	107	104	107	107	103	98	103	81
L-16		14,635	103	103	106	100	106	100	97	99	111	119	115	118	123	98	113	82
M-8		14,085	205	109	210	99	210	101	201	100	106	115	109	117	106	101	97	83
M-15		15,075	100	104	109	99	103	97	190	97	103	98	107	101	103	96	103	80
M-16		14,850	112	109	109	102	112	91	109	99	106	97	112	96	106	100	100	82
N-7		14,150	118	105	106	109	112	105	105	102	102	106	118	106	100	101	38	83
N-8		14,085	112	102	124	99	131	95	104	97	104	100	110	100	97	77	44	105
N-15		14,020	109	106	112	101	106	102	102	100	105	105	82	105	106	80	36	107
N-16		14,020	100	110	112	102	102	103	102	103	105	104	108	102	106	81	42	107
O-7		13,825	106	110	109	103	103	105	100	103	106	109	116	115	103	82	?	108
O-15		12,710	97	108	103	102	100	101	97	101	98	107	98	105	97	82	?	108
P-7		14,635	109	107	109	98	105	101	106	105	121	109	109	111	115	105	133	86

1. Loaded beams not tested for %E.
2. E and V^2 = 100% at start of test in July 1975.
3. Table extracted from WES Report.
4. Cycles of 33 in 1982 and 26 in 1983 are suspiciously low. An instrumentation problem may exist.
5. In 1983, V^2 readings that are over 100 are illogically high. Readings in 1985 should clarify this.

TABLE 2. Concrete Mixture Data

Mixture	Type	Fiber Length, mm	Slump, mm	Air Content %	Fiber Ratio by Volume
H	None	---	146	2.5	---
I	C	19	64	1.8	1.2
J	None	---	178	8.5	---
K	C	19	102	8.5	1.2
L	A	25	51	1.9	1.2
M	A	25	76	7.0	1.2
N	D	25	25	3.6	0.3
O	D	25	51	7.0	0.3
P	B	25	70	7.0	1.2

Interpretation of results to date.

To facilitate understanding the extensive data, Table 3 was assembled to permit a more direct comparison of beams that were exposed and subjected to flexure. The conclusions which possibly can be extracted, at this time, are as follows:

1. Beams with low air behaved the poorest.
2. Beams with high air and low % of fibers behaved the best.
3. Beams have not been autopsied to date, therefore depth of corrosion is not known.
4. There seems to be a trend showing that air entrained SFRC with a lower percentage of steel fibers performs better than with a high percentage, in this environment and for these concrete mixtures.

For the 152mm x 152mm (6 in. x 6 in.), beams that were not subjected to flexure the same conclusions as above are generally indicated. However, some test results are erratic and can not be explained at present.

LIGHTWEIGHT AGGREGATE SFRC EXPOSED IN THE TIDAL ZONE OF THE LONG ISLAND SOUND IN NORWALK, CONNECTICUT.

Description of specimens and exposure site.

This test program is a by-product of a SFRC product development project. In 1972, the author was asked to develop a lightweight fibrous concrete cable reel. Initially, it was my opinion that SFRC was not a viable medium either for performance or economic reasons. In the process of trying to develop this reel, a lightweight SFRC was developed and successfully cast into a usable product as shown in Fig. 2. Unfortunately, the economics of the reels could only be justified if they were mass produced.

TABLE 3. V^2 for Beams Exposed and Subject to Flexural Stresses

Beam 229mm x 229mm x 1143mm (9in. x 9in. x 45in.)

				Year:	July 1975		1976	1977	1978	1979	1980	1981	1982	1983
				Cycles:	0		146	223	370	463	612	677	710+	736+
Beam No.	Fiber Length mm	Slump mm	Entr'd Air %	Load Kg	fps	$\%V^2$	$\%V^2$	$\%V^2$	$\%V^2$	$\%V^2$	$\%V^2$	$\%V^2$	$\%V^2$	$\%V^2$
H-3	---	146	2.5	1233	16,095	100	103	102	101	106	107	105	95	*
I-1[A]	19	64	1.8	1969	15,560	100	97	97	97	94	102	100	79	70
I-3[A]	19	64	1.8	"	16,375	"	101	98	97	101	105	95	--	*
J-1	---	178	8.5	"	14,315	"	103	102	100	97	105	102	99	84
J-3	---	178	8.5	"	14,590	"	106	105	101	102	108	106	103	86
K-1[A]	19	102	8.5	"	14,590	"	104	102	100	101	108	104	103	85
K-3[A]	19	102	8.5	"	14,260	"	103	100	100	99	107	103	108	84
L-1[A]	25	51	1.9	"	16,520	"	94	93	94	82	95	*	--	--
L-3[A]	"	51	1.9	"	16,305	"	99	101	98	95	103	92	86	75
M-1[A]	"	76	7.0	"	14,590	"	98	99	98	100	114	103	99	82
M-3[A]	"	76	7.0	"	15,060	"	101	100	99	103	109	105	102	85
N-1[B]	"	25	3.6	"	14,765	"	108	103	104	106	108	108	106	88
N-3[B]	"	25	3.6	"	14,705	"	103	103	102	109	107	107	105	88
O-1[B]	"	51	7.0	"	14,150	"	108	107	105	110	110	108	105	89
O-3[B]	"	51	7.0	"	14,370	"	104	105	102	105	108	106	104	88
P-1[A]	"	70	7.0	"	14,940	"	106	103	102	107	106	106	98	85
P-2[A]	"	70	7.0	"	15,245	"	109	105	105	111	109	110	99	88

Beams loaded at third point to 35% of flexural ultimate strength.
A = 1.2% fibers by vol.
B = 0.3% " " "
* Failed + No. of cycles questionable (see Note 4 Table 1).

FIG 2. Reel with SFRC flanges - 1.2 diameter.

To establish the practicality of fabricating thin lightweight concrete SFRC, thin beam or slab sections were cast (7mm to 38mm). Different mixes and fibers were used to establish ranges of flexural strength and toughness. This development work proved that a tough and functional concrete cable reel could be made, with the major disadvantage of it being twice as heavy as a wooden reel.

Numerous flexural test specimens were made of the planned thicknesses of the drum and the flanges of the reel. The remains of these specimens were available for exposure tests. In the period of 1972 to 1974, the author placed a number of specimens into the tidal zone as shown in Fig. 3. The specimens were placed approximately at mid-tide, with some specimens partially inserted into the highly organic marsh silt. The site is located in a marsh of a tidal estuary, Village Creek, located approximately 500 meters from the Long Island Sound, in Norwalk, CT. The tidal range is, approximately, 2 to 3 meters. The salinity is about 28,000 parts/million. The water temperature in the Long Island Sound ranges from 0° Celsius (32° F) in the winter to about 22° Celsius (72° F) in the summer. Because of thermal pollution from a nearby thermal power plant, high temperatures at this site are estimated to be about 5° Celsius (8°F) higher. No measurements have been made at this site to determine actual cycles of freezing and thawing for specimens located at mid-tide. Crudely, it is estimated that 30 to 50 cycles occur per year.

FIG 3. Exposure test in a marine environment of SFRC previously tested flexural specimens.

The concrete mix used in most of these specimens is shown in Table 4. At the time of this development, after investigating many fibers, it was found that the greatest ductility and toughness was obtained with steel fibers 0.64mm (0.025") diameter x 64mm (2.5") long. These fibers created some difficulty in working the mix, but with the use of fairly small aggregate and a superplasticizer, the flanges of the reels could be cast by consolidating with external vibration.

TABLE 4. Typical lightweight concrete mix used in exposed specimens.

Constituent	Kg/Meter^3	Lbs./C.Y.
Solite sand SG=1.95	1025	1728
Cement type III	560	945
Water w/c = 0.46 (range: 0.42-0.55)	256	432
Steel fibers 1.5% by volume	118	200

Admixtures: Superplasticizer and air entraining agent.
Density: From 1750 to 2,000 Kg/M^3 (110 to 125 pcf)

Testing and findings.

The flexural strengths, after about ten years of exposure of uncracked specimens, seemed to be in the same general range as those shortly after fabrication. Using 1-1/2% fibers by volume, the average flexural strength was 17.3 MPa (2500 psi) for 25mm (1 in.) deep beams with 0.64mm diameter x 64mm fibers and 9.0 MPa (1300 psi) for 19mm (0.75 in.) deep beams with 0.25 x 0.50 x 19mm fibers. No doubt the forced orientation of the long fibers in this thickness had something to do with this. This also points out the need to make flexural specimens of the thickness to be actually used in the product being considered. Consolidation of specimens should also be equal to or similar to that to be used in fabrication.

Figure 4 shows a specimen tested in flexure after about 10 years exposure. The fibers are basically bright throughout except for the surface conditions. The specimens were visually inspected for cracks prior to testing and none were observed in this specimen. The flexural behavior seemed to support the visual observation. In none of the specimens, which seemed to be free of cracks, was corrosion found in any steel fibers immediately away from the exposed outer surface. As could be expected, fibers at outer surfaces were generally corroded, but did not seem to be disruptive to the matrix.

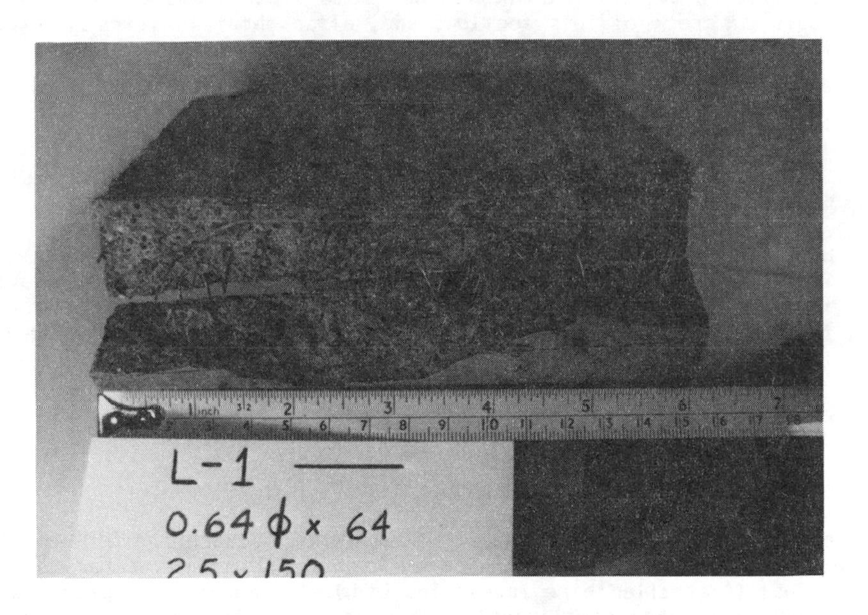

FIG 4. Lightweight SFRC beam - 25 x 150mm exposed in tidal site. Only surface fibers corroded.

Some of these uncracked specimens were pulverized to permit closer examination of the fibers. No fiber corrosion below the immediate surface was found. It is important to keep in mind that the cement rich mix of 560 kg per cubic meter of SFRC, with a water/cement ratio generally under 0.5, thoroughly consolidated, made a comparatively impervious concrete.

Since all specimens were remains of tested flexural specimens, a few had cracks away from the original failure plane. These cracked specimens did have corrosion progressing across the cross-section. Unfortunately, the original crack size is unknown.

Conclusions

For SFRC specimens exposed about 10 years in a marine environment,using the lightweight concrete mix described in Table 1, the following conclusions can be stated:

1. For uncracked concrete, for the steel fiber types used, the corrosion of fibers is limited to the surface of the specimen, even for specimens as thin as 19mm.

2. Uncracked specimens had no reduction in flexural strength.

3. For cracked specimens corrosion of some fibers does occur through the depth of crack of thin sections and, after about 10 years of exposure, there appears to be a reduction in flexural strength.

4. For air entrained SFRC at the exposure described, no freezing and thawing damage was observed.

5. Surface conditions were esthetically affected by surface corrosion but no significant deterioration in material behavior resulted.

6. Based on the deterioration of SFRC observed at Treat Island Exposure Station cited earlier, a comparative study is needed of the behavior of lightweight and natural aggregate concrete with different cement factors and consolidation methods, for a better understanding of the long term behavior of SFRC.

SFRC SPECIMENS EXPOSED IN THE TIDAL ZONE IN AUSTRALIA.

Description of specimens and exposure site.

In February 1971, four SFRC beams, 100mm (4 in.) square and 2134mm (84 in.) long, were placed in the upper Sidney Harbor two months after casting by Fibresteel (Australian Wire Industries Ltd.). The beams were placed so that the lower ends were continuously submerged, the top ends always exposed, and the center section subjected to tidal variations. Figure 5 shows the beams after 10 years of exposure.

The beams were cast in timber molds, internally vibrated and cured under polyethylene for 28 days. The design mix was:

	kg/m^3	pcf
Cement - Type A	580	978
Nepean River Sand (SSD)	847	1428
Cronulla Beach Sand (SSD)	455	767
25mm x 0.25mm diam. wire fibres	160	270
Water	298	503
Water Reducing Admixture	2730ml	

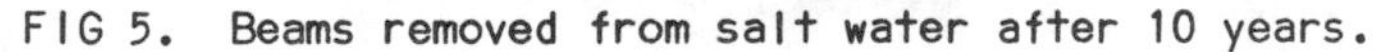

FIG 5. Beams removed from salt water after 10 years.

Testing and findings

Testing at 2-1/2, 5 and 10 years involved sawing a beam into 6 sections, as shown in Fig. 6. Specimens were examined for internal fiber corrosion and then tested for flexural strength. Average flexural strength in 2.5 years, 11.4 MPa; 5 years, 12.6 MPa and 10 years, 11.4 MPa.

At 5 years, the specimens were also tested to determine the depth of carbonation. The depth of carbonation was less than 1mm and fiber corrosion was limited to this thin surface layer. At 10 years, the specimens were again tested for carbonation. The depth of carbonation varied from 1mm at the base where the beam was normally immersed; 1.5mm to 2mm in the central section; and up to 3mm at the top of the beam which was continually exposed to the atmosphere. Internal examination showed that there was no rusting of fibers below the carbonated surface layer.

FIG 6. Sections of beam after 10 years

In addition, to determine the total soluble and insoluble chloride content at each position, the beams were sectioned and samples taken from various depths of the beam. The results are shown in Table 5.

TABLE 5. Total chloride* content of Fibrecrete specimens

Chloride Content - % By Weight of Concrete			
Specimen Position in Original Beams	Sample Position in Cross-section		
	0 to 5mm	8 to 13mm	Center
Top	0.54 (2.03)	0.47 (1.77)	0.08 (0.30)
Center	0.31 (1.17)	0.45 (1.70)	0.25 (0.94)
Bottom	0.62 (2.33)	0.67 (2.52)	0.56 (2.10)

* Includes both soluble and insoluble chlorides.

( ) Chloride content in brackets expressed as a % of cement content.

CONCLUSIONS

The conclusions that can be derived from SFRC exposure in a marine tidal zone in a sub-tropical region are:

1. After 10 years of exposure the corrosion of steel fibers seem to be limited to the depth of carbonation.

2. Even with chloride ion penetration, the fiber corrosion was limited to the concrete surface - no internal fiber corrosion was found even with the highest chloride penetration.

3. Surface staining occurs soon after exposure and is dependent on the concrete cover and density of fibers near the surface. Surface staining does not change significantly after the first months of exposure.

This data was supplied by Mr. W. Marsden, of Australian Wire Industries and it is acknowledged with thanks.

LABORATORY AND FIELD DURABILITY STUDIES OF BATTELLE DEVELOPMENT CORPORATION (Ref. 6) REPORT DATED JUNE 1978.

Exposure to deicing salts.

Two SFRC concrete slabs-on-grade containing 0.25 x 0.56 x 25.4mm (0.010 x 0.022 x 1.0 in.) and 0.640 diameter x 64mm (0.025 x 2.5 in.) steel fibers have been exposed to an outdoor environment (Columbus, Ohio) for 5 years. The exposure included the application of large quantities of deicing salts to the slab wearing surfaces. Similar exposure was given to 102 x 102 x 356mm (4 x 4 x 14 in.) SFRC concrete beams containing 0.410 diameter x 2.5mm (0.016 x 1.0 in.) and 0.640 diameter x 64mm (0.025 x 2.5 in.) steel fibers.

1. After 5 years of deicing salt application, slabs contained large quantities of chloride ion, up to 10.7 kg./c.m. (18 lb. per cu. yd.) at the 0 to 25mm level and the vast majority of steel fibers imbedded in this chloride contaminated concrete had not corroded. The only fibers exhibiting corrosion are those at or near an exposed surface (top, bottom, or edge of slab).

2. Over the 5 year salting period, the exposure of SFRC concretes to deicing salts has not had an adverse effect on flexural strength. Five year strengths are as good or better than pre-exposure 28 day strengths.

3. Concrete slab number 2 exhibited severe surface scaling and spalling. This freeze/thaw and deicer scaling damage is most severe in the portions of the slab containing concrete with a high water cement ratio and low entrained air content. Tests performed (ASTM C666) shortly after construction of the slab (August, 1972) showed this concrete to have a low freeze/thaw durability factor. This result suggests that steel fiber additions to concretes with a low resistance to freeze/thaw

damage and deicer scaling (concrete with an inadequate air void system, improperly finished or cured concretes, concretes with high water cement ratios) will not necessarily improve their performance.

4. As of June 1978, there was no report of failures or problems associated with any deicer salt-induced corrosion of steel fibers in SFRC used in the construction of highway pavement overlays, bridge deck overlays, sidewalks, curbs, and various slabs-on-grade. Some of these concretes have been in service for over 10 years. (Unfortunately, a follow-up to date was not available.)

5. The results of laboratory-controlled and normal field exposures of SFRC concrete flatwork strongly suggest that the corrosion of steel fibers in sound concrete will not occur even in the presence of chloride ion contents greater than the threshold value. The fact that the steel fibers are not electrically continuous probably has some bearing on this result.

Seawater exposure study.

SFRC specimens of plain mortar and conventionally reinforced mortar with a 9.5mm bar (one No. 3) have been exposed to laboratory seawater for up to 10 years. Cube specimens (for compressive strength determinations) were fully submerged and beam specimens (for flexural strength measurements) were partially submerged. The results after 10 years show that,

1. Carbon steel fibers contained in sound, uncracked mortar have shown no evidence of corrosion in fully or partially submerged situations.

2. As in other cases where exposed fibers come in contact with moisture, surface or near-surface fibers in the beams and cubes of steel fiber reinforced mortar have rusted with no apparent adverse affect on the structural integrity of the concrete.

3. Monel fibers in partially and fully submerged mortar specimens, have undergone no corrosion during a 9-year exposure period (including surface and near surface fibers).

4. The plain mortar and the SFRC (steel and Monel) have exhibited a loss in compressive strength over the 10-year exposure period as follows,

Plain Mortar	=	40%
Carbon SFRC	=	15%
Monel SFRC	=	4%

All of the mortars had compressive strengths in excess of 27.6 MPa (4000 psi) after 10 years submergence in seawater.

5. SFRC showed a loss in flexural strength over the 9 to 10 year seawater exposure period (31% for the steel and 26% for the Monel). The primary reduction in flexural strength of the steel fiber reinforced mortar occurred between the first and second year of exposure. The reason for the strength loss at this time is not obvious. It was not due to a loss of strength in the mortar or to corrosion of the fibers.

One possible explanation is that the steel fibers used in the specimens were brass coated. The continued exposure of these fibers to a moist, aggressive solution may have removed the coating during this time and reduced the fiber-matrix bond strength. Notes made the second year of testing established that the fibers in the outer perimeter of the beam cross-section (to a depth of about 19mm) were devoid of the brass coating, while those fibers in the interior of the beam were still brass coated. In any event, the flexural strength of the steel fiber reinforced mortar after 10 years exposure was still in excess of 103 MPa (1500 psi).

GENERAL CONCLUSIONS

The data presented in this paper is based generally on about ten years of exposure of various type specimens. Based on this information the following points concerning the durability of SFRC can be made:

1. Basic durable, reasonably impervious concrete will protect steel fiber from corroding.

2. Fibers located at or near the exposed surface, 1 to 5mm, can be expected to corrode with time.

3. Surface fiber corrosion has no significant effect on overall structural behavior except in very thin elements of possibly 20mm or less.

4. Different types and sizes of uncoated steel fiber seem to behave similarly in regard to corrosion susceptibility.

5. Concrete which is not resistant to freezing and thawing, will not have its durability enhanced when reinforced with steel fibers.

6. The behavior of air entrained SFRC subject to a severe marine freezing and thawing shows no improvement over non-reinforced reference specimens. There seems to be a trend, in the limited data, showing a decrease in properties with higher fiber contents. This is based only on the decay in pulse velocity readings without having made a physical in-depth examination of the specimens. Further investigation is warranted.

7. The durability of steel fibers passing through a crack is suspect. Exposure of relatively thin (up to 50mm) cracked specimens in a marine environment indicates that wires at cracks corrode. The severity of fiber corrosion at cracks needs further investigation and research.

8. Considering the fact that concrete cracking is likely to occur in some applications, the effects of possible fiber corrosion at cracks should be considered in design.

9. Considering some of the "surprises" in reinforcing and prestressing steel corrosion that has occurred and is occurring, it is prudent to continue to monitor SFRC corrosion performance both in the field and the laboratory.

REFERENCES.

1. "Investigation of Performance of Concrete and Concreting Materials Exposed to Natural Weathering", U. S. Army Engineers Waterway Experiment Station, Technical Report No. 6-553, Vicksburg, MS, June 1960.

2. Podvalnyi, A.M. , "Study of Concrete and Reinforced Concrete Durability in the Tidal Zone of the Northern Sea", Research Institute of Concrete and Reinforced Concrete (NIIZhb) Gosstroy USSR, Supplement to American Concrete Institute (ACI) SP65, August 1980.

3. Schupack, M., "The Behavior of Twenty Post-tensioned Test Beams Subject to up to 2200 Cycles of Freezing and Thawing in the Tidal Zone at Treat Island, Maine", American Concrete Institute SP65, August 1980.

4. Natural Weathering Exposure Station, Treat Island, Maine, U.S.A., "Review of Programs", U. S. Army Engineers Waterway Experiment Station, Vicksburg, MS, August 1984.

5. "Handbook for Concrete and Cement", Corps of Engineers, U. S. Army Waterways Experiment Station, CE, Vickburg, MS, August 1949.

6. Lankard, D.R. and Walker H. J., "Laboratory and Field Investigations of the Durability of Wirand Concrete Exposed to Various Service Environments", June 1978.

STEEL FIBER CONCRETE
US-SWEDEN joint seminar
(NSF-STU)
Stockholm 3-5 June, 1985
S P Shah - Å Skarendahl,
Editors

FIBRE REINFORCED SHOTCRETE-AN INTERESTING MATERIAL FOR LABORATORY EXPERIMENTS OR A MAJOR CONTRIBUTION TO BUILDING TECHNOLOGY OF TOMORROW?

by Bengt E Vretblad, DSc(Eng), Royal Swedish Fortifications Administration, Eskilstuna, Sweden

---

ABSTRACT

In Sweden the use of steelfiber reinforced concrete is very limited. One application is steelfiber reinforced shotcrete (SFRS). Lack of manuals and lack of data on longterm performance are two - interrelated - reasons for the limited use.

FortF has examined some structures where SFRS has been used and found that the few deficiences in performance that were discovered could be attributed to poor workmanship. However, the shotcreteing technique should have improved since the time when the structures were made.

Bengt E Vretblad is Director of Research and Head of the Research Department at FortF - Royal Swedish Fortifications Administration since 1980. At FortF he is primarily active in research on protective structures and risk analysis. Also the research at FortF on fiberreinforced concrete is primarily for applications to hardened structures.

INTRODUCTION

The fiber reinforcing technique is rather old. In the early 1900 primitiv structures based on clay reinforced with straw were used. During the sixties efforts were made to use fiber reinforced concrete (FRC). These efforts were continued and increased during the seventies. Comprehensive research led to a development of the FRC material for a variety of applications. Still, though, the application of FRC even within fields that have been considered promising has shown to be rather scarce.

FIBER REINFORCED CONCRETE FOR PROTECTIVE STRUCTURES

The development of FRC has been stimulated by the possibilities of getting a material with the favorable properties but without the disadvantages of concrete. In the first place this has to do with tensile strength and ductility e.g. the energy absorbing capacity.

For protective structures the FRC material should be of special interest. Some examples:

* High strength

* High ductility increases the applicability for linings where substantial deformations are to be foreseen e.g. from spalling or differential movements in rock installations

* High energyabsorbing capacity is a primary concern in resisting weapons effects

* Possibility to make thin structures of suitable shape for smaller shelters.

Still, though, the FRC has not been used very much for the constructing of protective structures. There might be many explanations for this:

* Lack of data on performance in applications
* Lack of codes and manuals
* Lack of experiences with the material
* Traditions
* Costs

While some data exist on FRC in impulsively and in impact loaded structures, /1/, /2/, more are required. Within FortF projects on steelfiber reinforced shotcrete (SFRS) for linings exposed to static loads have been performed, cfr. /3/. Also investigations on SFRS in linings subjected to dynamic loads have been executed, /4/. These investigations have among other things shown that a reliable performance of the shotcrete can be achieved. Therefor it should be of interest for support in rocks used for protective structures and in seismic areas, /5/.

Shotcrete for linings is used to a considerable extent at FortF. The latest revision of the code regulating the use of shotcrete is /6/. In it fiberreinforced shotcrete is not dealt with, however.

Traditions do not favor new materials. Lack of codes makes an extensive use of new materials difficult. This also works the other way around - with few actual applications few experiences will be gained that can be used to write new codes. The dilemma is that more use of FRC is required to give more data for the use of FRC. This is especially a case when it comes to longtime effects.

LONGTIME CONDITIONS

FortF installations in rock have to meet with various conditions. In some installations very dry air at room temperature is prevalent while in other installations the relative humidity is 100 %. Some have to sustain substantial variations both in temperature and humidity. Even problems with sulfurous water exist.

The corrosion of reinforcement in concrete is to be attributed to the carbonation process in the concrete or chloride initiation, /7/. A major way for chloride initiation to start is at wide cracks.

It is therefor of interest for FortF to study carbonation and crackwidths that are or might be relevant to the conditions in FortF underground installations.

Carbonation has for this reason been studied in shotcreted linings in up to 30 years old installations.

At a maximum more than 30 mm thick carbonation zones have been identified in an old installation with dry conditions and at room temperature. With dry conditions carbonation zones around 15 mm thick in 20 years old shotcrete are frequent. In more humid environments the carbonation zones are considerably smaller.

While the studies of shotcrete without fibers are relevant with respect to carbonation studies this is not the case for the study of cracks. In this field very few data seem to have been put together. FortF has therefor studied some structures that could give some additional information.

DATA ON LONGTIME EFFECTS

Mines

Two mines, one belonging to the Boliden Mine Ltd Company near Kristineberg, /8/, and a Zincmine close to Askersund, belonging to Vielle Montagne Ltd, both shotcreted in the mid seventies and with conditions comparable to many underground installations in Sweden were examined.

In the Kristineberg Mine the roof parts were shotcreted with fibers, the wall parts without. A difference in performance was clearly noticed: In the walls 1-2 mm wide cracks were visible while no large cracks were observed in the roof.

In the Zincmine 1-2 mm wide cracks were observed in the fiber-reinforced concrete. A closer examination, however, revealed that at the cracks the number of fibers was almost zero.

The carbonated zone was only a few millimeters thick and no corrosion could be observed on the fibers taken out of the concrete.

Petroleum Storage

At Scanraff, Brofjorden, in western Sweden about 4500 m^2 were shotcreted with FRC in 1974. Brofjorden is part of a heavily industrialized area with severe climatic conditions.

Only minor cracks 0.1-0.2 mm could be observed and no corrosion in the fibers could be identified. This also was true for some parts of the shotcrete close to the ground where the quality was poor.

The carbonation was also measured. The carbonated zone had not penetrated into the concrete more than 3-4 mm.

Lighthouse

Morup Tånge is an about 30 m high lighthouse off the Swedish west coast. It was built in the middle of the 19th century by stoneblocks jointed together with calcium mortar. As different means of protecting the surface from leakage and freezing had shown unsatisfying shotcreteing was tried in 1975, see /9/. First 50 mm of shotcrete was used. Thereafter additionally 30 mm of fiberreinforced shotcrete was placed around the lighthouse.

The lighthouse has been exposed to very severe effects from temperatur, humidity, water and ice.

It can be seen, FIG 1, that the lower part of the lighthouse has suffered the most. Wide cracks have opened up but closed

FIG 1. Part of Morup Tånge lighthouse.

almost completely. The fibers on the surface were corroded as could be expected - the fibers in the concrete, however, not.

It was clearly indicated that cracks opened up where there were few fibers in the concrete. Carbonation was minor.

Data in short

The data are far from comprehensive. They all are indicative, however, that

* Cracks are likely to occur in areas with bad workmanship
* Therefor clear instructions, skilled workers and qualified control is necessary
* The carbonation zones were small - as expected
* No corrosive effects on fibers in concrete were discovered

It should be pointed out that on the structures studied the cracks were very few. The work done was at least partly pioneer work. The means of applying SFRS have developed since.

More data, though, are desirable.

Conclusions

Fiberreinforced concrete has gained much attention through the latest 10-15 years. Steelfiber reinforced shotcrete has been regarded as one interesting application. Still, there are not an impressive amount of structures where it has been used, though.

A major increase in the use of SFRS is not likely until

* Manuals for the use of SFRS including
* Control specifications

are available.

For this more data on some properties must be available. The longtime effects naturally take longer to grasp. They are no less important, though. For SFRS to be a valuable contribution in the future a more extensive use is needed today. A way out of this dilemma might be provisional manuals for selected applications.

ACKNOWLEDGEMENTS

Dr Peter Balazs and Jörgen Asp, M Sc, both at FortF have been very helpful in the investigations. Their contributions are highly appreciated.

REFERENCES

/1/ Balazs, P, Fiber Reinforced Concrete. FortF report C 237. Eskilstuna 1984 (In Swedish).

/2/ Mullins, R K, & Baker C F, Interim Rapport on Use of Steel Fibers in Concrete Slab Construction to Resist Spall Caused by High-Explosive Blast Effects. Minutes of the Nineteenth Explosives Safety Seminar, Los Angeles, 9-11 September 1980.

/3/ Holmgren, J, Punch-loaded shotcrete linings on hard rock, Swedish Fortifications Administration, Research Dept, Report 121:6, Stockholm 1979. (Also Report no 7:2/79, Swedish Rock Engineering Research Foundation, Stockholm 1979. Dissertation Royal Inst of Technology, Stockholm 1979.)

/4/ Holmgren, J, Dynamiskt belastad bergförstärkning av sprutbetong, (Shotcrete linings subjected to dynamic loading), to be printed by the Swedish Fortifications Administration, Eskilstuna 1985. In Swedish with English summary.

/5/ Vretblad, B, Blast Effects on Underground Facilities. Adv Tunnel Technology & Subsurface Use. Vol 4. No 3. Great Britain. 1984.

/6/ Konstruktionsprinciper vid utformning av berganläggningar. FortF/Bk Publ 21. Stockholm 1978. In Swedish.

/7/ Tuutti, K, Corrosion of steel in concrete. CBI research Fo 4 Stockholm, 1982.

/8/ Bergssprängningskommitténs protokoll från diskussionsmöte i Stockholm den 17 februari 1977. In Swedish.

/9/ Sandell, N-O & Westerdahl, B, System BESAB for High Strength Steel Fiber Reinforced Shotcrete. USA-Sweden seminar on Steel Fibre Concrete. CBI. Stockholm, 1985.

STEEL FIBER CONCRETE
US-SWEDEN joint seminar
(NSF-STU)
Stockholm 3-5 June, 1985
S P Shah - Å Skarendahl,
Editors

PERFORMANCE CRITERIA, SAFETY AND QUALITY. SOME EXPERIENCES FROM THE SWEDISH STATE POWER BOARD

by Karl Åke Sjöborg, MSc(Eng), the Swedish State Power Board, Stockholm, Sweden

---

ABSTRACT

The use of steel-fibre-reinforced concrete by the Swedish State Power Board is relatively new. It is used by the Board in the form of shotcrete and almost exclusively for strengthening the rock in tunnels and caverns. Rock strengthening is an expensive operation, with stringent demands on performance, safety and quality.

It is important for the Board to obtain clarity on how interaction takes place between fibre-reinforced concrete and rock, and to determine the concrete properties that should be sought. Greater knowledge of this is required to be able to satisfy the performance and safety requirements of the finished structure. During the course of the studies and development work being carried out by the Board, the working theory that has been assumed is that adhesion, toughness and durability are the most important properties. The Board has therefore investigated the factors that give rise to these properties.

Karl-Åke Sjöborg is chief engineer at the Swedish State Power Board. Mr Sjöborg is head of Civil Engineering Development and is responsible for quality control and civil engineering research and development.

Mr Sjöborg earlier worked on feasibility studies, planning and the design of hydro-electric power stations. He has also been responsible for co-ordination of the civil engineering activities for a nuclear power plant.

INTRODUCTION

Shotcrete that does not contain fibres has been used by the Board for a long time. This involved the use of reinforced and unreinforced shotcrete and both the dry and wet methods of application have been employed. In recent years, however, the Board has started to investigate the possibility of using concrete containing steel fibres. This material has been used by the Board in the form of shotcrete and almost exclusively for the strengthening of rock in tunnels and rock caverns. This work is expensive, with stringent demands on performance, safety and quality.

Strengthening is carried out for safety reasons in the course of construction work and to ensure the performance of the installation during its entire service life. Safety during construction work requires speed, because the workforce must be able to work in complete safety the whole time and because disturbance of the normal work should be kept to a minimum, for financial reasons.

The performance of the installation during its service life places major demands on quality and durability, especially in parts of the installation such as tunnels carrying water, which are not accessible without very expensive measures.

To be able to evaluate a method or material - rock strengthening with fibre-reinforced concrete, in this case - a number of conditions must first be clarified. These include:

- How does rock that requires strengthening behave and what is to be achieved?

- What properties are required of the strengthening system?

The tests and development work being carried out by the Board are based on the following philosophy.

THEORETICAL APPROACH

In-situ rock contains fissures, joints and crushed zones. The rock in a driven tunnel or cavern will have lost some of its earlier vertical and horizontal support. The original compressive stresses are often of considerable magnitude, particularly the horizontal ones. Values of between 20 and 80 MPa have been measured in conjunction with planning and construction of Swedish rock installations. The blasting of tunnels and caverns involves transferring the forces to the rock surrounding the tunnel or cavern, which results in stress concentrations.

The roof and walls contain more or less firmly fixed boulders and blocks that can withstand the forces applied during scaling, but which may later work loose and fall as a result of direct gravitational forces and the rock stresses mentioned earlier (and possibly, water pressure). To counteract this, the rock may be strengthened by rock bolting and/or the application of shotcrete, which may be reinforced or unreinforced. In crushed zones it is often obviously necessary to carry out strengthening with shotcrete. The shotcrete performs a superficial support function, somewhat like that of a corrugated steel sheet pipe in a road embankment. In other cases, the amount of support required may be a matter of assessment. Unreinforced shotcrete is relatively brittle, but it may be regarded as being effective in preventing individual stones from falling.

Shotcrete containing fabric reinforcement is tough and also performs when the concrete has cracked, and may be regarded as effective in preventing boulders and blocks from falling.

It is important in both of these methods of strengthening to ensure that good adhesion is achieved between the shotcrete and the rock.

Shotcrete with fabric reinforcement and rock bolting, that is, reinforcing bars grouted fast in drillholes, both constitute high quality strengthening, with relatively reliable interaction between the shotcrete and the rock and between different blocks of rock. This method of strengthening is used when reliable adhesion between the concrete and rock cannot be achieved, due, for example, to water flows.

In all cases, the compressive forces in the rock must be transferred across the roof and round the walls by an arching action, for which the roof and wall rock surfaces must be kept intact because of the risk of injuring personnel and the need for reliability and efficiency of the installation.

It is conceivable to replace the fabric reinforcement mentioned above by fibre reinforcement that is applied at the same time as the concrete, because the fibres are mixed in it.

It is not at present possible for the Board to design the strength of the rock by means of calculation, as in the case of building structures above ground. The fissure systems and weakness zones of the rock in three dimensions are mainly unknown and vary greatly along the length of a tunnel, for instance. The shape of the rock surface, the geometry and thickness of the shotcrete and the position of the reinforcement within the concrete are factors that are unknown in advance and may vary a great deal. A concrete structure such as a bridge is designed with almost complete knowledge of the shape, dimensions, position of the reinforcement, strengths of materials and loads, whereas, in the case of shotcrete on rock, these factors, which are necessary for optimal design, are more or less unknown.

Rock strengthening with shotcrete containing different quantities of reinforcement, and rock bolting should, instead, be regarded as different classes of strengthening, the relative functions and load-bearing capacities of which are known or can be further investigated experimentally. Fibre-reinforced shotcrete must be classified in such a system, for which it is a prerequisite that its load-bearing function be compared with that of fabric-reinforced shotcrete by tests and other means.

The design of rock strengthening, that is, the selection of strengthening type, will then have to be carried out, even in the future, by means of assessment by knowledgable and experienced persons, in conjunction with on-site inspection, drawing on experience gained from earlier installations, whether or not they involved damage.

Fibre reinforcement is being regarded as a replacement for fabric reinforcement. The author is not completely clear of the manner in which fibre reinforcement should be compared to fabric reinforcement. It is possible that such a comparison should not be made, but that fibre-reinforced concrete should be regarded as a completely different material with properties that are different from those of fabric-reinforced concrete.

When there are no further problems with the mixing of fibres and with the shotcreting technique, fibres have the obvious advantage of eliminating the operation of placing and fixing the reinforcement. One advantage of the fibres appears to be that the concrete is not only reinforced against bending, but also against shear.

One special problem is the long-term strength and toughness of fibre-reinforced concrete. It is assumed that the fibres can easily rust away in a crack in the concrete. The Board has no experience of this very interesting point.

INSPECTION IN PRACTICE

In the applications of fibre-reinforced concrete in which the Board is interested, adhesion and toughness are probably the most interesting properties. For the moment, the Board has no quantitative requirements on these properties, but uses the requirements of relevant Swedish Standards, which include:

- The adhesive strength between layers of shotcrete shall be at least 1 MPa.

- The adhesive strength between a layer of shotcrete and the rock shall be as specified on drawings, but no less than 0.1 MPa.

- Adhesion to rock is not required in the case of shotcrete that is anchored to rock by rock bolts or supported by an arch effect or, in the case of a small section, that is supported by other sections.

Adhesion and compressive strengths are checked systematically at the Board's sites. Especially intensive checking is being carried out at present at two major sites, the Vietas Tunnel 3 and the Final Repository for Reactor Waste. Shotcrete is being applied at these two sites mostly using the wet method.

Compressive strengths used to be checked largely by preparing test panels of shotcrete, which were then sawn into cubes and subjected to compressive testing. This method has several disadvantages, such as the fact that personnel applying the shotcrete know that it is to be used for a test and can therefore be more careful. In terms of technique, it is also different to shotcrete panels, so the results obtained may be completely different from those achieved in an actual structure. For this reason, it is very important that sampling be carried out, as far as possible in practice, on a finished structure.

At the Final Repository for Reactor Waste at Forsmark, the adhesion and compressive strength are tested directly in the structure. The adhesion to rock is tested on a series of tests for each 1000 m^2 of surface for which the adhesion is specified. The compressive strength is tested for each 500 m^2, by removing three 72 mm dia. cores, which are then cut and subjected to compression testing.

Inspection of adhesion has been carried out for the past three years using an in-situ system developed by the Board. Equipment that was earlier available on the market has been found to have disadvantages and deficencies that were capable of having an adverse effect on test results. The principles on which the Board's present equipment operates differ radically from those of earlier equipment, and also permit considerably faster inspection.

The equipment consists of a small, conventional, hydraulically driven core drilling machine, with a triangular foot and a short column on which the motor slides. The triangular foot is securely fastened to the shotcrete using expanding sleeves. The short column results in a small bending moment and little risk of breaking cores during drilling.

Drilling is carried out in two stages. Drilling is first carried out with a double bit (Ø72/Ø86 mm inner bit and Ø 104/111 mm outer bit). Drilling is carried out with this bit until two grooves have been made in the surface of the concrete. The outer groove will later serve as the support for a jack.

The second stage involves continued drilling with a single bit, until the shotcrete core to be subjected to tensile testing has been drilled out. The tensile testing equipment has a built-in strain gauge. The load and course of load application is recorded on a plotter.

Tensile testing is carried out using a drawing sleeve with a clamping ring that grips the outside of the core that has been drilled out. The advantages are significant, because testing can be carried out quickly, irrespective of any water or low temperatures that can prevent adhesion.

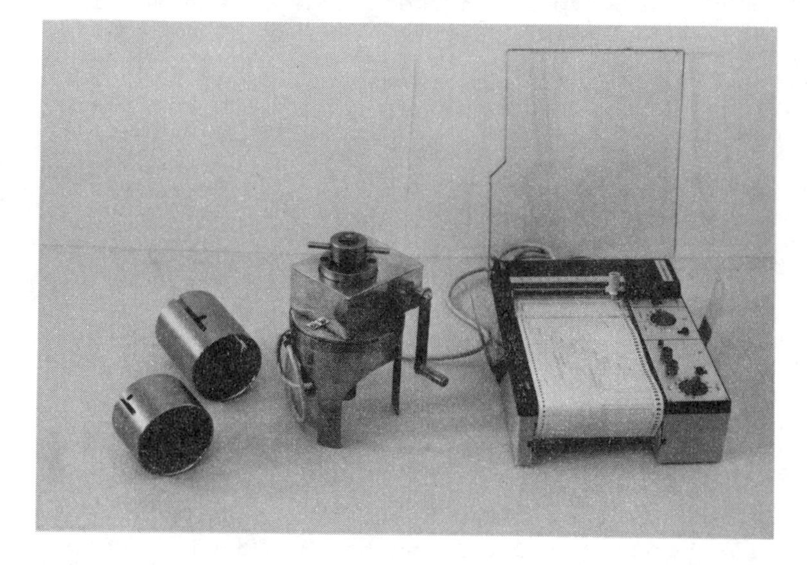

FIG 1. Tensile testing equipment with x-t-recorder The tensile force is made by hand with a crank.

DEVELOPMENT WORK

Development work being carried out on fibre-reinforced concrete is naturally being followed with great interest and the Board is carrying out tests and its own development work, which is mostly in the practical field. The Board wishes to find out how to satisfy specified requirements under site conditions - to clarify the method of determining reliably the required properties - and, at the same time, to work quickly and economically. As mentioned earlier, the Board also wishes to develop practical test equipment for use on site. Two of the trials performed are described briefly below, together with the results achieved so far or, rather, the problems encountered, which are regarded as requiring further work.

Trials are carried out preferably in conjunction with work in progress. Thus, at the Messaure Power Station, wet shotcreting, using steel-fibre reinforced concrete, has been carried out over a limited area for development purposes. The aim of the project was to study the properties of the concrete and the shotcreting capacity when using steel fibres, and silicon dust, plasticizer and accelerator additives. The investigation of the adhesion of the concrete to rock, which was performed about one month after the completion of shotcreting, showed that the adhesion was between 0.3 and 1.1 MPa. It is noteworthy that a "miss" was later detected and drilling of additional cores showed a loss of adhesion between the concrete and the rock in an area in which adhesion had been found during earlier trials.

During wet shotcreting, mostly fibres that are 18 mm long, with cross-sectional dimensions 0.3 x 0.6 mm, and known as E-E fibres have been used. The reason for selecting this type of fibre is that they can be mixed into the concrete to advantage, without noteworthy formation of balls.

The laboratory investigations showed that these steel fibres improved the mechanical properties of the concrete. The fibre reinforcement increased the energy required to induce failure and reduced the incidence of cracking. The orientation of the fibres is largely acceptable and the adhesion between the concrete and the fibres was generally good. The microstructure of the shotcrete without additives, as regards both cracking and pores, was better than that of concrete in which additives and silicon dust had been used.

To date, the Board has tested several types of fibre of varying quality. In dry shotcrete, 35-mm long, 0.35-mm dia. Beckerts steel fibres have been used mostly. The quality of this type of steel fibre is good and it provides very tough concrete, but it is, unfortunately, difficult to mix into the concrete. The method used in dry shotcreting with long fibres is to mix in the fibres at the nozzle, which results in a large amount of fibre spillage. Investigations carried out by the Board, involved the use of 35-mm long, hook-ended fibres of 0.35 mm dia. and called for 1% (by volume) of fibres in the finished structure. Inspection of cores drilled out of the finished concrete showed that the percentage of fibres was about 0.9% in walls and 0.4% in roofs.

Some of the results of the trials using 35-mm fibres have been summarized in the graph shown in Appendix 1

The graph shows that an increased quantity of fibres reduced the compressive strength and density of the concrete. This may be related to the distribution and orientation of the fibres, and elimination of voids around them, which may be dependent on the equipment and method used in the trials. It is therefore important to continue the trials with 30 - 35-mm long fibres, to find out the reason for this loss of compressive strength and density when using larger amounts of fibres.

It is our opinion that the consistency of the concrete should be semi-fluid. For practical reasons, it has been necessary for us to use fluid concrete in our wet-shotcrete field tests. This, in turn, has made it necessary to use an accelerator additive, so that the shotcrete will adhere to the surface to which it was applied. Large amounts of accelerator reduce the compressive strength of the concrete and also the adhesion of the shotcrete to the rock in the finished structure. Future tests will therefore involve trying to find out how to shotcrete using a less fluid consistency.

CONCLUSIONS

In summary, it may be said that the Board is very interested in fibre-reinforced shotcrete. Because of the application - rock support - the performance and quality requirements must be high. To be able to make these requirements, the Board should specify - preferably quantitatively - the characteristic properties and define the criteria to be used to identify them. The problem during the present development stage, however, is the inability to define these requirements with adequate reliability and accuracy. The basic reasons for this are the lack of both detailed qualitative knowledge or, even more so, quantitative knowledge of how rock requiring strengthening will behave. In other words, we know neither the nature nor the magnitude of the loads. In addition, it is unclear - at least to the author - how the shotcrete performs in conjunction with rock that exhibits properties which differ from those of a "manufactured" structural material. For example, after disturbance, the rock mass may require relatively large movements before it again achieves equilibrium.

In other words, we do not know (in the strictest meaning) what we wish to reinforce against and do not, therefore, really know how to do it. As a result, it is not completely clear which properties are the most important. But we have as a working theory that the following properties must be among the most significant.

- Adhesion - we are making studies to determine the factors that contribute to good adhesion.
- Toughness - we are interested in the effects of fibre contents, lengths and qualities.
- Durability - we are trying to achieve high strength and low permeability.

We are also performing tests to find out how to obtain these properties in practical work, with high outputs and uniform quality. Finally, it is important to emphasize the great importance of site personnel having good knowledge of the product and being skilful in the use of the equipment. When all is said and done, it is the hand of the man on the gun that determines whether or not the results will be good.

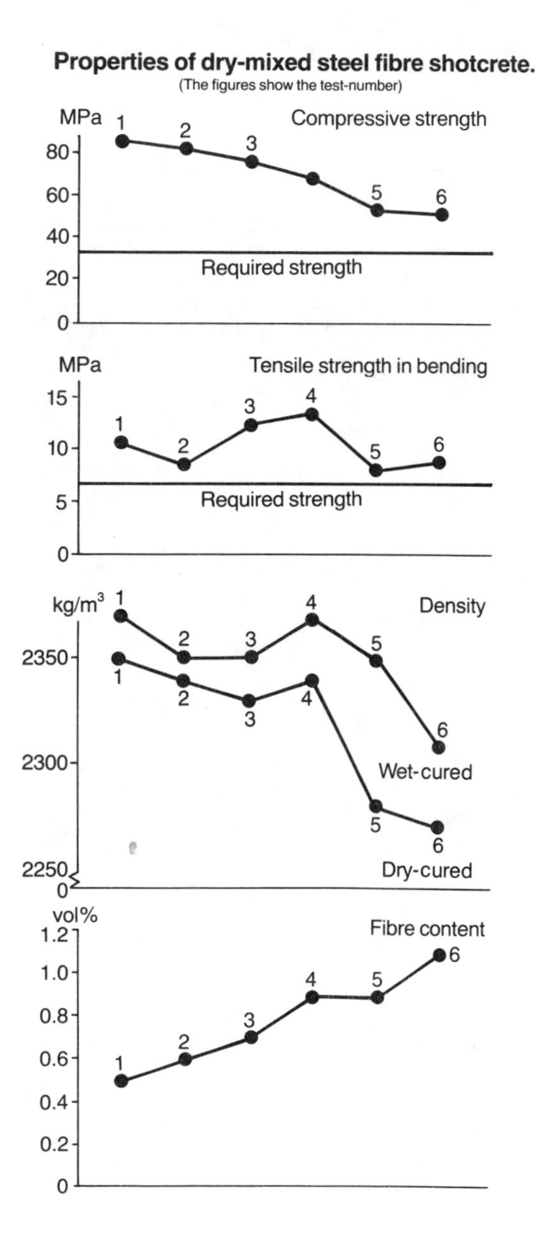
Properties of dry-mixed steel fibre shotcrete.
(The figures show the test-number)
MPa
Compressive strength
80
60
40
20
0
Required strength
1
2
3
5
6
MPa
Tensile strength in bending
15
10
5
0
Required strength
1
2
3
4
5
6
kg/m³
Density
2350
2300
2250
0
Wet-cured
Dry-cured
1
2
3
4
5
6
vol%
Fibre content
1.2
1.0
0.8
0.6
0.4
0.2
0
1
2
3
4
5
6

STEEL FIBER CONCRETE
US-SWEDEN joint seminar
(NSF-STU)
Stockholm 3-5 June, 1985
S P Shah - Å Skarendahl,
Editors

PRODUCTS OF STEEL FIBRE REINFORCED CONCRETE - QUALITY ASSURANCE

by Hans-Erik Gram, DSc(Eng), Swedish Cement and Concrete Research Institute, Stockholm, Sweden

---

ABSTRACT

The properties of steel fibre reinforced concrete open new applications in the use of concrete. In addition to new production techniques and labour training, the introduction of this "new" material also requires new ways of controlling the production in the factory and new ways of specifying and checking expected properties for the customer or user of the product. It may be necessary to introduce new test methods and test equipment. There is need of a quality assurance system. General reflections on quality assurance concerning steel fibre reinforced concrete are made and a system for quality assurance applied to the production of balconies made of fibre shotcrete is discussed.

Hans-Erik Gram received his MSc degree in civil engineering at the Royal Institute of Technology, Stockholm in 1977. In the same year he took up consultancy and research work at the Swedish Cement and Concrete Research Institute. His research work on the durability of natural fibres in concrete led to a DSc degree in 1983. He is now project leader for a research project titled "Concrete - new material combinations". He is a member of RILEM Committee 49-TFR Testing of fibre Concrete and FIP Commission on Concrete.

INTRODUCTION

The properties of steel fibre reinforced concrete open new applications in the use of concrete. Thin products made of concrete can replace products made of other materials. In addition to new production techniques and labour training, the introduction of a "new" material also requires new ways of controlling the production in the factory and new ways of specifying and checking expected properties for the customer or user of the product. It may be necessary to introduce new test methods and test equipment. These facts speak for the use of a quality assurance system. This paper contains reflections on what a quality assurance system for steel fibre reinforced concrete could include. The Swedish Cement and Concrete Research Institute has been contracted to give assistance in the development of a quality assurance system for the production of steel fibre reinforced concrete balconies made by Ekebro AB in Arboga/Sweden. The system is presented in this paper.

QUALITY ASSURANCE - GENERAL REFLECTIONS

Production technique

Control of constituents

The quality control of the constituents of steel fibre concrete is similar to that applied to ordinary concrete. It includes quality control of the binder, aggregates, water, admixtures and fibres. Test methods are available.

Control of the production equipment

As always, routine control programs for the equipment used in the production, i.e. tests of their function etc. are necessary. Plans for how to meet breakdowns must be prepared. Control programs for the formwork, if formwork is used, i.e. cleanness, dimensions, recesses etc. are also necessary.

Control of the fresh concrete

The control of the fresh concrete can include workability, segregation, bleeding, heat and strength development, air content, cement content, water/cement ratio, fibre content, distribution and orientation, density and temperature. In this field test methods must be developed if the fresh mix contains fibres.

Control of curing, demoulding properties and storing

A control system should include the curing conditions, demoulding strength, handling and storing in the factory. Such systems are well established.

Mechanical properties of the hardened product

The control of strength can be based on testing specially cast specimens and/or non-destructive or destructive testing methods. Depending on the required properties of the product, test routines and methods should be developed. The methods can include the compressive, flexural and tensile strength as well as fracture energy properties and impact strength.

Control of long term properties

Depending on the environment in which the actual product will be used, the long term properties must be ensured. Test methods for frost resistance, corrosion on the fibres in the matrix, abrasion resistance, surface appearance etc. can be used. In some cases new test methods are required.

Control of other properties

The control of other properties can include testing water tightness, shrinkage, creep, thermal properties, resistance to fire as well as the function of attachments.

QUALITY ASSURANCE - THE EKEBRO BALCONY

Introduction

The Swedish Cement and Concrete Research Institute (CBI) has given assistance in the development of a quality assurance system for the production of steel fibre reinforced balconies at Ekebro in Arboga/Sweden. The institute is also involved in production control at the factory. 500 balconies are being produced according to a new concept during a period of two years (1983-1985). The production process is based on a wet concrete spray method.

The matrix used is made of a specially composed pre-mixed dry mortar mixed with water in the factory. The reinforcement consists of a wire (tensile strength 1200 MPa) delivered on bobbins and cut to lengths of 40-50 mm in the gun.

During the development of the balcony the design loads were confirmed by full scale tests on balconies. A theoretical study of the corrosion process of embedded steel fibres was conducted and the behaviour during fire was examined.

Control of the concrete

The grading curve for the pre-mixed dry mortar is established weekly according to the Swedish standard method for particle size distribution by sieving (SS 13 21 23).

The consistency of the mortar is determined weekly with the aid of a slump cone according to Swedish Standard SS 13 71 10.

The temperature of the fresh mortar is determined in connection with the determination of the consistency.

The addition of admixtures to the mortar is noted.

The density of the fresh mortar is determined weekly according to Swedish Standard SS 13 71 25.

The water-to-binder ratio is determined weekly by drying out the water of a specified volume of fresh mortar at a temperature of +105 ^{o}C. The ratio is calculated under the assumption that the binder does not hydrate during this drying process.

The density of the hardened mortar as well as the compressive strength is determined weekly according to Swedish Standard SS 13 72 10 on cubes measuring 100x100x100 mm. In the first period, in which 10% of the total number of balconies have been produced, the cube strength at the age of 7 days was determined (in addition to the 28 day value).

Control of the composite

The volume of added fibres is determined for every single balcony in several ways for the three walls separately.

The amount of fibres added is calculated after the determination of the weight of fibres supplied to the gun (negative weighing of the spools). The fibre volume and distribution is also checked by using an electromagnetic covermeter on special spots on the three walls of the balcony. This method is described by Malmberg and Skarendahl (1) and special calibration diagrams have been prepared for the quality assurance program.

The flexural strength and the toughness of the composite is determined on specially prepared specimens, see FIG 1.

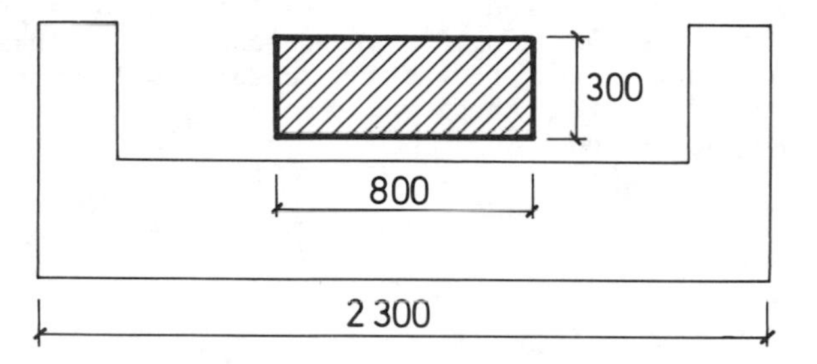

FIG 1. The balcony and specially prepared specimen for the flexural strength tests.

Beams measuring 60x360x40 mm have been sawn out and subjected to flexural strength tests at 28 day age with a load set-up shown in FIG 2.

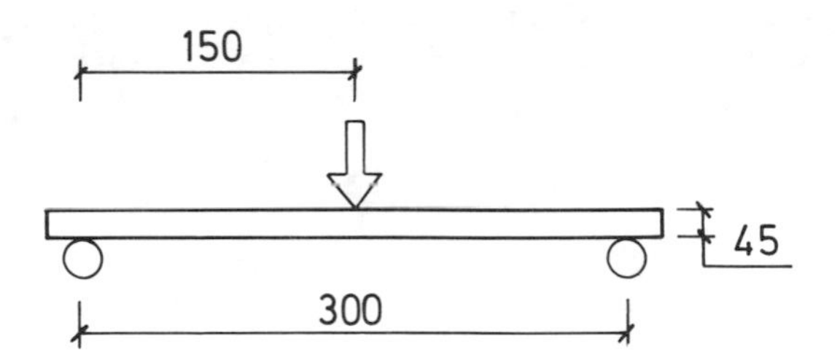

FIG 2. Load set-up.

The stress-strain curve is produced with the aid of a XY-plotter and the stress at the modulus of rupture and 1% deformation is calculated. The number of fibres at the crack surface is noted.

Control of the balcony

The sprayed layer thicknesses are measured with a gauge during the production.

The curing time (at 100% RH) and temperature are noted for each balcony.

The attachments are controlled for all balconies and each balcony is correctly marked.

External control of the production

Representatives from CBI have visited the factory and controlled the production of balconies after 50, 200, 350 produced units. On these occasions the internal control was checked and random samples were chosen. Checklists were inspected. Specimens from every tenth unit are sent to CBI for flexural testing.

CONCLUSIONS

One problem encountered when introducing products of steel fibre reinforced concrete is the lack of generally accepted methods for the evaluation of properties of great significance for the function of the product, for instance the ductility of the material under tensile and flexural loading or the variation in the material, for example the fibre distribution. At present our codes do not comprise test methods for evaluating some of the properties. The producer of steel fibre concrete needs new test methods for production control, and the customer or user of products made of this material needs instruments to specify and check expected properties. New test methods and new requirements for the properties of steel fibre reinforced concrete can also imply the development of new production equipment. This may be essential for a wider spread of the use of products made of steel fibre reinforced concrete.

INDEX